Graduate Texts in Mathematics 208

Springer
New York
Berlin
Heidelberg
Barcelona
Hong Kong
London
Milan
Paris
Singapore
Tokyo

Graduate Texts in Mathematics

1 TAKEUTI/ZARING. Introduction to Axiomatic Set Theory. 2nd ed.
2 OXTOBY. Measure and Category. 2nd ed.
3 SCHAEFER. Topological Vector Spaces. 2nd ed.
4 HILTON/STAMMBACH. A Course in Homological Algebra. 2nd ed.
5 MAC LANE. Categories for the Working Mathematician. 2nd ed.
6 HUGHES/PIPER. Projective Planes.
7 SERRE. A Course in Arithmetic.
8 TAKEUTI/ZARING. Axiomatic Set Theory.
9 HUMPHREYS. Introduction to Lie Algebras and Representation Theory.
10 COHEN. A Course in Simple Homotopy Theory.
11 CONWAY. Functions of One Complex Variable I. 2nd ed.
12 BEALS. Advanced Mathematical Analysis.
13 ANDERSON/FULLER. Rings and Categories of Modules. 2nd ed.
14 GOLUBITSKY/GUILLEMIN. Stable Mappings and Their Singularities.
15 BERBERIAN. Lectures in Functional Analysis and Operator Theory.
16 WINTER. The Structure of Fields.
17 ROSENBLATT. Random Processes. 2nd ed.
18 HALMOS. Measure Theory.
19 HALMOS. A Hilbert Space Problem Book. 2nd ed.
20 HUSEMOLLER. Fibre Bundles. 3rd ed.
21 HUMPHREYS. Linear Algebraic Groups.
22 BARNES/MACK. An Algebraic Introduction to Mathematical Logic.
23 GREUB. Linear Algebra. 4th ed.
24 HOLMES. Geometric Functional Analysis and Its Applications.
25 HEWITT/STROMBERG. Real and Abstract Analysis.
26 MANES. Algebraic Theories.
27 KELLEY. General Topology.
28 ZARISKI/SAMUEL. Commutative Algebra. Vol.I.
29 ZARISKI/SAMUEL. Commutative Algebra. Vol.II.
30 JACOBSON. Lectures in Abstract Algebra I. Basic Concepts.
31 JACOBSON. Lectures in Abstract Algebra II. Linear Algebra.
32 JACOBSON. Lectures in Abstract Algebra III. Theory of Fields and Galois Theory.
33 HIRSCH. Differential Topology.
34 SPITZER. Principles of Random Walk. 2nd ed.
35 ALEXANDER/WERMER. Several Complex Variables and Banach Algebras. 3rd ed.
36 KELLEY/NAMIOKA et al. Linear Topological Spaces.
37 MONK. Mathematical Logic.
38 GRAUERT/FRITZSCHE. Several Complex Variables.
39 ARVESON. An Invitation to C^*-Algebras.
40 KEMENY/SNELL/KNAPP. Denumerable Markov Chains. 2nd ed.
41 APOSTOL. Modular Functions and Dirichlet Series in Number Theory. 2nd ed.
42 SERRE. Linear Representations of Finite Groups.
43 GILLMAN/JERISON. Rings of Continuous Functions.
44 KENDIG. Elementary Algebraic Geometry.
45 LOÈVE. Probability Theory I. 4th ed.
46 LOÈVE. Probability Theory II. 4th ed.
47 MOISE. Geometric Topology in Dimensions 2 and 3.
48 SACHS/WU. General Relativity for Mathematicians.
49 GRUENBERG/WEIR. Linear Geometry. 2nd ed.
50 EDWARDS. Fermat's Last Theorem.
51 KLINGENBERG. A Course in Differential Geometry.
52 HARTSHORNE. Algebraic Geometry.
53 MANIN. A Course in Mathematical Logic.
54 GRAVER/WATKINS. Combinatorics with Emphasis on the Theory of Graphs.
55 BROWN/PEARCY. Introduction to Operator Theory I: Elements of Functional Analysis.
56 MASSEY. Algebraic Topology: An Introduction.
57 CROWELL/FOX. Introduction to Knot Theory.
58 KOBLITZ. p-adic Numbers, p-adic Analysis, and Zeta-Functions. 2nd ed.
59 LANG. Cyclotomic Fields.
60 ARNOLD. Mathematical Methods in Classical Mechanics. 2nd ed.
61 WHITEHEAD. Elements of Homotopy Theory.
62 KARGAPOLOV/MERLZJAKOV. Fundamentals of the Theory of Groups.
63 BOLLOBAS. Graph Theory.
64 EDWARDS. Fourier Series. Vol. I. 2nd ed.
65 WELLS. Differential Analysis on Complex Manifolds. 2nd ed.

(continued after index)

Ward Cheney

Analysis for Applied Mathematics

With 27 Illustrations

 Springer

Ward Cheney
Department of Mathematics
University of Texas at Austin
Austin, TX 78712-1082
USA

Editorial Board

S. Axler
Mathematics Department
San Francisco State
   University
San Francisco, CA 94132
USA

F.W. Gehring
Mathematics Department
East Hall
University of Michigan
Ann Arbor, MI 48109
USA

K.A. Ribet
Mathematics Department
University of California
   at Berkeley
Berkeley, CA 94720-3840
USA

Mathematics Subject Classification (2000): 46Bxx, 65L60, 32Wxx, 42B10

Library of Congress Cataloging-in-Publication Data
Cheney, E. W. (Elliott Ward), 1929–
    Analysis for applied mathematics / Ward Cheney.
        p.   cm. — (Graduate texts in mathematics ; 208)
    Includes bibliographical references and index.
    ISBN 0-387-95279-9 (alk. paper)
    1. Mathematical analysis.   I. Title.   II. Series.
QA300.C4437   2001
515—dc21   2001-1020440

Printed on acid-free paper.

Production managed by Terry Kornak; manufacturing supervised by Jerome Basma.
Photocomposed from the author's TeX files.
Printed and bound by Maple-Vail Book Manufacturing Group, York, PA.
Printed in the United States of America.

9 8 7 6 5 4 3 2 1

ISBN 0-387-95279-9   SPIN 10833405

Springer-Verlag   New York Berlin Heidelberg
A member of BertelsmannSpringer Science+Business Media GmbH

Preface

This book evolved from a course at our university for beginning graduate students in mathematics—particularly students who intended to specialize in applied mathematics. The content of the course made it attractive to other mathematics students and to graduate students from other disciplines such as engineering, physics, and computer science. Since the course was designed for two semesters duration, many topics could be included and dealt with in detail. Chapters 1 through 6 reflect roughly the actual nature of the course, as it was taught over a number of years. The content of the course was dictated by a syllabus governing our preliminary Ph.D. examinations in the subject of applied mathematics. That syllabus, in turn, expressed a consensus of the faculty members involved in the applied mathematics program within our department. The text in its present manifestation is my interpretation of that syllabus: my colleagues are blameless for whatever flaws are present and for any inadvertent deviations from the syllabus.

The book contains two additional chapters having important material not included in the course: Chapter 8, on measure and integration, is for the benefit of readers who want a concise presentation of that subject, and Chapter 7 contains some topics closely allied, but peripheral, to the principal thrust of the course.

This arrangement of the material deserves some explanation. The ordering of chapters reflects our expectation of our students: If they are unacquainted with Lebesgue integration (for example), they can nevertheless understand the examples of Chapter 1 on a superficial level, and at the same time, they can begin to remedy any deficiencies in their knowledge by a little private study of Chapter 8. Similar remarks apply to other situations, such as where some point-set topology is involved; Section 7.6 will be helpful here. To summarize: We encourage students to wade boldly into the course, starting with Chapter 1, and, where necessary, fill in any gaps in their prior preparation. One advantage of this strategy is that they will see the *necessity* for topology, measure theory, and other topics — thus becoming better motivated to study them. In keeping with this philosophy, I have not hesitated to make forward references in some proofs to material coming later in the book. For example, the Banach contraction mapping theorem is needed at least once prior to the section in Chapter 4 where it is dealt with at length.

Each of the book's six main topics could certainly be the subject of a year's course (or a lifetime of study), and many of our students indeed study functional analysis and other topics of the book in separate courses. Most of them eventually or simultaneously take a year-long course in analysis that includes complex analysis and the theory of measure and integration. However, the applied mathematics course is typically taken in the first year of graduate study. It seems to bridge the gap between the undergraduate and graduate curricula in a way that has been found helpful by many students. In particular, the course and the

v

book certainly do not presuppose a thorough knowledge of integration theory nor of topology. In our applied mathematics course, students usually enhance and reinforce their knowledge of undergraduate mathematics, especially differential equations, linear algebra, and general mathematical analysis. Students may, for the first time, perceive these branches of mathematics as being essential to the foundations of applied mathematics.

The book could just as well have been titled *Prolegomena to Applied Mathematics*, inasmuch as it is *not* about applied mathematics itself but rather about topics in analysis that impinge on applied mathematics. Of course, there is no end to the list of topics that could lay claim to inclusion in such a book. Who is bold enough to predict what branches of mathematics will be useful in applications over the next decade? A look at the past would certainly justify my favorite algorithm for creating an applied mathematician: Start with a pure mathematician, and turn him or her loose on real-world problems.

As in some other books I have been involved with, I owe a great debt of gratitude to Ms. Margaret Combs, our departmental TEX-pert. She typeset and kept up-to-date the notes for the course over many years, and her resourcefulness made my burden much lighter.

The staff of Springer–Verlag has been most helpful in seeing this book to completion. In particular, I worked closely with Dr. Ina Lindemann and Ms. Terry Kornak on editorial matters, and I thank them for their efforts on my behalf. I am indebted to David Kramer for his meticulous copy-editing of the manuscript; it proved to be very helpful in the final editorial process.

I thank my wife, Victoria, for her patience and assistance during the period of work on the book, especially the editorial phase. I dedicate the book to her in appreciation.

I will be pleased to hear from readers having questions or suggestions for improvements in the book. For this purpose, electronic mail is efficient: cheney@math.utexas.edu. I will also maintain a web site for material related to the book at http://www.math.utexas.edu/users/cheney/AAMbook

Ward Cheney
Department of Mathematics
University of Texas at Austin

Contents

Preface v

Chapter 1. Normed Linear Spaces 1

1.1 Definitions and Examples 1
1.2 Convexity, Convergence, Compactness, Completeness 6
1.3 Continuity, Open Sets, Closed Sets 15
1.4 More About Compactness 19
1.5 Linear Transformations 24
1.6 Zorn's Lemma, Hamel Bases, and the Hahn–Banach Theorem 30
1.7 The Baire Theorem and Uniform Boundedness 40
1.8 The Interior Mapping and Closed Mapping Theorems 47
1.9 Weak Convergence 53
1.10 Reflexive Spaces 58

Chapter 2. Hilbert Spaces 61

2.1 Geometry 61
2.2 Orthogonality and Bases 70
2.3 Linear Functionals and Operators 81
2.4 Spectral Theory 91
2.5 Sturm–Liouville Theory 105

Chapter 3. Calculus in Banach Spaces 115

3.1 The Fréchet Derivative 115
3.2 The Chain Rule and Mean Value Theorems 121
3.3 Newton's Method 125
3.4 Implicit Function Theorems 135
3.5 Extremum Problems and Lagrange Multipliers 145
3.6 The Calculus of Variations 152

Chapter 4. Basic Approximate Methods of Analysis 170

4.1 Discretization 170
4.2 The Method of Iteration 176
4.3 Methods Based on the Neumann Series 186
4.4 Projections and Projection Methods 191
4.5 The Galerkin Method 198
4.6 The Rayleigh–Ritz Method 205
4.7 Collocation Methods 213
4.8 Descent Methods 226
4.9 Conjugate Direction Methods 232
4.10 Methods Based on Homotopy and Continuation 237

Chapter 5. Distributions 246

  5.1 Definitions and Examples 246
  5.2 Derivatives of Distributions 253
  5.3 Convergence of Distributions 257
  5.4 Multiplication of Distributions by Functions 260
  5.5 Convolutions 268
  5.6 Differential Operators 273
  5.7 Distributions with Compact Support 280

Chapter 6. The Fourier Transform 287

  6.1 Definitions and Basic Properties 287
  6.2 The Schwartz Space 294
  6.3 The Inversion Theorems 301
  6.4 The Plancherel Theorem 305
  6.5 Applications of the Fourier Transform 310
  6.6 Applications to Partial Differential Equations 318
  6.7 Tempered Distributions 321
  6.8 Sobolev Spaces 325

Chapter 7. Additional Topics 333

  7.1 Fixed–Point Theorems 333
  7.2 Selection Theorems 339
  7.3 Separation Theorems 342
  7.4 The Arzelà–Ascoli Theorems 347
  7.5 Compact Operators and the Fredholm Theory 351
  7.6 Topological Spaces 361
  7.7 Linear Topological Spaces 367
  7.8 Analytic Pitfalls 373

Chapter 8. Measure and Integration 381

  8.1 Extended Reals, Outer Measures, Measurable Spaces 381
  8.2 Measures and Measure Spaces 386
  8.3 Lebesgue Measure 391
  8.4 Measurable Functions 394
  8.5 The Integral for Nonnegative Functions 399
  8.6 The Integral, Continued 404
  8.7 The L^p-Spaces 409
  8.8 The Radon–Nikodym Theorem 413
  8.9 Signed Measures 417
  8.10 Product Measures and Fubini's Theorem 420

References 429

Index 437

Symbols 443

Chapter 1

Normed Linear Spaces

1.1   Definitions and Examples   1
1.2   Convexity, Convergence, Compactness, Completeness   6
1.3   Continuity, Open Sets, Closed Sets   15
1.4   More about Compactness   19
1.5   Linear Transformations   24
1.6   Zorn's Lemma, Hamel Bases, and the Hahn–Banach Theorem   30
1.7   The Baire Theorem and Uniform Boundedness   40
1.8   The Interior Mapping and Closed Mapping Theorems   47
1.9   Weak Convergence   53
1.10 Reflexive Spaces   58

1.1 Definitions and Examples

This chapter gives an introduction to the theory of normed linear spaces. A skeptical reader may wonder why this topic in pure mathematics is useful in applied mathematics. The reason is quite simple: Many problems of applied mathematics can be formulated as a search for a certain function, such as the function that solves a given differential equation. Usually the function sought must belong to a definite family of acceptable functions that share some useful properties. For example, perhaps it must possess two continuous derivatives. The families that arise naturally in formulating problems are often linear spaces. This means that any linear combination of functions in the family will be another member of the family. It is common, in addition, that there is an appropriate means of measuring the "distance" between two functions in the family. This concept comes into play when the exact solution to a problem is inaccessible, while approximate solutions can be computed. We often measure how far apart the exact and approximate solutions are by using a norm. In this process we are led to a normed linear space, presumably one appropriate to the problem at hand. Some normed linear spaces occur over and over again in applied mathematics, and these, at least, should be familiar to the practitioner. Examples are the space of continuous functions on a given domain and the space of functions whose squares have a finite integral on a given domain. A knowledge of function spaces enables an applied mathematician to consider a problem from a more lofty viewpoint, from which he or she may have the advantage of being more aware of significant features as distinguished from less significant details.

We begin by reviewing the concept of a **vector space**, or **linear space.** (These terms are interchangeable.) The reader is probably already familiar with these spaces, or at least with the example of vectors in $\mathbb{R}^n$. However, many *function* spaces are also linear spaces, and much can be learned about these function spaces by exploiting their similarity to the more elementary examples. Here, as a reminder, we include the axioms for a vector space or linear space.

A **real vector space** is a triple $(X, +, \cdot)$, in which X is a set, and $+$ and $\cdot$ are binary operations satisfying certain axioms. Here are the axioms:

(i) If x and y belong to X then so does $x + y$ (closure axiom).

(ii) $x + y = y + x$ (commutativity).

(iii) $x + (y + z) = (x + y) + z$ (associativity).

(iv) X contains a unique element, 0, such that $x + 0 = x$ for all x in X.

(v) With each element x there is associated a unique element, $-x$, such that $x + (-x) = 0$.

(vi) If $x \in X$ and $\lambda \in \mathbb{R}$, then $\lambda \cdot x \in X$ ($\mathbb{R}$ denotes the set of real numbers.) (closure axiom)

(vii) $\lambda \cdot (x + y) = \lambda \cdot x + \lambda \cdot y$ ($\lambda \in \mathbb{R}$), (distributivity).

(viii) $(\lambda + \mu) \cdot x = \lambda \cdot x + \mu \cdot x$ ($\lambda, \mu \in \mathbb{R}$), (distributivity).

(ix) $\lambda \cdot (\mu \cdot x) = (\lambda\mu) \cdot x$ (associativity).

(x) $1 \cdot x = x$.

These axioms need not be intimidating. The essential feature of a linear space is that there is an *addition* defined among the elements of X, and when we add two elements, the result is again in the space X. One says that the space is *closed* (algebraically) under the operation of addition. A similar remark holds true for multiplication of an element by a real number. The remaining axioms simply tell us that the usual rules of arithmetic are valid for the two operations. Most rules that you expect to be true are indeed true, but if they do not appear among the axioms it is because they *follow* from the axioms. The effort to keep the axioms minimal has its rewards: When one must verify that a given system is a real vector space there will be a minimum of work involved!

In this set of axioms, the first five define an (additive) Abelian group. In axiom (iv), the uniqueness of 0 need not be mentioned, for it can be proved with the aid of axiom (ii). Usually, if $\lambda \in \mathbb{R}$ and $x \in X$, we write λx in place of $\lambda \cdot x$. The reader will note the ambiguity in the symbol $+$ and the symbol 0. For example, when we write $0x = 0$ two different zeros are involved, and in axiom (viii) the plus signs are not the same. We usually write $x - y$ in place of $x + (-y)$. Furthermore, we are not going to belabor elementary consequences of the axioms such as $\lambda \sum_1^n x_i = \sum_1^n \lambda x_i$. We usually refer to X as the linear space rather than $(X, +, \cdot)$. Observe that in a linear space, we have no way of assigning a meaning to expressions that involve a limiting process, such as $\sum_1^\infty x_i$. This drawback will disappear soon, upon the introduction of a norm.

From time to time we will prefer to deal with a **complex vector space.** In such a space $\lambda \cdot x$ is defined (and belongs to X) whenever $\lambda \in \mathbb{C}$ and $x \in X$. (The symbol $\mathbb{C}$ denotes the set of complex numbers.) Other fields can be employed in place of $\mathbb{R}$ and $\mathbb{C}$, but they are rarely useful in applied mathematics. The field elements are often termed **scalars,** and the elements of X are often called **vectors.**

Let X be a vector space. A **norm** on X is a real-valued function, denoted by $\|\ \|$, that fulfills three axioms:

(i) $\|x\| > 0$ for each nonzero element in X.

(ii) $\|\lambda x\| = |\lambda|\,\|x\|$ for each λ in $\mathbb{R}$ and each x in X.

(iii) $\|x + y\| \leqslant \|x\| + \|y\|$ for all $x, y \in X$. (Triangle Inequality)

A vector space in which a norm has been introduced is called a **normed linear space.** Here are eleven examples.

Example 1.   Let $X = \mathbb{R}$, and define $\|x\| = |x|$, the familiar absolute value function.   ∎

Example 2.   Let $X = \mathbb{C}$, where the scalar field is also $\mathbb{C}$. Use $\|x\| = |x|$, where $|x|$ has its usual meaning for a complex number x. Thus if $x = a + ib$ (where a and b are real), then $|x| = \sqrt{a^2 + b^2}$.   ∎

Example 3.   Let $X = \mathbb{C}$, and take the scalar field to be $\mathbb{R}$. The terminology we have adopted requires that this be called a *real* vector space, since the scalar field is $\mathbb{R}$.   ∎

Example 4.   Let $X = \mathbb{R}^n$. Here the elements of X are n-tuples of real numbers that we can display in the form $x = [x(1), x(2), \ldots, x(n)]$ or $x = [x_1, x_2, \ldots, x_n]$. A useful norm is defined by the equation

$$\|x\|_\infty = \max_{1 \leqslant i \leqslant n} |x(i)|$$

Note that an n-tuple is a function on the set $\{1, 2, \ldots, n\}$, and so the notation $x(i)$ is consistent with that interpretation. (This is the "sup" norm.)   ∎

Example 5.   Let $X = \mathbb{R}^n$, and define a norm by the equation $\|x\| = \sum_{i=1}^{n} |x(i)|$. Observe that in Examples 4 and 5 we have two *distinct* normed linear spaces, although each involves the *same* linear space. This shows the advantage of being more formal in the definition and saying that a normed linear space is a pair $(X, \|\ \|)$ etc. etc., but we refrain from doing this unless it is necessary.   ∎

Example 6.   Let X be the set of all real-valued continuous functions defined on a fixed compact interval $[a, b]$. The norm usually employed here is

$$\|x\|_\infty = \max_{a \leqslant s \leqslant b} |x(s)|$$

(The notation $\max_{a \leqslant s \leqslant b} |x(s)|$ denotes the maximum of the expression $|x(s)|$ as s runs over the interval $[a, b]$.) The space X described here is often denoted by $C[a, b]$. Sticklers would insist on $C([a, b])$, because $C(S)$ will be used for the continuous functions on some general domain S. (This again is the "sup" norm.)   ∎

Example 7. Let X be the set of all Lebesgue-integrable functions defined on a fixed interval $[a, b]$. The usual norm for this space is $\|x\| = \int_a^b |x(s)| ds$. In this space, the vectors are actually *equivalence classes* of functions, two functions being regarded as **equivalent** if they differ only on a set of measure 0. (The reader who is unfamiliar with the Lebesgue integral can substitute the Riemann integral in this example. The resulting spaces are different, one being complete and the other not. This is a rather complicated matter, best understood after the study of measure theory and Lebesgue integration. Chapter 8 is devoted to this branch of analysis. The notion of completeness of a space is taken up in the next section.) ∎

Example 8. Let $X = \ell$, the space of all sequences in $\mathbb{R}$

$$x = [x(1), x(2), \ldots]$$

in which only a finite number of terms are nonzero. (The number of nonzero terms is not fixed but can vary with different sequences.) Define $\|x\| = \max_n |x(n)|$. ∎

Example 9. Let $X = \ell_\infty$, the space of all real sequences x for which $\sup_n |x(n)| < \infty$. Define $\|x\|$ to be that supremum, as in Example 8. ∎

Example 10. Let $X = \Pi$, the space of all polynomials having real coefficients. A typical element of Π is a function x having the form

$$x(t) = a_0 + a_1 t + a_2 t^2 + \cdots + a_n t^n$$

One possible norm on Π is $x \mapsto \max_i |a_i|$. Others are $x \mapsto \max_{0 \leqslant t \leqslant 1} |x(t)|$ or $x \mapsto \int_0^1 |x(t)| \, dt$ or $x \mapsto (\sum_0^n |x|^3)^{1/3}$. ∎

Example 11. Let $X = \mathbb{R}^n$, and use the familiar **Euclidean norm**, defined by

$$\|x\|_2 = \left(\sum_{i=1}^n [x(i)]^2 \right)^{1/2}$$ ∎

In all of these examples (as well as in others to come) it is regarded as obvious how the algebraic structure is defined. A complete development would define $x + y$, λx, 0, and $-x$, and then verify the axioms for a linear space. After that, the alleged norm would be shown to satisfy the axioms for a norm. Thus, in Example 6, the zero element is the *function* denoted by 0 and defined by $0(s) = 0$ for all $s \in [a, b]$. The operation of addition is defined by the equation

$$(x + y)(s) = x(s) + y(s)$$

and so on.

The concept of linear independence is of central importance. Recall that a subset S in a linear space is **linearly independent** if it is **not** possible to find a finite, nonempty, set of distinct vectors $x_1, x_2, \ldots, x_m$ in S and nonzero scalars $c_1, c_2, \ldots, c_m$ for which

$$c_1 x_1 + c_2 x_2 + \cdots + c_m x_m = 0$$

(Linear independence is not a property of a point; it is a property of a *set* of points. Because of this, the usage "the vectors... are independent" is misleading.) The reader probably recalls how this notion enters into the theory of nth–order ordinary differential equations: A general solution must involve a linearly independent set of n solutions.

Some other basic concepts to recall from linear algebra are mentioned here. The **span** of a set S in a vector space X is denoted by span(S), and consists of all vectors in X that are expressible as linear combinations of vectors in S. Remember that linear combinations are always *finite* expressions of the form $\sum_{i=1}^{k} \lambda_i x_i$. We say that "$S$ spans X" when $X = \text{span}(S)$. A **base** or **basis** for a vector space X is any set that is linearly independent and spans X. Both properties are essential. Any set that is linearly independent is contained in a basis, and any set that spans the space contains a basis. A vector space is said to be **finite dimensional** if it has a finite basis. An important theorem states that if a space is finite dimensional, then every basis for that space has the same number of elements. This common number is then called the **dimension** of the space. (There is an infinite-dimensional version of this theorem as well.)

The material of this chapter is accessible in many textbooks and treatises, such as: [Au], [Av], [BN], [Ban], [Bea], [CP], [Day], [Dies], [Dieu], [DS], [Edw], [Frie2], [Fried], [GP], [Gre], [Gri], [HS], [HP], [Hol], [Horv], [Jam], [KA], [Kee], [KF], [Kre], [Lan1], [Lo], [Moo], [NaSn], [OD], [Ped], [Red], [RS], [RN], [Roy], [Ru1], [Sim], [Tay2], [Yo], and [Ze].

Problems 1.1

Here is a Chinese proverb that is pertinent to the problems: *I hear, I forget; I see, I remember; I do, I understand!*

1. Let X be a linear space over the complex field. Let X^r be the space obtained from X by restricting the scalars to the real field. Prove that X^r is a real linear space. Show by an example that *not* every real linear space is of the form X^r for some complex linear space X. *Caution:* When we say that a linear space is a *real* linear space, this has nothing to do with the elements of the space. It means only that the scalar field is $\mathbb{R}$ and not $\mathbb{C}$.

2. Prove the norm axioms for Examples 4–7.

3. Prove that in any normed linear space,

$$\|0\| = 0 \quad \text{and} \quad \Big| \|x\| - \|y\| \Big| \leqslant \|x - y\|$$

4. Denote the norms in Examples 4 and 5 by $\| \ \|_{\infty}$ and $\| \ \|_1$, respectively. Find the best constants in the inequality

$$\alpha \|x\|_1 \leqslant \|x\|_{\infty} \leqslant \beta \|x\|_1$$

   Prove that your constants are the best. (The "constants" α and β will depend on n but not x.)

5. In Examples 4, 5, 6, and 7 find the precise conditions under which we have $\|x + y\| = \|x\| + \|y\|$.

6. Prove that in any normed linear space, if $x \neq 0$, then $x/\|x\|$ is a vector of norm 1.

7. The Euclidean norm on $\mathbb{R}^n$ is defined in Example 11. Find the best constants in the inequality $\alpha \|x\|_{\infty} \leqslant \|x\|_2 \leqslant \beta \|x\|_{\infty}$.

8. What theorems in elementary analysis are needed to prove the closure axioms for Example 6?

9. What is the connection between the normed linear spaces ℓ and Π defined in Examples 8 and 10?

10. For any t in the open interval $(0, 1)$, let $\tilde{t}$ be the sequence $[t, t^2, t^3, \ldots]$. Notice that $\tilde{t} \in \ell_\infty$. Prove that the set $\{\tilde{t} : 0 < t < 1\}$ is linearly independent.

11. In the space Π we define special elements called **monomials**. They are given by $x_n(t) = t^n$ where $n = 0, 1, 2, \ldots$ Prove that $\{x_n : n = 0, 1, 2, 3 \ldots\}$ is linearly independent.

12. Let T be a set of real numbers. We say that T is **bounded above** if there is an M in $\mathbb{R}$ such that $t \leqslant M$ for all t in T. We say that M is an **upper bound** of T. The **completeness axiom** for $\mathbb{R}$ asserts that if a set T is bounded above, then the set of all its upper bounds is an interval of the form $[b, \infty)$. The number b is the **least upper bound**, or **supremum** of T, written $b = \mathrm{l.u.b.}(T) = \sup(T)$. Prove that if $x < b$, then $(x, \infty) \cap T$ is nonempty. Give examples to show that $[b, \infty) \cap T$ can be empty or nonempty. There are corresponding concepts of **bounded below, lower bound, greatest lower bound**, and **infimum**.

13. Which of these expressions define norms on $\mathbb{R}^2$? Explain.

    (a) $\max\{|x(1)|, |x(1) + x(2)|\}$

    (b) $|x(2) - x(1)|$

    (c) $|x(1)| + |x(2) - x(1)| + |x(2)|$

14. Prove that in any normed linear space the conditions $\|x\| = 1$ and $\|x - y\| < \varepsilon < 1$ imply that $\|x - y/\|y\| \| < 2\varepsilon$.

15. Prove that if N_1 and N_2 are norms on a linear space, then so are $\alpha_1 N_1 + \alpha_2 N_2$ (when $\alpha_1 > 0$ and $\alpha_2 > 0$) and $(N_1^2 + N_2^2)^{1/2}$.

16. Is the following set of axioms for a norm equivalent to the set given in the text? (a) $\|x\| \neq 0$ if $x \neq 0$, (b) $\|\lambda x\| = -\lambda \|x\|$ if $\lambda \leqslant 0$, (c) $\|x + y\| \leqslant \|x\| + \|y\|$.

17. Prove that in a normed linear space, if $\|x + y\| = \|x\| + \|y\|$, then $\|\alpha x + \beta y\| = \|\alpha x\| + \|\beta y\|$ for all nonnegative α and β.

18. Why is the word "distinct" essential in our definition of linear independence on page 4?

19. Is the set of functions $f_i(x) = |x - i|$, where $i = 1, 2 \ldots$, linearly independent?

20. One example of an "exotic" vector space is described as follows. Let X be the set of positive real numbers. We define an "addition", $\oplus$, by $x \oplus y = xy$ and a "scalar multiplication" by $a \odot x = x^a$. Prove that $(X, \oplus, \odot)$ is a vector space.

21. In Example 10, two norms (say N_1 and N_2) were suggested. Do there exist constants such that $N_1 \leqslant \alpha N_2$ or $N_2 \leqslant \beta N_1$?

22. In Examples 4 and 5, let $n = 2$, and draw sketches of the sets $\{x \in \mathbb{R}^2 : \|x\| = 1\}$. (Symmetries can be exploited.)

1.2 Convexity, Convergence, Compactness, Completeness

A subset K in a linear space is said to be **convex** if it contains every line segment connecting two of its elements. Formally, convexity is expressed as follows:

$$[x \in K \quad \& \quad y \in K \quad \& \quad 0 \leqslant \lambda \leqslant 1] \implies \lambda x + (1 - \lambda)y \in K$$

The notion of convexity arises frequently in optimization problems. For example, the theory of linear programming (optimization of linear functions) is based on the fact that a linear function on a convex polyhedral set must attain its extrema at the vertices of the set. Thus, to locate the maxima of a linear function over a convex polyhedral set, one need only test the vertices. The central idea of Dantzig's famous simplex method is to move from vertex to vertex, always improving the value of the objective function.

Another application of convexity occurs in studying deformations of a physical body. The "yield surface" of an object is generally convex. This is the surface in 6-dimensional space that gives the stresses at which an object will fail structurally. Six dimensions are needed to account for all the variables. See [Mar], pages 100–104.

Among examples of convex sets in a linear space X we have:

  (i) the space X itself;

 (ii) any set consisting of a single point;

(iii) the empty set;

(iv) any linear subspace of X;

 (v) any line segment; i.e. a set of the following form in which a and b are fixed:

$$\{\lambda a + (1 - \lambda)b \ : \ 0 \leqslant \lambda \leqslant 1\}$$

In a normed linear space, another important convex set is the **unit cell** or **unit ball**:

$$\{x \in X \ : \ \|x\| \leqslant 1\}$$

In order to see that the unit ball is convex, let $\|x\| \leqslant 1$, $\|y\| \leqslant 1$, and $0 \leqslant \lambda \leqslant 1$. Then, with $\mu = 1 - \lambda$,

$$\|\lambda x + \mu y\| \leqslant \|\lambda x\| + \|\mu y\| = \lambda \|x\| + \mu \|y\| \leqslant \lambda + \mu = 1$$

If we let $n = 2$ in Examples 4 and 5 of Section 1.1, then we can draw pictures of the unit balls. They are shown in Figures 1.1 and 1.2.

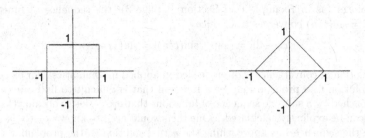

Figures 1.1 and 1.2. Unit balls

There is a family of norms on $\mathbb{R}^n$, known as the ℓ_p-norms, of which the norms in Examples 4 and 5 are special cases. The general formula, for $1 \leqslant p < \infty$, is

$$\|x\|_p = \left(\sum_{i=1}^{n} |x(i)|^p \right)^{1/p}$$

The case $p = \infty$ is special; for it we use the formula

$$\|x\|_\infty = \max_{1 \leqslant i \leqslant n} |x(i)|$$

It can be shown (Problem 1) that $\lim_{p \to \infty} \|x\|_p = \|x\|_\infty$. (This explains the notation.) The unit balls (in $\mathbb{R}^2$) for $\| \; \|_p$ are shown for $p = \frac{4}{3}$, 2, and 7, in Figure 1.3.

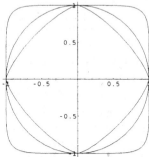

Figure 1.3. The unit balls in ℓ_p, for $p = \frac{4}{3}$, 2, and 7.

In any normed linear space there exists a metric (and its corresponding topology) that arises by defining the distance between two points as

$$d(x, y) = \|x - y\|$$

All the topological notions from the theory of metric spaces then become available in a normed linear space. (See Problem 23.) In Chapter 7, Section 6, the theory of general topological spaces is broached. But we shall discuss here topological concepts restricted to metric spaces or to normed linear spaces. A sequence $x_1, x_2, \ldots$ in a normed linear space is said to **converge** to a point x (and we write $x_n \to x$) if

$$\lim_{n \to \infty} \|x_n - x\| = 0$$

For example, in the space of continuous functions on $[0, 1]$ furnished with the max-norm (as in Example 6 of Section 1, page 3), the sequence of functions $x_n(t) = \sin(t/n)$ converges to 0, since

$$\|x_n - 0\| = \sup_{0 \leqslant t \leqslant 1} |\sin(t/n)| = \sin(1/n) \to 0$$

The notion of convergence is often needed in applied mathematics. For example, the solution to a problem may be a function that is difficult to find but can be approached by a suitable sequence of functions that are easier to obtain. (Maybe they can be explicitly calculated.) One then would need to know exactly in what sense the sequence was approaching the actual solution to the problem.

A subset K in a normed space is said to be **compact** if each sequence in K has a subsequence that converges to a point in K. (*Caution:* In general topology, this concept would be called sequential compactness. Refer to Section 7.6.) A subsequence of a sequence $x_1, x_2, \ldots$ is of the form $x_{n_1}, x_{n_2}, \ldots$, where the integers n_i satisfy $n_1 < n_2 < n_3 < \cdots$. Our notation for a sequence is $[x_n]$, or $[x_n : n \in \mathbb{N}]$, or $[x_1, x_2, \ldots]$. With this meagre equipment we can already prove some interesting results.

Theorem 1.   Let K be a compact set in a normed linear space X.
To each x in X there corresponds at least one point in K of minimum
distance from x.

Proof.   Let x be any member of X. The distance from x to K is defined to be
the number
$$\operatorname{dist}(x, K) = \inf_{z \in K} \|x - z\|$$
By the definition of an infimum (Problem 12 in Section 1.1, page 6), there exists
a sequence $[y_n]$ in K such that $\|x - y_n\| \to \operatorname{dist}(x, K)$. Since K is compact,
there is a subsequence converging to a point in K, say $y_{n_i} \to y \in K$. Since
$$\|x - y\| \leqslant \|x - y_{n_i}\| + \|y_{n_i} - y\|$$
we have in the limit $\|x - y\| \leqslant \operatorname{dist}(x, K) \leqslant \|x - y\|$. (The final inequality follows
from the definition of the distance function.)   ∎

The preceding theorem can be useful in problems involving noisy measure-
ments. For example, suppose that a noisy measurement of a single entity x is
available. If a set K of admissible noise-free values for x is prescribed, then
the best noise-free estimate of x can be taken to be a point of K as close as
possible to x. Theorem 1 is also important in approximation theory, a branch
of analysis that provides the theoretical underpinning for many areas of applied
mathematics.

Example 1.   On the real line, an open interval (a, b) is not compact, for we
can take a sequence in the interval that converges to the endpoint b, say. Then
every subsequence will also converge to b. Since b is not in the interval, the
interval cannot be compact. On the other hand, a closed and bounded interval,
say $[a, b]$, is compact. This is a special case of the Heine–Borel theorem. See the
discussion before Lemma 1 in Section 1.4, page 20.   ∎

Given a sequence $[x_n]$ in a normed linear space (or indeed in any metric
space), is it possible to determine, from the sequence alone, whether it converges? This is certainly an important matter for practical purposes, since we
often use algorithms to generate sequences that should converge to a solution
of a given problem. The answer to the posed question is that we *cannot* infer
convergence, in general, solely from the sequence itself. If we confine ourselves to
the information contained in the sequence, we can construct the doubly indexed
sequence $c_{nm} = \|x_n - x_m\|$. If $[c_{nm}]$ does not converge to zero, then the given
sequence $[x_n]$ cannot converge, as is easily proved: For any x in the space, write
$$c_{nm} = \|x_n - x_m\| = \|(x_n - x) - (x_m - x)\| \leqslant \|x_n - x\| + \|x_m - x\|$$
This shows that if c_{nm} does not converge to 0, then $[x_n]$ cannot converge. On
the other hand, if c_{nm} converges to zero, one intuitively thinks that the sequence
ought to converge, and if it does not, there must be a flaw in the space itself: The
limit of the sequence should exist, but the limiting point is somehow missing from
the space. Think of the rational numbers as an example. The missing ingredient
is *completeness* of the space, to which we now turn.

A sequence $[x_n]$ in a normed linear space X is said to have the **Cauchy property** or to be a **Cauchy sequence** if

$$\lim_{n\to\infty} \sup_{\substack{i\geqslant n \\ j\geqslant n}} \|x_i - x_j\| = 0$$

If every Cauchy sequence in the space X is convergent (to a point of X, of course), then the space X is said to be **complete**. A complete normed linear space is termed a **Banach** space, in honor of Stefan Banach, who lived from 1892 to 1945. His book [Ban] stimulated the study of functional analysis for several decades. Examples 1–7, 9, and 11, given previously, are all Banach spaces. The real number field $\mathbb{R}$ is complete, and so is the complex number field $\mathbb{C}$. The rational field $\mathbb{Q}$ is not complete. These facts are established in elementary analysis courses.

Completeness is important in constructing solutions to a problem by taking the limit of successive approximations. One often wants information about the limit (i.e., the solution). Does it have the same properties as the approximations? For example, if all the approximating functions are continuous, must the limit also be continuous? If all the approximating functions are bounded, is the limit also bounded? The answers to such questions depend on the sense in which the limit is achieved; in other words, they depend on the norm that has been chosen and the function space that goes with it. Typically, one wants a norm that leads to a complete normed linear space, i.e., a Banach space.

Here is an example of a normed linear space that is not a Banach space:

Example 2.   Let the space be the one described in Example 8 of Section 1.1, page 4. This is ℓ, the space of "finitely–nonzero sequences," with the "sup norm" $\|x\| = \max_i |x(i)|$. Define a sequence $[x_k]$ in ℓ by the equation

$$x_k = \left[1, \frac{1}{2}, \frac{1}{3}, \ldots, \frac{1}{k}, 0, 0, \ldots\right]$$

If $m > n$, then

$$x_m - x_n = \left[0, \ldots, 0, \frac{1}{n+1}, \ldots, \frac{1}{m}, 0, \ldots\right]$$

Since $\|x_m - x_n\| = 1/(n+1)$, we conclude that the sequence $[x_k]$ has the Cauchy property. If the space were complete, we would have $x_n \to y$, where $y \in \ell$. The point y would be finitely nonzero, say $y(n) = 0$ for $n > N$. Then for $m > N$, x_m would have as its Nth term the value $1/N$, while the Nth term of y is 0. Thus $\|x_m - y\| \geqslant 1/N$, and convergence cannot take place.   ∎

Theorem 2.   *The space $C[a,b]$ with norm $\|x\| = \max_s |x(s)|$ is a Banach space.*

Proof.   Let $[x_n]$ be a Cauchy sequence in $C[a,b]$. (This space is described in Example 6, page 3.) Then for each s, $[x_n(s)]$ is a Cauchy sequence in $\mathbb{R}$. Since $\mathbb{R}$ is complete, this latter sequence converges to a real number that we may denote by $x(s)$. The function x thus defined must now be shown to be continuous, and we must also show that $\|x_n - x\| \to 0$. Let t be fixed as the point at which continuity is to be proved. We write

$$(1) \qquad \left|x(s) - x(t)\right| \leqslant \left|x(s) - x_n(s)\right| + \left|x_n(s) - x_n(t)\right| + \left|x_n(t) - x(t)\right|$$

This inequality should suggest to the reader how the proof must proceed. Let $\varepsilon > 0$. Select N so that $\|x_n - x_m\| \leqslant \varepsilon/3$ whenever $m \geqslant n \geqslant N$ (Cauchy property). Then for $m \geqslant n \geqslant N$, $\left|x_n(s) - x_m(s)\right| \leqslant \varepsilon/3$. By letting $m \to \infty$ we get $\left|x_n(s) - x(s)\right| \leqslant \varepsilon/3$ for all s. This shows that $\|x_n - x\| \leqslant \varepsilon/3$ and that the sequence $\|x_n - x\|$ converges to 0. By the continuity of x_n there exists a $\delta > 0$ such that $\left|x_n(s) - x_n(t)\right| < \varepsilon/3$ whenever $|t - s| < \delta$. Inequality (1) now shows that $\left|x(s) - x(t)\right| < \varepsilon$ when $|t - s| < \delta$. (This proof illustrates what is sometimes called "an $\varepsilon/3$ argument.")   ∎

Remarks.   Theorem 2 is due to Weierstrass. It remains valid if the interval $[a, b]$ is replaced by any compact Hausdorff space. (For topological notions, refer to Section 7.6, starting on page 361.) The traditional formulation of this theorem states that a uniformly convergent sequence of continuous functions on a closed and bounded interval must have a continuous limit. A sequence of functions $[f_n]$ converges **uniformly** to f if

$$(2) \qquad \forall \varepsilon \;\; \exists n \;\; \forall k \;\; \forall s \;\; \left[\, k > n \;\; \implies \;\; |f_k(s) - f(s)| < \varepsilon \,\right]$$

(In this succinct description, it is understood that $\varepsilon > 0$, $n \in \mathbb{N}$, $k \in \mathbb{N}$, and s is in the domain of the functions.) By contrast, **pointwise** convergence is defined by

$$\forall s \;\; \forall \varepsilon \;\; \exists n \;\; \forall k \;\; \left[\, k > n \;\; \implies \;\; |f_k(s) - f(s)| < \varepsilon \,\right]$$

Our use of the austere and forbidding logical notation is to bring out clearly and to emphasize the importance of the *order* of the quantifiers. Thus, in the definition of uniform convergence, n does not (*cannot*) depend on s, while in the definition of pointwise convergence, n may depend on s. Notice that by the definition of the norm being used, (2) can be written

$$\forall \varepsilon \;\; \exists n \;\; \forall k \;\; \left[\, k > n \;\; \implies \;\; \|f_k - f\|_\infty \leqslant \varepsilon \,\right]$$

or simply as $\lim_{n \to \infty} \|f_n - f\|_\infty = 0$. The latter is conceptually rather simple, if one is already comfortable with this norm (called the "supremum norm" or the "maximum norm").

The (perhaps) simplest example of a sequence of continuous functions that converges pointwise but not uniformly to a continuous function is the sequence $[f_n]$ described as follows. The value of $f_n(x)$ is 1 everywhere except on the interval $[0, 2/n]$, where its value is given by $|nx - 1|$.

Problems 1.2

1. Prove that $\lim_{p \to \infty} \|x\|_p = \max_{1 \leqslant i \leqslant n} |x(i)|$ for every x in $\mathbb{R}^n$.

2. Is this property of a sequence equivalent to the Cauchy property?

$$\lim_{n \to \infty} \sup_{k \geqslant n} \|x_k - x_n\| = 0$$

   Answer the same question for this property: For every positive ε there is a natural number n such that $\|x_m - x_n\| < \varepsilon$ whenever $m \geqslant n$.

3. Prove that if a sequence $[x_n]$ in a Banach space satisfies $\sum_{n=1}^{\infty} \|x_n\| < \infty$, then the series $\sum_{n=1}^{\infty} x_n$ converges.

4. Prove that Theorem 2 is not true for the norm $\int |x(t)| \, dt$.

5. Prove that the union of a finite number of compact sets is compact. Give an example to show that the union of an infinite family of compact sets can fail to be compact.

6. Prove that $\| \; \|_p$ on $\mathbb{R}^n$ does not satisfy the triangle inequality if $0 < p < 1$ and $n \geqslant 2$.

7. Prove that if $x_n \to x$, then the set $\{x, x_1, x_2, \ldots\}$ is compact.

8. A **cluster point** (or **accumulation** point) of a sequence is the limit of any convergent subsequence. Prove that if a sequence lies in a compact set and has only one cluster point, then it is convergent.

9. Prove that the convergence in Problem 1 above is monotone.

10. Give an example of a countable compact set in $\mathbb{R}$ having infinitely many accumulation points. If your example has more than a countable number of accumulation points, give another example, having no more than a countable number.

11. Let x_0 and x_1 be any two points in a normed linear space. Define $x_2, x_3, \ldots$ inductively by putting

$$x_{n+2} = \tfrac{1}{2}(x_{n+1} + x_n) \qquad n = 0, 1, 2, \ldots$$

   Prove that the resulting sequence is a Cauchy sequence.

12. A particular Banach space of great importance is the space $\ell_\infty(S)$, consisting of all bounded real-valued functions on a given set S. For $x \in \ell_\infty(S)$ we define

$$\|x\|_\infty = \sup_{s \in S} |x(s)|$$

   Prove that this space is complete. *Cultural note:* The space $\ell_\infty(\mathbb{N})$ is of special interest. Every separable metric space can be embedded isometrically in it! You might enjoy trying to prove this, but that is not part of problem 12.

13. Prove that in a normed linear space a sequence cannot converge to two different points.

14. How does a sequence $[x_n : n \in \mathbb{N}]$ differ from a countable set $\{x_n : n \in \mathbb{N}\}$?

15. Is there a norm that makes the space of all real sequences a Banach space?

16. Let c_0 denote the space of all real sequences that converge to zero. Define $\|x\| = \sup_n |x(n)|$. Prove that c_0 is a Banach space.

17. If K is a convex set in a linear space, then these two sets are also convex:

$$u + K = \{u + x : x \in K\} \quad \text{and} \quad \lambda K = \{\lambda x : x \in K\}$$

18. Let A be a subset of a linear space. Put

$$A^c = \left\{ \sum_{i=1}^{n} \lambda_i a_i \; : \; n \in \mathbb{N} \,, \; \lambda_i \geqslant 0 \,, \; a_i \in A \,, \; \sum_{i=1}^{n} \lambda_i = 1 \right\}$$

   Prove that $A \subset A^c$. Prove that A^c is convex. Prove that A^c is the smallest convex set containing A. This latter assertion means that if A is contained in a convex set B, then A^c is also contained in B. The set A^c is the **convex hull** of A.

19. If A and B are convex sets, is their vector sum convex? The vector sum of these two sets is $A + B = \{a + b \, : \, a \in A, \, b \in B\}$.

20. Can a norm be recovered from its unit ball? Hint: If $x \in X$, then x/λ is in the unit ball whenever $|\lambda| \geqslant \|x\|$. (Prove this.) On the other hand, x/λ is not in the unit ball if $|\lambda| < \|x\|$. (Prove this.)

21. What are necessary and sufficient conditions on a set S in a linear space X in order that S be the unit ball for some norm on X?

22. Prove that the intersection of a family of convex sets (all contained in one linear space) is convex.

23. A **metric space** is a pair (X, d) in which X is a set and d is a function (called a *metric*) from $X \times X$ to $\mathbb{R}$ such that

    (i) $d(x, y) \geqslant 0$

    (ii) $d(x, y) = 0$ if and only if $x = y$

    (iii) $d(x, y) = d(y, x)$

    (iv) $d(x, y) \leqslant d(x, z) + d(z, y)$

    Prove that a normed linear space is a metric space if $d(x, y)$ is defined as $\|x - y\|$.

24. For this problem only, we use the following notation for a line segment in a linear space:

$$\langle a, b \rangle = \{\lambda a + (1 - \lambda) b : 0 \leqslant \lambda \leqslant 1\}$$

A **polygonal path** joining points a and b is any finite union of line segments $\bigcup_{i=1}^{n} \langle a_i, a_{i+1} \rangle$, where $a_1 = a$ and $a_{n+1} = b$. If the linear space has a norm, the **length** of the polygonal path is $\sum_{i=1}^{n} \|a_i - a_{i+1}\|$. Give an example of a pair of points a, b in a normed linear space and a polygonal path joining them such that the polygonal path is not identical to $\langle a, b \rangle$ but has the same length. A path of length $\|a - b\|$ connecting a and b is called a **geodesic** path. Prove that any geodesic polygonal path connecting a and b is contained in the set $\{x : \|x - a\| \leqslant \|b - a\|\}$.

25. If $x_n \to x$ and if the **Césarò means** are defined by $\sigma_n = (x_1 + \cdots + x_n)/n$, then $\sigma_n \to x$. (This is to be proved in an arbitrary normed linear space.)

26. Prove that a Cauchy sequence that contains a convergent subsequence must converge.

27. A compact set in a normed linear space must be bounded; i.e., contained in some multiple of the unit ball.

28. Prove that the equation $f(x) = \sum_{k=0}^{\infty} a^k \cos b^k x$ defines a continuous function on $\mathbb{R}$, provided that $0 \leqslant a < 1$. The parameter b can be any real number. You will find useful Theorem 2 and Problem 3. Cultural Note: If $0 < a < 1$ and if b is an odd integer greater than a^{-1}, then f is differentiable nowhere. This is the famous Weierstrass nondifferentiable function. (See Section 7.8, page 374, for more information about this function.)

29. Prove that a sequence $[x_n]$ in a normed linear space converges to a point x if and only if every subsequence of $[x_n]$ converges to x.

30. Prove that if ϕ is a strictly increasing function from $\mathbb{N}$ into $\mathbb{N}$, then $\phi(n) \geqslant n$ for all n.

31. Let S be a subset of a linear space. Let S_1 be the union of all line segments that join pairs of points in S. Is S_1 necessarily convex?

32. (continuation)   What happens if we repeat the process and construct $S_2, S_3, \ldots$? (Thus, for example, S_2 is the union of line segments joining points in S_1.)

33. Let I be a compact interval in $\mathbb{R}$, $I = [a, b]$. Let X be a Banach space. The notation $C(I, X)$ denotes the linear space of all continuous maps $f : I \to X$. We norm $C(I, X)$ by putting $\|f\| = \sup_{t \in I} \|f(t)\|$. Prove that $C(I, X)$ is a Banach space.

34. Define $f_n(x) = e^{-nx}$. Show that this sequence of functions converges pointwise on $[0, 1]$ to the function g such that $g(0) = 1$ and $g(t) = 0$ for $t \neq 0$. Show that in the L^2-norm on $[0, 1]$, f_n converges to 0. The L^2-norm is defined by $\|f\| = \{ \int_0^1 |f(t)|^2 dt \}^{1/2}$.

35. Let $[x_n]$ be a sequence in a Banach space. Suppose that for every $\varepsilon > 0$ there is a convergent sequence $[y_n]$ such that $\sup_n \|x_n - y_n\| < \varepsilon$. Prove that $[x_n]$ converges.

36. In any normed linear space, define $K(x, r) = \{y : \|x - y\| \leqslant r\}$. Prove that if $K(x, \frac{1}{2}) \subset K(0, 1)$ then $0 \in K(x, \frac{1}{2})$.

37. Show that the closed unit ball in a normed linear space cannot contain a disjoint pair of closed balls having radius $\frac{1}{2}$.

38. (Converse of Problem 3)   Prove that if every absolutely convergent series converges in a normed linear space, then the space is complete. (A series $\sum x_n$ is **absolutely convergent** if $\sum \|x_n\| < \infty$.)

39. Let X be a compact Hausdorff space, and let $C(X)$ be the space of all real-valued continuous functions on X, with norm $\|f\| = \sup |f(x)|$. Let $[f_n]$ be a Cauchy sequence in $C(X)$. Prove that

$$\lim_{x \to x_0} \lim_{n \to \infty} f_n(x) = \lim_{n \to \infty} \lim_{x \to x_0} f_n(x)$$

Give examples to show why compactness, continuity, and the Cauchy property are needed.

40. The space ℓ_1 consists of all sequences $x = [x(1), x(2), \ldots]$ in which $x(n) \in \mathbb{R}$ and $\sum |x(n)| < \infty$. The space ℓ_2 consists of sequences for which $\sum |x(n)|^2 < \infty$. Prove that $\ell_1 \subset \ell_2$ by establishing the inequality $\sum |x(n)|^2 \leqslant (\sum |x(n)|)^2$.

41. Let X be a normed linear space, and S a dense subset of X. Prove that if each Cauchy sequence in S has a limit in X, then X is complete. A set S is **dense** in X if each point of X is the limit of some sequence in S.

42. Give an example of a linearly independent sequence $[x_0, x_1, x_2, \ldots]$ of vectors in ℓ_∞ such that $\sum_{n=0}^{\infty} x_n = 0$. Don't forget to *prove* that $\sum x_n = 0$.

43. Prove, in a normed space, that if $x_n \to x$ and $\|x_n - y_n\| \to 0$, then $y_n \to x$. If $x_n \to x$ and $\|x_n - y_n\| \to 1$, what is $\lim y_n$?

44. Whenever we consider real-valued or complex-valued functions, there is a concept of absolute value of a function. For example, if $x \in C[0, 1]$, we define $|x|$ by writing $|x|(t) = |x(t)|$. A norm on a space of functions is said to be **monotone** if $\|x\| \geqslant \|y\|$ whenever $|x| \geqslant |y|$. Prove that the norms $\| \quad \|_\infty$ and $\| \quad \|_p$ are monotone norms.

45. (Continuation) Prove that there is no monotone norm on the space of all real-valued sequences.

46. Why isn't the example of this section a counterexample to Theorem 2?

47. Any normed linear space X can be embedded as a dense subspace in a complete normed linear space $\overline{X}$. The latter is fully determined by the former, and is called the **completion** of X. A more general assertion of the same sort is true for metric spaces. Prove that the completion of the space ℓ in Example 8 of Section 1.1 (page 4) is the space c_0 described in Problem 16. Further remarks about the process of completion occur in Section 1.8, page 60.

48. Metric spaces were defined in Problem 23, page 13. In a metric space, a Cauchy sequence is one that has the property $\lim_{n,m} d(x_n, x_m) = 0$. A metric space is **complete** if every Cauchy sequence converges to some point in the space. For the discrete metric space mentioned in Problem 11 (page 19), identify the Cauchy sequences and determine whether the space is complete.

1.3 Continuity, Open Sets, Closed Sets

Consider a function f, defined on a subset D of a normed linear space X and taking values in another normed linear space Y. We say that f is **continuous at a point** x in D if for every sequence $[x_n]$ in D converging to x, we have also $f(x_n) \to f(x)$. Expressed otherwise,

$$f(\lim x_n) = \lim f(x_n)$$

A function that is continuous at each point of its domain is said simply to be **continuous**. Thus a continuous function is one that preserves the convergence of sequences.

Example.   The norm in a normed linear space is continuous. To see that this is so, just use Problem 3, page 5, to write

$$\Big|\ \|x_n\| - \|x\|\ \Big| \leqslant \|x_n - x\|$$

Thus, if $x_n \to x$, it follows that $\|x_n\| \to \|x\|$.   ∎

With these definitions at our disposal, we can prove a number of important (yet elementary) theorems.

Theorem 1.   *Let f be a continuous mapping whose domain D is a compact set in a normed linear space and whose range is contained in another normed linear space. Then $f(D)$ is compact.*

Proof.   To show that $f(D)$ is compact, we let $[y_n]$ be any sequence in $f(D)$, and prove that this sequence has a convergent subsequence whose limit is in $f(D)$. There exist points $x_n \in D$ such that $f(x_n) = y_n$. Since D is compact, the sequence $[x_n]$ has a subsequence $[x_{n_i}]$ that converges to a point $x \in D$. Since f is continuous,

$$f(x) = f\big(\lim_i x_{n_i}\big) = \lim_i f(x_{n_i}) = \lim_i y_{n_i}$$

Thus the subsequence $[y_{n_i}]$ converges to a point in $f(D)$.   ∎

The following is a generalization to normed linear spaces of a theorem that should be familiar from elementary calculus. It provides a tool for optimization problems—even those for which the solution is a function.

Theorem 2.   *A continuous real-valued function whose domain is a compact set in a normed linear space attains its supremum and infimum; both of these are therefore finite.*

Proof.   Let f be a continuous real-valued function whose domain is a compact set D in a normed linear space. Let $M = \sup\{f(x) : x \in D\}$. Then there is a sequence $[x_n]$ in D for which $f(x_n) \to M$. (At this stage, we admit the possibility that M may be $+\infty$.) By compactness, there is a subsequence $[x_{n_i}]$ converging to a point $x \in D$. By continuity, $f(x_{n_i}) \to f(x)$. Hence $f(x) = M$, and of course $M < \infty$. The proof for the infimum is similar.   ∎

A function f whose domain and range are subsets of normed linear spaces is said to be **uniformly continuous** if there corresponds to each positive ε a positive δ such that $\|f(x) - f(y)\| < \varepsilon$ for *all* pairs of points (in the domain of f) satisfying $\|x - y\| < \delta$. The crucial feature of this definition is that δ serves simultaneously for all pairs of points. The definition is *global*, as distinguished from *local*.

Theorem 3.   *A continuous function whose domain is a compact subset of a normed space and whose values lie in another normed space is uniformly continuous.*

Proof.   Let f be a function (defined on a compact set) that is not uniformly continuous. We shall show that f is not continuous. There exists an $\varepsilon > 0$ for which there is no corresponding δ to fulfill the condition of uniform continuity. That implies that for each n there is a pair of points (x_n, y_n) satisfying the condition $\|x_n - y_n\| < 1/n$ and $\|f(x_n) - f(y_n)\| \geq \varepsilon$. By compactness the sequence $[x_n]$ has a subsequence $[x_{n_i}]$ that converges to a point x in the domain of f. Then $y_{n_i} \to x$ also because $\|y_{n_i} - x\| \leq \|y_{n_i} - x_{n_i}\| + \|x_{n_i} - x\|$. Now the continuity of f at x *fails* because

$$\varepsilon \leq \|f(x_{n_i}) - f(y_{n_i})\| \leq \|f(x_{n_i}) - f(x)\| + \|f(x) - f(y_{n_i})\| \qquad ∎$$

A subset F in a normed space is said to be **closed** if the limit of every convergent sequence in F is also in F. Thus, for all sequences this implication is valid:

$$[x_n \in F \ \& \ x_n \to x] \implies x \in F$$

As is true of the notion of completeness, the concept of a closed set is useful when the solution of a problem is constructed as a limit of an approximating sequence.

By Problem 4, the intersection of any family of closed sets is closed. Therefore, the intersection of all the closed sets containing a given set A is a closed set containing A, and it is the smallest such set. It is commonly written as $\overline{A}$ or $\mathrm{cl}(A)$, and is called the **closure** of A.

Theorem 4.   *The inverse image of a closed set by a continuous map is closed.*

Proof.   Recall that the inverse image of a set A by a map f is defined to be $f^{-1}(A) = \{x \, : \, f(x) \in A\}$. Let $f : X \to Y$, where X and Y are normed spaces and f is continuous. Let K be a closed set in Y. To show that $f^{-1}(K)$ is closed, we start by letting $[x_n]$ be a convergent sequence in $f^{-1}(K)$. Thus $x_n \to x$ and $f(x_n) \in K$. By continuity, $f(x_n) \to f(x)$. Since K is closed, $f(x) \in K$. Hence $x \in f^{-1}(K)$.   ∎

As an example, consider the unit ball in a normed space:

$$\{x : \|x\| \leqslant 1\}$$

This is the inverse image of the closed interval $[0, 1]$ by the function $x \mapsto \|x\|$. This function is continuous, as shown above. Hence, the unit ball is closed. Likewise, each of the sets

$$\{x : \|x - a\| \leqslant r\} \qquad \{x : \|x - a\| \geqslant r\} \qquad \{x : \alpha \leqslant \|x - a\| \leqslant \beta\}$$

is closed.

An **open** set is a set whose complement is closed. Thus, from the preceding remarks, the so-called "open unit ball," i.e., the set

$$U = \{x \, : \, \|x\| < 1\}$$

is open, because its complement is closed. Likewise, all of these sets are open:

$$\{x : \|x\| > 1\} \qquad \{x : \|x - a\| < r\} \qquad \{x : \alpha < \|x - a\| < \beta\}$$

An alternative way of describing the open sets, closer to the spirit of general topology, will now be discussed.

The open ε-**cell** or ε-**ball** about a point x_0 is the set

$$B(x_0, \varepsilon) = \{x \, : \, \|x - x_0\| < \varepsilon\}$$

Sometimes this is called the ε-**neighborhood** of x_0. A useful characterization of open sets is the following: A subset U in X is open if and only if for each $x \in U$ there is an $\varepsilon > 0$ such that $B(x, \varepsilon) \subset U$. The collection of open sets is called the **topology** of X. One can verify easily that the topology $\mathcal{T}$ for a normed linear space has these characteristic properties:

(1) the empty set, $\varnothing$, belongs to $\mathcal{T}$;

(2) the space itself, X, belongs to $\mathcal{T}$;

(3) the intersection of any two members of $\mathcal{T}$ belongs to $\mathcal{T}$;

(4) the union of any subfamily of $\mathcal{T}$ belongs to $\mathcal{T}$.

These are the axioms for *any* topology. One section of Chapter 7 provides an introduction to general topology.

A series $\sum_{k=1}^{\infty} x_k$ whose elements are in a normed linear space is **convergent** if the sequence of partial sums $s_n = \sum_{k=1}^{n} x_k$ converges. The given series is said to be **absolutely convergent** if the series of real numbers $\sum_{k=1}^{\infty} \|x_k\|$ is convergent. That means simply that $\sum_{k=1}^{\infty} \|x_k\| < \infty$. Problem 3, page 13, asks for a proof that absolute convergence implies convergence, provided that the space is complete. See also Problem 38, page 14. The following theorem gives another important property of absolutely convergent series.

Theorem 5.   *If a series in a Banach space is absolutely convergent,*
then all rearrangements of the series converge to a common value.

Proof.   Let $\sum_{i=1}^{\infty} x_i$ be such a series and $\sum_{i=1}^{\infty} x_{k_i}$ a rearrangement of it. Put
$x = \sum_{i=1}^{\infty} x_i$, $S_n = \sum_{i=1}^{n} x_i$, $s_n = \sum_{i=1}^{n} x_{k_i}$, and $M = \sum_{i=1}^{\infty} \|x_i\|$. Then
$\sum_{i=1}^{n} \|x_{k_i}\| \leqslant M$. This proves that $\sum_{i=1}^{\infty} x_{k_i}$ is absolutely convergent and hence
convergent. (Here we require the completeness of the space.) Put $y = \sum_{i=1}^{\infty} x_{k_i}$.
Let $\varepsilon > 0$. Select n such that $\sum_{i \geqslant n} \|x_i\| < \varepsilon$ and such that $\|S_m - x\| < \varepsilon$ when
$m \geqslant n$. Select r so that $\|s_r - y\| < \varepsilon$ and so that $\{1, \ldots, n\} \subset \{k_1, \ldots, k_r\}$.
Select m such that $\{k_1, \ldots, k_r\} \subset \{1, \ldots, m\}$. Then $m \geqslant n$ and

$$\|S_m - s_r\| = \|(x_1 + \cdots + x_m) - (x_{k_1} + \cdots + x_{k_r})\| \leqslant \sum_{i=n+1}^{m} \|x_i\| < \varepsilon$$

Hence

$$\|x - y\| \leqslant \|x - S_m\| + \|S_m - s_r\| + \|s_r - y\| < 3\varepsilon \qquad ■$$

In using a series that is not absolutely convergent, some caution must be
exercised. Even in the case of a series of real numbers, bizarre results can arise
if the series is randomly re-ordered. A good example of a series of real numbers
that converges yet is not absolutely convergent is the series $\sum_n (-1)^n/n$. The
series of corresponding absolute values is the divergent **harmonic** series. There
is a remarkable theorem that includes this example:

Riemann's Theorem.   *If a series of real numbers is convergent but*
not absolutely so, then for every real number, some rearrangement of
the series converges to that real number.

Proof.   Let the series $\sum x_n$ satisfy the hypotheses. Then $\lim x_n = 0$ and

$$\sum_{x_n > 0} x_n - \sum_{x_n < 0} x_n = \sum |x_n| = \infty$$

Since the series $\sum x_n$ converges, the two series on the left of the preceding
equation must diverge to $+\infty$ and $-\infty$, respectively. (See Problems 12 and 13.)
Now let r be any real number. Select positive terms (in order) from the series
until their sum exceeds r. Now add negative terms (chosen in order) until the
new partial sum is less than r. Continue in this manner. Since $\lim x_n = 0$, the
partial sums thus created differ from r by quantities that tend to zero.   ■

Problems 1.3

1. Prove that the sequential definition of continuity of f at x is equivalent to the "ε, δ"
definition, which is

$$\forall \varepsilon > 0 \quad \exists \delta > 0 \quad \forall u \, [\, \|x - u\| < \delta \Longrightarrow \|f(x) - f(u)\| < \varepsilon \,]$$

2. Let U be an arbitrary subset of a normed space. Prove that the function $x \mapsto \mathrm{dist}(x, U)$ is continuous. This function was defined in the proof of Theorem 1 in Section 1.2, page 9. Prove, in fact, that it is "nonexpansive":

$$| \mathrm{dist}(x, U) - \mathrm{dist}(y, U) | \leqslant \|x - y\|$$

3. Let X be a normed space. We make $X \times X$ into a normed linear space by defining $\|(x, y)\| = \|x\| + \|y\|$. Show that the map $(x, y) \mapsto x + y$ is continuous. Show that the norm is continuous. Show that the map $(\lambda, x) \mapsto \lambda x$ is continuous when $\mathbb{R} \times X$ is normed by $\|(\lambda, x)\| = |\lambda| + \|x\|$.

4. Prove that the intersection of a family of closed sets is closed.

5. If $x \neq 0$, put $\widetilde{x} = x/\|x\|$. This defines the *radial projection* of x onto the surface of the unit ball. Prove that if x and y are not zero, then

$$\|\widetilde{x} - \widetilde{y}\| \leqslant 2\|x - y\|/\|x\|$$

6. Use Theorem 2 and Problem 2 in this section to give a brief proof of Theorem 1 in Section 2, page 9.

7. Using the definition of an open set as given in this section, prove that a set U is open if and only if for each x in U there is a positive ε such that $B(x, \varepsilon) \subset U$.

8. Prove that the inverse image of an open set by a continuous map is open.

9. The (algebraic) sum of two sets in a linear space is defined by $A + B = \{a + b : a \in A, b \in B\}$. Is the sum of two closed sets (in a normed linear space) closed? (Cf. Problem 19, page 13.)

10. Prove that if the series $\sum_{i=1}^{\infty} x_i$ converges (in some normed linear space), then $x_i \to 0$.

11. A common misconception about metric spaces is that the closure of an open ball $S = \{x : d(a, x) < r\}$ is the closed ball $S^* = \{x : d(a, x) \leqslant r\}$. Investigate whether this is correct in a discrete metric space (X, d), where $d(x, y) = 1$ if $x \neq y$. What is the situation in a normed linear space? (Refer to Problem 23, page 13.)

12. Let $\sum x_n$ and $\sum y_n$ be two series of nonnegative terms. Prove that if one of these series converges but the other does not, then the series $\sum (x_n - y_n)$ diverges. Can you improve this result by weakening the hypotheses?

13. Let $\sum x_n$ be a convergent series of real numbers such that $\sum |x_n| = \infty$. Prove that the series of positive terms extracted from the series $\sum x_n$ diverges to ∞. It may be helpful to introduce $u_n = \max(x_n, 0)$ and $v_n = \min(x_n, 0)$. By using the partial sums of series, one reduces the question to matters concerning the convergence of sequences.

14. Refer to Problem 12, page 12, for the space $\ell_\infty(S)$. We write $\leqslant$ to signify a pointwise inequality between two members of this space. Let g_n and f_n be elements of this space, for $n = 1, 2, \ldots$ Let $g_n \geqslant 0$, $f_{n-1} - g_{n-1} \leqslant f_n \leqslant M$, and $\sum_1^n g_i \leqslant M$ for all n. Prove that the sequence $[f_n]$ converges pointwise. Give an example to show that convergence in norm may fail.

1.4 More About Compactness

We continue our study of compactness in normed linear spaces. The starting point for the next group of theorems is the Heine–Borel theorem, which states that every closed and bounded subset of the real line is compact, and conversely. We assume that the reader is familiar with that theorem.

Our first goal in this section is to show that the Heine–Borel theorem is true for a normed linear space if and only if the space is finite-dimensional. Since most interesting function spaces are infinite-dimensional, verifying the compactness of a set in these spaces requires information beyond the simple properties of being bounded and closed. Many important theorems in functional analysis address the question of identifying the compact sets in various normed linear spaces. Examples of such theorems will appear in Chapter 7.

Lemma 1.   *In the space $\mathbb{R}^n$ with norm $\|x\|_\infty = \max_{1 \leqslant i \leqslant n} |x(i)|$ each ball $\{x : \|x\|_\infty \leqslant c\}$ is compact.*

Proof.   Let $[x_k]$ be a sequence of points in $\mathbb{R}^n$ satisfying $\|x_k\|_\infty \leqslant c$. Then the components obey the inequality $-c \leqslant x_k(i) \leqslant c$. By the compactness of the interval $[-c, c]$, there exists an increasing sequence $I_1 \subset \mathbb{N}$ having the property that $\lim [x_k(1) : k \in I_1]$ exists. Next, there exists another increasing sequence $I_2 \subset I_1$ such that $\lim [x_k(2) : k \in I_2]$ exists. Then $\lim [x_k(1) : k \in I_2]$ exists also, because $I_2 \subset I_1$. Continuing in this way, we obtain at the nth step an increasing sequence I_n such that $\lim [x_k(i) : k \in I_n]$ exists for each $i = 1, \ldots, n$. Denoting that limit by $x^*(i)$, we have defined a vector x^* such that $\|x_k - x^*\|_\infty \to 0$ as k runs through the sequence of integers I_n.   ∎

Lemma 2.   *A closed subset of a compact set is compact.*

Proof.   If F is a closed subset of a compact set K, and if $[x_n]$ is a sequence in F, then by the compactness of K a subsequence converges to a point of K. The limit point must be in F, since F is closed.   ∎

A subset S in a normed linear space is said to be **bounded** if there is a constant c such that $\|x\| \leqslant c$ for all $x \in S$. Expressed otherwise, $\sup_{x \in S} \|x\| < \infty$.

Theorem 1.   *In a finite-dimensional normed linear space, each closed and bounded set is compact.*

Proof.   Let X be a finite-dimensional normed linear space. Select a basis for X, say $\{x_1, \ldots, x_n\}$. Define a mapping $T : \mathbb{R}^n \to X$ by the equation

$$Ta = \sum_{i=1}^{n} a(i)x_i \qquad a = \big(a(1), \ldots, a(n)\big) \in \mathbb{R}^n$$

If we assign the norm $\| \ \|_\infty$ to $\mathbb{R}^n$, then T is continuous because

$$\|Ta - Tb\| = \left\| \sum_{i=1}^{n} (a(i) - b(i))x_i \right\| \leqslant \sum_{i=1}^{n} |a(i) - b(i)| \, \|x_i\|$$

$$\leqslant \max_i |a(i) - b(i)| \cdot \sum_{j=1}^{n} \|x_j\| = \|a - b\|_\infty \sum_{j=1}^{n} \|x_j\|$$

Now let F be a closed and bounded set in X. Put $M = T^{-1}(F)$. Then M is closed by Theorem 4 in Section 1.3, page 17. Since $F = T(M)$, we can use Theorem 1 in Section 1.3, page 15, to conclude that F is compact, provided that M is compact. To show that M is compact, we can use Lemmas 1 and 2 above if we can show that for some c,

$$M \subset \{a \in \mathbb{R}^n : \|a\|_\infty \leqslant c\}$$

In other words, we have only to prove that M is bounded. To this end, define

$$\beta = \inf\{\|Ta\| : \|a\|_\infty = 1\}$$

This is the infimum of a continuous map on a compact set (prove that). Hence the infimum is *attained* at some point b. Thus $\|b\|_\infty = 1$ and

$$\beta = \|Tb\| = \left\|\sum_{i=1}^n b(i)x_i\right\|$$

Since the points x_i constitute a linearly independent set, and since $b \neq 0$, we conclude that $Tb \neq 0$ and that $\beta > 0$. Since F is bounded, there is a constant c such that $\|x\| \leqslant c$ for all $x \in F$. Now, if $a \in \mathbb{R}^n$ and $a \neq 0$, then $a/\|a\|_\infty$ is a vector of norm 1; consequently, $\|T(a/\|a\|_\infty)\| \geqslant \beta$, or

$$\|Ta\| \geqslant \beta\|a\|_\infty$$

This is obviously true for $a = 0$ also. For $a \in M$ we have $Ta \in F$, and $\beta\|a\|_\infty \leqslant \|Ta\| \leqslant c$, whence $\|a\|_\infty \leqslant c/\beta$. Thus, M is indeed bounded. ∎

Corollary 1.   *Every finite-dimensional normed linear space is complete.*

Proof.   Let $[x_n]$ be a Cauchy sequence in such a space. Let us prove that the sequence is bounded. Select an index m such that $\|x_i - x_j\| < 1$ whenever $i, j \geqslant m$. Then we have

$$\|x_i\| \leqslant \|x_i - x_m\| + \|x_m\| \leqslant 1 + \|x_m\| \qquad (i \geqslant m)$$

Hence for all i,

$$\|x_i\| \leqslant 1 + \|x_1\| + \cdots + \|x_m\| \equiv c$$

Since the ball of radius c is compact, our sequence must have a convergent subsequence, say $x_{n_i} \to x^*$. Given $\varepsilon > 0$, select N so that $\|x_i - x_j\| < \varepsilon$ when $i, j \geqslant N$. Then $\|x_j - x_{n_i}\| < \varepsilon$ when $i, j \geqslant N$, because $n_i > i$. By taking the limit as $i \to \infty$, we conclude that $\|x_j - x^*\| \leqslant \varepsilon$ when $j \geqslant N$. This shows that $x_j \to x$. ∎

Corollary 2.   *Every finite-dimensional subspace in a normed linear space is closed.*

Proof. Recall that a subset Y in a linear space is a **subspace** if it is a linear space in its own right. (The only axioms that require verification are the ones concerned with algebraic closure of Y under addition and scalar multiplication.) Let Y be a finite-dimensional subspace in a normed space. To show that Y is closed, let $y_n \in Y$ and $y_n \to y$. We want to know that $y \in Y$. The preceding corollary establishes this: The convergent sequence has the Cauchy property and hence converges to a point in Y, because Y is complete.   ∎

Riesz's Lemma.   *If U is a closed and proper subspace (U is neither 0 nor the entire space) in a normed linear space, and if $0 < \lambda < 1$, then there exists a point x such that $1 = \|x\|$ and $\operatorname{dist}(x, U) > \lambda$.*

Proof.   Since U is proper, there exists a point $z \in X \smallsetminus U$. Since U is closed, $\operatorname{dist}(z, U) > 0$. (See Problem 11.) By the definition of $\operatorname{dist}(z, U)$ there is an element u in U satisfying the inequality $\|z - u\| < \lambda^{-1} \operatorname{dist}(z, U)$. Put $x = (z - u)/\|z - u\|$. Obviously, $\|x\| = 1$. Also, with the help of Problem 7, we have
$$\operatorname{dist}(x, U) = \operatorname{dist}(z - u, U)/\|z - u\| = \operatorname{dist}(z, U)/\|z - u\| > \lambda \qquad ∎$$

Theorem 2.   *If the unit ball in a normed linear space is compact, then the space has finite dimension.*

Proof. If the space is not finite dimensional, then a sequence $[x_n]$ can be defined inductively as follows. Let x_1 be any point such that $\|x_1\| = 1$. If $x_1, \ldots, x_{n-1}$ have been defined, let U_{n-1} be the subspace that they span. By Corollary 2, above, U_{n-1} is closed. Use Riesz's Lemma to select x_n so that $\|x_n\| = 1$ and $\operatorname{dist}(x_n, U_{n-1}) > \frac{1}{2}$. Then $\|x_n - x_i\| > \frac{1}{2}$ whenever $i < n$. This sequence cannot have any convergent subsequence.   ∎

Putting Theorems 1 and 2 together, we have the following result.

Theorem 3.   *A normed linear space is finite dimensional if and only if its unit ball is compact.*

In any normed linear space, a compact set is necessarily closed and bounded. In a finite-dimensional space, these two conditions are also *sufficient* for compactness. In any infinite-dimensional space, some additional hypothesis is required to imply compactness. For many spaces, necessary and sufficient conditions for compactness are known. These invariably involve some uniformity hypothesis. See Section 7.4, page 347, for some examples, and [DS] (Section IV.14) for many others.

Problems 1.4

1. A real-valued function f defined on a normed space is said to be **lower semicontinuous** if each set $\{x : f(x) \leqslant \lambda\}$ is closed ($\lambda \in \mathbb{R}$). Prove that every continuous function is lower

semicontinuous. Prove that if f and $-f$ are lower semicontinuous, then f is continuous. Prove that a lower semicontinuous function attains its infimum on a compact set.

2. Prove that the collection of open sets (as we have defined them) in a normed linear space fulfills the axioms for a topology.

3. Two norms, N_1 and N_2, on a vector space X are said to be **equivalent** if there exist positive constants α and β such that $\alpha N_1 \leqslant N_2 \leqslant \beta N_1$. Show that this is an equivalence relation. Show that the topologies engendered by a pair of equivalent norms are identical.

4. Prove that a Cauchy sequence converges if and only if it has a convergent subsequence.

5. Let X be the linear subspace of all real sequences $x = [x(1), x(2), \ldots]$ such that only a finite number of terms are nonzero. Is there a norm for X such that $(X, \| \ \|)$ is a Banach space?

6. Using the notation in the proof of Theorem 1, prove in detail that $F = T(M)$.

7. Prove these properties of the distance function $\operatorname{dist}(x, U)$ (defined in Section 1.2, page 9) when U is a linear subspace in a normed linear space:

   (a) $\operatorname{dist}(\lambda x, U) = |\lambda| \operatorname{dist}(x, U)$

   (b) $\operatorname{dist}(x - u, U) = \operatorname{dist}(x, U) \qquad (u \in U)$

   (c) $\operatorname{dist}(x + y, U) \leqslant \operatorname{dist}(x, U) + \operatorname{dist}(y, U)$

8. Prove this version of Riesz's Lemma: If U is a finite-dimensional proper subspace in a normed linear space X, then there exists a point x for which $\|x\| = \operatorname{dist}(x, U) = 1$.

9. Prove that if the unit ball in a normed linear space is complete, then the space is complete.

10. Let U be a finite-dimensional subspace in a normed linear space X. Show that for each $x \in X$ there exists a $u \in U$ satisfying $\|x - u\| = \operatorname{dist}(x, U)$.

11. Let U be a closed subspace in a normed space X. Prove that the distance functional has the property that for $x \in X \smallsetminus U$, $\operatorname{dist}(x, U) > 0$.

12. In any infinite-dimensional normed linear space, the open unit ball contains an infinite disjoint family of open balls all having radius $\frac{1}{3}$ (!!) (Prove it, of course. While you're at it, try to improve the number $\frac{1}{3}$.)

13. In the proof of Theorem 1, show that M is bounded as follows. If it is not bounded, let $a_k \in M$ and $\|a_k\|_\infty \to \infty$. Put $a'_k = a_k / \|a_k\|_\infty$. Prove that the sequence $[a'_k]$ has a convergent subsequence whose limit is nonzero. By considering Ta'_k, obtain a contradiction of the injective nature of T.

14. Prove that the sequence $[x_n]$ constructed in the proof of Theorem 2 is linearly independent.

15. Prove that in any infinite-dimensional normed linear space there is a sequence $[x_n]$ in the unit ball such that $\|x_n - x_m\| > 1$ when $n \neq m$. If you don't succeed, prove the same result with the weaker inequality $\|x_n - x_m\| \geqslant 1$. (Use the proof of Theorem 2 and Problem 8 above.) Also prove that the unit ball in ℓ_∞ contains a sequence satisfying $\|x_n - x_m\| = 2$ when $n \neq m$. Reference: [Dies].

16. Let S be a subset of a normed linear space such that $\|x - y\| \geqslant 1$ when x and y are different points in S. Prove that S is closed. Prove that if S is an infinite set then it cannot be compact. Give an example of such a set that is bounded and infinite in the space $C[0, 1]$.

17. Let A and B be nonempty closed sets in a normed linear space. Prove that if $A + B$ is compact, then so are A and B. Why do we assume that the sets are nonempty? Prove that if A is compact, then $A + B$ is closed.

1.5 Linear Transformations

Consider two vector spaces X and Y over the same scalar field. A mapping $f : X \to Y$ is said to be **linear** if

$$f(\alpha u + \beta v) = \alpha f(u) + \beta f(v)$$

for all scalars α and β and for all vectors u, v in X. A linear map is often called a linear **transformation** or a linear **operator**. If Y happens to be the scalar field, the linear map is called a linear **functional**. By taking $\alpha = \beta = 0$ we see at once that a linear map f must have the property $f(0) = 0$. This meaning of the word "linear" differs from the one used in elementary mathematics, where a linear function of a real variable x means a function of the form $x \mapsto ax + b$.

Example 1. If $X = \mathbb{R}^n$ and $Y = \mathbb{R}^m$, then each linear map of X into Y is of the form $f(x) = y$,

$$y(i) = \sum_{j=1}^{n} a_{ij} x(j) \qquad (1 \leqslant i \leqslant m)$$

where the a_{ij} are certain real numbers that form an $m \times n$ matrix.   ∎

Example 2. Let $X = C[0,1]$ and $Y = \mathbb{R}$. One linear functional is defined by $f(x) = \int_0^1 x(s)\,ds$.   ∎

Example 3. Let X be the space of all functions on $[0,1]$ that possess n continuous derivatives, $x', x'', \ldots, x^{(n)}$. Let $a_0, a_1, \ldots, a_n$ be fixed elements of X. Then a linear operator D is defined by

$$Dx = \sum_{i=0}^{n} a_i x^{(i)}$$

Such an operator is called a **differential operator**.   ∎

Example 4. Let $X = C[0,1] = Y$. Let k be a continuous function on $[0,1] \times [0,1]$. Define K by

$$(Kx)(s) = \int_0^1 k(s,t)x(t)\,dt$$

This is a linear operator, in fact a **linear integral operator**.   ∎

Example 5. Let X be the set of all bounded continuous functions on $\mathbb{R}_+ = \{t \in \mathbb{R} : t \geqslant 0\}$. Put

$$(Lx)(s) = \int_0^{\infty} e^{-st} x(t)\,dt$$

This linear operator is called the **Laplace Transform**.   ∎

Example 6. Let X be the set of all continuous functions on $\mathbb{R}$ for which $\int_{-\infty}^{\infty} |x(t)|\,dt < \infty$. Define

$$(Fx)(s) = \int_{-\infty}^{\infty} e^{-2\pi i s t} x(t)\,dt$$

This linear operator is called the **Fourier Transform**.   ∎

If a linear transformation T acts between two normed linear spaces, then the concept of continuity becomes meaningful.

Theorem 1.   *A linear transformation acting between normed linear spaces is continuous if and only if it is continuous at zero.*

Proof.   Let $T : X \to Y$ be such a linear transformation. If it is continuous, then of course it is continuous at 0. For the converse, suppose that T is continuous at 0. For each $\varepsilon > 0$ there is a $\delta > 0$ such that for all x,

$$\|x\| < \delta \quad \Longrightarrow \quad \|Tx\| < \varepsilon$$

Hence

$$\|x - y\| < \delta \quad \Longrightarrow \quad \|Tx - Ty\| = \|T(x - y)\| < \varepsilon \qquad \blacksquare$$

A linear transformation T acting between two normed linear spaces is said to be **bounded** if it is bounded in the usual sense on the unit ball:

$$\sup\{\|Tx\| : \|x\| \leqslant 1\} < \infty$$

Example 7.   Let $X = C^1[0, 1]$, the space of all continuously differentiable functions on $[0, 1]$. Give X the norm $\|x\|_\infty = \sup |x(s)|$. Let f be the linear functional defined by $f(x) = x'(1)$. This functional is not bounded, as is seen by considering the vectors $x_n(s) = s^n$. On the other hand, the functional in Example 2 is bounded since $|f(x)| \leqslant \int_0^1 |x(s)| \, ds \leqslant \|x\|_\infty$. $\blacksquare$

Theorem 2.   *A linear transformation acting between normed linear spaces is continuous if and only if it is bounded.*

Proof.   Let $T : X \to Y$ be such a map. If it is continuous, then there is a $\delta > 0$ such that

$$\|x\| \leqslant \delta \Longrightarrow \|Tx\| \leqslant 1$$

If $\|x\| \leqslant 1$, then δx is a vector of norm at most δ. Consequently, $\|T(\delta x)\| \leqslant 1$, whence $\|Tx\| \leqslant 1/\delta$. Conversely, if $\|Tx\| \leqslant M$ whenever $\|x\| \leqslant 1$, then

$$\|x\| \leqslant \frac{\varepsilon}{M} \Longrightarrow \left\| \frac{Mx}{\varepsilon} \right\| \leqslant 1 \Longrightarrow \left\| T\left(\frac{Mx}{\varepsilon} \right) \right\| \leqslant M \Longrightarrow \|Tx\| \leqslant \varepsilon$$

This proves continuity at 0, which suffices, by the preceding theorem. $\blacksquare$

If $T : X \to Y$ is a bounded linear transformation, we define

$$\|T\| = \sup\{\|Tx\| : \|x\| \leqslant 1\}$$

It can be shown that this defines a norm on the family of all bounded linear transformations from X into Y; this family is a vector space, and it now becomes a normed linear space, denoted by $\mathcal{L}(X, Y)$.

The definition of $\|T\|$ leads at once to the important inequality

$$\|Tx\| \leqslant \|T\| \, \|x\|$$

To prove this, notice first that it is correct for $x = 0$, since $T0 = 0$. On the other hand, if $x \neq 0$, then $x/\|x\|$ is a vector of norm 1. By the definition of $\|T\|$, we have $\|T(x/\|x\|)\| \leqslant \|T\|$, which is equivalent to the inequality displayed above. That inequality contains three distinct norms: the ones defined on X, Y, and $\mathcal{L}(X, Y)$.

Theorem 3.   *A linear functional on a normed space is continuous if and only if its kernel ("null space") is closed.*

Proof.   Let $f : X \to \mathbb{R}$ be a linear functional. Its kernel is

$$\ker(f) = \{x : f(x) = 0\}$$

This is the same as $f^{-1}(\{0\})$. Thus if f is continuous, its kernel is closed, by Theorem 4 in Section 1.3, page 17. Conversely, if f is discontinuous, then it is not bounded. Let $\|x_n\| \leqslant 1$ and $f(x_n) \to \infty$. Take any x not in the kernel and consider the points $x - \varepsilon_n x_n$, where $\varepsilon_n = f(x)/f(x_n)$. These points belong to the kernel of f and converge to x, which is not in the kernel, so the latter is not closed.   ∎

Corollary 1.   *Every linear functional on a finite-dimensional normed linear space is continuous.*

Proof.   If f is such a functional, its null space is a subspace, which, by Corollary 2 in Section 1.4, page 22, must be closed. Then Theorem 3 above implies that f is continuous.   ∎

Corollary 2.   *Every linear transformation from a finite-dimensional normed space to another normed space is continuous.*

Proof.   Let $T : X \to Y$ be such a transformation. Let $\{b_1, \ldots, b_n\}$ be a basis for X. Then each $x \in X$ has a unique expression as a linear combination of basis elements. The coefficients depend on x, and so we write $x = \sum_{i=1}^{n} \lambda_i(x) b_i$. These functionals λ_i are in fact **linear**. Indeed, from the previous equation and the equation $u = \sum \lambda_i(u) b_i$ we conclude that

$$\alpha x + \beta u = \sum_{i=1}^{n} \left[\alpha \lambda_i(x) + \beta \lambda_i(u) \right] b_i$$

Since we have also

$$\alpha x + \beta u = \sum_{i=1}^{n} \lambda_i(\alpha x + \beta u) b_i$$

we may conclude (by the uniqueness of the representations) that

$$\lambda_i(\alpha x + \beta u) = \alpha \lambda_i(x) + \beta \lambda_i(u)$$

Now use the preceding corollary to infer that the functionals λ_i are continuous. Getting back to T, we have

$$Tx = T\left(\sum_{i=1}^{n} \lambda_i(x) b_i \right) = \sum_{i=1}^{n} \lambda_i(x) T b_i$$

and this is obviously continuous.   ∎

Corollary 3.   *All norms on a finite-dimensional vector space are equivalent, as defined in Problem 3, page 23.*

Proof.   Let X be a finite-dimensional vector space having two norms $\| \ \|_1$ and $\| \ \|_2$. The identity map I from $(X, \| \ \|_1)$ to $(X, \| \ \|_2)$ is continuous by the preceding result. Hence it is bounded. This implies that

$$\|x\|_2 = \|Ix\|_2 \leqslant \alpha \|x\|_1$$

By the symmetry in the hypotheses, there is a β such that $\|x\|_1 \leqslant \beta \|x\|_2$.   ∎

Recall that if X and Y are two normed linear spaces, then the notation $\mathcal{L}(X, Y)$ denotes the set of all bounded linear maps of X into Y. We have seen that boundedness is equivalent to continuity for linear maps in this context. The space $\mathcal{L}(X, Y)$ has, in a natural way, all the structure of a normed linear space. Specifically, we define

$$(\alpha A + \beta B)(x) = \alpha(Ax) + \beta(Bx)$$
$$\|A\| = \sup\{\|Ax\|_Y : x \in X \ , \ \|x\|_X \leqslant 1\}$$

In these equations, A and B are elements of $\mathcal{L}(X, Y)$, and x is any member of X.

Theorem 4.   *If X is a normed linear space and Y is a Banach space, then $\mathcal{L}(X, Y)$ is a Banach space.*

Proof.   The only issue is the completeness of $\mathcal{L}(X, Y)$. Let $[A_n]$ be a Cauchy sequence in $\mathcal{L}(X, Y)$. For each $x \in X$, we have

$$\|A_n x - A_m x\| = \|(A_n - A_m)x\| \leqslant \|A_n - A_m\| \ \|x\|$$

This shows that $[A_n x]$ is a Cauchy sequence in Y. By the completeness of Y we can define $Ax = \lim A_n x$. The linearity of A follows by letting $n \to \infty$ in the equation

$$A_n(\alpha x + \beta u) = \alpha A_n x + \beta A_n u$$

The boundedness of A follows from the boundedness of the Cauchy sequence $[A_n]$. If $\|A_n\| \leqslant M$ then $\|A_n x\| \leqslant M\|x\|$ for all x, and in the limit we have $\|Ax\| \leqslant M\|x\|$. Finally, we have $\|A_n - A\| \to 0$ because if $\|A_n - A_m\| \leqslant \varepsilon$ when $m, n \geqslant N$, then for all x of norm 1 we have $\|A_n x - A_m x\| \leqslant \varepsilon$ when $m, n \geqslant N$. Then we can let $m \to \infty$ to get $\|A_n x - Ax\| \leqslant \varepsilon$ and $\|A_n - A\| \leqslant \varepsilon$.   ∎

The composition of two linear mappings A and B is conventionally written as AB rather than $A \circ B$. Thus, $(AB)x = A(Bx)$. If AA is well-defined (i.e., the range of A is contained in its domain), then we write it as A^2. All nonnegative powers are then defined recursively by writing $A^0 = I$, $A^{n+1} = AA^n$.

Theorem 5. The Neumann Theorem.   *Let A be a bounded linear operator on a Banach space X (and taking values in X). If $\|A\| < 1$, then $I - A$ is invertible, and*

$$(I - A)^{-1} = \sum_{k=0}^{\infty} A^k$$

Proof.   Put $B_n = \sum_{k=0}^{n} A^k$. The sequence $[B_n]$ has the Cauchy property, for if $n > m$, then

$$\|B_n - B_m\| = \left\| \sum_{k=m+1}^{n} A^k \right\| \leqslant \sum_{k=m+1}^{n} \|A^k\| \leqslant \sum_{k=m}^{\infty} \|A\|^k$$

$$= \|A\|^m \sum_{k=0}^{\infty} \|A\|^k = \|A\|^m / \left(1 - \|A\|\right)$$

(In this calculation we used Problem 20.) Since the space of all bounded linear operators on X into X is complete (Theorem 4), the sequence $[B_n]$ converges to a bounded linear operator B. We have

$$(I - A)B_n = B_n - AB_n = \sum_{k=0}^{n} A^k - \sum_{k=1}^{n+1} A^k = I - A^{n+1}$$

Taking a limit, we obtain $(I - A)B = I$. Similarly, $B(I - A) = I$. Hence $B = (I - A)^{-1}$.   ∎

The Neumann Theorem is a powerful tool, having applications to many applied problems, such as integral equations and the solving of large systems of linear equations. For examples, see Section 4.3, which is devoted to this theorem, and Section 3.3, which has an example of a nonlinear integral equation.

Problems 1.5

1. Prove that the closure of a linear subspace in a normed linear space is also a subspace. (The closure operation is defined on page 16.)

2. Prove that the operator norm defined here has the three properties required of a norm.

3. Prove that the kernel of a linear functional is either closed or dense. (A subset in a topological space X is **dense** if its closure is X.)

4. Let $\{x_1, \ldots, x_k\}$ be a linearly independent finite set in a normed linear space. Show that there exists a $\delta > 0$ such that the condition

$$\max_{1 \leqslant i \leqslant k} \|x_i - y_i\| < \delta$$

implies that $\{y_1, \ldots, y_k\}$ is also linearly independent.

5. Prove directly that if T is an unbounded linear operator, then it is discontinuous at 0. (Start with a sequence $[x_n]$ such that $\|x_n\| \leqslant 1$ and $\|Tx_n\| \to \infty$.)

6. Let A be an $m \times n$ matrix. Let $X = \mathbb{R}^n$, with norm $\|x\|_\infty = \max_{1 \leqslant i \leqslant n} |x(i)|$. Let $Y = \mathbb{R}^m$, with norm $\|y\|_\infty = \max_{1 \leqslant i \leqslant m} |y(i)|$. Define a linear transformation T from X to Y by putting $(Tx)(i) = \sum_{j=1}^n a_{ij}x(j)$, $1 \leqslant i \leqslant m$. Prove that $\|T\| = \max_i \sum_{j=1}^n |a_{ij}|$.

7. Prove that a linear map is injective (i.e., one-to-one) if and only if its kernel is the 0 subspace. (The kernel of a map T is $\{x : Tx = 0\}$.)

8. Prove that the norm of a linear transformation is the infimum of all the numbers M that satisfy the inequality $\|Tx\| \leqslant M\|x\|$ for all x.

9. Prove the (surprising) result that a linear transformation is continuous if and only if it transforms every sequence converging to zero into a bounded sequence.

10. If f is a linear functional on X and N is its kernel, then there exists a one-dimensional subspace Y such that $X = Y \oplus N$. (For two sets in a linear space, we define $U + V$ as the set of all sums $u + v$ when u ranges over U and v ranges over V. If U and V are subspaces with only 0 in common we write this sum as $U \oplus V$.)

11. The space $\ell_\infty(S)$ was defined in Problem 12 of Section 1.2, page 12. Let $S = \mathbb{N}$, and define $T : \ell_\infty(\mathbb{N}) \to C[-\frac{1}{2}, \frac{1}{2}]$ by the equation $(Tx)(s) = \sum_{k=1}^\infty x(k)s^k$. Prove that T is linear and continuous.

12. Prove or disprove: A linear map from a normed linear space into a finite-dimensional normed linear space must be continuous.

13. Addition of sets in a vector space is defined by $A + B = \{a + b : a \in A, \; b \in B\}$. Better: $A + B = \{x : \exists a \in A \; \& \; \exists b \in B \text{ such that } x = a + b\}$. Scalar multiplication is $\lambda A = \{\lambda a : a \in A\}$. Does the family of all subsets of a vector space X form a vector space with these definitions?

14. Let Y be a closed subspace in a Banach space X. A "coset" is a set of the form $x + Y = \{x + y : y \in Y\}$. Show that the family of all cosets is a normed linear space if we use the norm $\|x + Y\| = \text{dist}(x, Y)$.

15. Refer to Problem 12 in the preceding section, page 23. Show that the assertion there is not true if $\frac{1}{3}$ is replaced by $\frac{1}{2}$.

16. Prove that for a bounded linear transformation $T : X \to Y$

$$\|T\| = \sup_{\|x\|=1} \|Tx\| = \sup_{x \neq 0} \|Tx\|/\|x\|$$

17. Prove that a bounded linear transformation maps Cauchy sequences into Cauchy sequences.

18. Prove that if a linear transformation maps some nonvoid open set of the domain space to a bounded set in the range space, then it is continuous.

19. On the space $C[0,1]$ we define "point-evaluation functionals" by $t^*(x) = x(t)$. Here $t \in [0,1]$ and $x \in C[0,1]$. Prove that $\|t^*\| = 1$. Prove that if $\phi = \sum_{i=1}^n \lambda_i t_i^*$, where $t_1, t_2, \ldots, t_n$ are distinct points in $[0,1]$, then $\|\phi\| = \sum_{i=1}^n |\lambda_i|$.

20. In the proof of the Neumann Theorem we used the inequality $\|A^k\| \leqslant \|A\|^k$. Prove this.

21. Prove that if $\{\phi_1, \ldots, \phi_n\}$ is a linearly independent set of linear functionals, then for suitable x_j we have $\phi_i(x_j) = \delta_{ij}$ for $1 \leqslant i, j \leqslant n$.

22. Prove that if a linear transformation is discontinuous at one point, then it is discontinuous everywhere.

23. Linear transformations on infinite-dimensional spaces do not always behave like their counterparts on finite-dimensional spaces. The space c_0 was defined in Problem 1.2.16 (page 12). On the space c_0 define

$$Ax = A[x(1), x(2), \ldots] = [x(2), x(3), \ldots]$$
$$Bx = B[x(1), x(2), \ldots] = [0, x(1), x(2), \ldots]$$

Prove that A is surjective but not invertible. Prove that B is injective but not invertible. Determine whether right or left inverses exist for A and B.

24. What is meant by the assertion that the behavior of a linear map at any point of its domain is exactly like its behavior at 0?

25. Prove that every linear functional f on $\mathbb{R}^n$ has the form $f(x) = \sum_{i=1}^n \alpha_i x(i)$, where $x(1), x(2), \ldots, x(n)$ are the coordinates of x. Let $\alpha = [\alpha_1, \alpha_2, \ldots, \alpha_n]$ and show that the relationship $f \mapsto \alpha$ is linear, injective, and surjective (hence, an isomorphism).

26. Is it true for linear operators in general that continuity follows from the null space being closed?

27. Let $\phi_0, \phi_1, \ldots, \phi_n$ be linear functionals on a linear space. Prove that if the kernel of ϕ_0 contains the kernels of all ϕ_i for $1 \leqslant i \leqslant n$, then ϕ_0 is a linear combination of $\phi_1, \ldots, \phi_n$.

28. If L is a bounded linear map from a normed space X to a Banach space Y, then L has a unique continuous linear extension defined on the completion of X and taking values in Y. (Refer to Problem 1.2.47, page 15.) Prove this assertion as well as the fact that the norm of the extension equals the norm of the original L.

29. Let A be a continuous linear operator on a Banach space X. Prove that the series $\sum_{n=0}^{\infty} A^n/n!$ converges in $\mathcal{L}(X, X)$. The resulting sum can be denoted by e^A. Is e^A invertible?

30. Investigate the continuity of the Laplace transform (in Example 5, page 24).

1.6 Zorn's Lemma, Hamel Bases, and the Hahn-Banach Theorem

This section is devoted to two results that require the Axiom of Choice for their proofs. These are a theorem on existence of Hamel bases, and the Hahn–Banach Theorem. The first of these extends to all vector spaces the notion of a base, which is familiar in the finite-dimensional setting. The Hahn–Banach Theorem is needed at first to guarantee that on a given normed linear space there can be defined continuous maps into the scalar field. There are many situations in applied mathematics where the Hahn–Banach Theorem plays a crucial role; convex optimization theory is a prime example.

The Axiom of Choice is an axiom that most mathematicians use unreservedly, but is nonetheless controversial. Its status was clarified in 1940 by a famous theorem of Gödel [Go]. His theorem can be stated as follows.

Theorem 1. *If a contradiction can be derived from the Zermelo–Fraenkel axioms of set theory (which include the Axiom of Choice), then a contradiction can be derived within the restricted set theory based on the Zermelo–Fraenkel axioms without the Axiom of Choice.*

In other words, the Axiom of Choice by itself cannot be responsible for introducing an inconsistency in set theory. That is why most mathematicians are willing to accept it. In 1963, Paul Cohen [Coh] proved that the Axiom of Choice is independent of the remaining axioms in the Zermelo–Fraenkel system. Thus it cannot be proved from them. The statement of this axiom is as follows:

Axiom of Choice. *If A is a set and f a function on A such that $f(\alpha)$ is a nonvoid set for each $\alpha \in A$, then f has a "choice function." That means a function c on A such that $c(\alpha) \in f(\alpha)$ for all $\alpha \in A$.*

For example, suppose that A is a finite set: $A = \{\alpha_1, \ldots, \alpha_n\}$. For each i in $\{1, 2, \ldots, n\}$ a nonempty set $f(\alpha_i)$ is given. In n steps, we can select "representatives" $x_1 \in f(\alpha_1)$, $x_2 \in f(\alpha_2)$, etc. Having done so, define $c(\alpha_i) = x_i$ for $i = 1, 2, \ldots, n$. Attempting the same construction for an infinite set such as $A = \mathbb{R}$, with accompanying infinite sets $f(\alpha)$, leads to an immediate difficulty. To get around the difficulty, one might try to order the elements of each set $f(\alpha)$ in such a way that there is always a "first" element in $f(\alpha)$. Then $c(\alpha)$ can be defined to be the first element in $f(\alpha)$. But the proposed ordering will require another axiom at least as strong as the Axiom of Choice! For a second example, see Problem 45, page 40.

A number of other set-theoretic axioms are equivalent to the Axiom of Choice. See [Kel] and [RR]. Among these equivalent axioms, we single out Zorn's Lemma as being especially useful. First, we require some definitions.

Definition 1. A **partially ordered set** is a pair $(X, \prec)$ in which X is a set and $\prec$ is a relation on X such that

(i) $x \prec x$ for all x
(ii) If $x \prec y$ and $y \prec z$, then $x \prec z$

Definition 2. A **chain**, or **totally ordered set**, is a partially ordered set in which for any two elements x and y, either $x \prec y$ or $y \prec x$.

Definition 3. In a partially ordered set X, an **upper bound** for a subset A in X is any point x in X such that $a \prec x$ for all $a \in A$.

Example 1. Let S be any set, and denote by 2^S the family of all subsets of S, including the empty set $\varnothing$ and S itself. This is often called the "power set" of S. Order 2^S by the inclusion relation $\subset$. Then $(2^S, \subset)$ is a partially ordered set. It is not totally ordered. An upper bound for any subset of 2^S is S. ∎

Example 2. In $\mathbb{R}^2$, define $x \prec y$ to mean that $x(i) \leqslant y(i)$ for $i = 1$ and 2. This is a partial ordering but not a total ordering. Which quadrants in $\mathbb{R}^2$ have upper bounds? ∎

Example 3. Let $\mathcal{F}$ be a family of functions (whose ranges and domains need not be specified). For f and g in $\mathcal{F}$ we write $f \prec g$ if two conditions are fulfilled:

(i) $\mathrm{dom}(f) \subset \mathrm{dom}(g)$

(ii) $f(x) = g(x)$ for all x in $\mathrm{dom}(f)$

When this occurs, we say that "g is an extension of f." Notice that this is equivalent to the assertion $f \subset g$, provided that we interpret (as ultimately we must) f and g as sets of pairs of elements.   ∎

Definition 4.   An element m in a partially ordered set X is said to be a **maximal element** if every x in X that satisfies the condition $m \prec x$ also satisfies $x \prec m$.

Zorn's Lemma.   *A partially ordered set contains a maximal element if each totally ordered subset has an upper bound.*

Definition 5.   Let X be a linear space. A subset H of X is called a **Hamel base**, or **Hamel basis**, if each point in X has a unique expression as a finite linear combination of elements of H.

Example 4.   Let X be the space of all polynomials defined on $\mathbb{R}$. A Hamel base for X is given by the sequence $[h_n]$ where $h_n(s) = s^n$, $n = 0, 1, 2, \ldots$.   ∎

Theorem 2.   *Every nontrivial vector space has a Hamel base.*

Proof.   Let X be a nontrivial vector space. To show that X has a Hamel base we first prove that X has a maximal linearly independent set, and then we show that any such set is necessarily a Hamel base. Consider the collection of all linearly independent subsets of X, and partially order this collection by inclusion, $\subset$. In order to use Zorn's Lemma, we verify that every chain in this partially ordered set has an upper bound. Let C be a chain. Consider $S^* = \bigcup \{S : S \in C\}$. This certainly satisfies $S \subset S^*$ for all $S \in C$. But is S^* linearly independent? Suppose that $\sum_{i=1}^{n} \alpha_i s_i = 0$ for some scalars α_i and for some distinct points s_i in S^*. Each s_i belongs to some $S_i \in C$. Since C is a chain (and since there are only finitely many s_i), one of these sets (say S_j) contains all the others. Since S_j is linearly independent, we conclude that $\sum |\alpha_i| = 0$. This establishes the linear independence of S^* and the fact that every chain in our partially ordered set has an upper bound. Now by Zorn's Lemma, the collection of all linearly independent sets in X has a maximal element, H. To see that H is a Hamel base, let x be any element of X. By the maximality of H, either $H \cup \{x\}$ is linearly dependent or $H \cup \{x\} \subset H$ (and then $x \in H$). In either case, x is a linear combination of elements of H. If x can be represented in two different ways as a linear combination of members of H, then by subtraction, we obtain 0 as a nontrivial linear combination of elements of H, contradicting the linear independence of H.   ∎

In the next theorem, when we say that one real-valued function, f, is **dominated** by another, p, we mean simply that $f(x) \leqslant p(x)$ for all x.

Hahn–Banach Theorem.   *Let X be a real linear space, and let p be a function from X to $\mathbb{R}$ such that $p(x + y) \leqslant p(x) + p(y)$ and $p(\lambda x) = \lambda p(x)$ if $\lambda \geqslant 0$. Any linear functional defined on a subspace of*

X and dominated by p has an extension that is linear, defined on X, and dominated by p.

Proof.   Let f be such a functional, and let X_0 be its domain. Thus X_0 is a linear subspace of X. In approaching the theorem for the first time and wondering how to discover a proof, one naturally asks how to extend the functional f to a domain containing X_0 that is only one dimension larger than X_0. If that is impossible, then the theorem itself cannot be true. Accordingly, let y be a point not in the original domain. To extend f to $X_0 + \mathrm{span}(y)$ it suffices to specify a value for $f(y)$ because of the necessary equation

$$f(x + \lambda y) = f(x) + \lambda f(y) \qquad (x \in X_0 \ , \ \lambda \in \mathbb{R})$$

The value of $f(y)$ must be assigned in such a way that

$$f(x) + \lambda f(y) \leqslant p(x + \lambda y) \qquad (x \in X_0 \ , \ \lambda \in \mathbb{R})$$

If $\lambda = 0$, this inequality is certainly valid. If $\lambda > 0$, we must have

$$f\left(\frac{x}{\lambda}\right) + f(y) \leqslant p\left(\frac{x}{\lambda} + y\right) \qquad (x \in X_0)$$

or

$$f(x_1) + f(y) \leqslant p(x_1 + y) \qquad (x_1 \in X_0)$$

If $\lambda < 0$, we must have

$$f(x_2) + f(y) \geqslant \frac{1}{\lambda} p(x + \lambda y) = -p(-x_2 - y) \qquad (x_2 \in X_0)$$

These two conditions on $f(y)$ can be written together as

$$-p(-x_2 - y) - f(x_2) \leqslant f(y) \leqslant p(x_1 + y) - f(x_1) \qquad (x_1, x_2 \in X_0)$$

In order to see that there is a number satisfying this inequality, we compute

$$f(x_1) - f(x_2) = f(x_1 - x_2) \leqslant p(x_1 - x_2) = p(x_1 + y - x_2 - y)$$
$$\leqslant p(x_1 + y) + p(-x_2 - y)$$

This completes the extension by one dimension.

Next, we partially order by the inclusion relation ($\subset$) all the linear extensions of f that are dominated by p. Thus $h \subset g$ if and only if the domain of g contains the domain of h, and $g(x) = h(x)$ on the domain of h. In order to use Zorn's Lemma, we must verify that each chain in this partially ordered set has an upper bound. But this is true, since the union of all the elements in such a chain is an upper bound for the chain. (Problem 2.)   By Zorn's Lemma, there exists a maximal element $\widetilde{f}$ in our partially ordered set. Then $\widetilde{f}$ is a linear functional that is an extension of f and is dominated by p. Finally, $\widetilde{f}$ must be defined on all of X, for if it were not, a further extension would be possible, as shown in the first part of the proof.   ∎

Corollary 1.   Let ϕ be a linear functional defined on a subspace Y in a normed linear space X and satisfying

$$|\phi(y)| \leqslant M\|y\| \qquad (y \in Y)$$

Then ϕ has a linear extension defined on all of X and satisfying the above inequality on X.

Proof.   Use the Hahn–Banach Theorem with $p(x) = M\|x\|$.   ∎

Corollary 2.   Let Y be a subspace in a normed linear space X. If $w \in X$ and $\operatorname{dist}(w, Y) > 0$, then there exists a continuous linear functional ϕ defined on X such that $\phi(y) = 0$ for all $y \in Y$, $\phi(w) = 1$, and $\|\phi\| = 1/\operatorname{dist}(w, Y)$.

Proof.   Let Z be the subspace generated by Y and w. Each element of Z has a unique representation as $y + \lambda w$, where $y \in Y$ and $\lambda \in \mathbb{R}$. It is clear that ϕ must be defined on Z by writing $\phi(y + \lambda w) = \lambda$. The norm of ϕ on Z is computed as follows, in which the supremum is over all nonzero vectors in Z:

$$\|\phi\| = \sup \; |\phi(y + \lambda w)/\|y + \lambda w\|| = \sup \; |\lambda|/\|y + \lambda w\| = \sup \; 1/\|y/\lambda + w\|$$
$$= 1/\inf \|y + w\| = 1/\operatorname{dist}(w, Y)$$

By Corollary 1, we can extend the functional ϕ to all of X without increase of its norm.   ∎

Corollary 3.   To each point w in a normed linear space there corresponds a continuous linear functional ϕ such that $\|\phi\| = 1$ and $\phi(w) = \|w\|$.

Proof.   In Corollary 2, take Y to be the 0-subspace.   ∎

At this juncture, it makes sense to associate with any normed linear space X a normed space X^* consisting of all continuous linear functionals defined on X. Corollary 3 shows that X^* is not trivial. The space X^* is called the **conjugate space** of X, or the **dual** space or the **adjoint** of X.

Example 4.   Let $X = \mathbb{R}^n$, endowed with the max-norm. Then X^* is (or can be identified with) $\mathbb{R}^n$ with the norm $\|\ \|_1$. To see that this is so, recall (Problem 1.5.25, page 30) that if $\phi \in X^*$, then $\phi(x) = \sum_{i=1}^n u(i)x(i)$ for a suitable $u \in \mathbb{R}^n$. Then

$$\|\phi\| = \sup_{\|x\|_\infty \leqslant 1} \left| \sum_{i=1}^n u(i)x(i) \right| = \sum_{i=1}^n |u(i)| = \|u\|_1 \qquad ∎$$

Example 5.   Let c_0 denote the Banach space of all real sequences that converge to zero, normed by putting $\|x\|_\infty = \sup |x(n)|$. Let ℓ_1 denote the Banach space of all real sequences u for which $\sum_{n=1}^\infty |u(n)| < \infty$, normed by putting $\|u\|_1 = \sum_{n=1}^\infty |u(n)|$. With each $u \in \ell_1$ we associate a functional $\phi_u \in c_0^*$ by means of the equation $\phi_u(x) = \sum_{n=1}^\infty u(n)x(n)$. (The connection between these two spaces is the subject of the next result.)   ∎

Proposition.   *The mapping $u \mapsto \phi_u$ is an isometric isomorphism between ℓ_1 and c_0^*. Thus we can say that c_0^* "is" ℓ_1.*

Proof. Perhaps we had better give a name to this mapping. Let $A : \ell_1 \rightarrow c_0^*$ be defined by $Au = \phi_u$. It is to be shown that for each u, Au is linear and continuous on c_0. Then it is to be shown that A is linear, surjective, and isometric. **Isometric** means $\|Au\| = \|u\|_1$. That ϕ_u is well-defined follows from the absolute convergence of the series defining $\phi_u(x)$:

$$\sum |x(n)|\,|u(n)| \leqslant \sum \|x\|_\infty |u(n)| = \|x\|_\infty \|u\|_1$$

The linearity of ϕ_u is obvious:

$$\phi_u(\alpha x + \beta y) = \sum u(n)\bigl[\alpha x(n) + \beta y(n)\bigr] = \alpha \sum u(n)x(n) + \beta \sum u(n)y(n)$$
$$= \alpha\phi_u(x) + \beta\phi_u(y)$$

The continuity or boundedness of ϕ_u is easy:

$$|\phi_u(x)| = \left|\sum u(n)x(n)\right| \leqslant \sum |u(n)|\,|x(n)| \leqslant \|x\|_\infty \|u\|_1$$

By taking a supremum in this last inequality, considering only x for which $\|x\|_\infty \leqslant 1$, we get

$$\|\phi_u\| \leqslant \|u\|_1$$

On the other hand, if $\varepsilon > 0$ is given, we can select N so that $\sum_{n=N+1}^\infty |u(n)| < \varepsilon$. Then we define x by putting $x(n) = \operatorname{sgn} u(n)$ for $n \leqslant N$, and by setting $x(n) = 0$ for $n > N$. Clearly, $x \in c_0$ and $\|x\|_\infty = 1$. Hence

$$\|\phi_u\| \geqslant \phi_u(x) = \sum_{n=1}^N x(n)u(n) = \sum_{n=1}^N |u(n)| > \|u\|_1 - \varepsilon$$

Since ε was arbitrary, $\|\phi_u\| \geqslant \|u\|_1$. Hence we have proved

$$\|Au\| = \|\phi_u\| = \|u\|_1$$

Next we show that A is surjective. Let $\psi \in c_0^*$. Let δ_n be the element of c_0 that has a 1 in the nth coordinate and zeros elsewhere. Then for any x,

$$x = \sum_{n=1}^\infty x(n)\delta_n$$

Since ψ is continuous and linear,

$$\psi(x) = \sum x(n)\psi(\delta_n)$$

Consequently, if we put $u(n) = \psi(\delta_n)$, then $\psi(x) = \phi_u(x)$ and $\psi = \phi_u$. To verify that $u \in \ell_1$, we define (as above) $x(n) = \text{sgn}\, u(n)$ for $n \leqslant N$ and $x(n) = 0$ for $n > N$. Then

$$\sum_{n=1}^{N} |u(n)| = \sum x(n)u(n) = \psi(x) \leqslant \|\psi\| \, \|x\| = \|\psi\|$$

Thus $\|u\|_1 \leqslant \|\psi\|$.

Finally, the linearity of A follows from writing

$$\phi_{\alpha u + \beta v}(x) = \sum (\alpha u + \beta v)(n)x(n) = \alpha \sum u(n)x(n) + \beta \sum v(n)x(n)$$
$$= (\alpha \phi_u + \beta \phi_v)(x) \qquad \blacksquare$$

Corollary 4.   For each x in a normed linear space X, we have

$$\|x\| = \max\{|\phi(x)| : \phi \in X^* \,, \, \|\phi\| = 1\}$$

Proof. If $\phi \in X^*$ and $\|\phi\| = 1$, then

$$|\phi(x)| \leqslant \|\phi\| \, \|x\| = \|x\|$$

Therefore,

$$\sup\{|\phi(x)| : \phi \in X^* \,, \, \|\phi\| = 1\} \leqslant \|x\|$$

For the reverse inequality, note first that it is trivial if $x = 0$. Otherwise, use Corollary 3. Then there is a functional $\psi \in X^*$ such that $\psi(x) = \|x\|$ and $\|\psi\| = 1$. Note that the supremum is attained. $\qquad \blacksquare$

A subset Z in a normed space X is said to be **fundamental** if the set of all linear combinations of elements in Z is dense in X. Expressed otherwise, for each $x \in X$ and for each $\varepsilon > 0$ there is a vector $\sum_{i=1}^{n} \lambda_i z_i$ such that $z_i \in Z$, $\lambda_i \in \mathbb{R}$, and

$$\left\| x - \sum \lambda_i z_i \right\| < \varepsilon$$

We could also state that $\text{dist}(x, \text{span}\, Z) = 0$ for all $x \in X$. As an example, the vectors

$$\delta_1 = (1, 0, 0, \dots)$$
$$\delta_2 = (0, 1, 0, \dots)$$
$$\text{etc.}$$

form a fundamental set in the space c_0.

Example 6. In the space $C[a, b]$, with the usual supremum norm, an important fundamental set is the sequence of monomials

$$u_0(t) = 1 \,, \; u_1(t) = t \,, \; u_2(t) = t^2 \,, \; \dots$$

The Weierstrass Approximation Theorem asserts the fundamentality of this sequence. Thus, for any $x \in C[a, b]$ and any $\varepsilon > 0$ there is a polynomial u for which $\|x - u\|_\infty < \varepsilon$. Of course, u is of the form $\sum_{i=0}^{n} \lambda_i u_i$. $\qquad \blacksquare$

Definition 5. If A is a subset of a normed linear space X, then the **annihilator** of A is the set

$$A^\perp = \{\phi \in X^* : \phi(a) = 0 \; \text{for all} \; a \in A\}$$

Theorem 3.   *A subset in a normed space is fundamental if and only if its annihilator is $\{0\}$.*

Proof.   Let X be the space and Z the subset in question. Let Y be the closure of the linear span of Z. If $Y \neq X$, let $x \in X \setminus Y$. Then by Corollary 2, there exists $\phi \in X^*$ such that $\phi(x) = 1$ and $\phi \in Y^\perp$. Hence $\phi \in Z^\perp$ and $Z^\perp \neq 0$. If $Y = X$, then any element of $Z^\perp$ annihilates the span of Z as well as Y and X. Thus it must be the zero functional; i.e., $Z^\perp = 0$.   ∎

Theorem 4.   *If X is a normed linear space (not necessarily complete) then its conjugate space X^* is complete.*

Proof.   This follows from Theorem 4 in Section 1.5, page 27, by letting $Y = \mathbb{R}$ in that theorem.   ∎

Problems 1.6

1. Let X and Y be sets. A function from a subset of X to Y is a subset f of $X \times Y$ such that for each $x \in X$ there is at most one $y \in Y$ satisfying $(x, y) \in f$. We write then $f(x) = y$. The set of all such functions is denoted by S. Prove or disprove the following: (a) S is partially ordered by inclusion. (b) The union of two elements of S is a member of S. (c) The intersection of two elements of S is a member of S. (d) The union of any chain in S is a member of S.

2. In the proof of the Hahn–Banach theorem, show that the union of the elements in a chain is an upper bound for the chain. (There are five distinct things to prove.)

3. Denote by c_0 the normed linear space of all functions $x : \mathbb{N} \to \mathbb{R}$ having the property $\lim_{n \to \infty} x(n) = 0$, with norm given by $\|x\| = \sup_n |x(n)|$. Do the vectors e_m defined by $e_m(n) = \delta_{nm}$ form a Hamel base for c_0?

4. If $\{h_\alpha : \alpha \in I\}$ is a Hamel base for a vector space X, then each element x in X has a representation $x = \sum_\alpha \lambda(\alpha) h_\alpha$ in which $\lambda : I \to \mathbb{R}$ and $\{\alpha : \lambda(\alpha) \neq 0\}$ is finite. (Prove this.)

5. Prove that every real vector space is isomorphic to a vector space whose elements are real-valued functions. ("Function spaces are all there are.")

6. Prove that any linearly independent set in a vector space can be extended to produce a Hamel base.

7. If U is a linear subspace in a vector space X, then U has an "algebraic complement," which is a subspace V such that $X = U + V$, $U \cap V = 0$. ("0" denotes the zero subspace.) (Prove this.)

FIVE EXERCISES (8–12) ON BANACH LIMITS

8. The space ℓ^∞ consists of all bounded sequences, with norm $\|x\|_\infty = \sup_n |x(n)|$. Define $T : \ell^\infty \to \ell^\infty$ by putting

$$Tx = [x(1),\ x(2) - x(1),\ x(3) - x(2),\ x(4) - x(3)\ \ldots]$$

Let M denote the range of T, and put $u = [1, 1, 1, \ldots]$. Prove that $\operatorname{dist}(u, M) = 1$.

9. Prove that there exists a continuous linear functional $\phi \in M^\perp$ such that $\|\phi\| = \phi(u) = 1$. The functional ϕ is called a Banach limit, and is sometimes written LIM.

10. Prove that if $x \in \ell^\infty$ and $x \geqslant 0$, then $\phi(x) \geqslant 0$.

11. Prove that $\phi(x) = \lim_n x(n)$ when the limit exists.

12. Prove that if $y = [x(2), x(3), \ldots]$ then $\phi(x) = \phi(y)$.

13. Let ℓ_∞ denote the normed linear space of all bounded real sequences, with norm given by $\|x\|_\infty = \sup_n |x(n)|$. Prove that ℓ_∞ is complete, and therefore a Banach space. Prove that $\ell_1^* = \ell_\infty$, where the equality here really means isometrically isomorphic.

14. A **hyperplane** in a normed space is any translate of the null space of a continuous, linear, nontrivial functional. Prove that a set is a hyperplane if and only if it is of the form $\{x : \phi(x) = \lambda\}$, where $\phi \in X^* \setminus 0$ and $\lambda \in \mathbb{R}$. A **translate** of a set S in a vector space is a set of the form $v + S = \{v + s : s \in S\}$.

15. A **half-space** in a normed linear space X is any set of the form $\{x : \phi(x) \geqslant \lambda\}$, where $\phi \in X^* \setminus 0$ and $\lambda \in \mathbb{R}$. Prove that for every x satisfying $\|x\| = 1$ there exists a half-space such that x is on the boundary of the half-space and the unit ball is contained in the half-space.

16. Prove that a linear functional ϕ is a linear combination of linear functionals $\phi_1, \ldots, \phi_n$ if and only if $N(\phi) \supset \bigcap_{i=1}^n N(\phi_i)$. Here $N(\phi)$ denotes the null space of ϕ. (Use induction and trickery.)

17. Prove that a linear map transforms convex sets into convex sets.

18. Prove that in a normed linear space, the closure of a convex set is convex.

19. Let Y be a linear subspace in a normed linear space X. Prove that

$$\text{dist}(x, Y) = \sup\{\phi(x) : \phi \in X^*\ ,\ \phi \perp Y\ ,\ \|\phi\| = 1\}$$

Here the notation $\phi \perp Y$ means that $\phi(y) = 0$ for all $y \in Y$.

20. Let Y be a subset of a normed linear space X. Prove that $Y^\perp$ is a closed linear subspace in X^*.

21. If Z is a linear subspace in X^*, where X is a normed linear space, we define

$$Z_\perp = \{x \in X : \phi(x) = 0\ \text{ for all }\ \phi \in Z\}$$

Prove that for any closed subspace Y in X, $(Y^\perp)_\perp = Y$. Generalize.

22. Let $f(z) = \sum_{n=0}^\infty a_n z^n$, where $[a_n]$ is a sequence of complex numbers for which $n a_n \to 0$. Prove the famous theorem of Tauber that $\sum a_n$ converges if and only if $\lim_{z \to 1} f(z)$ exists. (See [DS], page 78.)

23. Do the vectors δ_n defined just after Corollary 4 form a fundamental set in the space ℓ_∞ consisting of bounded sequences with norm $\|x\|_\infty = \max_n |x(n)|$?

THREE EXERCISES (24–26) ON SCHAUDER BASES (See [Sem] and [Sing].)

24. A **Schauder base** (or basis) for a Banach space X is a sequence $[u_n]$ in X such that each x in X has a unique representation

$$x = \sum_{n=1}^\infty \lambda_n u_n$$

This equation means, of course, that $\lim_{N \to \infty} \|x - \sum_{n=1}^N \lambda_n u_n\| = 0$. Show that one Schauder base for c_0 is given by $u_n(m) = \delta_{nm}$ $(n, m = 1, 2, 3, \ldots)$.

25. Prove that the λ_n in the preceding problem are functions of x and must be, in fact, linear and continuous.

26. Prove that if the Banach space X possesses a Schauder base, then X must be separable. That is, X must contain a countable dense set.

27. Prove that for any set A in a normed linear space all these sets are the same:
$A^\perp$, (closure $A)^\perp$, (span $A)^\perp$, [closure (span $A)]^\perp, \ldots$

28. Prove that for $x \in c_0$,

$$\|x\|_\infty = \sup\left\{ \sum_{n=1}^\infty x(n)u(n) : u \in \ell_1 \ , \ \|u\|_1 \leqslant 1 \right\}$$

29. Use the Axiom of Choice to prove that for any set S having at least 2 points there is a function $f : S \to S$ that does not have a fixed point.

30. An interesting Banach space is the space c consisting of all convergent sequences. The norm is $\|x\|_\infty = \sup_n |x(n)|$. Obviously, we have these set inclusions among the examples encountered so far:

$$\ell_1 \subset c_0 \subset c \subset \ell_\infty$$

Prove that c_0 is a hyperplane in c. Identify in concrete terms the conjugate space c^*.

31. Prove that if H is a Hamel base for a normed linear space, then so is $\{h/\|h\| : h \in H\}$.

32. Let X and Y be linear spaces. Let H be a Hamel base for X. Prove that a linear map from X to Y is completely determined by its values on H, and that these values can be arbitrarily–assigned elements of Y.

33. Prove that on every infinite-dimensional normed linear space there exist discontinuous linear functionals. (The preceding two problems can be useful here.)

34. Using Problem 33 and Problem 1.5.3, page 28, prove that every infinite-dimensional normed linear space is the union of a disjoint pair of dense convex sets.

35. Let two equivalent norms be defined on a single linear space. (See Problem 1.4.3, page 23.) Prove that if the space is complete with respect to one of the norms, then it is complete with respect to the other. Prove that this result fails (in general) if we assume only that one norm is less than or equal to a constant multiple of the other.

36. Let Y be a subspace of a normed space X. Prove that there is a norm-preserving injective map $J : Y^* \to X^*$ such that for each $\phi \in Y^*$, $J\phi$ is an extension of ϕ.

37. Let Y be a subspace of a normed space X. Prove that if $Y^\perp = 0$, then Y is dense in X.

38. Let T be a bounded linear map of c_0 into c_0. Show that T must have the form $(Tx)(n) = \sum_{i=1}^\infty a_{ni}x(i)$ for a suitable infinite matrix $[a_{ni}]$. Prove that $\sup_n \sum_{i=1}^\infty |a_{ni}| = \|T\|$.

39. Prove that if $\#S = n$, then $\#2^S = 2^n$.

40. What implications exist among these four properties of a set S in a normed linear space X? (a) S is fundamental in X; (b) S is linearly independent; (c) S is a Schauder base for X; (d) S is a Hamel base for X.

41. A "spanning set" in a linear space is a set S such that each point in the space is a linear combination of elements from S. Prove that every linear space has a minimal spanning set.

42. Let $f : \mathbb{R} \to \mathbb{R}$. Define $x \prec y$ to mean $f(x) \leqslant f(y)$. Under what conditions is this a partial order or a total order?

43. Criticize the following "proof" that if X and Y are any two normed linear spaces, then $X^* = Y^*$. We can assume that X and Y are subspaces of a third normed space Z. (For example, we could use $Z = X \oplus Y$, a direct sum.) Clearly, X^* is a subspace of Z^*, since the Hahn–Banach Theorem asserts that an element of X^* can be extended, without increasing its norm, to Z. Clearly, Z^* is a subspace of Y^*, since each element of Z^* can be restricted to become an element of Y^*. So, we have $X^* \subset Z^* \subset Y^*$. By symmetry, $Y^* \subset X^*$. So $X^* = Y^*$.

44. Let K be a subset of a linear space X, and let $f : K \to \mathbb{R}$. Establish necessary and sufficient conditions in order that f be the restriction to K of a linear functional on X.

45. For each α in a set A, let $f(\alpha)$ be a subset of $\mathbb{N}$. Without using the Axiom of Choice, prove that f has a choice function.

1.7 The Baire Theorem and Uniform Boundedness

This section is devoted to the first consequences of *completeness* in a normed linear space. These are stunning and dramatic results that distinguish Banach spaces from other normed linear spaces. Once we have these theorems (in this section and the next), it will be clear why it is always an advantage to be working with a complete space. The reader has undoubtedly seen this phenomenon when studying the real number system (which *is* complete). When we compare the real and the rational number systems, we notice that the latter has certain deficiencies, which indeed had already been encountered by the ancient Greeks. For example, they knew that no square could have rational sides and rational diagonal! Put another way, certain problems posed within the realm of rational numbers do not have solutions among the rational numbers; rather, we must expect solutions sometimes to be irrational. The simplest example, of course, is $x^2 = 2$. Our story begins with a purely metric-space result.

Theorem 1. Baire's Theorem. *In a complete metric space, the intersection of a countable family of open dense sets is dense.*

Proof. (A set is "dense" if its closure is the entire space.) Let $O_1, O_2, \dots$ be open dense sets in a complete metric space X. In order to show that $\bigcap_{n=1}^{\infty} O_n$ is dense, it is sufficient to prove that this set intersects an arbitrary nonvoid open ball S_1 in X. For each n we will define an open ball and a closed ball:

$$ S_n = \{x \in X : d(x, x_n) < r_n\} \qquad S_n' = \{x \in X : d(x, x_n) \leqslant r_n\} $$

Select any $x_1 \in X$ and let $r_1 > 0$. We want to prove that S_1 intersects $\bigcap_{n=1}^{\infty} O_n$. Since O_1 is open and dense, $O_1 \cap S_1$ is open and nonvoid. Take $S_2' \subset S_1 \cap O_1$.

Then take $S_3' \subset S_2 \cap O_2$, $S_4' \subset S_3 \cap O_3$, and so on. At the same time we can insist that $r_n \downarrow 0$. Then for all n,

$$S_{n+1}' \subset S_n \cap O_n \subset S_1 \cap O_n$$

The points x_n form a Cauchy sequence because $x_i, x_j \in S_n$ if $i, j > n$, and so

$$d(x_i, x_j) \leqslant d(x_i, x_n) + d(x_n, x_j) < 2r_n$$

Since X is complete, the sequence $[x_n]$ converges to some point x^*. Since for $i > n$,

$$x_i \in S_{n+1}' \subset S_1 \cap O_n$$

we can let $i \to \infty$ to conclude that $x^* \in S_{n+1}' \subset S_1 \cap O_n$. Since this is true for all n, the set $\bigcap_{n=1}^{\infty} O_n$ does indeed intersect S_1.   ∎

Corollary.   *If a complete metric space is expressed as a countable union of closed sets, then one of the closed sets must have a nonempty interior.*

Proof.   Let X be a complete metric space, and suppose that $X = \bigcup_{n=1}^{\infty} F_n$, where each F_n is a closed set having empty interior. The sets $O_n = X \smallsetminus F_n$ are open and dense. Hence by Baire's Theorem, $\bigcap_{n=1}^{\infty} O_n$ is dense. In particular, it is nonempty. If $x \in \bigcap_{n=1}^{\infty} O_n$, then $x \in X \smallsetminus \bigcup_{n=1}^{\infty} F_n$, a contradiction.   ∎

A subset in a metric space X (or indeed in any topological space) is said to be **nowhere dense** in X if its closure has an empty interior. Thus the set of irrational points on the horizontal axis in $\mathbb{R}^2$ is nowhere dense in $\mathbb{R}^2$. A set that is a countable union of nowhere dense sets is said to be of **category I** in X. A set that is not of category I is said to be of **category II** in X.

Observe that all three of these notions are dependent on the space. Thus one can have $E \subset X \subset Z$, where E is of category II in X and of category I in Z. For a concrete example, the one in the preceding paragraph will serve.

The Corollary implies that if X is a complete metric space, then X is of the second category in X.

Intuitively, we think of sets of the first category as being "thin," and those of the second category as "fat." (See Problems 5, 6, 7, for example.)

Theorem 2.   The Banach–Steinhaus Theorem.   Let $\{A_\alpha\}$ *be a family of continuous linear transformations defined on a Banach space X and taking values in a normed linear space. In order that $\sup_\alpha \|A_\alpha\| < \infty$, it is necessary and sufficient that the set $\{x \in X : \sup_\alpha \|A_\alpha x\| < \infty\}$ be of the second category in X.*

Proof.   Assume first that $c = \sup_\alpha \|A_\alpha\| < \infty$. Then every x satisfies $\|A_\alpha x\| \leqslant c\|x\|$, and every x belongs to the set $F = \{x : \sup_\alpha \|A_\alpha x\| < \infty\}$. Since $F = X$, the preceding corollary implies that F is of the second category in X.

For the sufficiency, define

$$F_n = \left\{ x \in X : \sup_\alpha \|A_\alpha x\| \leqslant n \right\}$$

and assume that F is of the second category in X. Notice that $F = \bigcup_{n=1}^{\infty} F_n$. Since F is of the second category, and each F_n is a closed subset of X, the definition of second category implies that some F_m contains a ball. Suppose that

$$B \equiv \{x \in X : \|x - x_0\| \leqslant r\} \subset F_m \qquad (r > 0)$$

For any x satisfying $\|x\| \leqslant 1$ we have $x_0 + rx \in B$. Hence

$$\|A_\alpha x\| = \|A_\alpha [r^{-1}(x_0 + rx - x_0)]\|$$
$$\leqslant r^{-1}\|A_\alpha(x_0 + rx)\| + r^{-1}\|A_\alpha x_0\| \leqslant 2r^{-1}m$$

Hence $\|A_\alpha\| \leqslant 2r^{-1}m$ for all α.   ∎

> **Theorem 3. The Principle of Uniform Boundedness.**   Let $\{A_\alpha\}$ be a collection of continuous linear maps from a Banach space X into a normed linear space. If $\sup_\alpha \|A_\alpha x\| < \infty$ for each $x \in X$, then $\sup_\alpha \|A_\alpha\| < \infty$.

Example 1.   Consider the familiar space $C[0,1]$. We are going to show that most members of $C[0,1]$ are not differentiable. Select a point ξ in the open interval $(0,1)$. For small positive values of h we define a linear functional ϕ_h by the equation

$$\phi_h(x) = \frac{x(\xi + h) - x(\xi - h)}{2h} \qquad (x \in C[0,1])$$

It is elementary to prove that ϕ_h is linear and that $\|\phi_h\| = h^{-1}$. Consequently, by the Banach-Steinhaus Theorem, the set of x such that $\sup_h |\phi_h(x)| < \infty$ is of the first category. Hence the set of x for which $\sup_h |\phi_h(x)| = \infty$ is of the second category in $C[0,1]$. In other words, the set of functions in $C[0,1]$ that are not differentiable at ξ is of the second category in $C[0,1]$.   ∎

Example 2.   The formal Fourier series of a function x is

$$\sum_{n=-\infty}^{\infty} \alpha_n(x)e^{int}$$

where the functionals α_n are defined by

$$\alpha_n(x) = \frac{1}{2\pi} \int_0^{2\pi} x(s)e^{-ins}\, ds$$

If x belongs to $C_{2\pi}$, the space of continuous 2π-periodic functions on $[0, 2\pi]$ (endowed with the sup-norm), then the coefficients $\alpha_n(x)$ certainly exist; (in fact, they exist if x is only Lebesgue integrable). A sequence of linear operators, called **Fourier projections**, is obtained by truncation of the series:

$$(A_n x)(t) = \sum_{k=-n}^{n} \alpha_k(x)e^{ikt}$$

It can be shown that the norm of A_n, considered as a map of $C_{2\pi}$ into itself, is roughly $(4/\pi^2) \log n$. In fact, the norm of each functional $t^* \circ A_n$ has this property. Recall from Problem 19 in Section 1.5 (page 29) that t^* denotes point–evaluation at t, so that

$$(t^* \circ A_n)(x) = t^*(A_n x) = (A_n x)(t)$$

Since $\sup_n \|t^* \circ A_n\| = +\infty$, the set of x in $C_{2\pi}$ whose Fourier series diverge at a specified point t is a set of the second category. Thus, for most periodic continuous functions, the Fourier series do not converge. ∎

Theorem 4.   *Let $[A_n]$ be a sequence of continuous linear transformations from a Banach space X into a normed linear space. In order that $\lim_n A_n x = 0$ for all $x \in X$ it is necessary and sufficient that $\sup_n \|A_n\| < \infty$ and that $A_n u \to 0$ for each u in some fundamental subset of X.*

Proof. If $A_n x \to 0$ for all x, then obviously $\sup_n \|A_n x\| < \infty$ for all x. Hence $\sup_n \|A_n\| < \infty$, by the Principle of Uniform Boundedness.

For the other half of the theorem, assume that $\|A_n\| < M$ for all n and that $A_n u \to 0$ for all u in a fundamental set F. It is elementary to prove that $A_n y \to 0$ for all y in the linear span of F. Now let $x \in X$. Let $\epsilon > 0$. Select y in the linear span of F so that $\|x - y\| < \epsilon/2M$. Select m so that $\|A_n y\| < \epsilon/2$ whenever $n \geqslant m$. Then for $n \geqslant m$ we have

$$\|A_n x\| \leqslant \|A_n(x - y)\| + \|A_n y\| < M\|x - y\| + \epsilon/2 < \epsilon \qquad ∎$$

Example 3.   The Riemann integral of a continuous function x defined on $[a, b]$ can be obtained as a limit as follows:

$$\int_a^b x(s)\,ds = \lim_{n \to \infty} \sum_{i=1}^{n} x\left(a + i\,\frac{b-a}{n}\right) \cdot \frac{b-a}{n}$$

This suggests that we consider the problem of approximating functionals ψ that have the form

$$(1) \qquad\qquad \psi(x) = \int_a^b x(s)\,w(s)\,ds \qquad x \in C[a, b]$$

in which w is a fixed integrable function called the **weight**. We seek to approximate ψ by a sequence of functionals ϕ_n having the form

$$(2) \qquad\qquad \phi_n(x) = \sum_{i=1}^{n} A_{ni} x(s_{ni}) \qquad x \in C[a, b]$$

Notice that ϕ_n is simply a linear combination of point-evaluation functionals. One can argue with some justification that from the practical, numerical, standpoint only such functionals are realizable. Other functionals, such as those

involving integrals, *must* be approximated by the simpler realizable ones. Functionals of this type were considered in Problem 1.5.19 (page 29), and a result of
that problem is the formula

$$\|\phi_n\| = \sum_{i=1}^{n} |A_{ni}|$$

Here it is necessary to assume that for each n, $\{s_{n1}, s_{n2}, \ldots, s_{nn}\}$ is a set of n
distinct points in $[a, b]$. We call these points the "nodes" of the functional ϕ_n.
An old theorem of Szegő, presented next, concerns this example.   ∎

Theorem 5.   Let ψ and ϕ_n be as in Equations (1) and (2) above.
In order that $\phi_n(x) \to \psi(x)$ for each $x \in C[a, b]$, it is necessary and
sufficient that these two conditions be fulfilled:

(i) $\displaystyle\sup_n \sum_{i=1}^{n} |A_{ni}| < \infty$

(ii) The convergence occurs for all the elementary monomial functions,
$s \mapsto s^k$, $k = 0, 1, 2, \ldots$.

Proof.   Consider the sequence of functionals $[\psi - \phi_n]$. The norm of ψ is

$$\|\psi\| = \sup_{\|x\| \leqslant 1} \left| \int_a^b x(s)\, w(s)\, dx \right| \leqslant \int_a^b |w(s)|\, ds$$

Consequently, condition (i) is equivalent to the condition

$$\sup_n \|\psi - \phi_n\| < \infty$$

Next observe that the functions e_k defined by the equation $e_k(s) = s^k$, where
$k = 0, 1, \ldots$, form a fundamental set in $C[a, b]$, by the Weierstrass Polynomial
Approximation Theorem. Now apply the preceding theorem.   ∎

Problems 1.7

1. Prove the equivalence of these properties of a set U in a normed linear space X:

   (a) U intersects each nonempty open set in X

   (b) U intersects each open ball in X

   (c) The closure of U is X

   (d) For each $x \in X$ and each $\varepsilon > 0$ there is a point $u \in U$ satisfying the inequality $\|x - u\| < \varepsilon$

   (e) The set $X \smallsetminus U$ contains no open ball.

2. An interesting metric space is obtained by taking any set X and defining $d(x, y)$ to be 1 if $x \neq y$ and 0 if $x = y$. In such a metric space identify the open sets, the closed sets, the convergent sequences, and the compact sets. Also determine whether the closure of $\{x : d(x, y) < r\}$ is the set $\{x : d(x, y) \leqslant r\}$. Is (X, d) complete?

3. Prove that the set of functions in $C[0, 1]$ that do not possess a right–derivative at a given point in $[0, 1)$ is dense.

4. Is every set of the second category the complement of a set of the first category?

5. Prove that in a complete metric space, the complement of a set of the first category is dense and of second category.

6. Prove that a closed, proper subspace in a normed linear space is nowhere dense (and hence of first category).

7. Prove that in a Banach space, a subspace of second category must be dense.

8. Prove that in a Banach space every nonempty open set is of the second category. Prove that this assertion is not true for normed linear spaces in general. (Give an example.)

9. Let $[x_n]$ be a sequence in a Banach space X. Assume that $\sup_n |\phi(x_n)| < \infty$ for each $\phi \in X^*$. Prove that $[x_n]$ is bounded. Does X have to be complete for this? If so, give a suitable example.

10. Determine the category of these sets: (a) the rationals in $\mathbb{R}$; (b) the irrationals in $\mathbb{R}$; (c) the union of all vertical lines in $\mathbb{R}^2$ that pass through a rational point on the horizontal axis; (d) the set of all polynomials in $C[0, 1]$.

11. Does a homeomorphism (continuous map having a continuous inverse) preserve the category of sets?

12. Give an example to show that a homeomorphic image of a complete metric space need not be complete.

13. Prove that any subset of a set of the first category is also of the first category. Prove that a set that contains a set of second category is also of second category.

14. Is the closure of a nowhere dense set also nowhere dense? Is the closure of a set of the first category also of the first category?

15. For each natural number n, let A_n be a continuous linear transformation of a Banach space X into a normed linear space Y. Suppose that for each $x \in X$ the sequence $[A_n x]$ is convergent. Define A by the equation $Ax = \lim_{n \to \infty} A_n x$. Prove that A is linear and continuous. Explain why completeness is needed.

16. Let X be the space of real sequences $x = [x(1), x(2), \ldots]$ in which only a finite number of terms are nonzero. Give X the supremum norm. Define functionals ϕ_n by the equation $\phi_n(x) = \sum_{i=1}^{n} x(i)$. Show that the sequence $[\phi_n(x)]$ is bounded for each x, that each ϕ_n is continuous, but that the sequence $[\phi_n]$ is not bounded. (Compare to the Uniform Boundedness Theorem.)

17. Prove that the set of reals whose decimal expansions do not contain the digit 7 is a set of the first category.

18. Select a function $x_0 \in C[0, 1]$ and a sequence of reals $[\alpha_n]$. Define recursively

$$x_{n+1}(t) = \alpha_n + \int_0^t x_n(s) \, ds \qquad n = 0, 1, \ldots$$

Assume that for each $t \in [0, 1]$ there is an n for which $x_n(t) = 0$. Prove that $x_0 = 0$.

19. Return to Problem 15, and suppose that Y is complete. Weaken the hypotheses on A_n so that A_n is not necessarily linear and the set of x for which $[A_n x]$ converges is of the second category. Prove that this set must be X and that A is continuous.

20. Let $[A_n]$ be a sequence of continuous linear maps from one Banach space X to another. Prove that the set of x for which $[A_n x]$ is a Cauchy sequence is either X or a set of first category.

21. Prove that in a complete metric space a set of the first category has empty interior.

22. Prove that in a complete metric space, if a countable intersection of open sets is dense, then it is of second category.

23. Give an example of a metric space having countably many points that contains no subset of second category.

24. Prove that a set V is nowhere dense if and only if each nonempty open set has a nonempty open subset that lies in the complement of V.

25. (**Principle of Condensation of Singularities**). For each n and m in $\mathbb{N}$, let A_{nm} be a bounded linear operator from a Banach space X into a normed linear space Y. Assume that $\sup_m \|A_{nm}\| = \infty$ for each n. Prove that the set

$$\left\{ x \in X : \sup_m \|A_{nm}x\| = \infty \quad \text{for each } n \right\}$$

is of second category.

26. (**The Cantor Set**). This famous set is $C = [0,1] \smallsetminus \bigcup_{n=1}^{\infty} A_n$, where

$$A_1 = \left(\tfrac{1}{3}, \tfrac{2}{3}\right), \quad A_2 = \left(\tfrac{1}{9}, \tfrac{2}{9}\right) \cup \left(\tfrac{7}{9}, \tfrac{8}{9}\right),$$
$$A_3 = \left(\tfrac{1}{27}, \tfrac{2}{27}\right) \cup \left(\tfrac{7}{27}, \tfrac{8}{27}\right) \cup \left(\tfrac{19}{27}, \tfrac{20}{27}\right) \cup \left(\tfrac{25}{27}, \tfrac{26}{27}\right), \quad \text{and so on.}$$

Draw pictures of $[0,1] \smallsetminus A_1$, $[0,1] \smallsetminus (A_1 \cup A_2)$, and so on to see that we are successively removing the middle thirds from intervals. Each A_n is open, so $\bigcup A_n$ is open. Hence C is closed. Prove that C is nowhere dense. Prove that the lengths of the removed intervals add up to 1. Explain how there can be anything left in C. Prove that C is a "perfect set," i.e., if $x \in C$, then $C \smallsetminus \{x\}$ is not closed.

27. Prove this theorem: Let X be a complete metric space. Let $\{f_\alpha\}$ be a family of continuous real-valued maps defined on X. Assume that for each x, $\sup_\alpha |f_\alpha(x)| < \infty$. Then for some nonvoid open set O, $\sup_{x \in O} \sup_\alpha |f_\alpha(x)| < \infty$.

28. Prove that a countable union of sets of the first category is also a set of the first category.

29. Prove that a nowhere dense set is of the first category.

30. Is a countable set in a metric space necessarily a set of the first category?

31. Answer the question in Problem 30 for countable subsets of a normed linear space.

32. Prove that the sets F_n occurring in the proof of the Banach–Steinhaus Theorem are closed.

33. In a complete metric space, is every nonempty open set of the second category?

34. A metric space (X, d) is said to be **discrete** if $d(x, y) = 1$ whenever $x \neq y$. In such a space identify the nowhere dense sets, sets of first category, sets of second category, and dense sets. (Cf. Problem 2.)

35. Can a normed linear space have any of the peculiar properties of discrete metric spaces?

36. Show that a countable discrete metric space can be embedded isometrically in the Banach space c_0.

37. Give an example of sets $S \subset F \subset X$, where X is a complete metric space, F is a closed set in X, and S is of Category II in F but of Category I in X.

38. The intersection of a countable family of open sets is called a G_δ-set. Prove that the set of rationals is not a G_δ-set in $\mathbb{R}$.

39. (Continuation)   Let $f : \mathbb{R} \to \mathbb{R}$ be continuous. Show that each set $f^{-1}(r)$ is a G_δ-set.

40. (Continuation)   Let $f : \mathbb{R} \to \mathbb{R}$. Define

$$\omega(x) = \inf_{\varepsilon > 0} \sup_{\substack{|x-u| < \varepsilon \\ |x-v| < \varepsilon}} |f(u) - f(v)|$$

Prove that $\omega(x) = 0$ for each x at which f is continuous. Prove that for $\varepsilon > 0$, the set $\{x : \omega(x) < \varepsilon\}$ is open.

41. (Continuation)   Prove that there is no function $f : \mathbb{R} \to \mathbb{R}$ that is continuous at each rational point and discontinuous at each irrational point.

42. Can an infinite–dimensional Banach space have a countable Hamel base?

43. Prove that the complement of a nowhere dense set is dense. What about the converse: is it true?

44. Let $f \in C^\infty(\mathbb{R})$. Thus f has derivatives of all orders on $\mathbb{R}$. Suppose that $0 \in \{f^{(n)}(t) : n = 0, 1, 2, \ldots\}$ for each t. Then f is a polynomial.

45. A point x in a metric space is **isolated** if for some $\varepsilon > 0$, the ball of radius ε centered at x contains no point of the space except x. Prove that a complete metric space in which there are no isolated points is uncountable.

46. If $f : \mathbb{R} \to \mathbb{R}$, then there is an interval (a, b) and a number M such that each point of (a, b) is the limit of a sequence $[x_n]$ such that $a < x_n < b$ and $|f(x_n)| \leqslant M$.

47. Let X be a complete metric space and for each n let F_n be a closed set having empty interior. Prove that $\bigcup_{n=1}^{\infty} F_n$ has empty interior.

48. Prove that a set E in a metric space X (or any topological space) is nowhere dense if and only if $X \smallsetminus \overline{E}$ is dense.

49. In a metric space, is a singleton $\{x\}$ always nowhere dense? Answer the same question for a normed linear space.

50. Prove that if A is of the second category and B is of the first category, then $A \smallsetminus B$ is of the second category.

51. Is a countable intersection of sets of the second category necessarily a set of the second category?

52. A subset of a metric space is called a *residual* set if its complement is of the first category. Prove that the intersection of countably many residual sets is a residual set.

1.8 The Interior Mapping and Closed Mapping Theorems

A function f from one normed linear space X to another Y is said to be **closed** (or to **have a closed graph**) if f is closed as a subset of $X \times Y$. Expressed otherwise, the set

$$\{(x, f(x)) : x \in X\}$$

is a closed set in $X \times Y$. In terms of sequences, the closed property of f is that the conditions $x_n \to x$ and $f(x_n) \to y$ imply that $y = f(x)$. It is clear that a continuous map is closed. For general topological spaces this is still true if Y is a Hausdorff space ([Ru1], page 29). The outstanding example of a linear transformation that is closed but *not* continuous is the derivative operator D acting on the differentiable functions in $C[a, b]$ and mapping into $C[a, b]$. If $x_n \to x$ and $Dx_n \to y$, then $y = Dx$. This is actually a theorem of calculus ([Wid], page 305). Let us stop to prove it. We denote by $C^1[a, b]$ the linear space of all functions on $[a, b]$ whose derivatives exist and are continuous on $[a, b]$.

Theorem 1.   Let $x_n \in C^1[a, b]$, $\left\| x_n - x \right\|_\infty \to 0$, and $\left\| x_n' - y \right\|_\infty \to 0$. Then $y \in C[a, b]$ and $x' = y$.

Proof. Since $x_n \in C^1[a, b]$, we have $x_n' \in C[a, b]$. Thus $y \in C[a, b]$, by Theorem 2 in Section 1.2, page 10. By the Fundamental Theorem of Calculus and the continuity of integration,

$$\int_a^t y(s)\, ds = \int_a^t \lim_n x_n'(s)\, ds = \lim_n \int_a^t x_n'(s)\, ds$$

$$= \lim_n [x_n(t) - x_n(a)] = x(t) - x(a)$$

Differentiation with respect to t now yields $y(t) = x'(t)$.   ∎

Of course, in general we may not infer that $x_n' \to x'$ from the sole hypothesis that $x_n \to x$, even if $x_n \in C^1[a, b]$ and the convergence is uniform. For example, the sequence $x_n(s) = \frac{1}{n} \sin ns$ converges (uniformly) to 0, but the sequence $x_n'(s) = \cos ns$ does not converge even pointwise.

Another property that a mapping $f : X \to Y$ may have is being an **interior** (or "open") mapping. That means that f maps open sets to open sets.

Theorem 2.   The Interior Mapping Theorem.   *If a closed linear transformation maps one Banach space onto another, then it is an interior map.*

Proof.   Let $L : X \twoheadrightarrow Y$, where L is linear and closed, and X and Y are Banach spaces. (This double arrow signifies a surjection.) Let S be the open unit ball in X or Y, depending on the context. Since L is surjective,

$$Y = L(X) = L\left(\bigcup_{n=1}^\infty nS\right) \subset \bigcup_{n=1}^\infty L(nS)$$

Since Y is complete, the Baire Theorem implies that one of the sets $\mathrm{cl}\,[L(nS)]$ has a nonempty interior. Suppose, then, that for some m in $\mathbb{N}$ and $r > 0$

$$v + rS \subset \mathrm{cl}\, L(mS)$$

It follows that $v \in \mathrm{cl}\, L(mS)$, and hence

$$rS \subset \mathrm{cl}\, L(mS) - v \subset \mathrm{cl}\, L(mS) - \mathrm{cl}\, L(mS) \subset \mathrm{cl}\, L(2mS)$$

Hence $S \subset \mathrm{cl}\, L(tS)$ for some $t > 0$, namely $t = 2m/r$.

We will now prove that $S \subset L(2tS)$. Let y be any point of S. Select a sequence of positive numbers δ_n such that $\sum_{n=1}^\infty \delta_n < 1$. Since $y \in \mathrm{cl}\, L(tS)$, there is an x_1 in tS such that $\|y - Lx_1\| < \delta_1$. Since

$$y - Lx_1 \in \delta_1 S \subset \mathrm{cl}\, L(\delta_1 tS)$$

there is a point $x_2 \in \delta_1 tS$ such that $\|y - Lx_1 - Lx_2\| < \delta_2$. We continue this construction, obtaining a sequence $x_1, x_2, \ldots$ whose partial sums $z_n = x_1 + \cdots + x_n$ have the property $\|y - Lz_n\| < \delta_n$. Also, we have

$$\|z_n\| \leqslant \|x_1\| + \cdots + \|x_n\| \leqslant t + \delta_1 t + \cdots + \delta_{n-1} t \leqslant t\left(1 + \sum_{k=1}^\infty \delta_k\right) < 2t$$

The sequence $[z_n]$ has the Cauchy property because

$$\|z_{n+i} - z_n\| = \|x_{n+1} + \cdots + x_{n+i}\| < t\delta_n + \cdots + t\delta_{n+i-1} < t \sum_{j \geqslant n} \delta_j$$

Since X is complete, $z_n \to z$ for some $z \in X$. Clearly, $\|z\| < 2t$, or $z \in 2tS$. Since L is closed and $Lz_n \to y$, we conclude that $y = Lz$; thus $y \in L(2tS)$ as claimed.

To complete the proof we show that $L(U)$ is open in Y whenever U is open in X. Let y be any point in $L(U)$. Then $y = Lx$ for some $x \in U$. Since U is open, there exists a $\theta > 0$ such that $x + \theta S \subset U$. Then $y + \theta L(S) \subset L(U)$. By our previous work, we know that $S \subset L(2tS)$. Hence $(\theta/2t)S \subset \theta L(S)$ and

$$y + (\theta/2t)S \subset L(U)$$

Thus $L(U)$ contains a neighborhood of y, and $L(U)$ is open.   ∎

Corollary 1.   *If an algebraic isomorphism of one Banach space onto another is continuous, then its inverse is continuous.*

Proof.   Let $L : X \twoheadrightarrow Y$ be such a map. (The two-headed arrow denotes a surjective map. Thus $L(X) = Y$.) Being continuous, L is closed. By the Interior Mapping Theorem, L is an interior map. Hence L^{-1} is continuous. (Recall that a map f is continuous if f^{-1} carries open sets to open sets.)   ∎

Corollary 2.   *If a linear space can be made into a Banach space with two norms, one of which dominates the other, then these norms are equivalent.*

Proof.   Let X be the space, and N_1, N_2 the two norms. The equivalence of two norms is explained in Problem 1.4.3, page 23. Let I denote the identity map acting from (X, N_2) to (X, N_1). Assume that the norms bear the relationship $N_1 \leqslant N_2$. Since $N_1(Ix) \leqslant N_2(x)$, we see that I is continuous. By the preceding corollary, I^{-1} is continuous. Hence for some α, $N_2(x) = N_2(I^{-1}x) \leqslant \alpha N_1(x)$.   ∎

Theorem 3. The Closed Graph Theorem.   *A closed linear map from one Banach space into another is continuous.*

Proof.   Let $L : X \to Y$ be closed and linear. In X, define a new norm $N(x) = \|x\| + \|Lx\|$. Then (X, N) is complete. Indeed, if $[x_n]$ is a Cauchy sequence with the norm N, then $[x_n]$ and $[Lx_n]$ are Cauchy sequences with the given norms in X and Y. Hence $x_n \to x$ and $Lx_n \to y$, since X and Y are complete. Since L is closed, $Lx = y$ and so

$$N(x - x_n) = \|x - x_n\| + \|Lx - Lx_n\| \to 0$$

By the preceding corollary, $N(x) \leqslant \alpha\|x\|$ for some α. Hence $\|Lx\| \leqslant \alpha\|x\|$ ∎

Theorem 4.   *A normed linear space that is the image of a Banach space by a bounded, linear, interior map is also a Banach space.*

Proof.   Let $L : X \twoheadrightarrow Y$ be the bounded, linear, interior map. Assume that X is a Banach space. By Problem 1.2.38 (page 14), it suffices to prove that each absolutely convergent series in Y is convergent. Let $y_n \in Y$ and $\sum \|y_n\| < \infty$. By Problem 2 (of this section), there exist $x_n \in X$ such that $Lx_n = y_n$ and (for some $c > 0$) $\|x_n\| \leqslant c\|y_n\|$. Then $\sum \|x_n\| \leqslant c \sum \|y_n\| < \infty$. By Problem 1.2.3, page 12, the series $\sum x_n$ converges. Since L is continuous and linear, $L(\sum x_n) = \sum Lx_n = \sum y_n$, and the latter series is convergent.   ∎

Let L be a bounded linear transformation from one normed linear space, X, to another, Y. The **adjoint** of L is the map $L^* : Y^* \to X^*$ defined by $L^*\phi = \phi \circ L$. Here ϕ ranges over Y^*. It is elementary to prove that L^* is linear. It is bounded because

$$\|L^*\| = \sup_{\phi} \|L^*\phi\| = \sup_{\phi} \sup_{x} |(L^*\phi)(x)|$$

$$= \sup_{x} \sup_{\phi} |\phi(Lx)| = \sup_{x} \|Lx\| = \|L\|$$

In this equation ϕ ranges over functionals of norm 1 in Y^*, and x ranges over vectors of norm 1 in X. We used Corollary 4 on page 36.

In a finite-dimensional setting, an operator L can be represented by a matrix A (which is not necessarily square). This requires the prior selection of bases for the domain and range of L. The adjoint operator L^* is represented by the complex conjugate matrix A^*. An elementary theorem asserts that A is surjective ("onto") if and only if A^* is injective ("one-to-one"). (See Problem 20.) The situation in an infinite-dimensional space is only slightly more complicated, as indicated in the next three theorems.

Theorem 5.   *Let L be a continuous linear transformation from one normed linear space to another. The range of L is dense if and only if L^* is injective.*

Proof.   Let $L : X \to Y$. By Theorem 3 in Section 1.6 (page 37), applied to $L(X)$, we have these equivalent assertions: (1) $L(X)$ is dense in Y. (2) $L(X)^{\perp} = 0$. (3) If $\phi \in L(X)^{\perp}$, then $\phi = 0$. (4) If $\phi(Lx) = 0$ for all x, then $\phi = 0$. (5) If $L^*\phi = 0$, then $\phi = 0$. (6) L^* is injective.   ∎

Theorem 6. The Closed Range Theorem.   *Let L be a bounded linear transformation defined on a normed linear space and taking values in another normed linear space. The range of L and the null space of L^*, denoted by $\mathcal{N}(L^*)$, are related by the fact that $[\mathcal{N}(L^*)]_{\perp}$ is the closure of the range of L.*

Proof.   Recall the notation $U_{\perp}$ for the set $\{x \in X : \phi(x) = 0 \text{ for all } \phi \in U\}$, where X is a normed linear space and U is a subset of X^*. (See Problems 1.6.20 and 1.6.21, on page 38, as well as Problem 13 in this section, page 52.) We denote by $\mathcal{R}(L)$ the range of L. To prove [closure $\mathcal{R}(L)] \subset [\mathcal{N}(L^*)]_{\perp}$, let y be an element of the set on the left. Then $y = \lim y_n$ for some sequence $[y_n]$ in $\mathcal{R}(L)$. Write $y_n = Lx_n$ for appropriate x_n. To show that $y \in [\mathcal{N}(L^*)]_\perp$ we must prove that $\phi(y) = 0$ for all $\phi \in \mathcal{N}(L^*)$. We have

$$\phi(y) = \phi(\lim y_n) = \lim \phi(y_n) = \lim \phi(Lx_n)$$
$$= \lim(\phi \circ L)(x_n) = \lim(L^*\phi)(x_n) = \lim 0 = 0$$

To prove the reverse inclusion, suppose that y is not in [closure $\mathcal{R}(L)$]. We shall show that y is not in $[\mathcal{N}(L^*)]_\perp$. By Corollary 2 of the Hahn–Banach Theorem (page 34), there is a continuous linear functional ϕ such that $\phi(y) \neq 0$ and ϕ annihilates each member of [closure $\mathcal{R}(L)$]. It follows that for all x, $(L^*\phi)(x) = (\phi \circ L)(x) = \phi(Lx) = 0$. Consequently, $\phi \in \mathcal{N}(L^*)$. Since $\phi(y) \neq 0$, we conclude that $y \notin [\mathcal{N}(L^*)]_\perp$.   ∎

Theorem 7   *Let L be a continuous, linear, injective map from one Banach space into another. The range of L is closed if and only if L is bounded below:* $\inf_{\|x\|=1} \|Lx\| > 0.$

Proof.   Assume first that $\|Lx\| \geqslant c > 0$ when $\|x\| = 1$. By homogeneity, $\|Lx\| \geqslant c\|x\|$ for all x. To prove that the range, $\mathcal{R}(L)$, is closed, let $y_n \in \mathcal{R}(L)$ and $y_n \to y$. It is to be shown that $y \in \mathcal{R}(L)$. Let $y_n = Lx_n$. The inequality

$$\|y_n - y_m\| = \|L(x_n - x_m)\| \geqslant c\|x_n - x_m\|$$

reveals that $[x_n]$ is a Cauchy sequence. By the completeness of the domain space, $x_n \to x$ for some x. Then, by continuity,

$$Lx = L(\lim x_n) = \lim Lx_n = \lim y_n = y$$

Hence $y \in \mathcal{R}(L)$.

Now assume that $\mathcal{R}(L)$ is closed. Then L maps the domain space X injectively onto the Banach space $\mathcal{R}(L)$. By Corollary 1 of the Interior Mapping Theorem (page 49), L has a continuous inverse. The equation $\|L^{-1}y\| \leqslant \|L^{-1}\| \|y\|$ is equivalent to $\|x\| \leqslant \|L^{-1}\| \|Lx\|$, showing that L is bounded below.   ∎

Problems 1.8

1. Use the notation in the proof of the Interior Mapping Theorem. Show that a linear map $L : X \to Y$ is interior if and only if $L(S) \supset r\,S$ for some $r > 0$.

2. Show that a linear map $L : X \twoheadrightarrow Y$ is interior if and only if there is a constant c such that for each $y \in Y$ there is an $x \in X$ satisfying $Lx = y$, $\|x\| \leqslant c\|y\|$.

3. Define $T : c_0 \to c_0$ by the equation $(Tx)(n) = x(n+1)$. Which of these properties does T have: injective, surjective, open, closed, invertible? Does T have either a right or a left inverse?

4. Prove that a closed (and possibly nonlinear) map of one normed linear space into another maps compact sets to closed sets.

5. Let L be a linear map from one Banach space into another. Suppose that the conditions $x_n \to 0$ and $Lx_n \to y$ imply that $y = 0$. Prove that L is continuous.

6. Prove that if a closed map has an inverse, then the inverse is also closed.

7. Let M and N be closed linear subspaces in a Banach space. Define $L : M \times N \to M + N$ by writing $L(x, y) = x + y$. Prove that $M + N$ is closed if and only if L is an interior map.

8. Adopt the hypotheses of Problem 7. Prove that $M + N$ is closed if and only if there is a constant c such that each $z \in M + N$ can be written $z = x + y$ where $x \in M$, $y \in N$, and $\|x\| + \|y\| \leqslant c\|z\|$.

9. Let $L : X \twoheadrightarrow Y$ be a continuous linear surjection, where X and Y are Banach spaces. Let $y_n \to y$ in Y. Prove that there exist points $x_n \in X$ and a constant $c \in \mathbb{R}$ such that $Lx_n = y_n$, the sequence $[x_n]$ converges, and $\|x_n\| \leqslant c\|y_n\|$.

10. Recall the space ℓ defined in Example 8 of Section 1.1. Define $L : \ell \to \ell$ by $(Lx)(n) = nx(n)$. Use the sup-norm in ℓ and prove that L is discontinuous, surjective, and closed.

11. Is the identity map from $(C[-1, 1], \| \ \|_\infty)$ into $(C[-1, 1], \| \ \|_1)$ an interior map? Is it continuous?

12. (Continuation) Denote the two spaces in Problem 11 by X and Y, respectively. Let

$$G = \left\{ y \in Y : \int_{-1}^{1} y(s)\, ds = 0 \right\}$$

Show that G is closed in Y. Define

$$g_n(s) = \begin{cases} nx & |x| \leqslant 1/n \\ x/|x| & |x| > 1/n \end{cases}$$

Show that $[g_n]$ is a Cauchy sequence in Y. Since the space $L^1[-1, 1]$ is complete, $g_n \to g$ in L^1. Since G is closed, g should be in G. But it is discontinuous. Explain.

13. Let X be a normed linear space, and let $K \subset X$ and $U \subset X^*$. Define

$$K^\perp = \{\phi \in X^* : \phi(x) = 0 \text{ for all } x \in K\}$$

$$U_\perp = \{x \in X : \phi(x) = 0 \text{ for all } \phi \in U\}$$

Prove that these are closed subspaces in X^* and X, respectively.

14. Prove that for any subset K in a normed linear space, $(K^\perp)_\perp$ is the closure of the linear span of K. The Hahn–Banach Theorem can be used as in the proof of the Closed Range Theorem. Problem 13 will also be helpful.

15. Prove that if L is a linear operator having closed range and acting between normed linear spaces, then the equation $Lx = y$ is solvable for x if and only if $y \in [\mathcal{N}(L^*)]_\perp$.

16. Prove that if L is a bounded linear operator from one normed space into another, and if $\|Lx\|/\operatorname{dist}(x, \mathcal{N}(L))$ is bounded away from 0 when $\|x\| = 1$, then the conclusion of Problem 15 is again valid.

17. Let T be a linear map of a Banach space X into itself. Suppose that there exists a continuous, linear, one-to-one map $L : X \to X$ such that LT is continuous. Does it follow that T is continuous?

18. Define an operator L by the equation

$$(Lx)(t) = \int_{-1}^{1} (t - s)^2 x(s)\, ds$$

Describe the range of L and prove that it does not contain the function $f(x) = e^t$.

19. (Continuation) Draw the same conclusion as in Problem 18 by invoking the Closed Range Theorem. Thus, find ϕ in the null space of L^* such that $\phi(f) \neq 0$.

20. For an $m \times n$ matrix A prove the equivalence of these assertions: (a) A^* is injective. (b) The null space of A^* is 0. (c) The columns of A^* form a linearly independent set. (d) The rows of A form a linearly independent set. (e) The row space of A has dimension m. (f) The column space of A has dimension m. (g) The column space of A is $\mathbb{R}^m$. (h) The range of A is $\mathbb{R}^m$. (i) A is surjective, as a map from $\mathbb{R}^n$ to $\mathbb{R}^m$.

1.9 Weak Convergence

A sequence $[x_n]$ in a normed linear space X is said to **converge weakly** to an element x if $\phi(x_n) \to \phi(x)$ for every ϕ in X^*. (Sometimes we write $x_n \rightharpoonup x$.)

The usual type of convergence can be termed **norm** convergence or **strong** convergence. It refers, of course, to $\|x_n - x\| \to 0$. Clearly, if $x_n \to x$, then $x_n \rightharpoonup x$, because each ϕ in X^* is continuous. This observation justifies the terms "strong" and "weak."

Example 1.   For an example of a sequence that converges weakly to zero yet does not converge strongly to any point, consider the vectors e_n in c_0 defined by $e_n(i) = \delta_{in}$. These are the "standard unit vectors" in the space c_0. (This space was defined in Problem 1.2.16, on page 12.) Recall from Section 1.6, particularly the proposition on page 34, that every continuous linear functional on c_0 is of the form

$$\phi(x) = \sum_{i=1}^{\infty} \alpha(i)x(i)$$

for a suitable point $\alpha \in \ell_1$. Thus $\phi(e_n) = \alpha(n) \to 0$. The sequence $[x_n]$ does not have the Cauchy property, because $\|x_n - x_m\| = 1$ when $n \neq m$. ∎

Lemma.   *A weakly convergent sequence is bounded.*

Proof.   Let X be the ambient space, and suppose that $x_n \rightharpoonup x$. Define functionals $\overline{x}_n$ on X^* by putting

$$\overline{x}_n(\phi) = \phi(x_n) \qquad (\phi \in X^*)$$

For each ϕ, the sequence $[\phi(x_n)]$ converges in $\mathbb{R}$; hence it is bounded. Thus $\sup_n |\overline{x}_n(\phi)| < \infty$. By the Uniform Boundedness Theorem (page 42), applied in the complete space X^*, $\|\overline{x}_n\| \leqslant M$ for some constant M. Hence, for all n,

$$\sup\left\{ |\overline{x}_n(\phi)| : \phi \in X^*, \ \|\phi\| \leqslant 1 \right\} \leqslant M$$

By Corollary 4 of the Hahn–Banach Theorem (page 36), $\|x_n\| \leqslant M$. ∎

Theorem 1.   *In a finite-dimensional normed linear space, weak and strong convergence coincide.*

Proof.   Let X be a k-dimensional space. Select a base $\{b_1, \ldots, b_k\}$ for X and let $\phi_1, \ldots, \phi_k$ be the linear functionals such that for each x,

$$x = \sum_{i=1}^{k} \phi_i(x)b_i$$

By Corollary 1 on page 26, each functional ϕ_i is continuous. Now if $x_n \rightharpoonup x$, then we have $\phi_i(x_n) \to \phi(x)$, and consequently,

$$\left\| x - x_n \right\| = \left\| \sum_{i=1}^{k} \phi_i(x - x_n)b_i \right\| \leqslant \sum_{i=1}^{k} |\phi_i(x - x_n)| \, \|b_i\| \to 0 \qquad \blacksquare$$

Theorem 2.   *If a sequence $[x_n]$ in a normed linear space converges weakly to an element x, then a sequence of linear combinations of the elements x_n converges strongly to x.*

Proof.   Another way of stating the conclusion is that x belongs to the closed subspace
$$Y = \text{closure}\,(\text{span}\{x_1, x_2, \ldots\})$$
If $x \notin Y$, then by Corollary 2 of the Hahn–Banach Theorem (page 34), there is a continuous linear functional ϕ such that $\phi \in Y^{\perp}$ and $\phi(x) = 1$. This clearly contradicts the assumption that $x_n \rightharpoonup x$. $\blacksquare$

A refinement of this theorem states that a sequence of *convex* linear combinations of $\{x_1, x_2, \ldots\}$ converges strongly to x. This can be proved with the aid of a separation theorem, such as Theorem 3 in Section 7.3, page 344.

Theorem 3.   *If the sequence $[x_0, x_1, x_2, \ldots]$ is bounded in a normed linear space X and if $\phi(x_n) \to \phi(x_0)$ for all ϕ in a fundamental subset of X^*, then $x_n \rightharpoonup x$.*

Proof.   (The term "fundamental" was defined in Section 1.6, page 36.) Let $\mathcal{F}$ be the fundamental subset of X^* mentioned in the theorem. Let ψ be any member of X^*. We want to prove that $\psi(x_n) \to \psi(x_0)$. By hypothesis, there is a constant M such that $\|x_i\| < M$ for $i = 0, 1, 2, \ldots$ Given $\varepsilon > 0$, select $\phi_1, \ldots, \phi_m$ in $\mathcal{F}$ and scalars $\lambda_1, \ldots, \lambda_m$ such that

$$\left\| \psi - \sum_{i=1}^{m} \lambda_i \phi_i \right\| < \frac{\varepsilon}{3M}$$

Put $\phi = \sum \lambda_i \phi_i$. It is easily seen that $\phi(x_n) \to \phi(x_0)$. Select N so that for all $n > N$ we have the inequality $|\phi(x_n) - \phi(x_0)| < \varepsilon/3$. Then for $n > N$,

$$\begin{aligned}
|\psi(x_n) - \psi(x_0)| &\leqslant |\psi(x_n) - \phi(x_n)| + |\phi(x_n) - \phi(x_0)| + |\phi(x_0) - \psi(x_0)| \\
&\leqslant \|\psi - \phi\| \, \|x_n\| + \varepsilon/3 + \|\phi - \psi\| \, \|x_0\| \\
&\leqslant \frac{\varepsilon}{3M} M + \frac{\varepsilon}{3} + \frac{\varepsilon}{3M} M = \varepsilon \qquad \blacksquare
\end{aligned}$$

Example 2.   Fix a real number p in the range $1 \leqslant p < \infty$. The space ℓ_p is defined to be the set of all real sequences x for which $\sum_{n=1}^{\infty} |x(n)|^p < \infty$. We define a norm on the vector space ℓ_p by the equation

$$\|x\|_p = \left(\sum_{n=1}^{\infty} |x(n)|^p \right)^{1/p}$$

For $p = \infty$, we take ℓ_∞ to be the space of bounded sequences, with norm $\|x\|_\infty = \sup_n |x(n)|$. We shall outline some of the theory of these spaces. (This theory is actually included in the theory of the L^p spaces as given in Chapter 8.) Notice that in these spaces there is a natural partial order: $x \geqslant y$ means that $x(n) \geqslant y(n)$ for all n. We also define $|x|$ by the equation $|x|(n) = |x(n)|$. ∎

Hölder Inequality. *Let* $1 < p < \infty$, $1/p + 1/q = 1$, $x \in \ell_p$, *and* $y \in \ell_q$. *Then*

$$\sum_{n=1}^{\infty} x(n)y(n) \leqslant \|x\|_p \|y\|_q$$

Minkowski Inequality. *If* x *and* y *are two members of* ℓ_p, *then*

$$\|x + y\|_p \leqslant \|x\|_p + \|y\|_p$$

Proof. For $p = 1$ an elementary proof goes as follows:

$$\|x + y\|_1 = \sum |x(n) + y(n)| \leqslant \sum |x(n)| + \sum |y(n)| = \|x\|_1 + \|y\|_1$$

Now assume $1 < p < \infty$. Then

$$\sum |x(n) + y(n)|^p \leqslant \sum \{|x(n)| + |y(n)|\}^p$$
$$\leqslant \sum \{2 \max[|x(n)|, |y(n)|]\}^p$$
$$= \sum 2^p \max\{|x(n)|^p, |y(n)|^p\}$$
$$\leqslant 2^p \sum \{|x(n)|^p + |y(n)|^p\} < \infty$$

This proves that $x + y \in \ell_p$. Now let $1/p + 1/q = 1$ and observe that

$$x \in \ell_p \implies |x|^{p-1} \in \ell_q$$

because

$$\sum \{|x(n)|^{p-1}\}^q = \sum |x(n)|^p < \infty$$

Therefore, by the Hölder inequality,

$$\|x + y\|_p^p = \sum |x(n) + y(n)|^p$$
$$\leqslant \sum |x(n) + y(n)|^{p-1}|x(n)| + \sum |x(n) + y(n)|^{p-1}|y(n)|$$
$$\leqslant \| \, |x + y|^{p-1}\|_q \{\|x\|_p + \|y\|_p\}$$
$$= \|x + y\|_p^{p/q} \{\|x\|_p + \|y\|_p\}$$

Thus, finally,

$$\|x + y\|_p \leqslant \|x\|_p + \|y\|_p$$

∎

Some theorems about these spaces are given here without proof.

Theorem 4.　　*The conjugate of ℓ_p is isometrically isomorphic to ℓ_q, where $p^{-1} + q^{-1} = 1$. (Here $1 \leqslant p < \infty$.) The isomorphism pairs each element ϕ in ℓ_p^* with the unique element y in ℓ_q such that $\phi(x) = \sum_k x(k)y(k)$.*

Theorem 5.　　*Let x and x_n be in ℓ_p. We have $x_n \rightharpoonup x$ if and only if $\|x_n\|_p$ is bounded and $x_n(k) \to x(k)$ for each k.*

Theorem 6.　　*Let S be a compact Hausdorff space, and suppose $x, x_n \in C(S)$. We have $x_n \rightharpoonup x$ if and only if $\|x_n\|_\infty$ is bounded and $x_n(s) \to x(s)$ for each $s \in S$.*

Theorem 7.　　(Schur's Lemma) *In the space ℓ_1, the concepts of weak and strong convergence of sequences coincide.*

A subset F in a normed linear space X is said to be **weakly sequentially closed** if the weak limit of any weakly convergent sequence in F is also in F. A weakly sequentially closed set F is necessarily closed in the norm topology, for if $x_n \in F$ and $x_n \to x$, then $x_n \rightharpoonup x \in F$. (A simple example of a closed set that is not weakly sequentially closed is the surface of the unit ball in the space c_0.)

Theorem 8.　　*A subspace of a normed linear space is closed if and only if it is weakly sequentially closed.*

Proof.　Let Y be a weakly sequentially closed subspace in the normed space X. If $y_n \in Y$ and $y_n \to y$, then $y_n \rightharpoonup y$ and $y \in Y$. Hence Y is norm-closed.

For the converse, suppose that Y is norm-closed, and let $y_n \in Y$, $y_n \rightharpoonup y$. If $y \notin Y$, then (because Y is closed) we have $\text{dist}(y, Y) > 0$. By Corollary 2 of the Hahn-Banach Theorem (page 34) there is a functional $\phi \in Y^\perp$ such that $\phi(y) = 1$. Hence $\phi(y_n)$ does not converge to $\phi(y)$, contradicting the assumed weak convergence.　　　　■

A refinement of this theorem states that a convex set is closed if and only if it is weakly sequentially closed. See [DS], page 422.

Theorem 9.　　*A linear continuous mapping between normed spaces is weakly sequentially continuous.*

Proof.　Let $A : X \to Y$ be linear and norm-continuous. In order to prove that A is weakly continuous, let $x_n \rightharpoonup x$. For all $\phi \in Y^*$, $\phi \circ A \in X^*$. Hence $\phi(Ax_n - Ax) \to 0$ for all $\phi \in Y^*$.　　　　■

In a conjugate space X^*, the concept of weak convergence is also available. Thus $\phi_n \rightharpoonup \phi$ if and only if $F(\phi_n) \to F(\phi)$ for each $F \in X^{**}$. There is another type of convergence, called weak* convergence. We say that $[\phi_n]$ converges to ϕ in the weak* sense if $\phi_n(x) \to \phi(x)$ for all $x \in X$.

Theorem 10.   Let X be a separable normed linear space, and $[\phi_n]$ a bounded sequence in X^*. Then there is a subsequence $[\phi_{n_i}]$ that converges in the weak* sense to an element of X^*.

Proof.   Since X is separable, it contains a countable dense set, $\{x_1, x_2, \ldots\}$. Since $[\phi_n]$ is bounded, so is the sequence $[\phi_n(x_1)]$. We can therefore find an increasing sequence $N_1 \subset N$ such that $\lim_{n \in N_1} \phi_n(x_1)$ exists. By the same reasoning there is an increasing sequence $N_2 \subset N_1$ such that $\lim_{n \in N_2} \phi_n(x_2)$ exists. Continuing in this way, we generate sequences

$$N \supset N_1 \supset N_2 \supset \cdots$$

Now use the Cantor diagonalization process: Define n_i to be the ith element of N_i. We claim that $\lim_{i \to \infty} \phi_{n_i}(x_k)$ exists for each k. This is true because $\lim_{n \in N_k} \phi_n(x_k)$ exists by construction, and if $i \geqslant k$, then $n_i \in N_i \subset N_k$. For any $x \in X$ we write

$$\left|\phi_{n_i}(x) - \phi_{n_j}(x)\right| \leqslant \left|\phi_{n_i}(x) - \phi_{n_i}(x_k)\right| + \left|\phi_{n_i}(x_k) - \phi_{n_j}(x_k)\right| + \left|\phi_{n_j}(x_k) - \phi_{n_j}(x)\right|$$

This inequality shows that $[\phi_{n_i}(x)]$ has the Cauchy property in $\mathbb{R}$ for each $x \in X$. Hence it converges to something that we may denote by $\phi(x)$. Standard arguments show that $\phi \in X^*$.   ∎

For Schur's Lemma, see [HP] page 37, or [Ban] page 137. The original source is [Schu]. See also [Jam] page 288, or [Hol] page 149.

Problems 1.9

1. Show that the Hölder Inequality remains true if we replace the left-hand side by $\sum |x(n)|\,|y(n)|$.

2. If $1 < p < q$, what inclusion relation exists between ℓ_p and ℓ_q?

3. Prove that if $x_n \rightharpoonup x$ and $\|x_n\| \to c$, then $\|x\| \leqslant c$. Why can we not conclude that $\|x\| = c$? Give examples. Explain in terms of weak continuity and weak semicontinuity of the norm.

4. Fix $p > 1$ and define a nonlinear map T on ℓ_p by the equation $Tx = |x|^{p-1}\,\mathrm{sgn}(x)$. Thus, $(Tx)(n) = |x(n)|^{p-1}\,\mathrm{sgn}(x(n))$ for all n. Prove that T maps ℓ_p into ℓ_q, where $1/p + 1/q = 1$. Then determine whether T is surjective.

5. Prove this theorem: In order that a sequence $[x_n]$ in a normed linear space X converge weakly to an element x it is necessary and sufficient that the sequence be bounded and that $\phi(x_n) \to \phi(x)$ for all functionals ϕ in a set that is dense on the surface of the unit ball in X^*.

6. Prove this characterization of weak convergence in the space c_0: In order that a sequence x_n converge weakly to an element x in the space c_0 it is necessary and sufficient that the sequence be bounded and that (for each i) we have $\lim_{n \to \infty} x_n(i) = x(i)$.

7. A Banach space X is said to be **weakly complete** if every sequence $[x_n]$ such that $\phi(x_n)$ converges for each ϕ in X^* must converge weakly to an element x in X. Prove that the space c_0 is not weakly complete.

1.10 Reflexive Spaces

Let X be a Banach space. It is possible to embed X isomorphically and iso-metrically as a subspace of X^{**}. There may be many ways to do this, but one embedding is called the **natural** or **canonical** embedding, denoted by J. Thus $J : X \to X^{**}$, and its definition is

$$(Jx)(\phi) = \phi(x) \qquad \phi \in X^*, \quad x \in X$$

The reader may wish to pause and prove that J is a linear isometry.

For an example of this embedding, let $X = c_0$; then $X^* = \ell_1$ and $X^{**} = \ell_\infty$. In this case, $J : c_0 \to \ell_\infty$, and J can be interpreted as the identity embedding, since $\phi(x) = \sum_{n=1}^{\infty} u(n)x(n)$ for an appropriate $u \in \ell_1$.

If the natural map of X into X^{**} is surjective, we say that X is **reflexive**. Thus if X is reflexive, it is isometrically isomorphic to X^{**}. The converse is *false*, however. A famous example of R.C. James exhibits an X that is isometrically isomorphic to X^{**}, but the isometry is *not* the canonical map J, and indeed the canonical image of $J(X)$ is a *proper* subspace of X^{**} in the example. See [Ja2].

Theorem 1.   *Each space ℓ_p, where $1 < p < \infty$, is reflexive.*

Proof.   If $p^{-1} + q^{-1} = 1$, then $\ell_p^* = \ell_q$ and $\ell_q^* = \ell_p$ by Theorem 4 of Section 1.9, page 56. Hence $\ell_p^{**} = \ell_p$. But we must be sure that the isometry involved in this statement is the *natural* one, J. Let $A : \ell_p \to \ell_q^*$ and $B : \ell_q \to \ell_p^*$ be the isometries that have already been discussed in a previous section. Thus, for example, if $x \in \ell_p$ then Ax is the functional on ℓ_q defined by

$$(Ax)(y) = \sum_{n=1}^{\infty} x(n)y(n) \qquad y \in \ell_q$$

Define $B^* : \ell_p^{**} \to \ell_q^*$ by the equation

$$B^*\phi = \phi \circ B \qquad \phi \in \ell_p^{**}$$

One of the problems asks for a proof of the fact that B^* is an isometric isomor-phism of ℓ_p^{**} onto ℓ_q^*. Thus $B^{*-1}A$ is an isometric isomorphism of ℓ_p onto ℓ_p^{**}. Now we wonder whether $B^{*-1}A = J$. Equivalent questions are these:

$$\begin{aligned}
B^{*-1}Ax &= Jx & (x \in \ell_p) \\
Ax &= B^*Jx & (x \in \ell_p) \\
(Ax)(y) &= (B^*Jx)(y) & (x \in \ell_p, \quad y \in \ell_q) \\
(Ax)(y) &= (Jx)(By) & (x \in \ell_p, \quad y \in \ell_q) \\
(Ax)(y) &= (By)(x) & (x \in \ell_p, \quad y \in \ell_q)
\end{aligned}$$

The final assertion is true because both sides of the equation are by definition $\sum_{n=1}^{\infty} x(n)y(n)$. ∎

Theorem 2.   *A closed linear subspace in a reflexive Banach space is reflexive.*

Proof.   Let Y be a closed subspace in a reflexive Banach space X. Let $J : X \twoheadrightarrow X^{**}$ be the natural map. Define $R : X^* \to Y^*$ by the equation $R\phi = \phi|Y$. (This is the *restriction map.*) Let $f \in Y^{**}$. Define $y = J^{-1}(f \circ R)$. We claim that $y \in Y$. Suppose that $y \notin Y$. By a corollary of the Hahn–Banach Theorem, there exists $\phi \in X^*$ such that $\phi(y) \neq 0$ and $\phi(Y) = 0$. Then it will follow that $R\phi = 0$ and that $\phi(y) = \phi(J^{-1}(f \circ R)) = (f \circ R)(\phi) = 0$, a contradiction. Next we claim that for all $\psi \in Y^*$, $f(\psi) = \psi(y)$. Let $\tilde{\psi}$ be a Hahn–Banach extension of ψ in X^*. Then $\psi = R\tilde{\psi}$ and $f(\psi) = f(R\tilde{\psi}) = (f \circ R)(\tilde{\psi}) = (Jy)(\tilde{\psi}) = \tilde{\psi}(y) = \psi(y)$. ∎

Theorem 3.   *A Banach space is reflexive if and only if its conjugate space is reflexive.*

Proof.   Let X be reflexive. Then the natural embedding $J : X \to X^{**}$ is surjective. Let $\Phi \in X^{***}$, and define $\phi \in X^*$ by the equation $\phi = \Phi \circ J$. Then for arbitrary $f \in X^{**}$ we have $f = Jx$ for some x, and consequently,

$$f(\phi) = (Jx)(\phi) = \phi(x) = (\Phi \circ J)(x) = \Phi(Jx) = \Phi(f)$$

Thus Φ is the image of ϕ under the natural map of X^* into X^{***}. This natural map is therefore surjective, and X^* is reflexive.

For the converse, suppose that X^* is reflexive. By what we just proved, X^{**} is reflexive. But $J(X)$ is a closed subspace in X^{**}, and by the preceding theorem, $J(X)$ is reflexive. Hence X is reflexive (being isometrically isomorphic to $J(X)$).   ∎

Eberlein–Smulyan Theorem.   *A Banach space is reflexive if and only if its unit ball is weakly sequentially compact.*

Proof.   (Partial) Let X be reflexive, S its unit ball, and $[y_n]$ a sequence in S. We wish to extract a subsequence $[y_{n_i}]$ such that $y_{n_i} \rightharpoonup y \in S$. To start, let Y be the closure of the linear span of $\{y_1, y_2, \ldots\}$. Then Y is a closed and separable subspace of X. By Theorem 2, Y is reflexive, and so $Y = Y^{**}$. Since Y^{**} is separable, so is Y^*. Let $\{\psi_1, \psi_2, \ldots\}$ be a countable dense set in Y^*. Since $[\psi_1(y_n)]$ is bounded, there exists an infinite set $\mathbb{N}_1 \subset \mathbb{N}$ such that $\lim_{n \in \mathbb{N}_1} \psi_1(y_n)$ exists. Proceeding as we did in the proof Theorem 10, Section 1.9, page 57, we find a subsequence y_{n_i} such that $\psi(y_{n_i})$ converges for all $\psi \in Y^*$. By a corollary of the uniform boundedness theorem, there is an element f of Y^{**} such that $\psi(y_{n_i}) \to f(\psi)$ for all $\psi \in Y^*$. Since Y is reflexive, $f(\psi) = \psi(y)$ for some $y \in Y$. Hence $\psi(y_{n_i}) \to \psi(y)$ for all $\psi \in Y^*$. Now if $\phi \in X^*$, then $\phi|Y \in Y^*$. Hence

$$\phi(y_{n_i}) = (\phi|Y)(y_{n_i}) \to (\phi|Y)(y) = \phi(y)$$

Thus $y_{n_i} \rightharpoonup y$. By a corollary of the Hahn–Banach Theorem, $\|y\| \leqslant 1$.

The converse is more difficult, and we do not give the proof. See [Yo], page 141, or [Tay2], page 230.   ∎

Theorem of James.   *A Banach space X is reflexive if and only if each continuous linear functional on X attains its supremum on the unit ball of X.*

Proof. (Partial) Suppose that X is reflexive. Let $\phi \in X^*$, and select $x_n \in X$ such that $\|x_n\| \leqslant 1$ and $\phi(x_n) \to \|\phi\|$. By the Eberlein–Smulyan Theorem, there is a subsequence $[x_{n_i}]$ that converges weakly to a point x satisfying $\|x\| \leqslant 1$. By the definition of weak convergence,

$$\phi(x) = \lim_i \phi(x_{n_i}) = \|\phi\|$$

The converse is more difficult, and we refer the reader to [Hol], page 157. ∎

One application of the second conjugate space occurs in the process of *completion*. If X is a normed linear space that is not complete, can we embed it linearly and isometrically as a dense set in a Banach space? If so, such a Banach space is termed a **completion** of X. The Cantor method of completion of a metric space is fully discussed in [KF]. The idea of that method is to create a new metric space whose elements are Cauchy sequences in the original metric space.

If X is a normed linear space, we can embed it, using the natural map J, into its second conjugate space X^{**}. The latter is *automatically* complete. Hence $\overline{J(X)}$ can be regarded as a completion of X. It can be proved that all completions of X are isometrically isomorphic to each other.

The Lebesgue spaces $L_p[a,b]$ can be defined without knowing anything about Lebesgue measure or integration. Here is how to do this. Consider the space $C[a,b]$ of all continuous real-valued functions on the interval $[a,b]$. For $1 \leqslant p < \infty$, we introduce the norm

$$\|x\|_p = \left[\int_a^b |x(s)|^p \, ds \right]^{1/p}$$

In this equation, the integration is with respect to the Riemann integral. The space $C[a,b]$, endowed with this norm, is denoted by $C_p[a,b]$. It is not complete. Its completion is $L_p[a,b]$. Thus if J is the natural map of $C_p[a,b]$ into its second conjugate space, then

$$L_p[a,b] = \overline{J(C_p[a,b])}$$

Problems 1.10

1. Use the fact that $c_0^* = \ell_1$ and $\ell_1^* = \ell_\infty$ to prove that the successive conjugate spaces of c_0 are all nonreflexive.

2. Find a sequence in the unit ball of c_0 that has no weakly convergent subsequence.

Chapter 2

Hilbert Spaces

2.1 Geometry 61
2.2 Orthogonality and Bases 70
2.3 Linear Functionals and Operators 81
2.4 Spectral Theory 91
2.5 Sturm–Liouville Theory 105

2.1 Geometry

Hilbert spaces are a special type of Banach space. In fact, the distinguishing characteristic is that the Parallelogram Law is assumed to hold:

$$\|x - y\|^2 + \|x + y\|^2 = 2\|x\|^2 + 2\|y\|^2$$

This succinct description gives no hint of the manifold implications of that assumption. The additional structure available in a Hilbert space makes it the preferred domain for much of applied mathematics! We pursue a more traditional approach to the subject, not basing everything on the Parallelogram Law, but using ideas that are undoubtedly already familiar to the reader, in particular the **dot product** or **inner product** of vectors. An **inner-product space** is a vector space X over the complex field in which an inner product $\langle x, y \rangle$ has been defined. We require these properties, for all x, y, and z in X:

(1) $\langle x, y \rangle$ is a complex number
(2) $\langle x, y \rangle = \overline{\langle y, x \rangle}$ (complex conjugate)
(3) $\langle \alpha x, y \rangle = \alpha \langle x, y \rangle$ $\alpha \in \mathbb{C}$
(4) $\langle x, x \rangle > 0$ if $x \neq 0$
(5) $\langle x + y, z \rangle = \langle x, z \rangle + \langle y, z \rangle$

The term "pre-Hilbert space" is also used for an inner-product space. Occasionally, we will employ real inner-product spaces and real Hilbert spaces. For them, the scalar field is $\mathbb{R}$, and the inner product is real-valued. However, some theorems to be proved later are valid only in the complex case.

Example 1.   Let $X = \mathbb{C}^n$ (the set of all complex n-tuples). If two points are given in $\mathbb{C}^n$, say $x = [x(1), x(2), \ldots, x(n)]$ and $y = [y(1), y(2), \ldots, y(n)]$, let $\langle x, y \rangle = \sum_{i=1}^{n} x(i)\overline{y(i)}$.   ∎

Example 2.   Let X be the set of all complex-valued continuous functions defined on $[0, 1]$. For x and y in X, define $\langle x, y \rangle = \int_0^1 x(t)\overline{y(t)}dt$.   ∎

In any inner-product space it is easy to prove that

$$\langle x + y, x + y \rangle = \langle x, x \rangle + 2\mathcal{R}\langle x, y \rangle + \langle y, y \rangle \qquad \mathcal{R} = \text{``real part''}$$
$$\langle x, \alpha y \rangle = \overline{\alpha}\langle x, y \rangle$$
$$\langle x, y + z \rangle = \langle x, y \rangle + \langle x, z \rangle$$
$$\left\langle \sum_{i=1}^{n} x_i, y \right\rangle = \sum_{i=1}^{n} \langle x_i, y \rangle$$

In an inner-product space, we define the **norm** of an element x to be $\|x\| = \sqrt{\langle x, x \rangle}$.

Theorem 1.   *The norm has these properties*

    a. $\|x\| > 0 \quad if \quad x \neq 0$

    b. $\|\alpha x\| = |\alpha|\,\|x\| \qquad (\alpha \in \mathbb{C})$

    c. $|\langle x, y \rangle| \leqslant \|x\|\,\|y\| \qquad$ *Cauchy–Schwarz Inequality*

    d. $\|x + y\| \leqslant \|x\| + \|y\| \qquad$ *Triangle Inequality*

    e. $\|x + y\|^2 + \|x - y\|^2 = 2\|x\|^2 + 2\|y\|^2 \qquad$ *Parallelogram Equality*

    f. *If* $\langle x, y \rangle = 0$, *then* $\|x + y\|^2 = \|x\|^2 + \|y\|^2 \qquad$ *Pythagorean Law.*

Proof.   Only **c** and **d** offer any difficulty. For **c**, let $\|y\| = 1$ and write

$$0 \leqslant \langle x - \lambda y, \ x - \lambda y \rangle = \langle x, x \rangle - \overline{\lambda}\langle x, y \rangle - \lambda\langle y, x \rangle + |\lambda|^2\langle y, y \rangle.$$

Now let $\lambda = \langle x, y \rangle$ to get $0 \leqslant \|x\|^2 - |\langle x, y \rangle|^2$. This establishes **c** in the case $\|y\| = 1$. By homogeneity, this suffices. To prove **d**, we use **c** as follows:

$$\|x + y\|^2 = \langle x + y, x + y \rangle = \langle x, x \rangle + \langle y, x \rangle + \langle x, y \rangle + \langle y, y \rangle$$
$$= \|x\|^2 + 2\mathcal{R}\langle x, y \rangle + \|y\|^2 \leqslant \|x\|^2 + 2|\langle x, y \rangle| + \|y\|^2$$
$$\leqslant \|x\|^2 + 2\|x\|\,\|y\| + \|y\|^2 = (\|x\| + \|y\|)^2 \qquad ∎$$

Item **e** in Theorem 1 is called the Parallelogram Equality (or "Law") because it states that the sum of the squares of the four sides of a parallelogram is equal to the sum of the squares of the two diagonals.

Lemma.   *In an inner-product space:*

    a. $x = 0$ *if and only if* $\langle x, v \rangle = 0$ *for all* v
    b. $x = y$ *if and only if* $\langle x, v \rangle = \langle y, v \rangle$ *for all* v
    c. $\|x\| = \sup\{|\langle x, v \rangle| : \|v\| = 1\}$

Proof. If $x = 0$, then $\langle x, v \rangle = 0$ for all v by Axiom 3 for the inner product. If $\langle x, v \rangle = 0$ for all v, then $\langle x, x \rangle = 0$, and so $x = 0$ by Axiom 4. The condition $x = y$ is equivalent to $x - y = 0$, to $\langle x - y, v \rangle = 0$ for all v, and to $\langle x, v \rangle = \langle y, v \rangle$ for all v. If $\|v\| = 1$, then by the Cauchy–Schwarz Inequality, $|\langle x, v \rangle| \leqslant \|x\|$. If $x = 0$, then $\|x\| \leqslant |\langle x, v \rangle|$ for all v. If $x \neq 0$, let $v = x/\|x\|$. Then $\|v\| = 1$ and $\langle x, v \rangle = \|x\|$.   ■

Definition.   A **Hilbert space** is a complete inner-product space.

Recall the definition of completeness from Section 1.2 (page 10): It means that every Cauchy sequence in the space converges to an element of the space.

Example 3.   The space of complex-valued continuous functions on $[0, 1]$ furnished with inner product

$$\langle x, y \rangle = \int_0^1 x(t)\overline{y(t)} \, dt$$

is *not* complete. Consider the sequence shown in Figure 2.1. The sequence of functions has the Cauchy property, but does not converge to a continuous function.

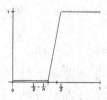

Figure 2.1

          ■

Example 4.   We write $L^2[a, b]$ for the set of all complex-valued Lebesgue measurable functions on $[a, b]$ such that

$$\int_a^b |x(t)|^2 \, dt < \infty$$

(The concept of measurability is explained in Chapter 8, Section 4, page 394.) In $L^2[a, b]$, put $\langle x, y \rangle = \int_a^b x(t)\overline{y(t)} \, dt$. This space is a Hilbert space, a fact known as the Riesz–Fischer Theorem (1906). See Chapter 8, Section 7, page 411 for the proof. This space contains many functions that have singularities. Thus, the function $t \mapsto t^{-1/3}$ belongs to $L^2[0, 1]$, but $t \mapsto t^{-2/3}$ does not.   ■

In $L^2[a, b]$, two functions f and g are regarded as equivalent if they differ only on a set of measure zero. Refer to Chapter 8 for an extended treatment of these matters. A set of measure 0 is easily described: For any $\varepsilon > 0$ we can cover the given set with a sequence of open intervals (a_n, b_n) whose total length satisfies $\sum_n (b_n - a_n) < \varepsilon$. An important consequence is that if f is an element of L^2, then $f(x)$ is meaningless! Indeed, f stands for an equivalence class of functions that can differ from each other at the point x, or indeed on any set of points having measure 0. When $f(x)$ appears under an integral sign, remember that the x is dispensable: The integration operates on the function as a whole, and no particular values $f(x)$ are involved.

Example 5. Let $(S, \mathcal{A}, \mu)$ be any measure space. The notation $L^2(S)$ then denotes the space of measurable complex functions on S such that $\int |f(s)|^2 \, d\mu < \infty$. In $L^2(S)$, define $\langle f, g \rangle = \int f(s)\overline{g(s)} \, d\mu$. Then $L^2(S)$ is a Hilbert space. See Theorem 3 in Section 8.7, page 411. ∎

Example 6. The space ℓ^2 (or ℓ_2) consists of all complex sequences $x = [x(1), x(2), \ldots]$ such that $\sum |x(n)|^2 < \infty$. The inner product is $\langle x, y \rangle = \sum x(n)\overline{y(n)}$. This is a Hilbert space, in fact a special case of Example 5. Just take $S = \mathbb{N}$ and use "counting" measure. (This is the measure that assigns to a set the number of elements in that set.) This example is also included in the general theory of the spaces ℓ_p, as outlined in Section 1.9, pages 54–56. ∎

Theorem 2.   *If K is a closed, convex, nonvoid set in a Hilbert space X, then to each x in X there corresponds a unique point y in K closest to x; that is,*

$$\|x - y\| = \operatorname{dist}(x, K) := \inf\{\|x - v\| : v \in K\}$$

Proof. Put $\alpha = \operatorname{dist}(x, K)$, and select $y_n \in K$ so that $\|x - y_n\| \to \alpha$. Notice that $\frac{1}{2}(y_n + y_m) \in K$ by the convexity of K. Hence $\|\frac{1}{2}(y_n + y_m) - x\| \geqslant \alpha$. By the Parallelogram Law,

$$\begin{aligned}
\|y_n - y_m\|^2 &= \|(y_m - x) - (y_n - x)\|^2 \\
&= 2\|y_n - x\|^2 + 2\|y_m - x\|^2 - \|y_n + y_m - 2x\|^2 \\
&= 2\|y_n - x\|^2 + 2\|y_m - x\|^2 - 4\|\tfrac{1}{2}(y_n + y_m) - x\|^2 \\
&\leqslant 2\|y_n - x\|^2 + 2\|y_m - x\|^2 - 4\alpha^2 \to 0
\end{aligned}$$

This shows that $[y_n]$ is a Cauchy sequence. Hence $y_n \to y$ for some $y \in X$. Since K is closed, $y \in K$. By continuity,

$$\|x - y\| = \|x - \lim_n y_n\| = \lim_n \|x - y_n\| = \alpha$$

For the uniqueness of the point y, suppose that y_1 and y_2 are points in K of distance α from x. By the previous calculation we have

$$\|y_1 - y_2\| \leqslant 2\|y_1 - x\|^2 + 2\|y_2 - x\|^2 - 4\alpha^2 = 0 \qquad ∎$$

In an inner-product space, the notion of orthogonality is important. If $\langle x, y \rangle = 0$, we say that the points x and y are **orthogonal** to each other, and we write $x \perp y$. (We *do not* say that the points are orthogonal, but we could say that the *pair of points* is orthogonal.) If Y is a set, the notation $x \perp Y$ signifies that $x \perp y$ for all $y \in Y$. If U and V are sets, $U \perp V$ means that $u \perp v$ for all $u \in U$ and all $v \in V$.

Theorem 3.   *Let Y be a subspace in an inner-product space X. Let $x \in X$ and $y \in Y$. These are equivalent assertions:*

    a. $x - y \perp Y$, *i.e.,* $\langle x - y, v \rangle = 0$ *for all $v \in Y$.*
    b. *y is the unique point of Y closest to x.*

Proof.   If **a** is true, then for any $u \in Y$ we have

$$\|x - u\|^2 = \|(x - y) + (y - u)\|^2 = \|x - y\|^2 + \|y - u\|^2 \geqslant \|x - y\|^2$$

Here we used the Pythagorean Law (part 6 of Theorem 1).

Now suppose that **b** is true. Let u be any point of Y and let λ be any scalar. Then (because y is the point closest to x)

$$0 \leqslant \|x - (y + \lambda u)\|^2 - \|x - y\|^2 = -2\mathcal{R}\langle x - y, \lambda u \rangle + |\lambda|^2 \|u\|^2$$

Hence

$$2\mathcal{R}\left\{\bar{\lambda}\langle x - y, u \rangle\right\} \leqslant |\lambda|^2 \|u\|^2$$

If $\langle x - y, u \rangle \neq 0$, then $u \neq 0$ and we can put $\lambda = \langle x - y, u \rangle / \|u\|^2$ to get a contradiction:

$$2\mathcal{R}\left\{\bar{\lambda}\lambda\|u\|^2\right\} \leqslant |\lambda|^2 \|u\|^2 \qquad \blacksquare$$

Definition.   The **orthogonal complement** of a subset Y in a inner-product space X is

$$Y^{\perp} = \left\{x \in X : \langle x, y \rangle = 0 \text{ for all } y \in Y\right\}$$

Theorem 4.   *If Y is a closed subspace of a Hilbert space X, then $X = Y \oplus Y^{\perp}$.*

Proof.   We have to prove that $Y^{\perp}$ is a subspace, that $Y \cap Y^{\perp} = 0$, and that $X \subset Y + Y^{\perp}$. If v_1 and v_2 belong to $Y^{\perp}$, then so does $\alpha_1 v_1 + \alpha_2 v_2$, since for $y \in Y$,

$$\langle y, \alpha_1 v_1 + \alpha_2 v_2 \rangle = \bar{\alpha}_1 \langle y, v_1 \rangle + \bar{\alpha}_2 \langle y, v_2 \rangle = 0$$

If $x \in Y \cap Y^{\perp}$, then $\langle x, x \rangle = 0$, so $x = 0$. If x is any element of X, let y be the element of Y closest to x. By the preceding theorem, $x - y \perp Y$. Hence the equation $x = y + (x - y)$ shows that $X \subset Y + Y^{\perp}$. $\blacksquare$

Theorem 5.   *If the Parallelogram Law is valid in a normed linear space, then that space is an inner-product space. In other words, an inner product can be defined in such a way that $\langle x, x \rangle = \|x\|^2$.*

Proof.   We define the inner product by the equation

$$4\langle x, y \rangle = \|x + y\|^2 - \|x - y\|^2 + i\|x + iy\|^2 - i\|x - iy\|^2$$

From the definition, it follows that

$$4\mathcal{R}\langle x,y\rangle = \|x+y\|^2 - \|x-y\|^2$$

From this equation and the Parallelogram Law we obtain

$$\begin{aligned}
4\mathcal{R}\langle u+v,y\rangle &= \|u+v+y\|^2 - \|u+v-y\|^2 \\
&= \{2\|u+y\|^2 + 2\|v\|^2 - \|u+y-v\|^2\} \\
&\quad - \{2\|u\|^2 + 2\|v-y\|^2 - \|u-v+y\|^2\} \\
&= \{\|u+y\|^2 - \|u-y\|^2\} + \{\|v+y\|^2 - \|v-y\|^2\} \\
&\quad + \{\|u+y\|^2 + \|u-y\|^2 - 2\|u\|^2 - 2\|y\|^2\} \\
&\quad + \{2\|y\|^2 + 2\|v\|^2 - \|v+y\|^2 - \|v-y\|^2\} \\
&= 4\mathcal{R}\langle u,y\rangle + 4\mathcal{R}\langle v,y\rangle
\end{aligned}$$

This proves that $\mathcal{R}\langle u+v,y\rangle = \mathcal{R}\langle u,y\rangle + \mathcal{R}\langle v,y\rangle$. Now by putting iy in place of y in the definition of $\langle x,y\rangle$ we obtain $\langle x,iy\rangle = -i\langle x,y\rangle$. Hence the imaginary parts of these complex numbers satisfy

$$\begin{aligned}
\mathcal{I}\langle u+v,y\rangle &= -\mathcal{R}i\langle u+v,y\rangle = \mathcal{R}\langle u+v,iy\rangle \\
&= \mathcal{R}\langle u,iy\rangle + \mathcal{R}\langle v,iy\rangle = -\mathcal{R}i\langle u,y\rangle - \mathcal{R}i\langle v,y\rangle \\
&= \mathcal{I}\langle u,y\rangle + \mathcal{I}\langle v,y\rangle
\end{aligned}$$

(In this equation, $\mathcal{I}$ denotes "the imaginary part of.") Thus we have fully established that $\langle u+v,y\rangle = \langle u,y\rangle + \langle v,y\rangle$. By induction, we can then prove that $\langle nx,y\rangle = n\langle x,y\rangle$ for all positive integers n. From this it follows, for any two positive integers m and n, that

$$\left\langle \frac{n}{m}x,y\right\rangle = \frac{n}{m}m\left\langle \frac{x}{m},y\right\rangle = \frac{n}{m}\langle x,y\rangle$$

By continuity, we obtain $\langle \lambda x,y\rangle = \lambda\langle x,y\rangle$ for any $\lambda \geqslant 0$. From the definition, we quickly verify that

$$\langle -x,y\rangle = -\langle x,y\rangle \quad \text{and} \quad \langle ix,y\rangle = i\langle x,y\rangle$$

Hence $\langle \lambda x,y\rangle = \lambda\langle x,y\rangle$ for all complex scalars λ. From the definition we obtain

$$\begin{aligned}
4\langle x,x\rangle &= \|2x\|^2 + i\|x+ix\|^2 - i\|x-ix\|^2 \\
&= 4\|x\|^2 + i|1+i|^2\|x\|^2 - i|1-i|^2\|x\|^2 = 4\|x\|^2
\end{aligned}$$

Finally, we have

$$\begin{aligned}
4\langle y,x\rangle &= \|y+x\|^2 - \|y-x\|^2 + i\|y+ix\|^2 - i\|y-ix\|^2 \\
&= \|x+y\|^2 - \|x-y\|^2 + i\|-i(y+ix)\|^2 - i\|i(y-ix)\|^2 \\
&= \|x+y\|^2 - \|x-y\|^2 + i\|x-iy\|^2 - i\|x+iy\|^2 \\
&= 4\overline{\langle x,y\rangle}
\end{aligned}$$

∎

In an inner-product space, the angle between two nonzero vectors can be defined. In order to see what a reasonable definition is, we recall the Law of Cosines from elementary trigonometry. In a triangle having sides a, b, c and angle θ opposite side c, we have

$$c^2 = a^2 + b^2 - 2ab\cos\theta$$

Notice that when $\theta = 90°$, this equation gives the Pythagorean rule. In an inner-product space, we consider a triangle as shown in Figure 2.2.

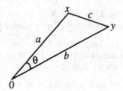

Figure 2.2

We have

$$\|x - y\|^2 = \langle x - y, x - y \rangle = \langle x, x \rangle - \langle x, y \rangle - \langle y, x \rangle + \langle y, y \rangle$$
$$= \|x\|^2 + \|y\|^2 - 2\mathcal{R}\langle x, y \rangle$$

On the other hand, we would like to have the law of cosines:

$$\|x - y\|^2 = \|x\|^2 + \|y\|^2 - 2\|x\| \, \|y\| \cos\theta$$

Therefore, we define $\cos\theta$ so that $\|x\| \, \|y\| \cos\theta = \mathcal{R}\langle x, y \rangle$ Thus

$$\theta = \text{Arccos} \, \frac{\mathcal{R}\langle x, y \rangle}{\|x\| \, \|y\|}$$

The "principal value" of Arccos is used; it is an angle in the interval $[0, \pi]$. Is the definition proper? Yes, because the number $\mathcal{R}\langle x, y \rangle \|x\|^{-1} \|y\|^{-1}$ lies in the interval $[-1, 1]$, by the Cauchy–Schwarz inequality. Other definitions for the angle between two vectors can be given. See [Ar], pages 87–90.

There are many sources for the theory of Hilbert spaces. In addition to the references indicated at the end of Section 1.1, there are these specialized texts: [AG], [Ar], [Berb], [Berb2], [DM], [Hal2], [Hal3], [St], and [Youn].

Problems 2.1

1. Verify that Example 1 is an inner-product space.

2. Verify that Example 2 gives an inner product. Give all details, especially for the fourth axiom.

3. Prove the four equations stated in the text just after Example 2.

4. Fix x and y in an inner-product space, and determine the value of λ for which $\|x - \lambda y\|$ is a minimum.

5. Prove the Parallelogram Law.

6. Let K be a convex set in an inner-product space. Let z be a point at distance α from K. Prove that the diameter of the set

$$\{x \in K : \|x - z\| \leqslant \alpha + \delta\}$$

is not greater than $2\sqrt{2\alpha\delta + \delta^2}$. The diameter of a set S is $\sup_{u,v \in S} \|u - v\|$.

7. Prove that in an inner-product space if $\|x\| = 1 < \|y\|$, then $\|(x - y/\|y\|)\| < \|x - y\|$.

8. Prove that in an inner-product space X, the mapping $x \mapsto \langle x, v \rangle$ is continuous. (Here v can be any fixed vector in the space.) Prove that on $X \times X$ the mapping $(x, y) \mapsto \langle x, y \rangle$ is continuous.

9. In an inner-product space X, let $M = \{x \in X \: : \: \langle x, v \rangle = 0\}$, where v is a fixed, nonzero vector. Show that M is a closed subspace. Prove that M has codimension 1.

10. For any subset M of an inner product space X, define $M^{\perp} = \{x \in X : \langle x, m \rangle = 0$ for all $m \in M\}$. Prove that $M^{\perp}$ is a closed subspace and that $M \cap M^{\perp}$ is either $\varnothing$ (the empty set) or 0. (Here 0 denotes the zero subspace, $\{0\}$.)

11. Prove that $|\langle x, y \rangle| = \|x\| \, \|y\|$ if and only if one of the vectors x and y is a multiple of the other.

12. Let X be any linear space, and let H be a Hamel basis for X. Show how to use H to define an inner product on X and thus create an inner-product space.

13. Let A be an $n \times n$ matrix. In the real space $\mathbb{R}^n$, define $\langle x, y \rangle = y^T A x$. (Here we interpret elements of $\mathbb{R}^n$ as $n \times 1$ matrices. Thus y^T is a $1 \times n$ matrix.) Find necessary and sufficient conditions on A in order that our definition shall produce a genuine inner product.

14. Let X be the space of all "finitely nonzero sequences" of complex numbers. Thus $x \in X$ if $x : \mathbb{N} \to \mathbb{R}$ and $\{n : x(n) \neq 0\}$ is finite. For x and y in X, define $\langle x, y \rangle = \sum_{n=1}^{\infty} x(n)\overline{y(n)}$. Prove that X is not a Hilbert space.

15. Let $X = \mathbb{R}^2$, and define an inner product between vectors $x = [x(1), x(2)]$ and $y = [y(1), y(2)]$ by the equation

$$\langle x, y \rangle = 2x(1)y(1) + x(2)y(2)$$

Prove that this makes X a real inner-product space. Let

$$Y = \{y \in X : y(1) - y(2) = 0\}$$

Find the point y of Y closest to $x = [0, 1]$. Draw an accurate sketch showing all of this. Explain why $x - y$ is not perpendicular to Y. Does this contradict Theorem 5? Draw a sketch of the unit ball.

16. Prove or disprove this analogue of Theorem 4: If Y is a subspace of a Hilbert space X, then $X = Y \oplus Y^{\perp}$.

17. Let x and y be points in a real inner-product space such that $\|x + y\|^2 = \|x\|^2 + \|y\|^2$. Show that $x \perp y$. Show that this is not always true in a complex inner-product space.

18. In an inner-product space, prove that if $\|x_n\| \to \|y\|$ and $\langle x_n, y \rangle \to \|y\|^2$, then $x_n \to y$.

19. Prove or disprove: In a Hilbert space, if $\sum_{n=1}^{\infty} \|x_n\|^2 < \infty$, then the series $\sum_{n=1}^{\infty} x_n$ converges.

20. Find all solutions to the equation $\langle x, a \rangle c = b$, assuming that a, b, and c are given vectors in an inner-product space.

21. Indicate how the equation $Ax = b$ can be solved if the operator A is defined by $Ax = \sum_{i=1}^{n} \langle x, a_i \rangle c_i$. Describe the set of all solutions.

22. Find all solutions to the equation $x + \langle x, a \rangle c = b$.

23. Use Problem 22 to solve the integral equation $x(s) + \int_0^1 x(t)t^2 s \, dt = \cos s$.

24. Let $v = [v(1), v(2), \ldots]$ be an element of ℓ^2. Prove that the set $\{x \in \ell^2 : |x(n)| \leqslant |v(n)|$ for all $n\}$ is compact in ℓ^2.

25. Prove that if M is a closed subspace in a Hilbert space, then $M^{\perp\perp} = M$.

26. Prove that if $M = M^{\perp\perp}$ for every closed linear subspace in an inner-product space, then the space is complete.

27. Prove that if M and N are closed subspaces of a Hilbert space and if $M \perp N$, then $M + N$ is closed.

28. Consider the mapping A in Problem 21. Find necessary and sufficient conditions on a_i and c_i in order that A have a fixed point other than 0.

29. In a Hilbert space, elements w, u_i, and v_i are given. Show how to find an x such that

$$x = w + \sum_{i=1}^{n} \langle x, v_i \rangle u_i$$

30. In a Hilbert space, let $\|x_n\| \to c$, $\|y_n\| \to c$, and $\langle x_n, y_n \rangle \to c^2$. Prove that $\|x_n - y_n\| \to 0$. Then make two generalizations. Is there any similar result for unbounded sequences?

31. If $M \subset N$, then $N^\perp \subset M^\perp$. Prove this.

32. Let K be a closed convex set in a Hilbert space X. Let $x \in X$ and let y be the point of K closest to x. Prove that $\mathcal{R}\langle x - y, v - y \rangle \leqslant 0$ for all $v \in K$. Interpret this as a separation theorem, i.e., an assertion about a hyperplane and a convex set. Prove the converse.

33. Prove that if $a_i \geqslant 0$ and $\sum_{i=1}^{\infty} a_i < \infty$, then

$$\left(\sum_{i=0}^{\infty} 2^{-i/2} a_i^{1/2} \right)^2 \leqslant \sum_{i=1}^{\infty} a_i$$

34. The Banach space ℓ^1 consists of sequences $[x_1, x_2, \ldots]$ for which $\sum |x_n| < \infty$. The norm is defined to be $\|x\| = \sum |x_n|$. Prove that ℓ^1 is dense in ℓ^2, and explain why this does not contradict the fact that ℓ^1 is complete.

35. Prove that in a real-inner product space

$$\|x - y\|^2 = \|x\|^2 - \|y\|^2 + 2\langle y - x, y \rangle$$

36. In a real inner-product space, does the equation $\|x + y + z\|^2 = \|x\|^2 + \|y\|^2 + \|z\|^2$ imply any orthogonality relations among the three points?

37. Find the necessary and sufficient conditions on the complex numbers $w_1, w_2, \ldots, w_n$ in order that the equation

$$\langle x, y \rangle = \sum_{k=1}^{n} x(k) \overline{y(k)}\, w_k$$

shall define an inner product on $\mathbb{C}^n$.

38. Prove that if x is an element of ℓ^2, then for all natural numbers n,

$$\inf_{k>n} |x(k)| \sum_{j=1}^{k} |x(j)| = 0$$

39. Let K be a closed convex set in a Hilbert space X. For each x in X, let Px be the point of K closest to x. Prove that $\|Px - Py\| \leqslant \|x - y\|$. (Cf. Problems 2.2.24, 2.1.32.)

40. (Continuation) Prove that each closed convex set K in a Hilbert space X is a "retract", i.e., the identity map on K has a continuous extension mapping X onto K.

41. Let F and G be two maps (not assumed to be linear or continuous) of an inner product space X into itself. Suppose that for all x and y in X, $\langle F(x), y \rangle = \langle x, G(y) \rangle$. Prove that if a sequence x_n converges to x, and $G(x_n)$ converges to y, then $y = G(x)$. Prove also that $F(0) = G(0) = 0$.

42. Prove that in an inner product space, if $\lambda > 0$, then

$$|\langle x, y \rangle| \leqslant \lambda \|x\|^2 + \frac{1}{4\lambda}\|y\|^2$$

2.2 Orthogonality and Bases

Definition.   A set $\mathcal{A}$ of vectors in an inner-product space is said to be **orthogonal** if $\langle x, y \rangle = 0$ whenever $x \in \mathcal{A}$, $y \in \mathcal{A}$, and $x \neq y$. Recall that we write $x \perp y$ to mean $\langle x, y \rangle = 0$, $x \perp S$ to mean that $x \perp y$ for all $y \in S$, and $U \perp V$ to mean that $x \perp y$ for all $x \in U$ and $y \in V$.

> **Theorem 1.   Pythagorean Law.**   *If $\{x_1, x_2, \ldots, x_n\}$ is a finite orthogonal set of n distinct elements in an inner-product space, then*

$$\left\| \sum_{j=1}^{n} x_j \right\|^2 = \sum_{j=1}^{n} \|x_j\|^2$$

Proof.   By our assumptions, $x_i \neq x_j$ if $i \neq j$, and consequently,

$$\left\| \sum_{j=1}^{n} x_j \right\|^2 = \left\langle \sum_{j=1}^{n} x_j, \sum_{i=1}^{n} x_i \right\rangle = \sum_{j=1}^{n}\sum_{i=1}^{n}\langle x_j, x_i \rangle = \sum_{j=1}^{n}\langle x_j, x_j \rangle = \sum_{j=1}^{n}\|x_j\|^2 \quad \blacksquare$$

This theorem has a counterpart for orthogonal sets that are not finite, but its meaning will require some explanation. What should we mean by the sum of the elements in an arbitrary subset $\mathcal{A}$ in X? If $\mathcal{A}$ is finite, we know what is meant. For an infinite set, we shall say that the sum of the elements of $\mathcal{A}$ is s if and only if the following is true: For each positive ϵ there exists a *finite* subset $\mathcal{A}_0$ of $\mathcal{A}$ such that for every larger finite subset F we have

$$\left| \sum \{x : x \in F\} - s \right| < \epsilon$$

When we say "larger set" we mean only that $\mathcal{A}_0 \subset F \subset \mathcal{A}$. Notice that the definition employs only finite subsets of $\mathcal{A}$. For the reader who knows all about "nets," "generalized sequences," or "Moore–Smith convergence," we remark that what is going on here is this: We partially order the finite subsets of $\mathcal{A}$ by inclusion. With each finite subset F of $\mathcal{A}$ we associate the sum $S(F)$ of all the elements in F. Then S is a net (i.e., a function on a directed set). The limit of this net, if it exists, is the sum s of all the elements of A. To be more precise, it is often called the *unordered* sum over A.

When dealing with an orthogonal indexed set of elements $[x_i]$ in an inner-product space, we always assume that $x_i \neq x_j$ if $i \neq j$. This assumption allows us to write $x_i \perp x_j$ when $i \neq j$.

Theorem 2. The General Pythagorean Law.   *Let $[x_j]$ be an orthogonal sequence in a Hilbert space. The series $\sum x_j$ converges if and only if $\sum \|x_j\|^2 < \infty$. If $\sum \|x_j\|^2 = \lambda < \infty$, then $\left\| \sum x_j \right\|^2 = \lambda$, and the sum $\sum x_j$ is independent of the ordering of the terms.*

Proof.   Put $S_n = \sum_1^n x_j$ and $s_n = \sum_1^n \|x_j\|^2$.

By the finite version of the Pythagorean Law, we have (for $m > n$)

$$\left\| S_m - S_n \right\|^2 = \left\| \sum_{n+1}^m x_j \right\|^2 = \sum_{n+1}^m \|x_j\|^2 = |s_m - s_n|$$

Hence $[S_n]$ is a Cauchy sequence in X if and only if $[s_n]$ is a Cauchy sequence in $\mathbb{R}$. This establishes the first assertion in the theorem.

Now assume that $\lambda < \infty$. By the Pythagorean Law, $\|S_n\|^2 = s_n$, and hence in the limit we have $\left\| \sum x_j \right\|^2 = \lambda$. Let u be a rearrangement of the original series, say $u = \sum x_{k_j}$. Let $U_n = \sum_1^n x_{k_j}$. By the theory of absolutely convergent series in $\mathbb{R}$, we have $\sum \|x_{k_j}\|^2 = \lambda$. Hence, by our previous analysis, $U_n \to u$ and $\|u\|^2 = \lambda$. Now compute

$$\langle U_n, S_m \rangle = \left\langle \sum_{j=1}^n x_{k_j}, \sum_{i=1}^m x_i \right\rangle = \sum_{j=1}^n \sum_{i=1}^m \|x_i\|^2 \delta_{ik_j}$$

We let $n \to \infty$ to get $\langle u, S_m \rangle = \sum_{i=1}^m \|x_i\|^2$. Then let $m \to \infty$ to get $\langle u, x \rangle = \lambda$, where $x = \lim S_m$. It follows that $x = u$, because

$$\|x - u\|^2 = \|x\|^2 - 2\mathcal{R}\langle x, u \rangle + \|u\|^2 = \lambda - 2\lambda + \lambda = 0 \qquad \blacksquare$$

Definition.   A set U in an inner-product space is said to be **orthonormal** if each element has norm 1 and if $\langle u, v \rangle = 0$ when $u, v \in U$ and $u \neq v$. If the set U is indexed in a one-to-one manner so that $U = [u_i : i \in I]$, then the condition of orthonormality is simply $\langle u_i, u_j \rangle = \delta_{ij}$, where, as usual, δ_{ij} is 1 when $i = j$ and is 0 otherwise. If an indexed set is asserted to be orthonormal, we shall always assume that the indexing is one-to-one, and that the equation just mentioned applies.

If $[v_i : i \in I]$ is an orthogonal set of nonzero vectors, then $[v_i/\|v_i\| : i \in I]$ is an orthonormal set.

Theorem 3.   If $[y_1, y_2, \ldots, y_n]$ is an orthonormal set in an inner-product space, and if Y is the linear span of $\{y_i : 1 \leqslant i \leqslant n\}$, then for any x, the point in Y closest to x is $\sum_{i=1}^{n} \langle x, y_i \rangle y_i$.

Proof.   Let $y = \sum_{i=1}^{n} \langle x, y_i \rangle y_i$. By Theorem 3 in Section 2.1, page 65, it suffices to verify that $x - y \perp Y$. For this it is enough to verify that $x - y$ is orthogonal to each basis vector y_k. We have

$$\langle x - y, y_k \rangle = \langle x, y_k \rangle - \Big\langle \sum_i \langle x, y_i \rangle y_i, y_k \Big\rangle = \langle x, y_k \rangle - \sum_i \langle x, y_i \rangle \langle y_i, y_k \rangle$$

$$= \langle x, y_k \rangle - \sum_i \langle x, y_i \rangle \delta_{ik} = \langle x, y_k \rangle - \langle x, y_k \rangle = 0 \ . \qquad \blacksquare$$

The vector y in the above proof is called the **orthogonal projection** of x onto Y. The coefficients $\langle x, y_i \rangle$ are called the **(generalized) Fourier** coefficients of x with respect to the given orthonormal system. The operator that produces y from x is called an **orthogonal projection** or an **orthogonal projector**. Look ahead to Theorem 7 for a further discussion.

Corollary 1.   If x is a point in the linear span of an orthonormal set $[y_1, y_2, \ldots, y_n]$ then $x = \sum_{i=1}^{n} \langle x, y_i \rangle y_i$.

Theorem 4. Bessel's Inequality.   If $[u_i : i \in I]$ is an orthonormal system in an inner-product space, then for every x,

$$\sum |\langle x, u_i \rangle|^2 \leqslant \|x\|^2$$

Proof.   For j ranging over a finite subset J of I, let $y = \sum \langle x, u_j \rangle u_j$. This vector y is the orthogonal projection of x onto the subspace $U = \text{span}[u_j : j \in J]$. By Theorem 3, $x - y \perp U$. Hence by the Pythagorean Law

$$\|x\|^2 = \|(x-y)+y\|^2 = \|x-y\|^2 + \|y\|^2 \geqslant \|y\|^2 = \sum \|\langle x, u_j \rangle u_j\|^2 = \sum |\langle x, u_j \rangle|^2$$

This proves our result for any finite set of indices. The result for I itself now follows from Problem 4. $\blacksquare$

Corollary 2.   If $[u_1, u_2, \ldots]$ is an orthonormal sequence in an inner-product space, then for each x, $\lim_{n \to \infty} \langle x, u_n \rangle = 0$.

Corollary 3.   If $[u_i : i \in I]$ is an orthonormal system, then for each x at most a countable number of the Fourier coefficients $\langle x, u_i \rangle$ are nonzero.

Proof.   Fixing x, put $J_n = \{i \in I : |\langle x, u_i \rangle| > 1/n\}$. By the Bessel Inequality,

$$\|x\|^2 \geqslant \sum_{j \in J_n} |\langle x, u_j \rangle|^2 \geqslant \sum_{j \in J_n} 1/n^2 = (\# J_n)/n^2$$

Hence J_n is a finite set. Since

$$\{i : \langle x, u_i \rangle \neq 0\} = \bigcup_{n=1}^{\infty} J_n$$

we see that this set must be countable, it being a union of countably many finite sets. ∎

Let X be any inner-product space. An **orthonormal basis** for X is any maximal orthonormal set in X. It is also called an "orthonormal base." In this context, "maximal" means *not properly contained in another orthonormal set*. In other words, it is a maximal element in the partially ordered family of all orthonormal sets, when the partial order is set inclusion, $\subset$. (Refer to Section 1.6, page 31, for a discussion of partially ordered sets.)

Theorem 5.   *Every nontrivial inner-product space has an orthonormal basis.*

Proof.   Call the space X. Since it is not 0, it contains a nonzero vector x. The set consisting solely of $x/\|x\|$ is orthonormal. Now order the family of all orthonormal subsets of X in the natural way (by inclusion). In order to use Zorn's Lemma, one must verify that each chain of orthonormal sets has an upper bound. Let $\mathcal{C}$ be such a chain, and put $A^* = \bigcup \{A : A \in \mathcal{C}\}$. It is obvious that A^* is an upper bound for $\mathcal{C}$, but is A^* orthonormal? Take x and y in A^* such that $x \neq y$. Say $x \in A_1 \in \mathcal{C}$ and $y \in A_2 \in \mathcal{C}$. Since $\mathcal{C}$ is a chain, either $A_1 \subset A_2$ or $A_2 \subset A_1$. Suppose the latter. Then $x, y \in A_1$. Since A_1 is orthonormal, $\langle x, y \rangle = 0$. Obviously, $\|x\| = 1$. Hence A^* is orthonormal. ∎

Theorem 6.   The Orthonormal Basis Theorem.   *For an orthonormal family $[u_i]$ (not necessarily finite or countable) in a Hilbert space X, the following properties are equivalent:*

   a. *$[u_i]$ is an orthonormal basis for X.*
   b. *If $x \in X$ and $x \perp u_i$ for all i, then $x = 0$.*
   c. *For each $x \in X$, $x = \sum \langle x, u_i \rangle u_i$.*
   d. *For each x and y in X, $\langle x, y \rangle = \sum \langle x, u_i \rangle \overline{\langle y, u_i \rangle}$.*
   e. *For each x in X, $\|x\|^2 = \sum |\langle x, u_i \rangle|^2$. (Parseval Identity)*

Proof.   To prove that **a** implies **b**, suppose that **b** is false. Let $x \neq 0$ and $x \perp u_i$ for all i. Adjoin $x/\|x\|$ to the family $[u_i]$ to get a larger orthonormal family. Thus the original family is not maximal and is not a basis.

To prove that **b** implies **c**, assume **b** and let x be any point in X. Let $y = \sum \langle x, u_i \rangle u_i$. By Bessel's inequality (Theorem 4), we have

$$\sum \|\langle x, u_i \rangle u_i\|^2 = \sum |\langle x, u_i \rangle|^2 \leqslant \|x\|^2$$

By Theorem 2, the series defining y converges. (Here the completeness of X is needed.) Then straightforward calculation (as in the proof of Theorem 3) shows that $x - y \perp u_i$ for all i. By **b**, $x - y = 0$.

To prove that **c** implies **d**, assume **c** and write

$$x = \sum \langle x, u_i \rangle u_i \qquad y = \sum \langle y, u_i \rangle u_i$$

Straightforward calculation then yields $\langle x, y \rangle = \sum \langle x, u_i \rangle \overline{\langle y, u_i \rangle}$.

To prove that **d** implies **e**, assume **d** and let $y = x$ in **d**. The result is the assertion in **e**.

To prove that **e** implies **a**, suppose that **a** is false. Then $[u_i]$ is not a maximal orthonormal set. Adjoin a new element, x, to obtain a larger orthonormal set. Then $1 = \|x\|^2 \neq \sum |\langle x, u_i \rangle|^2 = 0$, showing that **e** is false. ∎

Example 1.   One orthonormal basis in ℓ^2 is obtained by defining $u_n(j) = \delta_{nj}$. Thus

$$u_1 = [1, \ 0, \ 0, \ \ldots] \ , \quad u_2 = [0, \ 1, \ 0, \ \ldots] \ , \quad \text{etc.}$$

To see that this is actually an orthonormal base, use the preceding theorem, in particular the equivalence of **a** and **b**. Suppose $x \in \ell^2$ and $\langle x, u_n \rangle = 0$ for all n. Then $x(n) = 0$ for all n, and $x = 0$. ∎

Example 2.   An orthonormal basis for $L^2[0, 1]$ is provided by the functions $u_n(t) = e^{2\pi i n t}$, where $n \in \mathbb{Z}$. One verifies the orthonormality by computing the appropriate integrals. To show that $[u_n]$ is a base, we use Part **b** of Theorem 6. Let $x \in L^2[0, 1]$ and $x \neq 0$. It is to be shown that $\langle x, u_n \rangle \neq 0$ for some n. Since the set of continuous functions is dense in L^2, there is a continuous y such that $\|x - y\| < \|x\|/5$. Then $\|y\| \geqslant \|x\| - \|x - y\| > \frac{4}{5}\|x\|$. By the Weierstrass Approximation Theorem, the linear span of $[u_n]$ is dense in the space $C[0, 1]$, furnished with the supremum norm. Select a linear combination p of $[u_n]$ such that $\|p - y\|_\infty < \|x\|/5$. Then $\|p - y\| < \|x\|/5$. Hence $\|p\| > \|y\| - \|y - p\| > \frac{3}{5}\|x\|$. Then

$$|\langle x, p \rangle| \geqslant |\langle p, p \rangle| - |\langle y - p, p \rangle| - |\langle x - y, p \rangle|$$
$$\geqslant \|p\|^2 - \|y - p\| \, \|p\| - \|x - y\| \, \|p\| > 0$$

Thus it is not possible to have $\langle x, u_n \rangle = 0$ for all n. ∎

Recall that we have defined the **orthogonal projection** of a Hilbert space X onto a closed subspace Y to be the mapping P such that for each $x \in X$, Px is the point of Y closest to x.

Theorem 7.   The Orthogonal Projection Theorem.   *The orthogonal projection P of a Hilbert space X onto a closed subspace Y has these properties:*

    a. *It is well–defined; i.e., Px exists and is unique in Y.*
    b. *It is surjective, i.e., $P(X) = Y$.*
    c. *It is linear.*
    d. *If Y is not 0 (the zero subspace), then $\|P\| = 1$.*
    e. *$x - Px \perp Y$ for all x.*

f. P is Hermitian; i.e., $\langle Px, w \rangle = \langle x, Pw \rangle$ for all x and w.

g. If $[y_i]$ is an orthonormal basis for Y, then $Px = \sum \langle x, y_i \rangle y_i$.

h. P is idempotent; i.e., $P^2 = P$.

i. $Py = y$ for all $y \in Y$. Thus $P|Y = I_Y$.

j. $\|x\|^2 = \|Px\|^2 + \|x - Px\|^2$.

Proof. This is left to the problems. ∎

The Gram–Schmidt process, familiar from the study of linear algebra, is an algorithm for producing orthonormal bases. It is a recursive process that can be applied to any linearly independent sequence in an inner-product space, and it yields an orthonormal sequence, as described in the next theorem.

Theorem 8.   The Gram–Schmidt Construction.   *Let* $[v_1, v_2, v_3, \dots]$ *be a linearly independent sequence in an inner product space. Having set* $u_1 = v_1/\|v_1\|$, *define recursively*

$$u_n = \frac{v_n - \sum\limits_{i=1}^{n-1} \langle v_n, u_i \rangle u_i}{\left\| v_n - \sum\limits_{i=1}^{n-1} \langle v_n, u_i \rangle u_i \right\|} \qquad n = 2, 3, \dots$$

Then $[u_1, u_2, u_3, \dots]$ *is an orthonormal sequence, and for each* n, $\text{span}\{u_1, u_2, \dots, u_n\} = \text{span}\{v_1, v_2, \dots, v_n\}$.

Notice that in the equation describing this algorithm there is a normalization process: the dividing of a vector by its norm to produce a new vector pointing in the same direction but having unit length. The other action being carried out is the subtraction from the vector v_n of its projection on the linear span of the orthonormal set presently available, $u_1, u_2, \dots, u_{n-1}$. This action is obeying the equation in Theorem 3, and it produces a vector that is orthogonal to the linear span just described. These remarks should make the formulas easy to derive or remember.

Example 3.   (A nonseparable inner-product space). A normed linear space (or any topological space) is said to be **separable** if it contains a countable dense set. If an inner-product space is nonseparable, it cannot have a countable orthonormal base. For an example, we consider the uncountable family of functions $u_\lambda(t) = e^{i\lambda t}$, where $t \in \mathbb{R}$ and $\lambda \in \mathbb{R}$. This family of functions is linearly independent (Problem 5), and is therefore a Hamel basis for a linear space X. We introduce an inner product in X by defining the inner product of two elements in the Hamel base:

$$\langle u_\lambda, u_\sigma \rangle = \delta_{\lambda\sigma} = \begin{cases} 1 & \lambda = \sigma \\ 0 & \lambda \neq \sigma \end{cases}$$

This is the value that arises in the following integration:

$$\lim_{T \to \infty} \frac{1}{2T} \int_{-T}^{T} u_\lambda(t)\, \overline{u_\sigma(t)}\, dt = \lim_{T \to \infty} \frac{1}{2T} \int_{-T}^{T} e^{i(\lambda - \sigma)t}\, dt$$

If $\lambda = \sigma$, this calculation produces the result 1. If $\lambda \neq \sigma$, we get 0. Elements of X have the property of **almost periodicity**. (See Problem 1.) ∎

Example 4. (Other abstract Hilbert spaces). A higher level of abstraction can be used to generate further inner product spaces and Hilbert spaces. Let us create at one stroke a Hilbert space of any given dimension. Let S be any set. The notation $\mathbb{C}^S$ denotes the family of *all* functions from S to the field $\mathbb{C}$. This set of functions has a natural linear structure, for if x and y belong to $\mathbb{C}^S$, $x+y$ can be defined by

$$(x+y)(s) = x(s) + y(s)$$

A similar equation defines λx for $\lambda \in \mathbb{C}$. Within $\mathbb{C}^S$ we single out the subspace X of all $x \in \mathbb{C}^S$ such that

$$(1) \qquad\qquad \sum \big[|x(s)|^2 : s \in S \big] < \infty$$

(Here we are using the notion of unordered sum as defined previously.) This construction is familiar in certain cases. For example, if $S = \{1, 2, \ldots, n\}$, then the space X just constructed is the familiar space $\mathbb{C}^n$. On the other hand, if $S = \mathbb{N}$, then X is the familiar space ℓ^2. In the space X, addition and scalar multiplication are already defined, since $X \subset \mathbb{C}^S$. Naturally, we define the inner product by

$$(2) \qquad\qquad \langle x, y \rangle = \sum \big[x(s)\overline{y(s)} : s \in S \big]$$

Much of what we are doing here loses its mystery when we recall (from the Corollary to Theorem 4) that the sums in Equations (1) and (2) are always *countable*. The space discussed here is denoted by $\ell^2(S)$. ∎

Example 5. (Legendre polynomials.) An important example of an orthonormal basis is provided by the Legendre polynomials. We consider the space $C[-1, 1]$ and use the simple inner product

$$\langle f, g \rangle = \int_{-1}^{1} f(t)g(t)\, dt$$

Now apply the Gram–Schmidt process to the monomials $t \mapsto 1, t, t^2, t^3, \ldots$ The un-normalized polynomials that result can be described recursively, using the classical notation P_n:

$$P_0(t) = 1 \qquad P_1(t) = t$$

$$P_n(t) = \frac{2n-1}{n} t P_{n-1}(t) - \frac{n-1}{n} P_{n-2}(t) \qquad (n = 2, 3, \ldots)$$

The orthonormal system is, of course, $p_n = P_n / \|P_n\|$. The completion of the space $C[-1, 1]$ with respect to the norm induced by the inner product is the space $L^2[-1, 1]$. Every function f in this space is represented in the L^2–sense by the series

$$f = \sum_{k=0}^{\infty} \langle f, p_k \rangle p_k$$

We should be very cautious about writing

$$f(t) = \sum_{k=0}^{\infty} \langle f, p_k \rangle p_k(t)$$

because, in the first place, $f(t)$ is meaningless for an element $f \in L^2[-1, 1]$. In this context, f stands for an equivalence class of functions that differ from each other on sets of measure zero. In the second place, such an equation would seem to imply a pointwise convergence of the series, and that is questionable, if not false. Without more knowledge about the expansion of f in Legendre polynomials, we can write only

$$\int_{-1}^{1} \left[f(t) - \sum_{k=1}^{n} \langle f, p_k \rangle p_k(t) \right]^2 dt \to 0 \quad \text{as } n \to \infty$$

Consult [Davis] or [Sz] for the conditions on f that guarantee uniform convergence of the series to f.

Problems 2.2

1. A function $f : \mathbb{R} \to \mathbb{C}$ is said to be **almost periodic** if for every $\varepsilon > 0$ there is an $\ell > 0$ such that each interval of length ℓ contains a number τ for which

$$\sup_{s \in \mathbb{R}} |f(s + \tau) - f(s)| < \varepsilon$$

   Prove that every periodic function is almost periodic, and that the sum of two almost periodic functions is almost periodic. Refer to [Bes] and [Tay2] for further information.

2. Prove Theorem 7.

3. Prove Theorem 8. (Theorem 7 will help.)

4. Let $x : I \to \mathbb{R}_+$, where I is some index set. Suppose that there is a number M such that $\sum [x_j : j \in J] \leqslant M$ for every finite subset J in I. Prove that $\sum [x_i : i \in I]$ exists and does not exceed M. What happens if we drop the hypothesis $x_j \geqslant 0$?

5. Prove that the set of functions $\{u_\lambda : \lambda \in \mathbb{R}\}$, defined in Example 3, is linearly independent.

6. Using the inner product

$$\langle x, y \rangle = \int_{-1}^{1} x(t) y(t) \, dt$$

   construct an orthonormal set $\{u_0, u_1, u_2, u_3\}$ where (for each j) u_j is a polynomial of degree at most j. (One can apply the Gram–Schmidt process to the functions $v_j(t) = t^j$.)

7. Prove that the functions $u_n(t) = e^{int}$ $(n = 0, \pm 1, \pm 2, \ldots)$ form an orthonormal system with respect to the inner product

$$\langle x, y \rangle = \frac{1}{2\pi} \int_{-\pi}^{\pi} x(t) \, \overline{y(t)} \, dt$$

8. Prove that the functions

$$
u_n(t) = \begin{cases} \cos nt & n = -1, -2, -3, \ldots \\ \sin nt & n = 1, 2, 3, \ldots \\ 1/\sqrt{2} & n = 0 \end{cases}
$$

form an orthonormal system with respect to the inner product

$$
\langle x, y \rangle = \frac{1}{\pi} \int_{-\pi}^{\pi} x(t) y(t) \, dt
$$

9. Prove that the Chebyshev polynomials

$$
T_n(t) = \cos(n \operatorname{Arccos} t) \qquad (-1 \leqslant t \leqslant 1 \, ; \, n = 0, 1, 2, \ldots)
$$

form an orthogonal system with respect to the inner product

$$
\langle x, y \rangle = \int_{-1}^{1} x(t) y(t) (1 - t^2)^{-1/2} \, dt
$$

What is the corresponding orthonormal system? Hint: Make a change of variable $t = \cos \theta$ and apply Problem 8.

10. Let $v_1, v_2, \ldots$ be a sequence in a Hilbert space X such that $\operatorname{span}\{v_1, v_2, \ldots\} = X$. Show that X is finite dimensional.

11. Prove that any orthonormal set in an inner product space can be enlarged to form an orthonormal basis.

12. Let D be the open unit disk in the complex plane. The space $H^2(D)$ is defined to be the space of functions f analytic in D and satisfying $\int_D |f(z)|^2 \, dz < \infty$. In $H^2(D)$ we define $\langle f, g \rangle = \int_D f(z) \overline{g(z)} \, dz$. Prove that the functions $u_n(z) = z^n$ $(n = 0, 1, 2, \ldots)$ form an orthogonal system in $H^2(D)$. What is the corresponding orthonormal sequence?

13. If $0 < \alpha < \beta$, which of these implies the other?

(a) $\sum \|x_n\|^{\alpha} < \infty$,   (b) $\sum \|x_n\|^{\beta} < \infty$.

14. Prove that if $\{v_1, v_2, \ldots\}$ is linearly independent, then an orthogonal system can be constructed from it by defining $u_1 = v_1$ and

$$
u_n = v_n - \sum_{j=1}^{n-1} \langle v_n, u_j \rangle u_j / \|u_j\|^2 \qquad n = 2, 3, \ldots
$$

15. Illustrate the process in Problem 14 with the four vectors v_0, v_1, v_2, v_3, where $v_j(t) = t^j$ and the inner product is defined by $\langle x, y \rangle = \int_{-1}^{1} x(t) y(t) \, dt$.

16. Let $[u_i : i \in I]$ be an orthonormal basis for a Hilbert space X. Let $[v_i : i \in I]$ be an orthonormal set satisfying $\sum_i \|u_i - v_i\|^2 < 1$. Show that $[v_i]$ is also a basis for X.

17. Where does the proof of Theorem 6 fail if X is an incomplete inner-product space? Which equivalences remain true?

18. Prove that if P is the orthogonal projection of a Hilbert space X onto a closed subspace Y, then $I - P$ is the orthogonal projection of X onto $Y^\perp$.

19. (Cf. Problem 12.) Let Γ be the unit circle in the complex plane. For functions continuous on Γ define $\langle f, g \rangle = -i \int_\Gamma f(z)\overline{g(z)}\,\overline{z}\,dz$. Prove that this is an inner product and that the functions z^n form an orthogonal family.

20. Prove that an orthogonal projection P has the property that $\langle Px, x \rangle = \|Px\|^2$ for all x.

21. Let $[u_n]$ be an orthonormal sequence in an inner product space. Let $[\alpha_n] \subset \mathbb{C}$ and $\sum_{n=1}^\infty |\alpha_n|^2 < \infty$. Show that the sequence of vectors $y_n = \sum_{j=1}^n \alpha_j u_j$ has the Cauchy property.

22. Let $[u_1, u_2, \ldots, u_n]$ be an orthonormal set in an inner product space X. What choice of coefficients λ_j makes the expression $\|x - \sum_{j=1}^n \lambda_j u_j\|$ a minimum? Here x is a prescribed point in X.

23. Define $p_n(t) = \dfrac{d^n}{dt^n}(t^2 - 1)^n$ for $n = 0, 1, 2, \ldots$ Prove the orthogonality of $\{p_n : n \in \mathbb{N}\}$ with respect to the inner product $\langle x, y \rangle = \int_{-1}^1 x(t)y(t)\,dt$.

24. If K is a closed convex set in a Hilbert space X, there is a well-defined map $P : X \to K$ such that $\|x - Px\| = \text{dist}(x, K)$ for all x. Which properties $(a), \ldots, (j)$ in Theorem 7 does this mapping have? (Cf. Problem 2.1.39, page 70.)

25. Consider the real Hilbert space $X = L^2[-\pi, \pi]$, having its usual inner product, $\langle x, y \rangle = \int_{-\pi}^\pi x(t)y(t)\,dt$. Let U be the subspace of even functions in X; these are functions such that $u(-t) = u(t)$. Let V be the subspace of odd functions, $v(-t) = -v(t)$. Prove that $X = U + V$ and that $U \perp V$. Prove that the orthogonal projection of X onto U is given by $Px = u$, where $u(t) = \frac{1}{2}[x(t) + x(-t)]$. Find the orthogonal projection $Q : X \to V$. Give orthonormal bases for U and V, and express P and Q in terms of them.

26. Let $[e_n]$ be an orthonormal sequence in a Hilbert space. Let M be the linear span of this sequence. Prove that the closure of M is

$$\left\{ \sum_{n=1}^\infty \alpha_n e_n \; : \; \sum_{n=1}^\infty |\alpha_n|^2 < \infty \right\}$$

27. Let $[e_n : n \in \mathbb{N}]$ be an orthonormal basis in a Hilbert space. Let $[\alpha_n]$ be a sequence in $\mathbb{C}$. What are the precise conditions under which we can solve the infinite system of equations $\langle x, e_n \rangle = \alpha_n$ $(n \in \mathbb{N})$?

28. Find orthonormal bases for the Hilbert spaces in Examples 3 and 4.

29. What are necessary and sufficient conditions in order that an orthogonal set be linearly independent?

30. A linear map P is a **projection** if $P^2 = P$. Prove that if P is a projection defined on a Hilbert space and $\|P\| = 1$, then P is the orthogonal projection onto a subspace.

31. Let $[u_n : n \in \mathbb{N}]$ be an orthonormal sequence in a Hilbert space X. Define

$$Y = \left\{ \sum_{n=1}^\infty a_n u_n : \sum |a_n|^2 < \infty \right\}$$

Prove that the map $a \mapsto \sum a_n u_n$ is an isometry of ℓ^2 onto Y. Prove that Y is a closed subspace in X.

32. An indexed set $[u_i : i \in I]$ in a Hilbert space is said to be **stable** if there exist positive constants A and B such that

$$A \sum |a_i|^2 \leqslant \| \sum a_i u_i \|^2 \leqslant B \sum |a_i|^2$$

whenever $a \in \ell^2(I)$. Prove that a stable family is linearly independent. Prove that every orthonormal family is stable.

33. (Continuation) Let $[u_i : i \in \mathbb{Z}]$ be an orthonormal family. Define $v_i = u_i + u_{i+1}$. Prove that $[v_i : i \in \mathbb{Z}]$ is stable. Generalize.

34. (Continuation) Let $[u_i : i \in I]$ be a stable family. Let $a : I \to \mathbb{C}$. Prove that these properties of a are equivalent: (1) $\sum |a_i|^2 < \infty$; (2) $\sum a_i u_i$ converges; (3) $\sum a_i \langle x, u_i \rangle$ converges for each x in the Hilbert space.

35. (Continuation) Let $[u_i : i \in I]$ be an indexed family of vectors of norm 1 in a Hilbert space. Prove that if $\sum_{i \neq j} |\langle u_i, u_j \rangle|^2 < 1$, then the given family is stable.

36. (Continuation) Prove that if $[u_i : i \in I]$ is stable, then $\{ \sum a_i u_i : a \in \ell^2(I) \}$ is a closed subspace.

37. Let $[x_1, x_2, \ldots, x_n]$ be an ordered set in an inner-product space. Assume that it is orthogonal in this sense: If $x_i \neq x_j$, then $\langle x_i, x_j \rangle = 0$. Show by an example that the Pythagorean law in Theorem 1 may fail.

38. (Direct sums of Hilbert spaces). For $n = 1, 2, 3, \ldots$ let X_n be a Hilbert space over the complex field. The **direct sum** of these spaces is denoted by $\bigoplus_{n=1}^{\infty} X_n$, and its elements are sequences $[x_n : n \in \mathbb{N}]$, where $x_n \in X_n$ and $\sum_{n=1}^{\infty} \|x_n\|^2 < \infty$. Show how to make this space into a Hilbert space and prove the completeness.

39. This problem gives a pair of closed subspaces whose sum is not closed. Let X be an infinite-dimensional Hilbert space, and let $\{u_n\}$ be an orthonormal sequence in X. Put

$$v_n = u_{2n} \qquad w_n = u_{2n+1} \qquad z_n = \frac{1}{n} v_n + \frac{\sqrt{n^2-1}}{n} w_n \qquad x_0 = \sum_{n=1}^{\infty} \frac{1}{n} v_n$$

Let W and Z denote the closed linear spaces generated by $\{w_n\}$ and $\{z_n\}$. Prove that

(1) All three sequences $\{v_n\}$, $\{w_n\}$, $\{z_n\}$ are orthonormal.

(2) The vector x_0 is well-defined; i.e., its series converges.

(3) The vector x_0 is in the closure of $W + Z$.

(4) If $z \in Z$, then $\langle z, v_n \rangle = \langle z, z_n \rangle / n$.

(5) If $w \in W$, then $\langle w, v_n \rangle = 0$.

(6) If $x_0 = w + z$, where $w \in W$ and $z \in Z$, then

$$1 = n\langle x_0, v_n \rangle = n\langle w + z, v_n \rangle = \langle z, z_n \rangle \to 0$$

This contradiction will show that $x_0 \notin W + Z$.

40. Prove that an orthonormal set in a separable Hilbert space can have at most a countable number of elements. Hint: Consider the open balls of radius $\frac{1}{2}$ centered at the points in the orthonormal set.

41. Let $[u_n]$ be an orthonormal base in a Hilbert space. Define $v_n = 2^{-1/2}(u_{2n} + u_{2n+1})$. Prove that $[v_n]$ is orthonormal. Define another sequence $[w_n]$ by the same formula, except $+$ is replaced by $-$. Show that the v–sequence and the w–sequence together provide an orthonormal basis for the space.

42. Let X and Y be measure spaces, and $f \in L^2(X \times Y)$. Let $[u_i]$ be an orthonormal basis for $L^2(X)$. Prove that for suitable $v_i \in L^2(Y)$, we have $f(x, y) = \sum u_i(x)v_i(y)$.

2.3 Linear Functionals and Operators

Recall from Section 1.5, page 24, that a **linear functional** on a vector space X is a mapping ϕ from X into the scalar field such that for vectors x, y and scalars a, b,

$$\phi(ax + by) = a\phi(x) + b\phi(y)$$

If the space X has a norm, and if

(1) $$\sup_{\|x\|=1} |\phi(x)| < \infty$$

we say that ϕ is **bounded**, and we denote by $\|\phi\|$ the supremum in the inequality (1). (Boundedness is equivalent to continuity, by Theorem 2 on page 25.)

The bounded linear functionals on a Hilbert space have a very simple form, as revealed in the following important result.

Theorem 1. Riesz Representation Theorem.   *Every continuous linear functional defined on a Hilbert space is of the form $x \mapsto \langle x, v \rangle$ for an appropriate vector v that is uniquely determined by the given functional.*

Proof.   Let X be the Hilbert space, and ϕ a continuous linear functional. Define $Y = \{x \in X : \phi(x) = 0\}$. (This is the **null space** or **kernel** of ϕ). If $Y = X$, then $\phi(x) = 0$ for all x and $\phi(x) = \langle x, 0 \rangle$. If $Y \neq X$, then let $0 \neq u \in Y^{\perp}$. (Use Theorem 4 in Section 2.1, page 65.) We can assume that $\phi(u) = 1$. Observe that $X = Y \oplus \mathbb{C}u$, because $x = x - \phi(x)u + \phi(x)u$, and $x - \phi(x)u \in Y$. Define $v = u/\|u\|^2$. Then

$$\langle x, v \rangle = \langle x - \phi(x)u, v \rangle + \langle \phi(x)u, v \rangle = \phi(x)\langle u, v \rangle = \phi(x)\langle u, u \rangle / \|u\|^2 = \phi(x) \quad \blacksquare$$

Example 1.   Let X be a finite-dimensional Hilbert space with a basis $[u_1, u_2, \ldots, u_n]$, not necessarily orthonormal. Each point x of X can be represented uniquely in the form $x = \sum_j \lambda_j(x)u_j$, and the λ_j are continuous linear functionals. (Refer to Corollary 2 in Section 1.5, page 26.) Hence by Theorem 1 there exist points $v_j \in X$ such that

$$x = \sum_{j=1}^{n} \langle x, v_j \rangle u_j \qquad x \in X$$

Since $u_i = \sum_{j=1}^{n} \langle u_i, v_j \rangle u_j$, we must have $\langle u_i, v_j \rangle = \delta_{ij}$. In this situation, we say that the two sets $[u_1, u_2, \ldots, u_n]$ and $[v_1, v_2, \ldots, v_n]$ are mutually **biorthogonal** or that they form a biorthogonal pair. See [Brez]. ∎

Before reading further about linear operators on a Hilbert space, the reader may wish to review Section 1.5 (pages 24–30) concerning the theory of linear transformations acting between general normed linear spaces.

Example 2. The orthogonal projection P of a Hilbert space X onto a closed subspace Y is a bounded linear operator from X into X. Theorem 7 in Section 2.2 (page 74) indicates that P has a number of endearing properties. For example, $\|P\| = 1$. ∎

Example 3. It is easy to create bounded linear operators on a Hilbert space X. Take any orthonormal system $[u_i]$ (it may be finite, countable, or uncountable), and define $Ax = \sum_i \sum_j a_{ij} \langle x, u_j \rangle u_i$. If the coefficients a_{ij} have the property $\sum_i \sum_j |a_{ij}|^2 < \infty$, then A will be continuous. ∎

> **Theorem 2. Existence of Adjoints.** *If A is a bounded linear operator on a Hilbert space X (thus $A : X \to X$), then there is a uniquely defined bounded linear operator A^* such that*
>
> $$\langle Ax, y \rangle = \langle x, A^*y \rangle \qquad (x, y \in X)$$
>
> *Furthermore, $\|A^*\| = \|A\|$.*

Proof. For each fixed y, the mapping $x \mapsto \langle Ax, y \rangle$ is a bounded linear functional on X:

$$\langle A(\lambda x + \mu z), y \rangle = \langle \lambda Ax + \mu Az, y \rangle = \lambda \langle Ax, y \rangle + \mu \langle Ax, y \rangle$$
$$|\langle Ax, y \rangle| \leqslant \|Ax\| \, \|y\| \leqslant \|A\| \, \|x\| \, \|y\|$$

Hence by the Riesz Representation Theorem (Theorem 1 above) there is a unique vector v such that $\langle Ax, y \rangle = \langle x, v \rangle$. Since v depends on A and y, we are at liberty to denote it by A^*y. It remains to be seen whether the mapping A^* thus defined is linear and bounded. We ask whether

$$A^*(\lambda y + \mu z) = \lambda A^*y + \mu A^*z$$

By the Lemma in Section 2.1, page 63, it would suffice to prove that for all x,

$$\langle x, A^*(\lambda y + \mu z) \rangle = \langle x, \lambda A^*y + \mu A^*z \rangle$$

For this it will be sufficient to prove

$$\langle x, A^*(\lambda y + \mu z)\rangle = \overline{\lambda}\langle x, A^*y\rangle + \overline{\mu}\langle x, A^*z\rangle$$

By the definition of A^*, this equation can be transformed to

$$\langle Ax, \lambda y + \mu z\rangle = \overline{\lambda}\langle Ax, y\rangle + \overline{\mu}\langle Ax, z\rangle$$

This we recognize as a correct equation, and the steps we took can be reversed.

For the boundedness of A^* we use the lemma in Section 2.1 (page 63) and Problem 15 of this section (page 90) to write

$$\|A^*\| = \sup_{\|y\|=1} \|A^*y\| = \sup_{\|y\|=1}\ \sup_{\|x\|=1} |\langle x, A^*y\rangle|$$

$$= \sup_{\|x\|=1}\ \sup_{\|y\|=1} |\langle Ax, y\rangle| = \sup_{\|x\|=1} \|Ax\| = \|A\|$$

The uniqueness of A^* is left as a problem. (Problem 11, page 89)   ∎

The operator A^* described in Theorem 2 is called the **adjoint** of A. For an operator A on a Banach space X, A^* is defined on X^* by the equation $A^*\phi = \phi \circ A$. If X is a Hilbert space, X^* can be identified with X by the Riesz Theorem: $\phi(x) = \langle x, y\rangle$. Then $(A^*\phi)(x) = (\phi \circ A)(x) = \phi(Ax) = \langle Ax, y\rangle$. Thus, the Hilbert space adjoint is almost the same, and no shame attaches to this innocent blurring of the distinction.

Example 4.   Let an operator T on $L^2(S)$ be defined by the equation

$$(Tx)(s) = \int_S k(s,t)x(t)\, dt$$

Here, S can be any measure space, as in Example 5, page 64. Assume that the kernel of this integral operator satisfies the inequality

$$\int_S \int_S |k(s,t)|^2\, dt\, ds < \infty$$

Then T is bounded, and its adjoint is an integral operator of the same type, whose kernel is $(s,t) \mapsto \overline{k(t,s)}$. Such operators have other attractive properties. (See Theorem 5, below.) They are special cases of **Hilbert–Schmidt operators**, defined in Section 2.4, page 98.

If A is a bounded linear operator such that $A = A^*$, we say that A is **self-adjoint**. A related concept is that of being **Hermitian**. A linear map A on an inner product space is said to be Hermitian if $\langle Ax, y\rangle = \langle x, Ay\rangle$ for all x and y. This definition does not presuppose the boundedness of A. However, the following theorem indicates that the Hermitian property (together with the completeness of the space) implies self-adjointness.

Theorem 3.   *If a linear map A on a Hilbert space satisfies $\langle Ax, y \rangle = \langle x, Ay \rangle$ for all x and y, then A is bounded and self-adjoint.*

Proof.   For each y in the unit ball, define a functional ϕ_y by writing $\phi_y(x) = \langle Ax, y \rangle$. It is obvious that ϕ_y is linear, and we see also that it is bounded, since by the Cauchy–Schwarz inequality

$$|\phi_y(x)| = |\langle Ax, y \rangle| = |\langle x, Ay \rangle| \leqslant \|x\| \, \|Ay\|$$

Notice also that by the Lemma in Section 2.1, page 63,

$$\sup_{\|y\| \leqslant 1} |\varphi_y(x)| = \sup_{\|y\| \leqslant 1} |\langle Ax, y \rangle| = \|Ax\|$$

By the Uniform Boundedness Principle, (Section 1.7, page 42),

$$
\begin{aligned}
\infty > \sup_{\|y\| \leqslant 1} \|\phi_y\| &= \sup_{\|y\| \leqslant 1} \sup_{\|x\| \leqslant 1} |\phi_y(x)| \\
&= \sup_{\|x\| \leqslant 1} \sup_{\|y\| \leqslant 1} |\langle Ax, y \rangle| |\phi_y(x)| \\
&= \sup_{\|x\| \leqslant 1} \sup_{\|y\| \leqslant 1} |\langle Ax, y \rangle| |\phi_y(x)| \\
&= \sup_{\|x\| \leqslant 1} \sup_{\|y\| \leqslant 1} |\langle Ax, y \rangle| |\phi_y(x)| \\
&= \sup_{\|x\| \leqslant 1} \sup_{\|y\| \leqslant 1} |\langle Ax, y \rangle| = \sup_{\|x\| \leqslant 1} \|Ax\| = \|A\|
\end{aligned}
$$

The equation $\langle Ax, y \rangle = \langle x, Ay \rangle = \langle x, A^*y \rangle$, together with the uniqueness of the adjoint, shows that $A = A^*$.   ∎

With any bounded linear transformation A on an inner product space we can associate a quadratic form $x \mapsto \langle Ax, x \rangle$. We define

$$\|\!|A|\!\| = \sup_{\|x\| = 1} |\langle Ax, x \rangle|$$

Lemma 1. Generalized Cauchy–Schwarz Inequality.   *If A is a Hermitian operator, then*

$$|\langle Ax, y \rangle| \leqslant \|\!|A|\!\| \, \|x\| \, \|y\|$$

Proof.   Consider these two elementary equations:

$$\langle A(x+y), x+y \rangle = \langle Ax, x \rangle + \langle Ax, y \rangle + \langle Ay, x \rangle + \langle Ay, y \rangle$$
$$-\langle A(x-y), x-y \rangle = -\langle Ax, x \rangle + \langle Ax, y \rangle + \langle Ay, x \rangle - \langle Ay, y \rangle$$

By adding these equations and using the Hermitian property of A, we get

$$(1) \qquad \langle A(x+y), x+y \rangle - \langle A(x-y), x-y \rangle = 4\mathcal{R}\langle Ax, y \rangle$$

From the definition of $\|A\|$ and a homogeneity argument, we obtain

$$(2) \qquad\qquad |\langle Ax, x\rangle| \leqslant \|A\| \|x\|^2 \qquad (x \in X)$$

Using Equation (1), then (2), and finally the Parallelogram Law, we obtain

$$\begin{aligned}
|4\mathcal{R}\langle Ax, y\rangle &= |\langle A(x+y), x+y\rangle - \langle A(x-y), x-y\rangle| \\
&\leqslant |\langle A(x+y), x+y\rangle| + |\langle A(x-y), x-y\rangle| \\
&\leqslant \|A\| \|x+y\|^2 + \|A\| \|x-y\|^2 \\
&= \|A\|(2\|x\|^2 + 2\|y\|^2)
\end{aligned}$$

Letting $\|x\| = \|y\| = 1$ in the preceding equation establishes that

$$|\mathcal{R}\langle Ax, y\rangle| \leqslant \|A\| \qquad (\|x\| = \|y\| = 1)$$

For a fixed pair x, y we can select a complex number θ such that $|\theta| = 1$ and $\theta\langle Ax, y\rangle = |\langle Ax, y\rangle|$. Then

$$|\langle Ax, y\rangle| = |\langle A(\theta x), y\rangle| = |\mathcal{R}\langle A(\theta x), y\rangle| \leqslant \|A\|$$

By homogeneity, this suffices to prove the lemma.   ∎

Lemma 2.   *If A is Hermitian, then $\|A\| = \|A\|$.*

Proof.   By the Cauchy–Schwarz inequality,

$$\|A\| = \sup_{\|u\|=1} |\langle Au, u\rangle| \leqslant \sup_{\|u\|=1} \|Au\| \|u\| = \sup_{\|u\|=1} \|Au\| = \|A\|$$

For the reverse inequality, use the preceding lemma to write

$$\begin{aligned}
\|A\| &= \sup_{\|x\|=1} \|Ax\| = \sup_{\|x\|=1} \sup_{\|y\|=1} |\langle Ax, y\rangle| \\
&\leqslant \sup_{\|x\|=1} \sup_{\|y\|=1} \|A\| \|x\| \|y\| = \|A\|
\end{aligned}$$

    ∎

Definition.   An operator A, mapping one normed linear space into another, is said to be **compact** if it maps the unit ball of the domain to a set whose closure is compact.

When we recall that a continuous operator is one that maps the unit ball to a bounded set, it becomes evident that compactness of an operator is stronger than continuity. It is certainly not equivalent if the spaces involved are infinite dimensional. For example, the identity map on an infinite–dimensional space is continuous but not compact.

Lemma 3.   *Every continuous linear operator (from one normed linear space into another) having finite-dimensional range is compact.*

Proof.   Let A be such an operator, and let Σ be the unit ball. Since A is continuous, $A(\Sigma)$ is a bounded set in a finite-dimensional subspace, and its closure is compact, by Theorem 1 in Section 1.4, page 20.   ∎

Theorem 4.   *If X and Y are Banach spaces, then the set of compact operators in $\mathcal{L}(X,Y)$ is closed.*

Proof. Let $[A_n]$ be a sequence of compact operators from X to Y. Suppose that $\|A_n - A\| \to 0$. To prove that A is compact, let $[x_i]$ be a sequence in the unit ball of X. We wish to find a convergent subsequence in $[Ax_i]$. Since A_1 is compact, there is an increasing sequence $I_1 \subset \mathbb{N}$ such that $[A_1 x_i : i \in I_1]$ converges. Since A_2 is compact, there is an increasing sequence $I_2 \subset I_1$ such that $[A_2 x_i : i \in I_2]$ converges. Note that $[A_1 x_i : i \in I_2]$ converges. Continue this process, and use Cantor's diagonal process. Thus we let I be the sequence whose ith member is the ith member of I_i, for $i = 1, 2, \ldots$ By the construction, $[A_n x_i : i \in I]$ converges. To prove that $[Ax_i : i \in I]$ converges, it suffices to show that it is a Cauchy sequence. This follows from the inequality

$$\|Ax_i - Ax_j\| \leqslant \|Ax_i - A_n x_i\| + \|A_n x_i - A_n x_j\| + \|A_n x_j - Ax_j\|$$
$$\leqslant \|A - A_n\|\,\|x_i\| + \|A_n x_i - A_n x_j\| + \|A_n - A\|\,\|x_j\| \qquad \blacksquare$$

Theorem 5.   *Let S be any measure space. In the space $L^2(S)$, consider the integral operator T defined by the equation*

$$(Tx)(s) = \int_S k(s,t)x(t)\,dt$$

If the kernel k belongs to the space $L^2(S \times S)$, then T is a compact operator from $L^2(S)$ into $L^2(S)$.

Proof. Select an orthonormal basis $[u_n]$ for $L^2(S)$, and define $a_{nm} = \langle Tu_m, u_n\rangle$. This is the "matrix" for T relative to the chosen basis. In fact, we have for any x in $L^2(S)$, $x = \sum_n \langle x, u_n\rangle u_n$, whence

(3)
$$Tx = \sum_n \langle Tx, u_n\rangle u_n = \sum_n \Big\langle \sum_m \langle x, u_m\rangle Tu_m, u_n \Big\rangle u_n$$
$$= \sum_n \Big[\sum_m a_{nm}\langle x, u_m\rangle \Big] u_n$$

Using the notation k_s for the univariate function $t \mapsto k(s,t)$, we have

$$\|k\|^2 = \iint |k(s,t)|^2\, dt\, ds = \int \|k_s\|^2\, ds = \int \sum_n |\langle k_s, u_n\rangle|^2\, ds$$

$$= \int \sum_n \Big| \int k_s(t)u_n(t)\, dt \Big|^2 ds = \int \sum_n |(Tu_n)(s)|^2\, ds$$

$$= \sum_n \int |(Tu_n)(s)|^2\, ds = \sum_n \|Tu_n\|^2$$

$$= \sum_n \sum_m |\langle Tu_n, u_m\rangle|^2 = \sum_n \sum_m |a_{mn}|^2$$

$$= \sum_{m=1}^{\infty} \beta_m \quad \text{where} \quad \beta_m = \sum_n |a_{mn}|^2$$

Equation (3) suggests truncating the series that defines T in order to obtain operators of finite rank that approximate T. Hence, we put

$$T_n x = \sum_{i=1}^{n} \sum_{j=1}^{\infty} a_{ij} \langle x, u_j \rangle u_i$$

By subtraction,

$$Tx - T_n x = \sum_{i>n} \sum_{j=1}^{\infty} a_{ij} \langle x, u_j \rangle u_i$$

whence, by the Cauchy–Schwarz inequality (in ℓ^2!) and the Bessel inequality,

$$\left\| Tx - T_n x \right\|^2 = \sum_{i>n} \left| \sum_{j=1}^{\infty} a_{ij} \langle x, u_j \rangle \right|^2 \leqslant \sum_{i>n} \sum_{j=1}^{\infty} |a_{ij}|^2 \sum_{k=1}^{\infty} |\langle x, u_k \rangle|^2$$

$$\leqslant \|x\|^2 \sum_{i>n} \sum_{j=1}^{\infty} |a_{ij}|^2 = \|x\|^2 \sum_{i>n} \beta_i$$

This shows that $\left\| T - T_n \right\| \to 0$. Since each T_n is compact, so is the limit T, by Theorem 4.   ∎

Theorem 6.   *The null space of a bounded linear operator on a Hilbert space is the orthogonal complement of the range of its adjoint.*

Proof.   Let A be the operator and $\mathcal{N}(A)$ its null space. Denote the range of A^* by $\mathcal{R}(A^*)$. If $x \in \mathcal{N}(A)$ and z is arbitrary, then

$$\langle x, A^* z \rangle = \langle Ax, z \rangle = \langle 0, z \rangle = 0$$

Hence $x \in \mathcal{R}(A^*)^{\perp}$ and $\mathcal{N}(A) \subset \mathcal{R}(A^*)^{\perp}$. Conversely, if $x \in \mathcal{R}(A^*)^{\perp}$, then

$$\langle Ax, Ax \rangle = \langle x, A^*(Ax) \rangle = 0$$

whence $Ax = 0$, $x \in \mathcal{N}(A)$, and $\mathcal{R}(A^*)^{\perp} \subset \mathcal{N}(A)$.   ∎

Corollary.   *A Hermitian operator whose range is dense is injective (one-to-one).*

A sequence $[x_n]$ in a Hilbert space is said to **converge weakly** to a point x if, for all y,

$$\langle x_n, y \rangle \to \langle x, y \rangle$$

A convenient notation for this is $x_n \rightharpoonup x$. Notice that this definition is in complete harmony with the definition of weak convergence in an arbitrary normed linear space, as in Chapter 1, Section 9, (page 53). Of course, the Riesz Representation Theorem, proved earlier in this section (page 81), is needed to connect the two concepts.

Example 5.   If $[u_n]$ is an orthonormal sequence, then $u_n \rightharpoonup 0$. This follows from Bessel's inequality,

$$\sum |\langle u_n, y \rangle|^2 \leqslant \|y\|^2$$

which shows that $\langle u_n, y \rangle \to 0$ for all y. ∎

We say that a sequence $[x_n]$ in an inner-product space is a **weakly Cauchy** sequence if, for each y in the space, the sequence $[\langle x_n, y \rangle]$ has the Cauchy property in $\mathbb{C}$.

Lemma 4.   *A weakly Cauchy sequence in a Hilbert space is weakly convergent to a point in the Hilbert space.*

Proof.   Let $[x_n]$ be such a sequence. For each y, the sequence $[\langle y, x_n \rangle]$ has the Cauchy property, and is therefore bounded in $\mathbb{C}$. The linear functionals ϕ_n defined by $\phi_n(y) = \langle y, x_n \rangle$ have the property

$$\sup_n |\phi_n(y)| < \infty \qquad (y \in X)$$

By the Uniform Boundedness Principle (Section 1.7, page 42), we infer that $\|\phi_n\| \leqslant M$ for some constant M. Since

$$\|x_n\| = \sup_{\|y\|=1} |\langle y, x_n \rangle| = \|\phi_n\| \leqslant M$$

we conclude that $[x_n]$ is bounded. Put $\phi(y) = \lim_n \langle y, x_n \rangle$. Then ϕ is a bounded linear functional on X. By the Riesz Representation Theorem, there is an x for which $\phi(y) = \langle y, x \rangle$. Hence $\lim_n \langle y, x_n \rangle = \langle y, x \rangle$ and $x_n \rightharpoonup x$. ∎

Many problems in applied mathematics can be cast as solving a linear equation, $Ax = b$. For our discussion here, A can be any linear operator on a Hilbert space, X, and $b \in X$. Does the equation have a solution, and if it does, can we calculate it? The first question is the same as asking whether b is in the range of A. Here is a basic theorem, called the "Fredholm Alternative." It is the Hilbert space version of the Closed Range Theorem in Section 1.8, page 50. Other theorems called the Fredholm Alternative occur in Section 7.5.

Theorem 7.   *Let A be a continuous linear operator on a Hilbert space. If the range of A is closed, then it is the orthogonal complement of the null space of A^*; in symbols,*

$$\mathcal{R}(A) = [\mathcal{N}(A^*)]^\perp$$

Proof.   This is similar to Theorem 6, and is therefore left to the problems. (Half of the theorem does not require the closed range.) ∎

Problems 2.3

1. Let X be a Hilbert space and let $A : X \to X$ be a bounded linear operator. Let $[u_i : i \in I]$ be any orthonormal basis for X. (The index set may be uncountable.) Show that there exists a "matrix" (a function α on $I \times I$) such that for all x, $Ax = \sum_i \sum_j \alpha_{ij} \langle x, u_j \rangle u_i$.

2. Adopt the hypotheses of Problem 1. Show that there exist vectors v_i such that $Ax = \sum_i \langle x, v_i \rangle u_i$. Show also that the vectors v_i can be chosen so that $\|v_i\| \leqslant \|A\|$.

3. Let $[u_n]$ be an orthonormal sequence and let $Ax = \sum \lambda_n \langle x, u_n \rangle u_n$, where $[\lambda_n]$ is bounded. Prove that $A = A^*$ if and only if $[\lambda_n] \subset \mathbb{R}$.

4. Prove that a bounded linear transformation on a Hilbert space is completely determined by its values on an orthonormal basis. To what extent can these images be arbitrary?

5. Let X be a complex Hilbert space. Let $A : X \to X$ be bounded and linear. Prove that if $Ax \perp x$ for all x, then $A = 0$. Show that this is not true for real Hilbert spaces.

6. Let A be an operator on a Hilbert space having the form $Ax = \sum \lambda_n \langle x, u_n \rangle u_n$, where $[u_n]$ is an orthonormal sequence and $[\lambda_n]$ is a bounded sequence in $\mathbb{C}$. If f is analytic on a domain containing $[\lambda_n]$, then we define $f(A)(x)$ to be $\sum f(\lambda_n) \langle x, u_n \rangle u_n$. For the function $f(z) = e^z$ prove that $f(A + B) = f(A)f(B)$, *provided that* $AB = BA$.

7. Prove, without using the Hahn–Banach theorem, that a bounded linear functional defined on a closed subspace of a Hilbert space has an extension (of the same norm) to the whole Hilbert space.

8. Let Y be a subspace (possibly not closed) in a Hilbert space X. Let L be a linear map from X to Y such that $x - Lx \perp Y$ for all $x \in X$. Prove that L is continuous and idempotent. Prove that Y is closed and that L is the orthogonal projection of X onto Y.

9. Let A be a bounded linear operator mapping a Banach space X into X. Prove that if

$$\sum_{n=0}^{\infty} |c_n| \, \|A\|^n < \infty$$

then $\sum_{n=0}^{\infty} c_n A^n$ is also a bounded linear operator from X into X.

10. An operator A whose adjoint has dense range is injective.

11. Prove the uniqueness of A^* and that $A^{**} = A$.

12. Prove the continuity assertion in Example 3.

13. Let $[e_n]$ be an orthonormal sequence, $[\lambda_n]$ a bounded sequence in $\mathbb{C}$, and $Ax = \sum \lambda_n \langle x, e_n \rangle e_n$. Show that the operators defined by the partial sums $\sum_1^n \lambda_k \langle x, e_k \rangle e_k$ need not converge (in operator norm) to A. Find the exact conditions on $[\lambda_n]$ for which this operator convergence is valid. Prove that if the partial sum operators converge to A, then A is compact.

14. Let X be a separable Hilbert space, and $[u_n]$ an orthonormal basis for X. Define $A : X \to X$ by

$$Ax = \sum_{n=1}^{\infty} \frac{1}{n} \langle x, u_n \rangle u_n$$

Notice that A is a compact Hermitian operator. Prove that the range of A is the set

$$\left\{ y \in X \; : \; \sum_{n=1}^{\infty} |n \langle y, u_n \rangle|^2 < \infty \right\}$$

Prove that the range of A is not closed. Hint: Consider the vector $v = \sum_{n=1}^{\infty} u_n / n$.

15. Let X and Y be two arbitrary sets. For a function $f : X \times Y \to \mathbb{R}$, prove that

$$\sup_{x \in X} \sup_{y \in Y} f(x, y) = \sup_{y \in Y} \sup_{x \in X} f(x, y)$$

Show that this equation is not generally true if we replace $\sup_{x \in X}$ by $\inf_{x \in X}$ on both sides.

16. Prove that if $A_n \to A$, then $A_n^* \to A^*$. (This is continuity of the map $A \mapsto A^*$.)

17. Prove that the range of a Hermitian operator is orthogonal to its kernel. Can this phenomenon occur for an operator that is not Hermitian?

18. Prove that for a Hermitian operator A, the function $x \mapsto \text{dist}(x, \mathcal{R}(A))$ is a norm on $\ker(A)$. Here $\mathcal{R}(A)$ denotes the range of A.

19. Let A be a bounded linear operator on a Hilbert space. Define $[x, y] = \langle Ax, y \rangle$. Which properties of an inner product does $[\,,\,]$ have? What takes the place of the Cauchy–Schwarz inequality? What additional assumptions must be made in order that $[\,,\,]$ be an inner product?

20. Give an example of a nontrivial operator A on a real Hilbert space such that $Ax \perp x$ for all x. You should be able to find an example in $\mathbb{R}^2$. Can you do it with a Hermitian operator? (Cf. Problem 5.)

21. Let $[u_n]$ be an orthonormal sequence in an inner product space. Let $[\lambda_n]$ be a sequence of scalars such that the series $\sum \lambda_n u_n$ converges. Prove that $\sum |\lambda_n|^2 < \infty$.

22. Let $[u_n]$ be an orthonormal sequence in a Hilbert space. Let $Ax = \sum_{n=1}^{\infty} \alpha_n \langle x, u_n \rangle u_n$, where $[\alpha_n]$ is a bounded sequence in $\mathbb{C}$. Prove that A is continuous. Prove that if $[\alpha_n]$ is a bounded sequence in $\mathbb{R}$, then A is Hermitian. Prove that if $[\alpha_n]$ is a sequence in $\mathbb{C}$ such that $\sum |\alpha_n|^2 < \infty$, then A is compact. Suggestion: Use Lemma 3 and Theorem 4.

23. If $[u_n]$ is an orthonormal sequence and $Ax = \sum \lambda_n \langle x, u_n \rangle u_n$ where $\lambda_n \in \mathbb{C}$ and $\lambda_n \not\to 0$, then A is not compact.

24. Let v be a point in a Hilbert space X. Define $\phi(x) = \langle x, v \rangle$ for all $x \in X$. Show that the mapping T such that $Tv = \phi$ is one-to-one, onto X^*, norm-preserving, and **conjugate linear**: $T(\alpha_1 v_1 + \alpha_2 v_2) = \bar{\alpha}_1 Tv_1 + \bar{\alpha}_2 Tv_2$.

25. Prove that if X is an infinite–dimensional Hilbert space, then a compact operator on X cannot be invertible.

26. Let X be a Hilbert space. Let $A : X \to X$ be linear and let $B : X \to X$ be any map such that $\langle Ax, y \rangle = \langle x, By \rangle$ for all x and y. Prove that A is continuous, that B is linear, and that B is continuous.

27. Adopt the hypotheses of Problem 3, and prove that $\|A\| \leqslant \sup_n |\lambda_n|$.

28. Illustrate the Fredholm Alternative with this example. In a real Hilbert space, let A be defined by the equation $Ax = x - \lambda \langle v, x \rangle w$, where v and w are prescribed elements of the space, and $\langle v, w \rangle \neq 0$. The scalar λ is arbitrary. What are A^*, $\mathcal{N}(A^*)$, $\mathcal{R}(A)$? (The answers depend on the value of λ.)

29. Refer to Theorem 5, and assume that $S = [0, 1]$. Prove that if the kernel k is continuous, then Tx is continuous, for each x in $L^2(S)$.

30. Let A be a bounded linear operator on a Hilbert space, and let $[u_n]$ and $[v_n]$ be two orthonormal bases for the space. Prove that if $\sum_n \sum_m |\langle Au_n, v_m \rangle|^2 < \infty$, then A is compact. Suggestion: Base the proof on Lemma 3 and Theorem 4. Write

$$Ax = \sum \langle x, u_n \rangle Au_n = \sum \langle x, u_n \rangle \sum \langle Au_n, v_m \rangle$$

31. Define the operator T as in Theorem 5, page 86, and assume that

$$c = \int_S \int_S |k(s, t)|^2 \, ds \, dt < \infty$$

Prove that if $[u_n]$ is an orthonormal sequence and if $Tu_n = \lambda_n u_n$ for each n, then $\sum_n |\lambda_n|^2 \leqslant c$.

32. Prove Theorem 7.

33. Prove the assertion made in Example 3.

2.4 Spectral Theory

In this section we shall study the structure of linear operators on a Hilbert space. Ideally, we would dissect an operator into a sum of simple operators or perhaps an infinite sum of simple operators. In the latter scenario, the terms of the series should decrease in magnitude in order to achieve convergence and to make feasible the truncation of the series for actual computation.

What qualifies as a "simple" operator? Certainly, we would call this one "simple": $Qx = \langle x, u \rangle v$, where u and v are two prescribed elements of the space. The range of Q is the subspace generated by the single vector v. Thus, Q is an operator of rank 1 (*rank* = dimension of range). We may assume that $\|v\| = 1$, since we can compensate for this by redefining u. Every operator of rank one is of this form.

Another example of a simple operator (again of rank 1) is $Tx = a\langle x, u \rangle u$. Notice that in defining the operator T there is no loss of generality in assuming that $\|u\| = 1$, because one can adjust the scalar a to compensate. Next, having adopted this slight simplification, we notice that T has the property $Tu = au$. Thus, a is an eigenvalue of T and u is an accompanying eigenvector. From such primitive building blocks we can construct very general operators, such as

$$Lx = \sum_{j=1}^{\infty} a_j \langle x, u_j \rangle u_j$$

This goal, of representing a given operator L in the form shown, is beautifully achieved when the operator L is compact and Hermitian. (These terms are defined later.) We even have the serendipitous bonus of orthonormality in the sequence $[u_n]$. Each u_n will then be an eigenvector, since

$$Lu_n = \sum_{j=1}^{\infty} a_j \langle u_n, u_j \rangle u_j = a_n u_n$$

Definition. An **eigenvalue** of an operator A is a complex number λ such that $A - \lambda I$ has a nontrivial null space. The set of all eigenvalues of A is denoted here by $\Lambda(A)$. (Caution: $\Lambda(A)$ is defined differently in many books.)

If X is a finite-dimensional space, and if $A : X \to X$ is a linear map, then A certainly has some eigenvalues. To see that this is so, introduce a basis for X so that A can be identified with a square matrix. The following conditions on a complex number λ are then equivalent:

   (i) $A - \lambda I$ has a nontrivial null space

(ii) $A - \lambda I$ is singular

(iii) $\det(A - \lambda I) = 0$ (det is the determinant function)

Since the map $\lambda \mapsto \det(A - \lambda I)$ is a polynomial of degree n (if A is an $n \times n$ matrix), we see that there exist exactly n eigenvalues, it being understood that each is counted a number of times equal to its multiplicity as a root of the polynomial. This argument obviously fails for an infinite–dimensional space. Indeed, an operator with no eigenvalues is readily at hand in Problem 1.

If λ is an eigenvalue of an operator A, then any nontrivial solution of the equation $Ax = \lambda x$ is called an **eigenvector** of A belonging to the eigenvalue λ.

> **Lemma 1.**   *If A is a Hermitian operator on an inner-product space, then:*
>
>   *(1) All eigenvalues of A are real.*
>
>   *(2) Any two eigenvectors of A belonging to different eigenvalues are orthogonal to each other.*
>
>   *(3) The quadratic form $x \mapsto \langle Ax, x \rangle$ is real–valued.*

Proof.   Let $Ax = \lambda x$, $Ay = \mu y$, $x \neq 0$, $y \neq 0$, $\lambda \neq \mu$. Then

$$\lambda \langle x, x \rangle = \langle \lambda x, x \rangle = \langle Ax, x \rangle = \langle x, Ax \rangle = \langle x, \lambda x \rangle = \overline{\lambda} \langle x, x \rangle$$

Thus λ is real. To see that $\langle x, y \rangle = 0$, use the fact that λ and μ are real and write

$$(\lambda - \mu)\langle x, y \rangle = \langle \lambda x, y \rangle - \langle x, \mu y \rangle = \langle Ax, y \rangle - \langle x, Ay \rangle = 0$$

For (3), note that $\langle Ax, x \rangle = \overline{\langle x, Ax \rangle} = \overline{\langle Ax, x \rangle}$.   ∎

> **Lemma 2.**   *A compact Hermitian operator A on an inner-product space has at least one eigenvalue λ such that $|\lambda| = \|A\|$.*

Proof.   Since the case $A = 0$ is trivial, we assume that $A \neq 0$. Put $\|A\| = \sup\{|\langle Ax, x \rangle| : \|x\| = 1\}$. By Lemma 2 in Section 2.3 (page 85), $\|A\| = \|A\|$. Take a sequence of points x_n such that $\|x_n\| = 1$ and $\lim |\langle Ax_n, x_n \rangle| = \|A\|$. Since A is compact, there is a sequence of integers $n_1, n_2, \ldots$ such that $\lim_i Ax_{n_i}$ exists. Put $y = \lim_i Ax_{n_i}$. Then $y \neq 0$ because $|\langle Ax_{n_i}, x_{n_i} \rangle| \to \|A\| \neq 0$. By taking a further subsequence we can assume that the limit $\lambda = \lim \langle Ax_{n_i}, x_{n_i} \rangle$ exists. By Lemma 1, λ is real. Then

$$\left\| Ax_{n_i} - \lambda x_{n_i} \right\|^2 = \left\| Ax_{n_i} \right\|^2 - \lambda \langle Ax_{n_i}, x_{n_i} \rangle - \lambda \langle x_{n_i}, Ax_{n_i} \rangle + \lambda^2 \|x_{n_i}\|^2$$

Hence

$$0 \leqslant \lim \left\| Ax_{n_i} - \lambda x_{n_i} \right\|^2 = \|y\|^2 - \lambda^2 - \lambda^2 + \lambda^2 = \|y\|^2 - \lambda^2$$

This proves that $|\lambda| \leqslant \|y\|$. On the other hand, from the above work we also have

$$\|y\| = \lim \left\| Ax_{n_i} \right\| \leqslant \lim \|A\| \|x_{n_i}\| = \|A\| = |\lambda|$$

Thus our previous inequality shows that $0 \leqslant \lim \left\| Ax_{n_i} - \lambda x_{n_i} \right\| \leqslant 0$, and that

$$\|y - \lambda x_{n_i}\| \leqslant \|y - Ax_{n_i}\| + \|Ax_{n_i} - \lambda x_{n_i}\| \to 0$$

Thus $x_{n_i} \to y/\lambda$. Finally, $Ay = A(\lim \lambda x_{n_i}) = \lambda \lim Ax_{n_i} = \lambda y$, so y is in the null space of $A - \lambda I$, and λ is an eigenvalue.   ∎

Theorem 1. The Spectral Theorem.   *If A is a compact Hermitian operator defined on an inner-product space, then A is of the form $Ax = \sum \lambda_k \langle x, e_k \rangle e_k$ for an appropriate orthonormal sequence $\{e_k\}$ (possibly finite) and appropriate real numbers λ_k satisfying $\lim \lambda_k = 0$. Furthermore, the equations $Ae_k = \lambda_k e_k$ hold.*

Proof.   If $A = 0$, the conclusion is trivial. If $A \neq 0$, we let $X_1 = X$. Let λ_1 and e_1 be an eigenvalue and unit eigenvector determined by the preceding lemma. Thus, $|\lambda_1| = \|A\|$. Let $X_2 = \{x : \langle x, e_1 \rangle = 0\}$. Then X_2 is a subspace of X_1, and A maps X_2 into itself, since $\langle Ax, e_1 \rangle = \langle x, Ae_1 \rangle = \langle x, \lambda_1 e_1 \rangle = \overline{\lambda}_1 \langle x, e_1 \rangle = 0$ for any $x \in X_2$. (Thus X_2 is an **invariant** subspace of A.) We consider the restriction of A to the inner product space X_2, denoted by $A|X_2$. This operator is also compact and Hermitian. Also, $\|A|X_2\| \leqslant \|A\|$. If $A|X_2 \neq 0$, then the preceding lemma produces λ_2 and e_2, where $\|e_2\| = 1$, $|\lambda_2| = \|A|X_2\| \leqslant |\lambda_1|$, $e_2 \perp X_1$, $Ae_2 = \lambda_2 e_2$. We continue this process. At the nth stage we have $|\lambda_1| \geqslant |\lambda_2| \geqslant \cdots \geqslant |\lambda_n| > 0$, $\{e_1, \ldots, e_n\}$ orthonormal, and $Ae_k = \lambda_k e_k$ for $k = 1, \ldots, n$. We define X_{n+1} to be the orthogonal complement of the linear span of $[e_1, \ldots, e_n]$. If $A|X_{n+1} = 0$, the process stops. Then the range of A is spanned by $e_1, \ldots, e_n$. Indeed, for any x, the vector $x - \sum_1^n \langle x, e_k \rangle e_k$ is orthogonal to $\{e_1, \ldots, e_n\}$; hence it lies in X_{n+1}, and so A maps it to zero. In other words,

$$Ax = \sum_{k=1}^n \langle x, e_k \rangle Ae_k = \sum_{k=1}^n \lambda_k \langle x, e_k \rangle e_k$$

If $A|X_{n+1} \neq 0$, we apply the preceding lemma to get λ_{n+1} and e_{n+1}. It remains to be proved that if the above process does not terminate, then $\lim \lambda_k = 0$. Suppose on the contrary that $|\lambda_n| \geqslant \epsilon > 0$ for all $n \in \mathbb{N}$. Then e_n/λ_n is a bounded sequence, and by the compactness of A, the sequence $A(e_n/\lambda_n)$ must contain a convergent subsequence. But this is not possible, since $A(e_n/\lambda_n) = e_n$ and $\{e_n\}$, being orthonormal, satisfies $\|e_n - e_m\| = \sqrt{2}$. In the infinite case let $y_n = x - \sum_{k=1}^n \langle x, e_k \rangle e_k$. Since $y_n \perp \sum_{k=1}^n \langle x, e_k \rangle e_k$,

$$\|x\|^2 = \left\| y_n + \sum_{k=1}^n \langle x, e_k \rangle e_k \right\|^2 = \|y_n\|^2 + \sum_{k=1}^n |\langle x, e_k \rangle|^2 \geqslant \|y_n\|^2$$

Since $|\lambda_{n+1}|$ is the norm of $\|A|X_{n+1}\|$, we have

$$\|Ay_n\| \leqslant |\lambda_{n+1}| \, \|y_n\| \leqslant |\lambda_{n+1}| \, \|x\| \to 0$$

Since $Ay_n = Ax - \sum_1^n \lambda_k \langle x, e_k \rangle e_k$, we have $Ax = \lim_n \sum_1^n \lambda_k \langle x, e_k \rangle e_k$.   ∎

Remark.   *Every nonzero eigenvalue of A is in the sequence $[\lambda_n]$.*

Proof.   Suppose $Ax = \lambda x$, $x \neq 0$, $\lambda \neq 0$, $\lambda \notin \{\lambda_n : n \in \mathbb{N}\}$. Then $x \perp e_n$ for all n by Lemma 1. But then $Ax = \sum \lambda_n \langle x, e_n \rangle e_n = 0$, a contradiction.   ∎

Remark.   *Each nonzero eigenvalue λ of A occurs in the sequence $[\lambda_n]$ repeated a number of times equal to $\dim\{x : (A - \lambda I)x = 0\}$. Each of these numbers is finite.*

Proof.   Since $\lambda_n \to 0$, a nonzero eigenvalue λ can be repeated only a finite number of times in the sequence. If it is repeated p times, then the subspace $\{x : (A - \lambda I)x = 0\}$ contains an orthonormal set of p elements and so has dimension at least p. If the dimension were greater than p, there would exist $x \neq 0$ such that $Ax = \lambda x$ and $\langle x, e_n \rangle = 0$ for all n (again impossible). ∎

The next theorem gives an application of the spectral resolution of an operator, namely, a formula for inverting the operator $A - \lambda I$ when A is compact and λ is not an eigenvalue of A. (The Hermitian property is not assumed.)

Theorem 2.   *Let A be a compact operator (on an inner-product space) having spectral decomposition $Ax = \sum \lambda_n \langle x, e_n \rangle e_n$. (We allow λ_n to be complex.) If $0 \neq \lambda \notin \{\lambda_n\}$, then $A - \lambda I$ is invertible, and*

$$(A - \lambda I)^{-1}x = -\lambda^{-1}x + \lambda^{-1}\sum \lambda_n \frac{\langle x, e_n \rangle}{\lambda_n - \lambda} e_n$$

Proof.   If the series converges, then our formula is correct. Indeed, by the continuity of $A - \lambda I$ we have by straightforward calculation

$$(A - \lambda I)Bx = B(A - \lambda I)x = x$$

where Bx is defined by the right side of the equation in the statement of the theorem. In order to prove that the series converges, define the partial sums

$$v_n = \sum_{k=1}^{n} \frac{\langle x, e_k \rangle}{\lambda_k - \lambda} e_k$$

The sequence $[v_n]$ is bounded, because with an application of the Pythagorean law and Bessel's inequality we have

$$\|v_n\|^2 = \sum_{k=1}^{n} \left| \frac{\langle x, e_k \rangle}{\lambda_k - \lambda} \right|^2 \leqslant \sup_j \left| \frac{1}{\lambda_j - \lambda} \right|^2 \sum_{k=1}^{\infty} |\langle x, e_k \rangle|^2 \leqslant \beta \cdot \|x\|^2$$

Since A is compact, $\lambda_n \to 0$, by Problem 15. Thus $\beta < \infty$. Also, the sequence $[Av_n]$ contains a convergent subsequence. But $[Av_n]$ is a Cauchy sequence, and a Cauchy sequence having a convergent subsequence is convergent (Problem 1.2.26, page 13). To see that $[Av_n]$ is a Cauchy sequence, write

$$Av_n = \sum_{k=1}^{n} \lambda_k \frac{\langle x, e_k \rangle}{\lambda_k - \lambda} e_k$$

and

$$\|Av_n - Av_m\|^2 = \sum_{k=n+1}^{m} \left| \lambda_k \frac{\langle x, e_k \rangle}{\lambda_k - \lambda} \right|^2 \leqslant \sup_{1 \leqslant j < \infty} \left| \frac{\lambda_j}{\lambda_j - \lambda} \right|^2 \cdot \sum_{k=n+1}^{m} |\langle x, e_k \rangle|^2 \quad ∎$$

If an operator A is not necessarily compact but has a known spectral resolution (in the form of an orthonormal series), then certain conclusions can be drawn, as illustrated in the next three theorems.

Theorem 3.   *Let A be an operator on an inner–product space having the form $Ax = \sum_{n=1}^{\infty} \lambda_n \langle x, e_n \rangle e_n$, where $\{e_n\}$ is an orthonormal sequence and $[\lambda_n]$ is a bounded sequence of nonzero complex numbers. Let M be the linear span of $\{e_n : n \in \mathbb{N}\}$. Then $M^{\perp} = \ker(A)$.*

Proof.   The following are equivalent properties of a vector x:

    (a) $x \in \ker(A)$
    (b) $\|Ax\|^2 = 0$
    (c) $\sum |\lambda_n \langle x, e_n \rangle|^2 = 0$
    (d) $\langle x, e_n \rangle = 0$ for all n.   ∎

Theorem 4.   *Adopt the hypotheses of Theorem 3. The orthonormal set $\{e_n\}$ is maximal if and only if $\ker(A) = 0$.*

Proof.   By Theorem 3, $\ker(A) = 0$ if and only if $M^{\perp} = 0$. (In these equations, 0 denotes the 0 subspace.) The condition $M^{\perp} = 0$ is equivalent to the maximality of $\{e_n\}$. Here refer to Theorem 6 in Section 2.2, page 73, and observe that the equivalence of (a) and (b) in that theorem does not require the completeness of the space.   ∎

Theorem 5.   *Let A be an operator on a Hilbert space such that $Ax = \sum_{n=1}^{\infty} \lambda_n \langle x, e_n \rangle e_n$, where $[e_n]$ is an orthonormal sequence and $[\lambda_n]$ is a bounded sequence of nonzero complex numbers. If v is in the range of A, then one solution of the equation $Ax = v$ is $x = \sum_{n=1}^{\infty} \lambda_n^{-1} \langle v, e_n \rangle e_n$.*

Proof.   Since v is in the range of A, $v = Az$ for some z. Hence

$$\langle v, e_m \rangle = \langle Az, e_m \rangle = \left\langle \sum_{n=1}^{\infty} \lambda_n \langle z, e_n \rangle e_n, e_m \right\rangle = \lambda_m \langle z, e_m \rangle$$

From this we have

$$\sum_{n=1}^{\infty} |\lambda_n^{-1} \langle v, e_n \rangle|^2 = \sum_{n=1}^{\infty} |\langle z, e_n \rangle|^2 \leqslant \|z\|^2$$

This implies the convergence of the series $x = \sum_{n=1}^{\infty} \lambda_n^{-1} \langle v, e_n \rangle e_n$, by Theorem 2 in Section 2.2, page 71. It follows that

$$Ax = \sum \lambda_n^{-1} \langle v, e_n \rangle A e_n = \sum \langle v, e_n \rangle e_n = \sum \lambda_n \langle z, e_n \rangle e_n = Az = v \quad ∎$$

Example 1.   Consider the operator A defined on $L^2[0,1]$ by the equation

$$(Ax)(t) = \int_0^1 G(s,t) x(s)\, ds$$

where

$$G(s,t) = \begin{cases} (1-s)t & \text{when } 0 \leqslant t \leqslant s \leqslant 1 \\ (1-t)s & \text{when } 0 \leqslant s \leqslant t \leqslant 1 \end{cases}$$

The eigenvalues of A are $\lambda_n = n^2\pi^2$, and the corresponding eigenfunctions are $e_n(t) = \sqrt{2}\sin(2n\pi t)$. This example is discussed also in Section 2.5 (page 107) and in Section 4.7 (page 215). Theorem 5 shows how to solve the equation $Ax = v$ when v is a prescribed function in the range of A.   ∎

We turn now to the topic of Fredholm integral equations of the first kind. These have the form

$$(2) \qquad\qquad \int_S K(s,t)x(t)\,dt = f(s)$$

In this equation, the functions K and f are prescribed. The unknown function x is to be determined. A natural setting is the space $L^2(S)$, described on page 64. Let us assume that the kernel K is in the class $L^2(S \times S)$, so that Theorem 5 of Section 2.3 (page 86) is applicable.

If the integral on the left side of Equation (2) were a Riemann integral on the interval $[0,1]$, it would be a limit of linear combinations of sections of the bivariate function K. That is,

$$\lim_{n\to\infty} \sum_{j=1}^{n} \left(\frac{1}{n}\right) K\left(s,\frac{j}{n}\right) x\left(\frac{j}{n}\right)$$

The sections of K are functions of s parametrized by the variable t:

$$s \mapsto K^t(s) = K(s,t)$$

Thus, we must expect the integral equation to have a solution only if f is in the L^2–closure of the linear span of the sections K^t. This argument is informal, but nevertheless alerts us to the possibility of there being no solution.

Adopting the notation of Theorem 5 in Section 2.3 (and its proof), we have

$$(Tx)(s) = \int_S K(s,t)x(t)\,dt$$

The operator T thus defined from $L^2(S)$ to $L^2(S)$ is compact. (It is an example of a Hilbert–Schmidt operator.) Its range cannot be all of $L^2(S)$, except in the special case when $L^2(S)$ is finite dimensional. Equation (2) will have a solution if and only if f is in the range of T. Now, as in Section 2.3,

$$Tx = \sum_{i=1}^{\infty}\sum_{j=1}^{\infty} a_{ij}\langle x, u_j\rangle u_i$$

On the other hand, if f is in the range of T, we have

$$(3) \qquad\qquad f = \sum_{i=1}^{\infty}\langle f, u_i\rangle u_i$$

Hence the equation $Tx = f$ will be true if and only if Equation (3) holds and

$$\sum_{j=1}^{\infty} a_{ij}\langle x, u_j\rangle = \langle f, u_i\rangle \qquad (i = 1,2,\ldots)$$

Putting $\xi_j = \langle x, u_j \rangle$ and $\beta_i = \langle f, u_i \rangle$, we have the following infinite system of linear equations in an infinite number of unknowns:

$$\sum_{j=1}^{\infty} a_{ij}\xi_j = \beta_i \qquad (i = 1, 2, \ldots)$$

A pragmatic approach is to "truncate" the system by choosing a large integer n and considering the finite matrix problem

$$\sum_{j=1}^{n} a_{ij}\xi_j^{(n)} = \beta_i \qquad (i = 1, 2, \ldots, n)$$

Here the notation $\xi_j^{(n)}$ serves to remind us that we must not expect $\xi_j^{(n)}$ to equal $\langle x, u_j \rangle$. One can define $x_n = \sum_{j=1}^{n} \xi_j^{(n)} u_j$ and examine the behavior of the sequences $[x_n]$ and $[Tx_n]$. Will this procedure succeed always? Certainly not, for the integral equation may have no solution, as previously mentioned.

Other approaches to the solution of integral equations are explored in Chapter 4. The case of Equation (2) in which the kernel is separable or "degenerate," i.e., of the form

$$K(s, t) = \sum_{i=1}^{n} u_i(s)v_i(t)$$

is easily handled:

$$(Tx)(s) = \int_S K(s, t)x(t)\, dt = \int_S \sum_{i=1}^{n} u_i(s)v_i(t)x(t)\, dt$$

$$= \sum_{i=1}^{n} u_i(s)\langle v_i, x \rangle$$

This shows that the range of T is the finite-dimensional space spanned by the functions $u_1, u_2, \ldots, u_n$. Hence, in order that there exist a solution to the given integral equation it is necessary that f be in that same space: $f = \sum_{i=1}^{n} c_i u_i$. Any x such that $\langle v_i, x \rangle = c_i$ will be a solution.

Spectral methods can also be applied to Equation (2). Here, one assumes the kernel to be Hermitian: $K(s, t) = \overline{K(t, s)}$. Then the operator T is Hermitian, and consequently has a spectral form

$$Tx = \sum_{n=1}^{\infty} \lambda_n \langle x, u_n \rangle u_n$$

in which $[u_n]$ is an orthonormal sequence. If f is in the span of that orthonormal sequence, we write $f = \sum_{n=1}^{\infty} \langle f, u_n \rangle u_n$. The solution, if it exists, must then be the function x whose Fourier coefficients are $\langle f, u_n \rangle / \lambda_n$. If this sequence is not an ℓ^2 sequence, we are out of luck! Here we are following Theorem 5 above. This procedure succeeds if f is in the range of T.

For compact operators that are not self-adjoint or even normal there is still a useful canonical form that can be exploited. It is described in the next theorem.

Theorem 6. Singular-Value Decomposition for Compact Operators. *Every compact operator on a separable Hilbert space is expressible in the form*

$$Ax = \sum_{n=1}^{\infty} \langle x, u_n \rangle v_n$$

in which $[u_n]$ is an orthonormal basis for the space and $[v_n]$ is an orthogonal sequence tending to zero. (The sequences $[u_n]$ and $[v_n]$ depend on A.)

Proof. The operator A^*A is compact and Hermitian. Its eigenvalues are nonnegative, because if $A^*Ax = \beta x$, then

$$0 \leqslant \langle Ax, Ax \rangle = \langle x, A^*Ax \rangle = \langle x, \beta x \rangle = \beta \langle x, x \rangle$$

Now apply the spectral theorem to A^*A, obtaining

$$A^*Ax = \sum_{n=1}^{\infty} \lambda_n^2 \langle x, u_n \rangle u_n$$

where $[u_n]$ is an orthonormal basis for the space and $\lambda_n^2 \to 0$. Since we are assuming that $[u_n]$ is a base, we permit some (possibly an infinite number) of the λ_n to be zero. In the spectral representation above, each nonzero eigenvalue λ_n^2 is repeated a number of times equal to its geometric multiplicity. Define $v_n = Au_n$. Then we have

$$\langle v_m, v_n \rangle = \langle Au_m, Au_n \rangle = \langle u_m, A^*Au_n \rangle = \langle u_m, \lambda_n^2 u_n \rangle = \lambda_n^2 \delta_{nm}$$

Hence $[v_n]$ is orthogonal, and $\|v_n\| = \lambda_n \to 0$. Since $[u_n]$ is a base, we have for arbitrary x,

$$x = \sum_{n=1}^{\infty} \langle x, u_n \rangle u_n$$

Consequently,

$$Ax = \sum_{n=1}^{\infty} \langle x, u_n \rangle Au_n = \sum_{n=1}^{\infty} \langle x, u_n \rangle v_n \qquad \blacksquare$$

A general class of compact operators that has received much study is the **Hilbert–Schmidt class**, consisting of operators A such that

$$\sum_{\alpha} \|Au_\alpha\|^2 < \infty$$

for some orthonormal basis $[u_\alpha]$. It turns out that if this sum is finite for one orthonormal base, then it is finite for all. In fact, there is a better result:

Theorem 7.   Let $[u_\alpha]$ and $[v_\beta]$ be two orthonormal bases for a Hilbert space. Every linear operator A on the space satisfies

$$\sum_\alpha \left\| A u_\alpha \right\|^2 = \sum_\beta \left\| A v_\beta \right\|^2$$

Proof.   By the Orthonormal Basis Theorem, Section 2.2 (page 73), we have

$$\sum_\alpha \left\| A u_\alpha \right\|^2 = \sum_\alpha \sum_\beta |\langle A u_\alpha, v_\beta \rangle|^2 = \sum_\beta \sum_\alpha |\langle A u_\alpha, v_\beta \rangle|^2$$

$$= \sum_\beta \sum_\alpha |\langle u_\alpha, A^* v_\beta \rangle|^2 = \sum_\beta \left\| A^* v_\beta \right\|^2$$

Letting $\{u_\alpha\} = \{v_\beta\}$ in this calculation, we obtain $\sum_\beta \left\| A v_\beta \right\|^2 = \sum_\beta \left\| A^* v_\beta \right\|^2$. By combining these equations, we obtain the required result.   ∎

Example 2.   An example of a Hilbert–Schmidt operator arises in the following integral equation from scattering theory:

$$u(x) = f(x) \int_{\mathbb{R}^n} G(|x - y|) h(y) u(y)\, dy$$

Here, f, G, and h are prescribed functions, and u is the unknown function. The function h often has compact support. (Thus it vanishes on the complement of a compact set.) It models the sound speed in the medium, and in a simple case could be a constant on its support. The function f in the integral equation represents the incident wave in a scattering experiment. An important concrete case is

$$u(x) = e^{ip \cdot x} \int_{\mathbb{R}^3} \frac{e^{i|x-y|}}{4\pi|x - y|} h(y) u(y)\, dy$$

In this equation, p is a unit vector (prescribed). Notice the singularity in the kernel of this integral equation. Unfortunately, in the real world, such singularities are the rule rather than the exception.   ∎

References for operator theory in general are [DS, vol.II], [RS], [AG], [Hal2].

Problems 2.4

1. Let X be a Hilbert space having a countable orthonormal base $[u_1, u_2, \ldots]$. Define an operator A by the equation

$$Ax = \sum_{n=1}^{\infty} \langle x, u_n \rangle u_{n+1}$$

What are the eigenvalues of A? Is A compact? Is A Hermitian? What is the norm of A?

2. Repeat Problem 1 for the operator

$$Ax = \sum_{n=1}^{\infty} \alpha_n \langle x, u_n \rangle u_n$$

in which $[\alpha_n]$ is some prescribed bounded real sequence. Find the conditions under which A^{-1} exists as a bounded linear operator.

3. Repeat Problem 1 for the operator

$$Ax = \sum_{n=1}^{\infty} \langle x, u_{n+1} \rangle u_n$$

4. Repeat Problem 1 if the basis is $[\ldots, u_{-2}, u_{-1}, u_0, u_1, \ldots]$ and $A = \sum_{n=-\infty}^{\infty} \langle x, u_n \rangle u_{n+1}$. What is A^{-1}?

5. Let Y be a subspace of a Hilbert space X, and let $A : Y \to X$ be a (possibly unbounded) linear map such that $A^{-1} : X \to Y$ exists and is a compact linear operator. Prove that if $(A - \lambda I)^{-1}$ exists, then it is compact.

6. Prove that for a compact Hermitian operator A on a Hilbert space these properties are equivalent:

   (a) $\langle Ax, x \rangle \geqslant 0$ for all x

   (b) All eigenvalues of A are nonnegative

7. Prove these facts about the spectral sets: (Λ is defined on page 91.)

   (a) $\Lambda(A) = \overline{\Lambda(A^*)}$

   (b) If A is invertible, $\Lambda(A^{-1}) = \{\lambda^{-1} : \lambda \in \Lambda(A)\}$

   (c) $\Lambda(A^n) \supset \{\lambda^n : \lambda \in \Lambda(A)\}$ for $n = 1, 2, 3, \ldots$

8. Let $\{e_1, e_2, \ldots\}$ be an orthonormal system (countable or finite). Let $\lambda_1, \lambda_2, \ldots$ be complex numbers such that $\lim \lambda_n = 0$. Define $Ax = \sum \lambda_n \langle x, e_n \rangle e_n$. Prove that the series converges, that A is a bounded linear operator, and that A is compact. Prove also that if the λ_k are real, then A is Hermitian. Suggestion: Exploit the facts that operators of finite rank are compact and limits of compact operators are compact.

9. In the spectral theorem, when is the following equation true?

$$x = \sum_{n=1}^{\infty} \langle x, e_n \rangle e_n$$

10. Let P be the orthogonal projection of a Hilbert space onto a closed subspace. What are the eigenvalues of P? Give the spectral form of P and $I - P$.

11. Let A be a bounded linear operator on a Hilbert space. Prove that:

    (1) A commutes with A^n for $n = 0, 1, 2, \ldots$.

    (2) A commutes with $p(A)$ for any polynomial p.

    (3) If A^{-1} exists, then A commutes with A^{-n} for $n = 0, 1, 2, 3, \ldots$.

    (4) If $(A - \lambda I)^{-1}$ exists, then it commutes with A.

12. An operator A is said to be **normal** if $AA^* = A^*A$. Give an example of an operator that is not normal. (The eminent mathematician Olga Tausky once observed that most counterexamples in matrix theory are of size 2×2.) Are there any real 2×2 normal matrices that are not self-adjoint? (Other problems on normal operators: 29, 39, 40, 41.)

13. Establish the first equation in the proof of Theorem 4.

14. If A is a bounded linear operator on a Hilbert space, then $A + A^*$ and $i(A - A^*)$ are self-adjoint. Hence A is of the form $B + iC$, where B and C are self-adjoint.

15. Let $\{e_1, e_2, \ldots\}$ be an orthonormal sequence. Let $Ax = \sum \lambda_n \langle x, e_n \rangle e_n$, in which $0 < \inf |\lambda_n| < \sup |\lambda_n| < \infty$. Prove that the series defining Ax converges. Prove that A is not compact. Prove that A is bounded. What are the eigenvalues and eigenvectors of A?

16. Find the eigenvalues and eigenvectors for the operator $Ax = -x''$ acting on the space $X = \{x \in L^2[0,1] : x(0) = 0 \text{ and } x'(1) + \gamma x(1) = 0\}$. Here γ is a prescribed real number. How can the eigenvalues be computed numerically? Find the first one accurate to 3 digits when $\gamma = -\frac{1}{2}$. Newton's method, described in Section 3.3, can be used.

17. Prove that if $Ax = \lambda x$, $A^*y = \mu y$, and $\lambda \neq \overline{\mu}$, then $x \perp y$.

18. If A is Hermitian and x is a vector such that $Ax \neq 0$, then $A^n x \neq 0$ for $n = 0, 1, 2, \ldots$.

19. Every compact Hermitian operator is a limit of a sequence of linear combinations of orthogonal projections.

20. If λ is an eigenvalue of A^2 and $\lambda > 0$, then either $+\sqrt{\lambda}$ or $-\sqrt{\lambda}$ is an eigenvalue of A. (Here, A is any bounded operator.) Hint: If $A^2 x = \lambda x$, then for suitable c, $x \pm cAx$ is an eigenvector of A.

21. Consider the problem $x'' + (\lambda^2 - q)x = 0$, $x(0) = 1$, $x'(0) = 0$. Show that this initial-value problem can be solved by solving instead the integral equation

$$x(t) - \frac{1}{\lambda} \int_0^t q(s) \sin(\lambda(t-s)) x(s)\, ds = \cos(\lambda t)$$

22. If λ is an eigenvalue of A, then $\|A\| \geqslant |\lambda|$.

23. If A is Hermitian and p is a polynomial having real coefficients, then $p(A)$ is Hermitian.

24. A bounded linear operator A on a Hilbert space is said to be **unitary** if $AA^* = A^*A = I$. Prove that for such an operator, $\langle Ax, Ay \rangle = \langle x, y \rangle$ and $\|Ax\| = \|x\|$.

25. (Continuation)   All eigenvalues of a unitary operator satisfy $|\lambda| = 1$.

26. If $Ax = \sum_{n=1}^{\infty} \lambda_n \langle x, e_n \rangle e_n$, what is a formula for A^k ($k = 0, 1, 2, \ldots$)? (The e_n form an orthonormal sequence.)

27. Let A and B be compact Hermitian operators on a Hilbert space. Assume that $AB = BA$. Prove that there is an orthonormal sequence $[u_n]$ such that

$$Ax = \sum \lambda_n \langle x, u_n \rangle u_n \qquad Bx = \sum \mu_n \langle x, u_n \rangle u_n$$

Hint: If λ is an eigenvalue of A, put $E = \{x : Ax = \lambda x\}$, and show that $B(E) \subset E$. Apply the spectral theorem to $B|E$.

28. An operator A on a Hilbert space is said to be **skew-Hermitian** if $A^* = -A$. Prove a spectral theorem for compact skew-Hermitian operators. (Hint: Consider iA.)

29. Assume that A is "normal" ($AA^* = A^*A$) and compact. Prove a spectral theorem for A. Use $A = \frac{1}{2}(A + A^*) + \frac{1}{2}(A - A^*)$, Problem 28, and Problem 27.

30. Let A be a compact Hermitian operator on a Hilbert space X. Assume that all eigenvalues of A are positive, and prove that $\langle Ax, x \rangle > 0$ for all nonzero x.

31. Prove that a compact operator on an infinite-dimensional normed linear space cannot be invertible.

32. Let $[u_n]$ be an orthonormal sequence in a Hilbert space and let $[\lambda_n]$ be a bounded sequence in $\mathbb{C}$. The operator $Ax = \sum \lambda_n \langle x, u_n \rangle u_n$ is compact if and only if $\lambda_n \to 0$.

33. Criticize this argument: Let A be defined as in Problem 32. We show that A is surjective, provided that $\lambda_n \neq 0$ for all n. Take y arbitrarily. To find an x such that $Ax = y$ we write the equivalent equation $\sum \lambda_n \langle x, u_n \rangle u_n = y$. Take the inner product on both sides with u_m, obtaining $\lambda_m \langle x, u_m \rangle = \langle y, u_m \rangle$. Thus

$$x = \sum \frac{\langle y, u_m \rangle}{\lambda_m} u_m$$

34. Let A be a bounded linear operator on a Hilbert space. Suppose that the spectral decomposition of A is known:

$$Ax = \sum_{n=1}^{\infty} \lambda_n \langle x, e_n \rangle e_n$$

where $[e_n]$ is an orthonormal sequence. Show how this information can be used to solve the equation $Ax - \mu x = b$. Make modest additional assumptions if necessary.

35. Prove that the eigenvalues of a bounded linear operator A on a normed linear space all lie in the disk of radius $\|A\|$ in the complex plane.

36. Prove that if P is an orthogonal projection of a Hilbert space onto a subspace, then for any scalars α and β the operator $\alpha P + \beta(I - P)$ is normal (i.e., commutes with its adjoint).

37. Prove that an operator in the Hilbert–Schmidt class is necessarily compact.

38. Prove that every operator having the form described in Theorem 6 is compact, thus establishing a necessary and sufficient condition for compactness.

39. Find all normal 2×2 real matrices. Repeat the problem for complex matrices.

40. Prove that for a normal operator, eigenvectors corresponding to different eigenvalues are mutually orthogonal.

41. Prove that a normal operator and its adjoint have the same null space.

Appendix to Section 2.4

In this appendix we consider a finite-dimensional vector space X, and discuss the relationship between linear transformations and matrices.

Let $L : X \to X$ be a linear transformation. If an ordered basis is selected for X, then a matrix can be associated with L in a certain standard way. (If L is held fixed while the basis or its ordering is changed, then the matrix associated with L *will change*.) The association we use is very simple. Let $[u_1, \ldots, u_n]$ be an ordered basis for X. Then there must exist scalars a_{ij} such that

$$(1) \qquad\qquad Lu_j = \sum_{i=1}^{n} a_{ij} u_i \qquad (1 \leqslant j \leqslant n)$$

The $n \times n$ array of scalars

$$A = \begin{pmatrix} a_{11} & \cdots & a_{1n} \\ \vdots & & \vdots \\ a_{n1} & \cdots & a_{nn} \end{pmatrix}$$

is called the **matrix of L relative to the ordered basis** $[u_1, \ldots, u_n]$.

With the aid of the matrix A it is easy to describe the effect of L on *any* vector x. Write $x = \sum_{j=1}^{n} c_j u_j$. The n-tuple $(c_1, \ldots, c_n)$ is called the **coordinate vector** of x relative to the ordered basis $[u_1, \ldots, u_n]$. Then

$$(2) \qquad Lx = \sum_{j=1}^{n} c_j Lu_j = \sum_{j=1}^{n} c_j \sum_{i=1}^{n} a_{ij} u_i = \sum_{i=1}^{n} \left(\sum_{j=1}^{n} a_{ij} c_j \right) u_i$$

The coordinate vector of Lx therefore has as its entries

$$\sum_{j=1}^{n} a_{ij} c_j \qquad (1 \leqslant i \leqslant n)$$

This n-tuple is obtained from the matrix product

$$Ac = \begin{pmatrix} a_{11} & \cdots & a_{1n} \\ \vdots & & \vdots \\ a_{n1} & \cdots & a_{nn} \end{pmatrix} \begin{pmatrix} c_1 \\ \vdots \\ c_n \end{pmatrix}$$

If the basis for X is changed, what will be the new matrix for L? Let $[v_1, \ldots, v_n]$ be another ordered basis for X. Write

$$(3) \qquad v_j = \sum_{i=1}^{n} p_{ij} u_i \qquad (1 \leqslant j \leqslant n)$$

The $n \times n$ matrix P thus introduced is nonsingular. Now let B denote the matrix of L relative to the new ordered basis. Thus

$$(4) \qquad Lv_j = \sum_{k=1}^{n} b_{kj} v_k = \sum_{k=1}^{n} b_{kj} \sum_{i=1}^{n} p_{ik} u_i \qquad (1 \leqslant j \leqslant n)$$

Another expression for Lv_j can be obtained by use of Equations (2) and (3):

$$(5) \qquad Lv_j = \sum_{i=1}^{n} \left(\sum_{k=1}^{n} a_{ik} p_{kj} \right) u_i \qquad (1 \leqslant j \leqslant n)$$

Upon comparing (4) with (5) we conclude that

$$(6) \qquad \sum_{k=1}^{n} p_{ik} b_{kj} = \sum_{k=1}^{n} a_{ik} p_{kj} \qquad (1 \leqslant i, j \leqslant n)$$

Thus in matrix terms,

$$(7) \qquad PB = AP \qquad \text{or} \qquad B = P^{-1}AP$$

Any two matrices A and B are said to be **similar** to each other if there exists a nonsingular matrix P such that $B = P^{-1}AP$. The matrices for a given linear transformation relative to different ordered bases form an equivalence class under the similarity relation. What is the simplest matrix that can be obtained for a linear transformation by changing the basis? This is a difficult question, to which one answer is provided by the Jordan canonical form. Another answer can be given in the context of a finite-dimensional Hilbert space when the linear transformation L is Hermitian.

Let $[u_1, \ldots, u_n]$ be an ordered orthonormal basis for the n-dimensional Hilbert space X. Let L be Hermitian. The spectral theorem asserts the existence of an ordered orthonormal basis $[v_1, \ldots, v_n]$ and an n-tuple of real numbers $(\lambda_1, \ldots, \lambda_n)$ such that

$$(8) \qquad\qquad Lx = \sum_{i=1}^{n} \lambda_i \langle x, v_i \rangle v_i$$

As above, we introduce matrices A and P such that

$$(9) \qquad Lu_j = \sum_{i=1}^{n} a_{ij}\, u_i \qquad v_j = \sum_{i=1}^{n} p_{ij}\, u_i \qquad (1 \leqslant j \leqslant n)$$

The matrix B that represents L relative to the v-basis is the diagonal matrix $\mathrm{diag}(\lambda_1, \ldots, \lambda_n)$, as we see from Equation (8). Thus from Equation (7) we conclude that A is similar to a diagonal matrix having real entries. More can be said, however, because P has a special structure. Notice that

$$(I)_{jk} = \delta_{jk} = \langle v_k, v_j \rangle = \left\langle \sum_{i=1}^{n} p_{ik}\, u_i, \ \sum_{r=1}^{n} p_{rj}\, u_r \right\rangle = \sum_{i=1}^{n}\sum_{r=1}^{n} p_{ik}\, \bar{p}_{rj} \langle u_i, u_r \rangle$$

$$= \sum_{i=1}^{n} p_{ik}\, \bar{p}_{ij} = \sum_{i=1}^{n} (P^*)_{ji}(P)_{ik} = (P^*P)_{jk}$$

This shows that

$$(10) \qquad\qquad P^*P = I$$

(It follows by elementary linear algebra that $PP^* = I$.) Matrices having the property (10) are said to be **unitary**. We can therefore state that the matrix A (representing the Hermitian operator L with respect to an orthonormal base) is unitarily similar to a real diagonal matrix.

Finally, we note that if an $n \times n$ complex matrix A is such that $A = A^*$, then A is the matrix of a Hermitian transformation relative to an orthonormal basis. Indeed, we have only to select any orthonormal base $[u_1, \ldots, u_n]$ and *define* L by

$$Lu_j = \sum_{i=1}^{n} a_{ij}\, u_i \qquad (1 \leqslant j \leqslant n)$$

Then, of course,

$$Lx = L\left(\sum_{j=1}^{n} c_j\, u_j \right) = \sum_{j=1}^{n}\sum_{i=1}^{n} c_j\, a_{ij}\, u_i$$

By straightforward calculation we have

$$\langle Lx, y \rangle = \langle x, Ly \rangle$$

A matrix A satisfying $A^* = A$ is said to be **Hermitian**. We have proved therefore the following important result, regarded by many as the capstone of elementary matrix theory:

Theorem. *Every complex Hermitian matrix is unitarily similar to a real diagonal matrix.*

2.5 Sturm–Liouville Theory

In this section differential equations are attacked with the weapons of Hilbert space theory. Recall that in elementary calculus we interpret integration and differentiation as mutually inverse operations. So it is here, too, that differential operators and integral operators can be inverse to each other. We find that a differential operator is usually ill-behaved, whereas the corresponding integral operator may be well-behaved, even to the point of being compact. Thus, we often try to recast a differential equation as an equivalent integral equation, hoping that the transformed problem will be less troublesome. (This theme will reappear many times in Chapter 4.) This strategy harmonizes with our general impression that differentiation emphasizes the roughness of a function, whereas integration is a smoothing operation, and is thus applicable to a broader class of functions.

Definition.   The Sturm–Liouville operator is defined by

$$(Ax)(t) = [p(t)x'(t)]' + q(t)x(t) \qquad \text{i.e.,} \qquad Ax = (px')' + qx$$

where x is two–times continuously differentiable, p is real-valued and continuously differentiable, and q is real-valued continuous. The domain of the functions x, p, and q is an interval $[a, b]$. We permit x to be complex-valued. Let eight real numbers α_{ij}, β_{ij} be specified $1 \leqslant i, j \leqslant 2$. Assume that

$$p(a)(\beta_{11}\beta_{22} - \beta_{12}\beta_{21}) = p(b)(\alpha_{11}\alpha_{22} - \alpha_{12}\alpha_{21})$$

Let X be the subspace of $L^2[a, b]$ consisting of all twice continuously differentiable functions x such that

$$\alpha_{11}x(a) + \alpha_{12}x'(a) + \beta_{11}x(b) + \beta_{12}x'(b) = 0$$
$$\alpha_{21}x(a) + \alpha_{22}x'(a) + \beta_{21}x(b) + \beta_{22}x'(b) = 0$$

Assume also that $\beta_{11}\beta_{22} \neq \beta_{12}\beta_{21}$ or $\alpha_{11}\alpha_{22} \neq \alpha_{12}\alpha_{21}$.

Theorem 1.   *Under the preceding hypotheses, A is a Hermitian operator on X.*

Proof.   Let $x, y \in X$. We want to prove that $\langle Ax, y \rangle = \langle x, Ay \rangle$. We compute

$$\langle Ax, y \rangle - \langle x, Ay \rangle = \int_a^b [\overline{y}Ax - xA\overline{y}] = \int_a^b \left[\overline{y}(px')' + \overline{y}qx - x(p\overline{y}')' - xq\overline{y} \right]$$

$$= \int_a^b \left[\overline{y}(px')' - x(p\overline{y}')' \right]$$

$$= \int_a^b \left[\overline{y}(px')' + \overline{y}'px' - x(p\overline{y}')' - x'p\overline{y}' \right]$$

$$= \int_a^b [px'\overline{y} - px\overline{y}']' = [px'\overline{y} - px\overline{y}']_a^b$$

$$= p(b)\left[x'(b)\overline{y}(b) - x(b)\overline{y}'(b) \right] - p(a)\left[x'(a)\overline{y}(a) - x(a)\overline{y}'(a) \right]$$

$$= -p(b)\left[\det w(b) \right] + p(a)\left[\det w(a) \right]$$

where $w(t)$ is the Wroński matrix

$$w(t) = \begin{bmatrix} x(t) & \overline{y}(t) \\ x'(t) & \overline{y}'(t) \end{bmatrix}$$

Put also

$$\alpha = \begin{bmatrix} \alpha_{11} & \alpha_{12} \\ \alpha_{21} & \alpha_{22} \end{bmatrix} \qquad \beta = \begin{bmatrix} \beta_{11} & \beta_{12} \\ \beta_{21} & \beta_{22} \end{bmatrix}$$

Our hypothesis on p is that $p(a)\det\beta = p(b)\det\alpha$. The fact that $x, y \in X$ gives us $\alpha\, w(a) + \beta\, w(b) = 0$. This yields $(\det\alpha)[\det w(a)] = (\det\beta)[\det w(b)]$. Note that $\det(-\beta) = \det(\beta)$ because β is of even order. Multiplying this by $p(b)$ gives us $p(b)\det\alpha\det w(a) = p(b)\det\beta\det w(b)$. By a previous equation, this is $p(a)\det\beta\det w(a) = p(b)\det\beta\det w(b)$. If $\det\beta \neq 0$, we have $p(b)\det w(b) = p(a)\det w(a)$. If $\det\alpha \neq 0$, a similar calculation can be used. ∎

Lemma. *A second-order linear differential equation*

$$a(t)x''(t) + b(t)x'(t) + c(t)x(t) = d(t) \qquad (a \leqslant t \leqslant b)$$

can be put into the form of a Sturm–Liouville equation $(px')' + qx = f$, provided that the functions a, b, c are continuous and $a(t) \neq 0$ for all t in the interval $[a, b]$.

Proof. We transform the equation $ax'' + bx' + cx = d$ by multiplying by the integrating factor $\dfrac{1}{a}\exp\displaystyle\int \big(b(t)/a(t)\big)\, dt$. Thus

$$x''e^{\int b/a} + (b/a)x'e^{\int b/a} + (c/a)xe^{\int b/a} = (d/a)e^{\int b/a}$$

or

$$\left(x'e^{\int b/a}\right)' + (c/a)e^{\int b/a}x = f$$

Let

$$p = e^{\int b/a}, \qquad q = (c/a)e^{\int b/a} \qquad\qquad ∎$$

Example 1.   If $Ax = -x''$ (i.e., $p(t) = -1$ and $q(t) = 0$), what are the eigenvalues and eigenfunctions? The solutions to $-x'' = \lambda x$ are of the form $c_1 \sin\sqrt{\lambda}\,t + c_2 \cos\sqrt{\lambda}\,t$. Hence every complex number λ is an eigenvalue, and each eigenspace is of dimension 2. ∎

Example 2.   Let $Ax = -x''$ as before, but let the inner-product space be the subspace of $L^2[0, \pi]$ consisting of twice continuously differentiable functions that satisfy $x(0) = x(\pi) = 0$. The eigenvalues are n^2 for $n = 0, 1, 2, \ldots$, and the eigenfunctions are $\sin n^2 t$. ∎

The next theorem illustrates one case of the Sturm–Liouville Problem. We take $p(t) = 1$ in the differential equation and let $\beta_{11} = \beta_{12} = \alpha_{21} = \alpha_{22} = 0$. We assume that $|\alpha_{11}| + |\alpha_{12}| > 0$ and $|\beta_{21}| + |\beta_{22}| > 0$. It is left to Problem 8 to prove that the differential operator is Hermitian on the subspace of functions that satisfy the boundary conditions.

Our goal is to develop a method for solving the equation $Ax = y$, where y is a given function, and x is to be determined. The plan of attack is to find a right inverse of A (say $AB = I$) and to give $x = By$ as the solution to the problem. It will turn out that the spectral theorem is applicable to B.

We assume that there exist functions u and v such that

(1) $\qquad\qquad u'' = qu \qquad \beta_{21}u(b) + \beta_{22}u'(b) = 0$

(2) $\qquad\qquad v'' = qv \qquad \alpha_{11}v(a) + \alpha_{12}v'(a) = 0$

(3) $\qquad\qquad u'(a)v(a) - u(a)v'(a) = 1$

From (3) we see that $u \neq 0$ and $v \neq 0$. The left side of (3) is the Wrońskian of u and v evaluated at a.

In practical terms, u and v can be obtained by solving two initial-value problems. This is often done as follows. Find u_0 and v_0 such that

$$u_0'' = qu_0 \qquad u_0(b) = 1 \qquad u_0'(b) = 0$$
$$v_0'' = qv_0 \qquad v_0(a) = 0 \qquad v_0'(a) = 1$$

The u and v required will then be suitable linear combinations of u_0 and v_0.

Now we observe that for *all* s,

$$u'(s)v(s) - u(s)v'(s) = 1$$

This is true because the left side takes the value 1 at $s = a$ and is constant. Indeed,

$$\frac{d}{ds}\left[u'v - uv'\right] = u''v + u'v' - uv'' - u'v' = quv - uqv = 0$$

Next we construct a function g called the **Green's function** for the problem:

$$g(s,t) = \begin{cases} u(s)v(t) & a \leqslant t \leqslant s \leqslant b \\ v(s)u(t) & a \leqslant s \leqslant t \leqslant b \end{cases}$$

The operator A in this case is defined by

(4) $\qquad\qquad Ax = x'' - qx$

and the domain of A is the closure in $L_2[a,b]$ of the set of all twice continuously differentiable functions x such that

$$a_{11}x(a) + a_{12}x'(a) = \beta_{21}x(b) + \beta_{22}x'(b) = 0$$

Theorem 2. *A right inverse of A in Equation (4) is the operator B defined by*

(5) $\qquad\qquad (By)(s) = \int_a^b g(s,t)y(t)\,dt$

Proof.   It is to be proved that $AB = I$. Let $y \in C[a,b]$ and put $x = By$. We show first that $Ax = y$. From the equation

$$x(s) = \int_a^b g(s,t)y(t)\,dt$$

$$= \int_a^s u(s)v(t)y(t)\,dt + \int_s^b v(s)u(t)y(t)\,dt$$

$$= u(s) \int_a^s v(t)y(t)\,dt + v(s) \int_s^b u(t)y(t)\,dt$$

we have

$$x'(s) = u'(s) \int_a^s v(t)y(t)\,dt + u(s)v(s)y(s)$$

$$+ v'(s) \int_s^b u(t)y(t)\,dt - v(s)u(s)y(s)$$

$$= u'(s) \int_a^s v(t)y(t)\,dt + v'(s) \int_s^b u(t)y(t)\,dt$$

Another differentiation gives us

$$x''(s) = u''(s) \int_a^s v(t)y(t)\,dt + u'(s)v(s)y(s) + v''(s) \int_s^b u(t)y(t)\,dt - v'(s)u(s)y(s)$$

$$= q(s)u(s) \int_a^s v(t)y(t)\,dt + q(s)v(s) \int_s^b u(t)y(t)\,dt + y(s)\big[u'(s)v(s) - u(s)v'(s)\big]$$

$$= q(s)x(s) + y(s)$$

In the last step, the constant value of the Wroński an was substituted. Our calculation shows that $x'' - qx = y$ or $Ax = y$, as asserted. Hence $AB = I$.

It remains to prove that $x \in X$, i.e., that x satisfies the boundary conditions. We have, from previous equations,

$$x(a) = v(a) \int_a^b u(t)y(t)\,dt = cv(a)$$

and

$$x'(a) = v'(a) \int_a^b u(t)y(t)\,dt = cv'(a)$$

Hence

$$\alpha_{11}x(a) + \alpha_{12}x'(a) = \alpha_{11}cv(a) + \alpha_{12}cv'(a) = 0$$

Similarly we verify that

$$\beta_{21}x(b) + \beta_{22}x'(b) = 0 \qquad \blacksquare$$

Remark.   If it is known that the homogeneous boundary-value problem has only the trivial solution, then B is also a *left* inverse of A. In order to verify this, let $x \in X$, $y = Ax$, and $By = z$. The previous theorem shows that $y = ABy = Az$ and that $z \in X$. Hence $x - z \in X$ and $A(x - z) = 0$. It follows that $x - z = 0$, so that $x = By = BAx$.

Remark.   The operator B in the previous theorem is Hermitian, because (by Problem 9) g satisfies the equation

$$g(s,t) = g(t,s)$$

Now we apply the Spectral Theorem to the operator B. Notice that B is compact by Theorem 5 in Section 2.3, page 86. There exist an orthonormal sequence $[u_n]$ in $L^2[a,b]$ and real numbers λ_n such that

$$By = \sum_{n=1}^{\infty} \lambda_n \langle y, u_n \rangle u_n$$

Since $Bu_k = \lambda_k u_k$, we have $u_k = \lambda_k A u_k$, and u_k satisfies the boundary conditions. This equation shows that u_k is an eigenvector of A corresponding to the eigenvalue $1/\lambda_k$. Since $\lambda_k \to 0$, $1/\lambda_k \to \infty$. Consequently, a solution to the problem $Ax = y$, where y is given and x must satisfy the boundary conditions, is

$$x = By = \sum_{n=1}^{\infty} \lambda_n \langle y, u_n \rangle u_n$$

Example 3.   Consider the boundary-value problem

$$Ax \equiv x'' + x = y \qquad x'(0) = x(\pi) = 0$$

We shall solve it by means of a Green's function. For the functions u and v we can take $u(t) = \sin t$ and $v(t) = \cos t$. In this case the Green's function is

$$g(s,t) = \begin{cases} \sin s \cos t & 0 \leqslant t \leqslant s \leqslant \pi \\ \cos s \sin t & 0 \leqslant s \leqslant t \leqslant \pi \end{cases}$$

The compact Hermitian integral operator B is given by

$$(By)(s) = \sin s \int_0^s \cos t\, y(t)\, dt + \cos s \int_s^\pi \sin t\, y(t)\, dt \qquad \blacksquare$$

Example 4.   Let us solve the problem in Example 3 by using the Spectral Theorem. The eigenvalues and eigenvectors of the differential operator A are obtained by solving $x'' + x = \mu x$. The general solution of the differential equation is

$$x(t) = c_1 \sin \sqrt{1 - \mu}\, t + c_2 \cos \sqrt{1 - \mu}\, t$$

Imposing the conditions $x'(0) = x(\pi) = 0$, we find that the eigenvalues are $\mu_n = 1 - \left(n - \frac{1}{2}\right)^2$ and the eigenfunctions are $v_n(t) = \cos(2n - 1)t/2$. The v_n are also eigenfunctions of B, corresponding to eigenvalues $\lambda_n = 1/\mu_n = \left(n - n^2 + \frac{3}{4}\right)^{-1}$.

Observe that the eigenfunctions v_n are not of unit norm. If $\alpha_n = 1/\|v_n\|$, then $[\alpha_n v_n]$ is an orthonormal system, and the spectral resolution of B is

$$By = \sum_{n=1}^{\infty} \lambda_n \langle y, \alpha_n v_n \rangle (\alpha_n v_n)$$

A computation reveals that $\alpha_n = (2/\pi)^{1/2}$. Hence we can write

$$By = (2/\pi) \sum_{n=1}^{\infty} \lambda_n \langle y, v_n \rangle v_n$$

Use of this formula is equivalent to the traditional method for solving the boundary-value problem

$$x'' + x = y \qquad x'(0) = x(\pi) = 0$$

The traditional method starts with the functions $v_n(t) = \cos \dfrac{2n-1}{2} t$, which satisfy the boundary conditions. Then we build a function of the form $x = \sum_{n=1}^{\infty} c_n v_n$. This also satisfies the boundary conditions. We hope that with a correct choice of the coefficients we will have $Ax = y$. Since $Av_n = \mu_n v_n$, this equation reduces to $\sum_{n=1}^{\infty} c_n \mu_n v_n = y$. To discover the values of the coefficients, take the inner product of both sides with v_m:

$$\sum c_n \mu_n \langle v_n, v_m \rangle = \langle y, v_m \rangle$$

By orthogonality, we get $c_m \mu_m \alpha_m^{-2} = \langle y, v_m \rangle$ and $c_m = \langle y, v_m \rangle \mu_m^{-1} \alpha_m^2$.

Notice that Theorem 2 has given us an alternative method for solving the inhomogeneous boundary-value problem. Namely, we simply use the Green's function to get x:

$$x(s) = (By)(s) = \int_0^{\pi} g(s,t) y(t) \, dt \qquad \blacksquare$$

Our next task is to find out how to determine a Green's function for the more general Sturm–Liouville problem. The differential equation and its boundary conditions are as follows:

(6)
$$\begin{cases} Ax \equiv (px')' + qx = y \qquad x \in C^2[a,b] \\ \alpha_{11}x(a) + \alpha_{12}x'(a) + \beta_{11}x(b) + \beta_{12}x'(b) = 0 \\ \alpha_{21}x(a) + \alpha_{22}x'(a) + \beta_{21}x(b) + \beta_{22}x'(b) = 0 \end{cases}$$

We are looking for a function g defined on $[a,b] \times [a,b]$. As usual, the t-sections of g are given by $g^t(s) = g(s,t)$.

Theorem 3.   The Green's function for the above problem is characterized by these five properties:

(i) g is continuous in $[a, b] \times [a, b]$.

(ii) $\dfrac{\partial g}{\partial s}$ is continuous in $a < s < t < b$ and in $a < t < s < b$.

(iii) For each t, g^t satisfies the boundary conditions.

(iv) $Ag^t = 0$ in the two open triangles described in (ii).

(v) $\lim\limits_{t \downarrow s} \dfrac{\partial g}{\partial s}(s, t) - \lim\limits_{t \uparrow s} \dfrac{\partial g}{\partial s}(s, t) = -1/p(s)$.

Proof.   As in the previous proof, we take $y \in C[a, b]$ and define

$$x(s) = \int_a^b g(s, t)y(t)\, dt$$

It is to be shown that x is in the domain of A and that $Ax = y$. The domain of A
consists of twice-continuously differentiable functions that satisfy the boundary
conditions. Let us use $'$ to denote partial differentiation with respect to s. Since

$$x(s) = \int_a^s g(s, t)y(t)\, dt + \int_s^b g(s, t)y(t)\, dt$$

we have (as in the previous proof)

$$x'(s) = \int_a^s g'(s, t)y(t)\, dt + \int_s^b g'(s, t)y(t)\, dt$$

It follows that

$$x(a) = \int_a^b g(a, t)y(t)\, dt \qquad x(b) = \int_a^b g(b, t)y(t)\, dt$$

$$x'(a) = \int_a^b g'(a, t)y(t)\, dt \qquad x'(b) = \int_a^b g'(b, t)y(t)\, dt$$

Any linear combination of $x(a)$, $x(b)$, $x'(a)$, and $x'(b)$ is obtained by an integration of the same linear combination of $g(a, t)$, $g(b, t)$, $g'(a, t)$, and $g'(b, t)$. Since
g^t satisfies the boundary conditions, so does x. We now compute $x''(s)$ from the
equation for $x'(s)$:

$$x''(s) = g'(s, s-)y(s) + \int_a^s g''(s, t)y(t)\, dt - g'(s, s+)y(s) + \int_s^b g''(s, t)y(t)\, dt$$

$$= y(s)/p(s) + \int_a^b g''(s, t)y(t)\, dt$$

Here the following notation has been used:

$$g'(s, s+) = \lim\limits_{t \downarrow s} g'(s, t) \qquad g'(s, s-) = \lim\limits_{t \uparrow s} g'(s, t)$$

Now it is easy to verify that $Ax = y$. We have

$$Ax = (px')' + qx = p'x' + px'' + qx$$

Hence

$$(Ax)(s) = p'(s) \int_a^b g'(s,t)y(t)\, dt + y(s) + p(s) \int_a^b g''(s,t)y(t)\, dt$$

$$+ q(s) \int_a^b g(s,t)y(t)\, dt$$

$$= y(s) + \int_a^b \left[\left(p(s)g'(s,t) \right)' + q(s)g(s,t) \right] y(t)\, dt = y(s)$$

because g^t is a solution of the differential equation.   ∎

Example 5.   Find the Green's function for this Sturm–Liouville problem:

$$x'' = y \qquad x(0) = x'(0) = 0 \qquad x \in C^2[0,1]$$

The preceding theorem asserts that g^t should solve the homogeneous differential equation in the intervals $0 < s < t < 1$ and $0 < t < s < 1$. Furthermore, g^t should be continuous, and it should satisfy the boundary conditions. Lastly, $g'(s,t)$ should have a jump discontinuity of magnitude -1 as t passes through the value s. One can guess that g is given by

$$g(s,t) = \begin{cases} 0 & 0 \leqslant s \leqslant t \leqslant 1 \\ s - t & 0 \leqslant t \leqslant s \leqslant 1 \end{cases}$$

If we proceed systematically, it will be seen that this is the only solution. In the triangle $0 < s < t < 1$, $Ag^t = 0$, and therefore g^t must be a linear function of s. We write $g(s,t) = a(t) + b(t)s$. Since g^t must satisfy the boundary conditions, we have $g(0,t) = (\partial g/\partial s)(0,t) = 0$. Thus $a(t) = b(t) = 0$ and $g(s,t) = 0$ in this triangle. In the second triangle, $0 < t < s < 1$. Again g^t must be linear, and we write $g(s,t) = \alpha(t) + \beta(t)s$. Continuity of g on the diagonal implies that $\alpha(t) + \beta(t)t = 0$, and we therefore have $g(s,t) = -\beta(t)t + \beta(t)s = \beta(t)(s - t)$. The condition $(\partial g/\partial s)(s, s+) - (\partial g/\partial s)(s, s-) = -1/p$ leads to the equation $0 - \beta(t) = -1$. Hence $g(s,t) = s - t$ in this triangle. The solution to the inhomogeneous boundary-value problem $x'' = y$ is therefore given by

$$x(s) = \int_0^s (s - t)y(t)\, dt \qquad\qquad ∎$$

Example 6.   Find the Green's function for the problem

$$x'' - x' - 2x = y \qquad x(0) = 0 = x(1)$$

We tentatively set

(7) $$g(s,t) = \begin{cases} u(s)v(t) & 0 \leqslant s \leqslant t \leqslant 1 \\ v(s)u(t) & 0 \leqslant t \leqslant s \leqslant 1 \end{cases}$$

and try to determine the functions u and v. The homogeneous differential equation has as its general solution the function

$$x(s) = \alpha e^{-s} + \beta e^{2s}$$

The solution satisfying the condition $x(0) = 0$ is

$$u(s) = \alpha e^{-s} - \alpha e^{2s}$$

The solution satisfying the condition $x(1) = 0$ is

$$v(s) = -\beta e^3 e^{-s} + \beta e^{2s}$$

With these choices, the function g in Equation (7) satisfies the first four requirements in Theorem 3. With a suitable choice of the parameters α and β, the fifth requirement can be met as well. The calculation produces the following equation involving the Wrońskian of u and v:

$$g'(s, s+) - g'(s, s-) = u'(s)v(s) - v'(s)u(s)$$
$$= \alpha\beta(3 - 3e^3)e^s$$

In this problem, the function p is $p(s) = e^{-s}$, because

$$x''(s) - x'(s) = \left(e^{-s}x'(s)\right)'$$

Hence condition 5 in Theorem 3 requires us to choose α and β such that $\alpha\beta = -(3 - 3e^3)^{-1} \approx .0017465$. Then

$$g(s, t) = \begin{cases} \alpha\beta(e^{-s} - e^{2s})(e^{2t} - e^{3-t}) & 0 \leqslant s \leqslant t \leqslant 1 \\ \alpha\beta(e^{2s} - e^{3-s})(e^{-t} - e^{2t}) & 0 \leqslant t \leqslant s \leqslant 1 \end{cases}$$

Example 7. Find the Green's function for this Sturm–Liouville problem:

$$x'' + 9x = y \qquad x(0) = x(\pi/2) = 0$$

According to the preceding theorem, g should be a continuous function on the square $0 \leqslant s, t \leqslant \pi/2$, and g^t should solve the homogeneous problem in the intervals $0 \leqslant s \leqslant t$ and $t \leqslant s \leqslant \pi/2$. Finally, $\partial g/\partial s$ should have a jump of magnitude -1 as t increases through the value s. These considerations lead us to define

$$g(s, t) = \begin{cases} -\frac{1}{3}\sin 3s \cos 3t & 0 \leqslant s \leqslant t \leqslant \pi/2 \\ -\frac{1}{3}\cos 3s \sin 3t & 0 \leqslant t \leqslant s \leqslant \pi/2 \end{cases}$$

Problems 2.5

1. Find the eigenvalues and eigenfunctions for the Sturm–Liouville operator when $p = q = 1$ and

$$\alpha = \begin{bmatrix} 1 & 0 \\ 0 & 1 \end{bmatrix} = \beta$$

2. Prove that an operator of the form

$$(Ax)(s) = \int_a^b k(s,t)x(t)\,dt$$

   is Hermitian if and only if $k(s,t) = \overline{k(t,s)}$.

3. Find the Green's function for the problem

$$x'' - 3x' + 2x = y \qquad x(0) = 0 = x(1)$$

4. Find the Green's function for the problem

$$x'' - 9x = y \qquad x(0) = 0 = x(1)$$

5. Prove that if u and v are in $C^2[a,b]$, then the function

$$g(s,t) = \begin{cases} u(s)v(t) & a \leqslant s \leqslant t \leqslant b \\ v(s)u(t) & a \leqslant t \leqslant s \leqslant b \end{cases}$$

   has properties (i) and (ii) mentioned in Theorem 3.

6. (Continuation) Show that if
$$pv^2(u/v)' = -1$$

   then g (in Problem 5) will have property (v) in Theorem 3.

8. Prove that if $p = 1$ in the Sturm–Liouville problem and $\beta_{11} = \beta_{12} = \alpha_{21} = \alpha_{22} = 0$ then A is Hermitian.

9. Prove that the function g in Equation (4) is symmetric: $g(s,t) = g(t,s)$.

10. Let $Ax = (px')' - qx$. Prove Lagrange's identity:

$$xAy - yAx = [p(xy' - yx')]'$$

11. (Continuation) Prove Green's formula:

$$\int_a^b (xAy - yAx) = p(xy' - yx')\big|_a^b$$

12. Show that the Wrońskian for any two solutions of the equation $(px')' - qx = 0$ is a scalar multiple of $1/p$, and so is either identically zero or never zero. (Here we assume $p(t) \neq 0$ for $a \leqslant t \leqslant b$.)

13. Find the eigenvalues and eigenfunctions for the operator A defined by the equation $Ax = -x'' + 2x' - x$. Assume that the domain of A is the set of twice continuously differentiable functions on $[0,1]$ that have boundary values $x(0) = x(1) = 0$.

Chapter 3

Calculus in Banach Spaces

3.1 The Fréchet Derivative   115
3.2 The Chain Rule and Mean Value Theorems   121
3.3 Newton's Method   125
3.4 Implicit Function Theorems   135
3.5 Extremum Problems and Lagrange Multipliers   145
3.6 The Calculus of Variations   152

3.1 The Fréchet Derivative

In this chapter we develop the theory of the derivative for mappings between Banach spaces. Partial derivatives, Jacobians, and gradients are all examples of the general theory, as are the Gâteaux and Fréchet differentials. Kantorovich's theorem on Newton's method is proved. Following that there is a section on implicit function theorems in a general setting. Such theorems can often be used to prove the existence of solutions to integral equations and other similar problems. Another section, devoted to extremum problems, illustrates how the methods of calculus (in Banach spaces) can lead to solutions. A section on the "calculus of variations" closes the chapter.

The first step is to transfer, with as little disruption as possible, the elementary ideas of calculus to the more general setting of a normed linear space.

Definition. Let $f : D \to Y$ be a mapping from an open set D in a normed linear space X into a normed linear space Y. Let $x \in D$. If there is a bounded linear map $A : X \to Y$ such that

$$(1) \qquad \lim_{h \to 0} \frac{\|f(x + h) - f(x) - Ah\|}{\|h\|} = 0$$

then f is said to be **Fréchet differentiable** at x, or simply **differentiable** at x. Furthermore, A is called the (Fréchet) **derivative** of f at x.

115

Theorem 1.   *If f is differentiable at x, then the mapping A in the definition is uniquely defined. (It depends on x as well as f.)*

Proof.   Suppose that A_1 and A_2 are two linear maps having the required property, expressed in Equation (1). Then to each $\varepsilon > 0$ there corresponds a $\delta > 0$ such that

$$\|f(x+h) - f(x) - A_i h\| < \varepsilon\|h\| \qquad (i = 1, 2)$$

whenever $\|h\| < \delta$. By the triangle inequality, $\|A_1 h - A_2 h\| < 2\varepsilon\|h\|$ whenever $\|h\| < \delta$. Since $A_1 - A_2$ is homogeneous, the preceding inequality is true for all h. Hence $\|A_1 - A_2\| \leqslant 2\varepsilon$. Since ε was arbitrary, $\|A_1 - A_2\| = 0$. ∎

Notation.   If f is differentiable at x, its derivative, denoted by A in the definition, will usually be denoted by $f'(x)$. Notice that with this notation $f'(x) \in \mathcal{L}(X, Y)$. This is *NOT* the same as saying $f' \in \mathcal{L}(X, Y)$. It will be necessary to distinguish carefully between f' and $f'(x)$.

Theorem 2.   *If f is bounded in a neighborhood of x and if a linear map A has the property in Equation (1), then A is a bounded linear map; in other words, A is the Fréchet derivative of f at x.*

Proof.   Choose $\delta > 0$ so that whenever $\|h\| \leqslant \delta$ we will have

$$\|f(x+h)\| \leqslant M \quad \text{and} \quad \|f(x+h) - f(x) - Ah\| \leqslant \|h\|$$

Then for $\|h\| \leqslant \delta$ we have $\|Ah\| \leqslant 2M + \|h\| \leqslant 2M + \delta$. For $\|u\| \leqslant 1$, $\|\delta u\| \leqslant \delta$, whence $\|A(\delta u)\| \leqslant 2M + \delta$. Thus $\|A\| \leqslant (2M + \delta)/\delta$. ∎

Example 1.   Let $X = Y = \mathbb{R}$. Let f be a function whose derivative (in the elementary sense) at x is λ. Then the Fréchet derivative of f at x is the linear map $h \mapsto \lambda h$, because

$$\lim_{h \to 0} \frac{|f(x+h) - f(x) - \lambda h|}{|h|} = \lim_{h \to 0} \left| \frac{f(x+h) - f(x)}{h} - \lambda \right| = 0$$

Thus, the terminology adopted here is slightly different from the elementary notion of derivative in calculus. ∎

Example 2.   Let X and Y be arbitrary normed linear spaces. Define $f : X \to Y$ by $f(x) = y_0$, where y_0 is a fixed element of Y. (Naturally, such an f is called a **constant** map.) Then $f'(x) = 0$. (This is the 0 element of $\mathcal{L}(X, Y)$.) ∎

Example 3.   Let f be a bounded linear map of X into Y. Then $f'(x) = f$. Indeed, $\|f(x+h) - f(x) - f(h)\| = 0$. Observe that the equation $f' = f$ is *not* true. This illustrates again the importance of distinguishing carefully between $f'(x)$ and f'. ∎

Theorem 3. *If f is differentiable at x, then it is continuous at x.*

Proof. Let $A = f'(x)$. Then $A \in \mathcal{L}(X,Y)$. Given $\varepsilon > 0$, select $\delta > 0$ so that $\delta < \varepsilon/(1 + \|A\|)$ and so that the following implication is valid:

$$\|h\| < \delta \quad \Longrightarrow \quad \|f(x+h) - f(x) - Ah\|/\|h\| < 1$$

Then for $\|h\| < \delta$, we have by the triangle inequality

$$\|f(x+h) - f(x)\| \leqslant \|f(x+h) - f(x) - Ah\| + \|Ah\|$$
$$< \|h\| + \|Ah\| \leqslant \|h\| + \|A\|\,\|h\|$$
$$< \delta(1 + \|A\|) < \varepsilon \qquad \blacksquare$$

Example 4. Let $X = Y = C[0,1]$ and let $\phi : \mathbb{R} \to \mathbb{R}$ be continuously differentiable. Define $f : X \to Y$ by the equation $f(x) = \phi \circ x$, where x is any element of $C[0,1]$. What is $f'(x)$? To answer this, we undertake a calculation of $f(x+h) - f(x)$, using the classical mean value theorem:

$$\big[f(x+h) - f(x)\big](t) = \phi\big(x(t) + h(t)\big) - \phi\big(x(t)\big) = \phi'\Big(x(t) + \theta(t)h(t)\Big)h(t)$$

where $0 < \theta(t) < 1$. This suggests that we define A by

$$Ah = (\phi' \circ x)h$$

With this definition, we shall have at every point t,

$$\big[f(x+h) - f(x) - Ah\big](t) = \phi'\Big(x(t) + \theta(t)h(t)\Big)h(t) - \phi'\big(x(t)\big)h(t)$$

Hence, upon taking the supremum norm, we have

$$\|f(x+h) - f(x) - Ah\| \leqslant \|\phi' \circ (x + \theta h) - \phi' \circ x\|\,\|h\|$$

By comparing this to Equation (1) and invoking the continuity of ϕ', we see that A is indeed the derivative of f at x. Hence $f'(x) = \phi' \circ x$. $\qquad \blacksquare$

Theorem 4. *Let $f : \mathbb{R}^n \to \mathbb{R}$. If each of the partial derivatives $D_i f \ (= \partial f/\partial x_i)$ exists in a neighborhood of x and is continuous at x then $f'(x)$ exists, and a formula for it is*

$$f'(x)h = \sum_{i=1}^{n} D_i f(x) \cdot h_i \qquad h = (h_1, h_2, \ldots, h_n) \in \mathbb{R}^n$$

Speaking loosely, we say that the Fréchet derivative of f is given by the gradient of f.

Proof. We must prove that

$$\lim_{h \to 0} \frac{1}{\|h\|}\left[f(x+h) - f(x) - \sum_{i=1}^{n} h_i D_i f(x)\right] = 0$$

We begin by writing

$$f(x+h) - f(x) = f(v^n) - f(v^0) = \sum_{i=1}^{n} [f(v^i) - f(v^{i-1})]$$

where the vectors v^i and v^{i-1} differ in only one coordinate. Thus we put $v^0 = x$ and $v^i = v^{i-1} + h_i e^i$, where e^i is the ith standard unit vector. By the mean value theorem for functions of one variable,

$$f(v^i) - f(v^{i-1}) = f(v^{i-1} + h_i e^i) - f(v^{i-1}) = h_i D_i f(v^{i-1} + \theta_i h_i e^i)$$

where $0 < \theta_i < 1$. Putting this together, and using the Cauchy–Schwarz inequality, we have

$$\|h\|^{-1} \left| f(x+h) - f(x) - \sum h_i D_i f(x) \right|$$

$$= \|h\|^{-1} \left| \sum h_i \left[D_i f(v^{i-1} + \theta_i h_i e^i) - D_i f(x) \right] \right|$$

$$\leqslant \|h\|^{-1} \|h\| \sqrt{\sum \left[D_i f(v^{i-1} + \theta_i h_i e^i) - D_i f(x) \right]^2} \to 0$$

as $\|h\| \to 0$, by the continuity of $D_i f$ at x. Note that

$$\|v^{i-1} + \theta_i h_i e^i - x\| = \|(h_1, \ldots, h_{i-1}, \theta_i h_i, 0, 0, \ldots, 0)\| \leqslant \|h\| \qquad \blacksquare$$

Theorem 5.   Let $f : \mathbb{R}^n \to \mathbb{R}^m$, and let $f_1, \ldots, f_m$ be the component functions of f. If all partial derivatives $D_j f_i$ exist in a neighborhood of x and are continuous at x, then $f'(x)$ exists, and

$$(f'(x)h)_i = \sum_{j=1}^{n} D_j f_i(x) \cdot h_j \qquad \text{for all} \quad h \in \mathbb{R}^n$$

Speaking informally, we say that the Fréchet derivative of f is given by the Jacobian matrix J of f at x: $J_{ij} = (D_j f_i(x))$.

Proof.   By the definition of the Euclidean norm,

$$\frac{1}{\|h\|^2} \|f(x+h) - f(x) - Jh\|^2 = \frac{1}{\|h\|^2} \sum_{i=1}^{m} \left[f_i(x+h) - f_i(x) - \sum_{j=1}^{n} D_j f_i(x) \cdot h_j \right]^2$$

Each of the m terms in the sum (including the divisor $\|h\|^2$) converges to 0 as $h \to 0$. This is exactly the content of the preceding theorem. $\blacksquare$

Example 5.   Let $f(x) = \sqrt{|x_1 x_2|}$. Then the two partial derivatives of f exist at $(0,0)$, but $f'(0,0)$ does not exist. Details are left to Problem 16. $\blacksquare$

Example 6. Let L be a bounded linear operator on a real Hilbert space X. Define $F : X \to \mathbb{R}$ by the equation $F(x) = \langle x, Lx \rangle$. In order to discover whether F is differentiable at x, we write

$$F(x+h) - F(x) = \langle x + h, Lx + Lh \rangle - \langle x, Lx \rangle$$
$$= \langle x, Lh \rangle + \langle h, Lx \rangle + \langle h, Lh \rangle$$

Since the derivative is a linear map, we guess that A should be $Ah = \langle x, Lh \rangle + \langle h, Lx \rangle$. With that choice, $|Ah| \leqslant 2\|x\|\,\|L\|\,\|h\|$, showing that $\|A\| \leqslant 2\|x\|\,\|L\|$. Thus A is a bounded linear functional. Furthermore,

$$|F(x+h) - F(x) - Ah| = |\langle h, Lh \rangle| \leqslant \|L\|\,\|h\|^2 = o(h)$$

(The notation $o(h)$ is explained in Problem 6.) This establishes that $A = F'(x)$. Notice that

$$Ah = \langle L^* x + Lx, h \rangle$$ ∎

References for the material in this chapter are [Av1], [Av2], [Bart], [Bl], [Bo], [Car], [Cart], [CS], [Cou], [Dieu], [Els], [Ewi], [Fox], [FM], [GF], [Gold], [Hes1], [Hes2], [JLJ], [Lan1], [NSS], [PBGM], [Sag], [Schj], [Wein], and [Youl].

Problems 3.1

1. Let g be a function of two real variables such that g_{22} is continuous. (This notation means second partial derivative with respect to the second argument.) Define $f : C[0,1] \to C[0,1]$ by the equation $(f(x))(t) = \int_0^1 g(t, x(s))\, ds$. Compute the Fréchet derivative of f. You may need Taylor's Theorem.

2. Let f be a Fréchet-differentiable function from a Hilbert space X into $\mathbb{R}$. The gradient of f at x is a vector $v \in X$ such that $f'(x)h = \langle h, v \rangle$ for all $h \in X$. Prove that such a v exists. (It depends on x.) Illustrate with $f(x) = \langle a, x \rangle^2$, $a \in X$ and fixed.

3. Prove that if f and g are differentiable at x, then so is $f+g$, and $(f+g)'(x) = f'(x) + g'(x)$.

4. Let X, Y and Z be normed linear spaces. Prove that if $f : X \to Y$ is differentiable and if $A : Y \to Z$ is a bounded linear map, then $(A \circ f)' = A \circ f'$.

5. Let $f : X \to X$ be differentiable, X a real Hilbert space, and $v \in X$. Define $g : X \to \mathbb{R}$ by $g(x) = \langle f(x), v \rangle$. Prove that g is differentiable, and determine g'.

6. We write $h \mapsto o(h)$ for a generic function that has the property

$$\lim_{h \to 0} \frac{o(h)}{\|h\|} = 0$$

Thus $f'(x)$ is characterized by the equation $f(x+h) - f(x) - f'(x)h = o(h)$. Prove that the family of all such functions o from X to Y is a vector space.

7. Find the derivative of the map $f : C[0,1] \to C[0,1]$ defined by $f(x) = g \cdot x$. Here the dot signifies ordinary multiplication, and $g \in C[0,1]$.

8. Supply the missing details in Example 4. For example, you should establish the fact that $\|\phi' \circ (x + \theta h) - \phi' \circ x\|$ converges to 0 when h converges to 0. Quote any theorems from real analysis that you use.

9. Let X and Y be two normed linear spaces, and let $x \in X$. Let f and g be functions defined on a neighborhood of x and taking values in Y. Following Dieudonné, we say that "f and g are tangent at x" if

$$\lim_{h \to 0} \frac{\|f(x+h) - g(x+h)\|}{\|h\|} = 0$$

Prove that this is an equivalence relation. Prove that the relationship is preserved if the norms in X and Y are changed to equivalent ones. Prove that $x \mapsto f(x_0) + f'(x_0)(x - x_0)$ is the unique affine map tangent to f at x_0. (An affine map is a constant plus a linear map.)

10. Show that these two functions are tangent at $x = 2$:

$$f(x) = x^2 \qquad g(x) = 3 + \sqrt{17 - (x - 6)^2}$$

Draw a picture to illustrate.

11. Prove that if f and g are tangent at x and if both are differentiable at x, then $f'(x) = g'(x)$. Here f and g should be as general as in Problem 9.

12. Let $X = C[0,1]$, with its usual sup-norm. Select $t_i \in [0,1]$ and $v_i \in C[0,1]$, and define $f(x) = \sum_{i=1}^{n} [x(t_i)]^2 v_i$. Prove that f is differentiable at all points of X and give a formula for f'.

13. Prove that the supremum norm on the space $C[0,1]$ is *not* differentiable at any element x for which there are two or more points t in $[0,1]$ where $|x(t)| = \|x\|$.

14. Recall that c_0 is the space of sequences converging to 0 and that the norm is $\|x\| = \max_n |x(n)|$. Prove that the norm is differentiable at x if and only if there is a unique n such that $|x(n)| = \|x\|$.

15. Let y_0 be a point in a normed linear space Y. Define $f : \mathbb{R} \to Y$ by the equation $f(t) = t y_0$. Compute f'. Now define $g(t) = (\sin t) y_0$ and compute g'.

16. Supply the missing details for Example 5.

17. Define $f : C[0,1] \to C[0,1]$ by the equation $[f(x)](t) = x(t) + \int_0^1 [x(st)]^2 \, ds$. Compute $f'(x)$.

18. Prove that if f is differentiable at x, then f is Lipschitz continuous at x. This means that $\|f(y) - f(x)\| \leqslant \lambda \|y - x\|$ for some λ and all y in a neighborhood of x.

19. Let a_n $(n = 0, 1, 2, \ldots)$ be real numbers such that $\sum_{n=0}^{\infty} a_n z^n$ converges for all $z \in \mathbb{C}$. Let X be a Banach space. Define $f : \mathcal{L}(X, X) \to \mathcal{L}(X, X)$ by the equation $f(A) = \sum_{n=0}^{\infty} a_n A^n$. What is the Fréchet derivative of f?

20. Explain the difference between these statements:
    (i) f' is continuous at x.
    (ii) $f'(x)$ is continuous.
    Prove that if $f'(x)$ exists, then it is continuous and differentiable. Give an example of a mapping f such that f' is continuous but not differentiable.

21. Refer to the definition of the Fréchet derivative. If the bounded linear map A satisfies the weaker condition

$$\lim_{\lambda \to 0} \frac{1}{\lambda} \|f(x + \lambda h) - f(x) - \lambda A h\| = 0$$

for every $h \in X$, then f is said to be **Gâteaux differentiable** at x, and A is the **Gâteaux derivative** at x. Prove that if f is Fréchet differentiable at x, then it is Gâteaux differentiable at x, and the two derivatives are equal.

22. Let f be a differentiable map from one normed linear spaces into another. Let y be a point such that $f^{-1}(\{y\})$ contains no point x for which $f'(x) = 0$. Prove that $f^{-1}(\{y\})$ contains no nonvoid open set.

23. If $f : \mathbb{R} \to \mathbb{R}^n$, what is the formula for $f'(x)$?

24. Prove that in an inner-product space the functions $f(x) = \|x\|^2$ and $g(x) = \langle a, x \rangle$ are differentiable. Give formulas for the derivatives.

3.2 The Chain Rule and Mean Value Theorems

We continue to work with a function $f : D \to Y$, where D is an open set in a normed linear space X, and Y is another normed linear space. In the next theorem, we have another mapping g defined on an open set in Y and taking values in a third normed space. In the proof we use notation explained in Problem 3.1.6, page 119.

Theorem 1. The Chain Rule.   *If f is differentiable at x and if g is differentiable at $f(x)$, then $g \circ f$ is differentiable at x, and*

$$(g \circ f)'(x) = g'(f(x)) \circ f'(x)$$

Proof.   Define $F = g \circ f$, $A = f'(x)$, $y = f(x)$, $B = g'(y)$, and

$$o_1(h) = f(x + h) - f(x) - Ah \qquad (h \in X)$$
$$o_2(k) = g(y + k) - g(y) - Bk \qquad (k \in Y)$$
$$\phi(h) = Ah + o_1(h)$$

It is to be shown that $F'(x) = BA$. This requires a calculation as follows:

$$
\begin{aligned}
F(x + h) - F(x) - BAh &= g(f(x + h)) - g(f(x)) - BAh \\
&= g[f(x) + Ah + o_1(h)] - g(y) - BAh \\
&= g[y + \phi(h)] - g(y) - BAh \\
&= g(y) + B\phi(h) + o_2(\phi(h)) - g(y) - BAh \\
&= B[Ah + o_1(h)] + o_2(\phi(h)) - BAh \\
&= Bo_1(h) + o_2(\phi(h))
\end{aligned}
$$

In order to see that this last expression is $o(h)$, notice first that $\|Bo_1(h)\| \leqslant \|B\| \, \|o_1(h)\|$. Hence this term is $o(h)$. Now let $\varepsilon > 0$. Select $\delta_1 > 0$ so that

$$\|k\| < \delta_1 \quad \Longrightarrow \quad \|o_2(k)\| < \varepsilon \|k\| / (\|A\| + 1)$$

Select $\delta > 0$ so that $\delta < \delta_1 / (\|A\| + 1)$ and so that

$$\|h\| < \delta \quad \Longrightarrow \quad \|o_1(h)\| < \|h\|$$

Now let $\|h\| < \delta$. Then we have

$$\|\phi(h)\| = \|Ah + o_1(h)\| \leqslant \|A\| \, \|h\| + \|o_1(h)\|$$
$$< (\|A\| + 1)\|h\| < (\|A\| + 1)\delta < \delta_1$$

Consequently, using $k = \phi(h)$, we conclude that

$$\|o_2(\phi(h))\| < \varepsilon \|\phi(h)\| / (\|A\| + 1) < \varepsilon \|h\| \qquad \blacksquare$$

The mean value theorem of elementary calculus does not have an exact analogue for mappings between general normed linear spaces. (An exception to this assertion occurs in the case when $f : X \to \mathbb{R}$. See Theorem 2, below.) Even for functions $f : \mathbb{R} \to X$, the expected mean-value theorem fails, as we now illustrate.

Example. Define $f : \mathbb{R} \to \mathbb{R}^2$ by the equation $f(t) = (\cos t, \sin t)$. We ask: Is the equation

$$f(2\pi) - f(0) = f'(t)2\pi$$

true for some $t \in (0, 2\pi)$? The answer is "No," because the left side of the equation is $(0,0)$, while $f'(t) = (-\sin t, \cos t) \neq (0,0)$. ∎

However, the mean value theorem of elementary calculus does have a generalization to real-valued functions on a normed linear space. We present this first.

Theorem 2. Mean Value Theorem I. *Let f be a real-valued mapping defined on an open set D in a normed linear space. Let $a, b \in D$. Assume that the line segment*

$$[a, b] = \{a + t(b - a) : 0 \leqslant t \leqslant 1\}$$

lies in D. If f is continuous on $[a, b]$ and differentiable on the open line segment (a, b), then for some ξ in (a, b),

$$f(b) - f(a) = f'(\xi)(b - a)$$

Proof. Put $g(t) = f(a + t(b - a))$. Then g is continuous on the interval $[0, 1]$ and differentiable on $(0, 1)$. By the chain rule,

$$g'(t) = f'(a + t(b - a))(a - b)$$

By the mean value theorem of elementary calculus,

$$f(b) - f(a) = g'(\tau) = f'(a + \tau(b - a))(b - a)$$
$$= f'(\xi)(b - a)$$ ∎

Theorem 3. Mean Value Theorem II. *Let f be a continuous map of a compact interval $[a, b]$ of the real line into a normed linear space Y. If, for each x in (a, b), $f'(x)$ exists and satisfies $\|f'(x)\| \leqslant M$, then $\|f(b) - f(a)\| \leqslant M(b - a)$.*

Proof. It suffices to prove that if $a < \alpha < \beta < b$, then $\|f(\beta) - f(\alpha)\| \leqslant M(b-a)$ because, the desired result would follow from this by continuity. Also, it suffices to prove $\|f(\beta) - f(\alpha)\| \leqslant (M + \varepsilon)(b - a)$ for an arbitrary positive ε. Let S be the set of all x in $[\alpha, \beta]$ such that

$$\|f(x) - f(\alpha)\| \leqslant (M + \varepsilon)(x - a)$$

By continuity, S is a closed set. Let $x_0 = \sup S$. Since S is compact, $x_0 \in S$. To complete the proof, the main task is to show that $x_0 = \beta$. Suppose that $x_0 < \beta$ and look for a contradiction. Since f is differentiable at x_0, there is a positive δ such that $\delta < \beta - x_0$ and

$$|h| < \delta \implies \|f(x_0 + h) - f(x_0) - f'(x_0)h\| < \varepsilon|h|$$

Put $h = \delta/2$ and $u = x_0 + \delta/2$. Then

$$\|f(u) - f(x_0) - f'(x_0)(u - x_0)\| < \varepsilon(u - x_0)$$

Hence

$$\|f(u) - f(x_0)\| < \|f'(x_0)(u - x_0)\| + \varepsilon(u - x_0) \leqslant (M + \varepsilon)(u - x_0)$$

Since $x_0 \in S$, we have also

$$\|f(x_0) - f(\alpha)\| \leqslant (M + \varepsilon)(x_0 - a)$$

Hence

$$\|f(u) - f(\alpha)\| \leqslant \|f(u) - f(x_0)\| + \|f(x_0) - f(\alpha)\| \leqslant (M + \varepsilon)(u - a)$$

This proves that $u \in S$. Since $u > x_0$, we have a contradiction. Thus $x_0 = \beta$, $\beta \in S$, and

$$\|f(\beta) - f(\alpha)\| \leqslant (M + \varepsilon)(\beta - a) < (M + \varepsilon)(b - a) \qquad ■$$

Theorem 4. Mean Value Theorem III.   *Let f be a map from an open set D in one normed linear space into another normed linear space. If the line segment*

$$S = \{ta + (1 - t)b \; : \; 0 \leqslant t \leqslant 1\}$$

lies in D and if $f'(x)$ exists at each point of S, then

$$\|f(b) - f(a)\| \leqslant \|b - a\| \sup_{x \in S} \|f'(x)\|$$

Proof.   Define $g(t) = f\big(ta + (1 - t)b\big)$ for $0 \leqslant t \leqslant 1$. By the chain rule, g' exists and $g'(t) = f'(ta + (1 - t)b)(a - b)$. By the second Mean Value Theorem

$$\|f(b) - f(a)\| = \|g(1) - g(0)\| \leqslant \sup_{0 \leqslant t \leqslant 1} \|g'(t)\| \leqslant \|b - a\| \sup_{x \in S} \|f'(x)\|$$

Notice that $g = f \circ \ell$, where $\ell(t) = ta + (1 - t)b$. Thus $\ell'(t) \in \mathcal{L}(\mathbb{R}, X)$. Hence in the formula for g', the term $(a - b)$ is interpreted as a mapping from $\mathbb{R}$ to X defined by $t \mapsto t \cdot (a - b)$. ■

Theorem 5.   *Let X and Y be normed spaces, D a connected open set in X, and f a differentiable map of D into Y. If $f'(x) = 0$ for all $x \in D$, then f is a constant function.*

Proof.   Since $f'(x)$ exists for all $x \in D$, f is continuous on D (by Theorem 3 of Section 3.1, page 117). Select $x_0 \in D$ and define $A = \{x \in D : f(x) = f(x_0)\}$. This is a closed subset of D (i.e., the intersection of D with a closed set in X). But we can prove that A is also open. Indeed, if $x \in A$, then there is a ball $B(x, r) \subset D$, because D is open. If $y \in B(x, r)$, then the line segment from x to y lies in $B(x, r)$. By the Mean Value Theorem II,

$$\|f(x) - f(y)\| \leqslant \|x - y\| \sup_{0 \leqslant t \leqslant 1} \|f'(tx + (1 - t)y)\| = 0$$

So $f(y) = f(x) = f(x_0)$. This means that $y \in A$. Hence $B(x, r) \subset A$. Thus A is open (it contains a neighborhood of each of its points). A set is connected if it contains no proper subset that is open and closed. Since A is open and closed and nonempty, $A = D$. ∎

The connectedness of D is essential in the preceding theorem, even if $D \subset \mathbb{R}$. For example, suppose that $D = (0, 1) \cup (2, 3)$ and that $f(x) = 1$ on $(0, 1)$ while $f(x) = 2$ on $(2, 3)$. Then f is certainly not constant, although $f'(x) = 0$ at each point of D.

Problems 3.2

1. Let $X = C[0, 1]$ and let $f(x) = \|x\|_1 = \int_0^1 |x(t)|\, dt$. Is f differentiable?

2. Prove that the norm in a real Hilbert space is differentiable except at 0. Hint: Find the derivative of $\|x\|^2$ first.

3. Let X be a real Hilbert space and $v \in X$. Define $f(x) = \|x\|^2 v$. What is $f'(x)$?

4. Let f be a continuous real-valued map on a Hilbert space. If $f'(x_0)$ exists, then there is a direction of steepest descent at x_0. This means that there exists a vector u of norm 1 for which $(d/dt)f(x_0 + tu)|_{t=0}$ is a maximum. What is u?

5. Let f be a differentiable and continuous real-valued function defined on an open set D in a normed linear space. Suppose that $x_0 \in D$ and that $f(x_0) \geqslant f(x)$ for all $x \in D$. Prove that $f'(x_0) = 0$.

6. Let D be a bounded open set in a finite-dimensional normed linear space. Let $\overline{D}$ be the closure of D. Let $f : \overline{D} \to \mathbb{R}$ be continuous. Assume f differentiable in D and that f is constant on $\overline{D} \smallsetminus D$ (the boundary of D). Show that $f'(x) = 0$ for some $x \in D$. (*Hint*: A continuous real-valued function on a compact set achieves its maximum and minimum. Use Problem 5.)

7. Let K be a closed convex set contained in an open set D contained in a Banach space X. Let $f : D \to X$. Assume that $f'(x)$ exists for each $x \in K$ and that $f(K) \subset K$. Assume also that $\sup\{\|f'(x)\| : x \in K\} < 1$. Show that f has a unique fixed point in K. (Banach's Theorem, page 177, is helpful.)

8. The mean value theorem for functions $f : \mathbb{R} \to \mathbb{R}$ states that $f(x+h) - f(x) = hf'(x+\theta h)$ for some $\theta \in (0, 1)$. Show that this is not valid for complex functions. Try e^z, $z = 0$, $h = 2\pi i$, and at least one other function.

9. Let f be a differentiable map from a normed space X to a normed space Y. Let y_0 be a point of Y such that f' is invertible at each point of $f^{-1}(y_0)$. Prove that $f^{-1}(y_0)$ is a discrete set.

10. Write out the conclusion of Theorem 2 in the case that $X = \mathbb{R}^n$, using the partial derivatives $\partial f/\partial x_i$.

3.3 Newton's Method

The elementary form of Newton's method is used to find a zero of a function $f : \mathbb{R} \to \mathbb{R}$ (or "root" of the equation $f(x) = 0$). The method is iterative and employs the formula $x_{n+1} = x_n - f(x_n)/f'(x_n)$. Its rationale is as follows: Suppose that x_n is an approximation to a zero of f. We try to find a suitable correction to x_n so as to obtain the nearby root. That is, we try to determine h so that $f(x_n + h) = 0$. By Taylor's Theorem,

$$0 = f(x_n + h) = f(x_n) + hf'(x_n) + o(h)$$

So, by ignoring the $o(h)$ term, we are led to $h = -f(x_n)/f'(x_n)$. If f is now a mapping of one Banach space, X, into another, Y, the same rationale leads us to $x_{n+1} = x_n - [f'(x_n)]^{-1}f(x_n)$. Of course $f'(x_n)$ is a linear operator from X into Y, and the inverse $[f'(x_n)]^{-1}$ will have to be *assumed* to exist as a bounded linear operator from Y to X. First, we examine the simple case, when $f : \mathbb{R} \to \mathbb{R}$.

> **Theorem 1.**   Let f be a function from $\mathbb{R}$ to $\mathbb{R}$. Assume that f'' is bounded, that $f(r) = 0$, and that $f'(r) \neq 0$. Let δ be a positive number such that
>
> $$\rho \equiv \frac{1}{2}\delta \max_{|x-r|\leqslant\delta} |f''(x)| \div \min_{|x-r|\leqslant\delta} |f'(x)| < 1$$
>
> If Newton's method is started with $x_0 \in [r - \delta, r + \delta]$, then for all n,
>
> $$|x_{n+1} - r| \leqslant \frac{\rho}{\delta}|x_n - r|^2 \leqslant \rho|x_n - r|$$

Proof. Define $e_n = x_n - r$. Then

$$0 = f(r) = f(x_n - e_n) = f(x_n) - e_n f'(x_n) + \frac{1}{2}e_n^2 f''(\xi_n)$$

In this equation, the point ξ_n is between x_n and r. Hence $|\xi_n - r| \leqslant |x_n - r| = |e_n|$. Using this we have

$$e_{n+1} = x_{n+1} - r = x_n - \frac{f(x_n)}{f'(x_n)} - r = e_n - \frac{f(x_n)}{f'(x_n)}$$
$$= \frac{e_n f'(x_n) - f(x_n)}{f'(x_n)} = e_n^2 \frac{f''(\xi_n)}{f'(x_n)}$$

Since $|x_0 - r| \leqslant \delta$ by hypothesis, we have $|e_0| \leqslant \delta$ and $|\xi_0 - r| \leqslant \delta$. Hence $|e_1| \leqslant \frac{1}{2}e_0^2|f''(\xi_0)|/|f'(x_0)| \leqslant \frac{1}{2}e_0^2 \cdot 2\rho/\delta \leqslant \rho|e_0|$. By repeating this we establish that $|x_{n+1} - r| \leqslant \rho |x_n - r|$ (convergence). Similarly, we have $|e_1| \leqslant (\rho/\delta)e_0^2$ and $|e_{n+1}| \leqslant (\rho/\delta)e_n^2$. (quadratic convergence).   ∎

The successive errors e_n in the preceding theorem obey an inequality $|e_{n+1}| \leqslant C|e_n|^2$. Suppose, for example, that $C = 1$ and $|e_0| \leqslant 10^{-1}$. Then $|e_1| \leqslant 10^{-2}$, $|e_2| \leqslant 10^{-4}$, $|e_3| \leqslant 10^{-8}$, and so on. For an iterative process, this is an extraordinarily favorable state of affairs, as it indicates a doubling of the number of significant digits in the numerical solution at each step.

Example 1. For finding the square root of a given positive number a, one can solve the equation $x^2 - a = 0$ by Newton's method. The iteration formula turns out to be

$$x_{n+1} = \frac{1}{2}\left(x_n + \frac{a}{x_n}\right)$$

This formula was known to the ancient Greeks and is called Heron's formula. In order to see how well it performs, we can use a computer system such as Mathematica, Maple, or Matlab to obtain the Newton approximations to $\sqrt{2}$. The iteration function is $g(x) = (x + 2/x)/2$, and a reasonable starting point is $x_0 = 1$. Mathematica is capable of displaying x_n with any number of significant figures; we chose 60. The input commands to Mathematica are shown here. (Each one should be separated from the following one by a semicolon, as shown.) The output, not shown, indicates that the seventh iterate has at least 60 correct digits!

```
g[x_]:=(x+(2/x))/2;    g[1];    N[%,60];    g[%];    g[%];    ...
```
                    ∎

Example 2. We illustrate the mechanics of Newton's method in higher dimensions with the following problem:

$$\begin{cases} x - y + 1 = 0 \\ x^2 + y^2 - 4 = 0 \end{cases}$$

where x and y are real variables. We have here a mapping $f : \mathbb{R}^2 \to \mathbb{R}^2$, and we seek one or more zeros of f. The Newton iteration is $u_{n+1} = u_n - [f'(u_n)]^{-1}f(u_n)$, where $u_n = (x_n, y_n) \in \mathbb{R}^2$. The derivative $f'(u)$ is given by the Jacobian matrix J. We find that

$$J = \begin{bmatrix} 1 & -1 \\ 2x & 2y \end{bmatrix} \qquad J^{-1} = \frac{1}{2x + 2y}\begin{bmatrix} 2y & 1 \\ -2x & 1 \end{bmatrix}$$

Hence the iteration formula, in detail, is this:

$$\begin{bmatrix} x_{n+1} \\ y_{n+1} \end{bmatrix} = \begin{bmatrix} x_n \\ y_n \end{bmatrix} - \frac{1}{2x_n + 2y_n}\begin{bmatrix} 2y_n & 1 \\ -2x_n & 1 \end{bmatrix}\begin{bmatrix} x_n - y_n + 1 \\ x_n^2 + y_n^2 - 4 \end{bmatrix}$$

If we start at $u_0 = (0,2)^T$, the next vectors are $u_1 = (1,2)^T$ and $u_2 = (5/6, 11/6)$. A symbolic computation system such as those mentioned above can be used here, too. The problem is chosen intentionally as one easily visualized: One seeks the points where a line intersects a circle. See Figure 3.1.   ∎

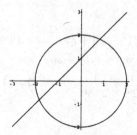

Figure 3.1

The remarkable theorem of Kantorovich is presented next. This theorem: (1) Proves the existence of a zero of a function from suitable hypotheses, and (2) Establishes the quadratic convergence of the Newton algorithm. When it was published in 1948, this theorem gave new information about Newton's method even when the domain space X was two-dimensional.

Theorem 2.   Kantorovich Theorem on Newton's Method.
Let $f : X \to Y$ be a map from a Banach space X into a Banach space Y. Let x_0 be a point of X where $f'(x_0)$ exists and is invertible. Define

$$a_0 = \left\| f'(x_0)^{-1} f(x_0) \right\| \qquad b_0 = \left\| f'(x_0)^{-1} \right\|$$
$$S = \{ x \in X : \| x - x_0 \| \leqslant 2a_0 \}$$
$$k = 2 \sup \{ \| f'(x) - f'(v) \| / \| x - v \| : x, v \in S, x \neq v \}$$

If f is differentiable in S and if $a_0 b_0 k \leqslant \frac{1}{2}$, then f has a zero in S. Newton's iteration started at x_0 converges quadratically to the zero.

Proof.   At the nth step we will have x_n, a_n, b_n such that
  (1) $x_n \in S$
  (2) $f'(x_n)^{-1}$ exists
  (3) $\| f'(x_n)^{-1} f(x_n) \| \leqslant a_n$
  (4) $\| f'(x_n)^{-1} \| \leqslant b_n$
  (5) $a_n b_n k \leqslant \frac{1}{2}$
  (6) $a_n \leqslant a_0 / 2^n$
Observe that $1 - a_n b_n k \geqslant \frac{1}{2}$, and that properties (1)–(6) are true for $n = 0$. Now define

$$x_{n+1} = x_n - f'(x_n)^{-1} f(x_n)$$
$$b_{n+1} = b_n (1 - a_n b_n k)^{-1}$$
$$a_{n+1} = \frac{1}{2} k b_{n+1} a_n^2$$

We will prove properties (1)–(6) for $n + 1$.

(I) x_{n+1} is well-defined because of (2).

(II) $\left\| x_n - x_{n+1} \right\| = \left\| f'(x_n)^{-1} f(x_n) \right\| \leqslant a_n$.

(III)

$$
\begin{aligned}
\left\| x_{n+1} - x_0 \right\| &\leqslant \left\| x_{n+1} - x_n \right\| + \left\| x_n - x_{n-1} \right\| + \cdots + \left\| x_1 - x_0 \right\| \\
&\leqslant a_n + a_{n-1} + \cdots + a_0 \\
&\leqslant a_0 \left(\frac{1}{2^n} + \frac{1}{2^{n-1}} + \cdots + \frac{1}{2} + 1 \right) \leqslant 2a_0
\end{aligned}
$$

Thus $x_{n+1} \in S$.

(IV)

$$
\begin{aligned}
a_{n+1} b_{n+1} k &= \left(\frac{1}{2} k b_{n+1} a_n^2 \right) b_{n+1} k \\
&= \frac{1}{2} (a_n b_{n+1} k)^2 \\
&= \frac{1}{2} (a_n b_n k)^2 (1 - a_n b_n k)^{-2} \\
&\leqslant 2 (a_n b_n k)^2 \leqslant \frac{1}{2}
\end{aligned}
$$

Observe that $a_n b_{n+1} k \leqslant 1$.

(V) Let $H = f'(x_n)^{-1} f'(x_{n+1})$. Then H is invertible because

$$
\begin{aligned}
\left\| I - H \right\| &= \left\| f'(x_n)^{-1} \{ f'(x_n) - f'(x_{n+1}) \} \right\| \\
&\leqslant \left\| f'(x_n)^{-1} \right\| \left\| f'(x_n) - f'(x_{n+1}) \right\| \\
&\leqslant b_n \frac{1}{2} k \left\| x_n - x_{n+1} \right\| \leqslant b_n \frac{1}{2} k a_n \leqslant \frac{1}{4}
\end{aligned}
$$

We know also that $\left\| H^{-1} \right\| \leqslant (1 - \frac{1}{2} a_n b_n k)^{-1}$. It follows that $f'(x_{n+1})$ is invertible since it is a product of invertible operators, $f'(x_{n+1}) = f'(x_n) H$.

(VI) From (V) we have

$$
\begin{aligned}
\left\| f'(x_{n+1})^{-1} \right\| &= \left\| H^{-1} f'(x_n)^{-1} \right\| \leqslant \left\| H^{-1} \right\| \left\| f'(x_n)^{-1} \right\| \\
&\leqslant (1 - \frac{1}{2} a_n b_n k)^{-1} b_n \leqslant b_n (1 - a_n b_n k)^{-1} = b_{n+1}
\end{aligned}
$$

(VII) Define $g(x) = x - f'(x_n)^{-1} f(x)$. Then $g(x_n) = x_{n+1}$. If $x \in S$, then

$$
\begin{aligned}
\left\| g'(x) \right\| &= \left\| I - f'(x_n)^{-1} f'(x) \right\| = \left\| f'(x_n)^{-1} \{ f'(x_n) - f'(x) \} \right\| \\
&\leqslant \left\| f'(x_n)^{-1} \right\| \left\| f'(x_n) - f'(x) \right\| \leqslant b_n \frac{1}{2} k \left\| x_n - x \right\|
\end{aligned}
$$

(VIII) Using the Mean Value Theorem and parts (VII) and (II), we have

$$
\begin{aligned}
\left\| f'(x_n)^{-1} f(x_{n+1}) \right\| &= \left\| x_{n+1} - g(x_{n+1}) \right\| = \left\| g(x_n) - g(x_{n+1}) \right\| \\
&\leqslant \left\| g'(\overline{x}) \right\| \left\| x_n - x_{n+1} \right\| \leqslant \left\| g'(\overline{x}) \right\| a_n
\end{aligned}
$$

Here $\bar{x}$ is some point on the line segment joining x_n to x_{n+1}. From part (VII), it follows that

$$\|f'(x_n)^{-1}f(x_{n+1})\| \leqslant b_n\frac{1}{2}k\|x_n - \bar{x}\|a_n$$

$$\leqslant b_n\frac{1}{2}k\|x_n - x_{n+1}\|a_n \leqslant b_n\frac{1}{2}ka_n^2$$

(IX)

$$\|f'(x_{n+1})^{-1}f(x_{n+1})\| = \|H^{-1}f'(x_n)^{-1}f(x_{n+1})\|$$

$$\leqslant \|H^{-1}\|\,\|f'(x_n)^{-1}f(x_{n+1})\|$$

$$\leqslant (1 - \frac{1}{2}a_nb_nk)^{-1}b_n\frac{1}{2}ka_n^2$$

$$\leqslant (1 - a_nb_nk)^{-1}b_n\frac{1}{2}ka_n^2 = \frac{1}{2}b_{n+1}a_n^2k = a_{n+1}$$

(X) Using the observation made in (IV), we have

$$a_{n+1} = \frac{1}{2}kb_{n+1}a_n^2 = (a_nb_{n+1}k)(\frac{1}{2}a_n) \leqslant \frac{1}{2}a_n \leqslant \frac{1}{2}a_0/2^n = a_0/2^{n+1}$$

This completes the induction phase of the proof.

(XI) If $m > n$, then from parts II and X we have

$$\|x_n - x_m\| \leqslant \|x_n - x_{n+1}\| + \|x_{n+1} - x_{n+2}\| + \cdots + \|x_{m-1} - x_m\|$$

$$\leqslant a_n + a_{n+1} + \cdots \leqslant a_n\left(1 + \frac{1}{2} + \frac{1}{4} + \cdots\right) = 2a_n$$

Notice that $a_n \to 0$ by (6). Hence $[x_n]$ is a Cauchy sequence. By the completeness of X there is a point x^* such that $x_n \to x^*$. We have $x^* \in S$ by (1).

(XII) Define $h_n = a_nb_nk$. From part (IV), we have

$$h_{n+1} = a_{n+1}b_{n+1}k = \frac{1}{2}(a_nb_nk)^2(1 - a_nb_nk)^{-2} \leqslant 2(a_nb_nk)^2 = 2h_n^2$$

Therefore,

$$h_n \leqslant 2h_{n-1}^2 \leqslant 2(2h_{n-2}^2)^2 = 8h_{n-2}^4 \leqslant \cdots \leqslant \frac{1}{2}(2h_0)^{2^n}$$

(XIII)

$$a_n = \frac{1}{2}kb_na_{n-1}^2 = \frac{1}{2}kb_{n-1}(1 - a_{n-1}b_{n-1}k)^{-1}a_{n-1}^2$$

$$\leqslant (kb_{n-1}a_{n-1})a_{n-1} = h_{n-1}a_{n-1}$$

Repeating this inequality, we obtain

$$a_n \leqslant h_{n-1}h_{n-2}\cdots h_0 a_0$$

$$\leqslant \left[\frac{1}{2}(2h_0)^{2^{n-1}}\right]\left[\frac{1}{2}(2h_0)^{2^{n-2}}\right]\cdots\left[\frac{1}{2}(2h_0)^{2^0}\right]a_0$$

$$= a_0\left(\frac{1}{2}\right)^n(2h_0)^{2^{n-1}+2^{n-2}+\cdots+1}$$

$$= a_0 2^{-n}(2h_0)^{2^n-1}$$

(XIV) $\|x_n - x^*\| \leqslant 2a_n \leqslant 2a_0 2^{-n}(2a_0 b_0 k)^{2^n-1} = c\cdot 2^{-n}\cdot\theta^{2^n}$, where $c = 4a_0^2 b_0 k$. If $\theta \equiv 2a_0 b_0 k < 1$, this is *quadratic* convergence. Here $x^* = \lim x_n$. The sequence $[x_n]$ has the Cauchy property, by part (XI).

(XV) In order to prove that $f(x^*) = 0$, write $\|f(x_n)\| = \|f'(x_n)(x_n - x_{n+1})\| \leqslant \|f'(x_n)\|\,\|x_n - x_{n+1}\|$. Now $\|x_n - x_{n+1}\| \to 0$ and $\|f'(x_n)\|$ is bounded as a function of n because $\|f'(x_n)\| \leqslant \|f'(x_n) - f'(x_0)\| + \|f'(x_0)\| \leqslant \frac{k}{2}\|x_n - x_0\| + \|f'(x_0)\|$. Since f is continuous, $f(x^*) = f(\lim x_n) = \lim f(x_n) = 0$. ∎

For the preceding theorem and the one to follow (for which we do not give the proof), we refer the reader to [Gold] and to [KA].

Theorem 3. Kantorovich's Theorem on the Simplified Newton Method. *Assume the hypothesis of Kantorovich's Theorem except that the radius of S is set equal to the quantity $[1 - \sqrt{1 - 2a_0 b_0 k}]/(b_0 k)$. Then the simplified Newton iteration*

$$x_{n+1} = x_n - f'(x_0)^{-1}f(x_n)$$

converges at least geometrically to a zero of f in S.

The next theorem concerns a variant of Newton's method due to R.E. Moore [Moo]. In this theorem, we have two normed linear spaces X and Y. An open set Ω in X is given, and a mapping $F : \Omega \to Y$ is prescribed. It is known that F has a zero x^* in Ω and that $F'(x^*)$ exists. We wish to determine x^*. For this purpose we set up an iterative scheme of the form

(7) $x_{n+1} = G(x_n), \qquad G(x) = x - A(x)F(x)$

Here, $A(x) \in \mathcal{L}(Y, X)$, and we assume that

(8) $\sup_{x\in\Omega}\|A(x)\| = M < \infty$

It is intended that $A(x)$ be an approximate inverse of $F'(x^*)$. We assume that

(9) $\sup_{x\in\Omega}\|I - A(x)F'(x^*)\| = \lambda < 1$

Theorem 4. *There is a neighborhood of x^* such that the iteration sequence defined in Equation (7) converges to x^* for arbitrary starting points in that neighborhood.*

Proof. Select $\varepsilon > 0$ such that

$$(10) \qquad\qquad \theta \equiv \lambda + M\varepsilon < 1$$

By the definition of the Fréchet derivative $F'(x^*)$, we can write

$$(11) \qquad F(x) = F(x^*) + F'(x^*)(x - x^*) + \eta(x)$$

where $\eta(x)$ is $o(\|x - x^*\|)$. In particular, we can select $\delta > 0$ so that

$$(12) \qquad \|x - x^*\| \leqslant \delta \Longrightarrow \Big[x \in \Omega \ \text{ and } \ \|\eta(x)\| \leqslant \varepsilon \|x - x^*\| \Big]$$

From (11), using the fact that $F(x^*) = 0$ and the definition of G, we have

$$\begin{aligned}
G(x) - x^* &= x - x^* - A(x)F(x) \\
&= x - x^* - A(x)\big[F'(x^*)(x - x^*) + \eta(x)\big] \\
&= x - x^* - A(x)F'(x^*)(x - x^*) - A(x)\eta(x) \\
&= \big[I - A(x)F'(x^*)\big](x - x^*) - A(x)\eta(x)
\end{aligned}$$

If we assume further that $\|x - x^*\| \leqslant \delta$, then

$$\begin{aligned}
\|G(x) - x^*\| &\leqslant \lambda \|x - x^*\| + M\|\eta(x)\| \\
&\leqslant \lambda \|x - x^*\| + M\varepsilon \|x - x^*\| \\
&= (\lambda + M\varepsilon)\|x - x^*\| = \theta \|x - x^*\|
\end{aligned}$$

If the starting point x_0 for the iteration is within distance δ of x^*, then

$$\|x_1 - x^*\| = \|G(x_0) - x^*\| \leqslant \theta \|x_0 - x^*\| \leqslant \theta\delta$$

Continuing, we have

$$\|x_2 - x^*\| = \|G(x_1) - x^*\| \leqslant \theta \|x_1 - x^*\| \leqslant \theta^2\delta$$

In general, $\|x_n - x^*\| \leqslant \theta^n\delta$, and hence $x_n \to x^*$. ∎

Corollary 1. *If there is an $r > 0$ such that*

$$\sup_{\|x-x^*\|\leqslant r} \|F'(x)^{-1}\| < \infty \quad \text{and} \quad \sup_{\|x-x^*\|\leqslant r} \|I - F'(x)^{-1}F'(x^*)\| < 1$$

then there is a neighborhood of x^* in which Newton's method converges from arbitrary starting points.

Corollary 2. *If $X = Y$ and if $\|I - F'(x^*)\| < 1$, then the iteration $x_{n+1} = x_n - F(x_n)$ will converge to x^* if started sufficiently near to x^*.*

Corollary 3. *If $\|I - F'(x_0)^{-1}F'(x^*)\| < 1$, then the simplified Newton iteration $x_{n+1} = x_n - F'(x_0)^{-1}F(x)$ converges to x^* if started sufficiently near to x^*.*

Applications to nonlinear integral equations. In the following paragraphs we shall discuss the application of Newton's method and the Neumann

Theorem to nonlinear integral equations. A rather general model problem is
considered:

(13) $$x(s) - \lambda \int_0^1 g\big(s, t, x(t)\big)\, dt = v(s)$$

Here λ, v, and g are given. We assume that $v \in C[0, 1]$ and that g is continuous
on the 3-dimensional set

$$D = \big\{(s, t, u) : 0 \leqslant s \leqslant 1,\; 0 \leqslant t \leqslant 1,\; -\infty < u < \infty\big\}$$

Also, we assume that $\big|g(s, t, u_1) - g(s, t, u_2)\big| \leqslant k\big|u_1 - u_2\big|$ in the domain D.

Theorem 5.   *If $\big|\lambda\big|k < 1$, then the integral equation (13) above has
a unique solution.*

Proof.   Apply the Contraction Mapping Theorem (Chapter 4, Section 2, page
177) to the mapping F defined on $C[0, 1]$ by $(Fx)(s) = v(s) + \lambda \int_0^1 g(s, t, x(t))\, dt$.
We see easily that

$$\|Fx_1 - Fx_2\| = \sup_s \big|(Fx_1)(s) - (Fx_2)(s)\big|$$
$$\leqslant |\lambda| \sup_s \int_0^1 \big|g\big(s, t, x_1(t)\big) - g\big(s, t, x_2(t)\big)\big|$$
$$\leqslant |\lambda| \int_0^1 k\big|x_1(t) - x_2(t)\big|\, dt$$
$$\leqslant |\lambda|k\|x_1 - x_2\| \qquad\blacksquare$$

If $|\lambda|k < 1$, then the sequence $x_{n+1} = F(x_n)$ will converge, in the space $C[0, 1]$,
to a solution of the integral equation. In this process, x_0 can be an arbitrary
starting point in $C[0, 1]$. Newton's method can also be used, provided that we
start at a point sufficiently close to the solution. For Newton's method, we define
the mapping f by

$$(f(x))(s) = x(s) - \lambda \int g\big(s, t, x(t)\big)\, dt - v(s)$$

We require f', which is given by

$$[f'(x)h](s) = h(s) - \lambda \int_0^1 g_3\big(s, t, x(t)\big)h(t)\, dt$$

where g_3 is the partial derivative of g with respect to its third argument, i.e.,
$g_3(s, t, u) = (\partial/\partial u)g(s, t, u)$.

Lemma.   *If g_3 exists on the domain*

$$Q = \{(s,t,u) \;:\; 0 \leqslant s \leqslant 1, \; 0 \leqslant t \leqslant 1, \; |u| \leqslant \|x\|\}$$

and if

$$\lim_{r \to 0} \frac{1}{r}\big[g(s,t,u+r) - g(s,t,u) - rg_3(s,t,u)\big] = 0$$

uniformly in Q, then $f'(x)$ is as given above.

The next step in using Newton's method is to compute $f'(x)^{-1}$. Observe that $f'(x) = I - \lambda A$, where A is the integral operator whose kernel is $g_3(s,t,x(t))$. Explicitly,

$$(Ah)(s) = \int_0^1 g_3\big(s,t,x(t)\big)h(t)\,dt$$

This is a linear operator, since x is *fixed*. If $|\lambda|\,\|A\| < 1$, then $I - \lambda A$ is invertible, and by the Neumann Theorem in Chapter 1, Section 5, page 28, we have

$$f'(x)^{-1} = \sum_0^\infty (\lambda A)^k = I + \lambda A + \lambda^2 A^2 + \lambda^3 A^3 + \cdots = I + \lambda B$$

where $B = A + \lambda A^2 + \cdots$.

If A is any integral operator of the form

$$(Ah)(s) = \int_0^1 k(s,t)h(t)\,dt$$

then we can define a companion operator B depending on a real parameter λ by the equation

(14) $$B = \lambda^{-1}\big[(I - \lambda A)^{-1} - I\big]$$

Theorem 6.   *The operator B, as just defined, is also an integral operator, having the form*

(15) $$(Bh)(s) = \int_0^1 r(s,t)h(t)\,dt$$

The kernel satisfies these two integral equations

(16) $$\begin{cases} r(s,t) = k(s,t) + \lambda \int_0^1 k(s,u)r(u,t)\,du \\[2mm] r(s,t) = k(s,t) + \lambda \int_0^1 k(u,t)r(s,u)\,du \end{cases}$$

Proof. From the definition of B we have $\lambda B = (I - \lambda A)^{-1} - I$ or $I + \lambda B = (I - \lambda A)^{-1}$. Consequently, we have

$$(I + \lambda B)(I - \lambda A) = (I - \lambda A)(I + \lambda B) = I$$

$$I + \lambda B - \lambda A - \lambda^2 BA = I - \lambda A + \lambda B - \lambda AB = I$$

$$\lambda B - \lambda A - \lambda^2 BA = \lambda B - \lambda A - \lambda^2 AB = 0$$

and

(17) $$B = (I + \lambda B)A = A(I + \lambda B)$$

Conversely, from Equation (17) we can prove Equation (14). Thus Equation (17) serves to characterize B. Now assume that r satisfies Equations (16) and that B is defined by Equation (15). We will show that B must satisfy Equation (17), and hence Equation (14). We have

$$
\begin{aligned}
\left[(A + \lambda BA)h \right](s) &= \int_0^1 k(s,t)h(t)\,dt + \lambda \int r(s,u)(Ah)(u)\,du \\
&= \int_0^1 k(s,t)h(t)\,dt + \lambda \int_0^1 r(s,u) \int_0^1 k(u,t)h(t)\,dt\,du \\
&= \int_0^1 \left\{ k(s,t) + \lambda \int_0^1 r(s,u)k(u,t)\,du \right\} h(t)\,dt \\
&= \int_0^1 r(s,t)h(t)\,dt = (Bh)(s)
\end{aligned}
$$

This proves that $B = A + \lambda BA$. Similarly, $B = A + \lambda AB$. ∎

Example 2. ([Gold], page 160.) Solve the integral equation

$$x(s) - \int_0^1 st\,\operatorname{Arctan} x(t)\,dt = 1 + s^2 - 0.485s$$

This conforms to the general theory outlined above.   We have as kernel $g(s,t,u) = st\,\operatorname{Arctan} u$, and $g_3(s,t,u) = st/(1 + u^2)$. We take as starting point for the Newton iteration the constant function $x_0(t) = 3/2$. Then

$$g(s,t,x_0(t)) = st / \left(1 + \frac{9}{4} \right) = \frac{4}{13} st = \alpha\, st$$

Then $f'(x_0) = I - A$, where $(Ax)(s) = \int_0^1 \alpha st\,x(t)\,dt$. Also we can express $f'(x_0)^{-1} = (I - A)^{-1} = I + B$, as in the preceding proof. We know that B is an integral operator whose kernel, r, satisfies the equations

$$
\begin{cases}
r(s,t) = \alpha\,s\,t + \int_0^1 \alpha\,s\,u\,r(u,t)\,du \\
r(s,t) = \alpha\,s\,t + \int_0^1 \alpha\,t\,u\,r(s,u)\,du
\end{cases}
$$

From these equations it is evident that $r(s,t)/st$ is on the one hand a function of t only, and on the other hand a function of s only. Thus $r(s,t)/st$ is constant, say β, and $r(s,t) = \beta\,s\,t$. Substituting in the integral equation for r and solving gives us $\beta = 12/35$. One step in the Newton algorithm will be $x_1 = x_0 - f'(x_0)^{-1}f(x_0)$. We compute $y = f(x_0)$ as follows:

$$
\begin{aligned}
y(s) &= x_0(s) - \int_0^1 g(s,t,x_0(t))\,dt - v(s) \\
&= \frac{3}{2} - \int_0^1 st\,\operatorname{Arctan} \frac{3}{2}\,dt - 1 - s^2 + .485s \\
&= \frac{1}{2} - .0063968616s - s^2
\end{aligned}
$$

Then $x_1 = x_0 - f'(x_0)^{-1}y = x_0 - (I+B)y = x_0 - y - By$. Hence

$$x_1(s) = \frac{3}{2}\left[\frac{1}{2} - \gamma s - s^2\right] - \int_0^1 \beta st\left[\frac{1}{2} - \gamma t - t^2\right] dt \qquad (\gamma \approx .0063968616)$$
$$= 1 + s^2 + (.0071279315)\,s \qquad\qquad\qquad\qquad \blacksquare$$

Problems 3.3

1. For the one-dimensional version of Newton's method, prove that if r is a root of multiplicity m, then quadratic convergence in the algorithm can be preserved by defining

$$x_{n+1} = x_n - mf(x_n)/f'(x_n)$$

2. Prove the corollaries, giving in each case the precise assumptions that must be made concerning the starting points.

3. Let $e_0, e_1, \dots$ be a sequence of positive numbers satisfying $e_{n+1} \leqslant ce_n^2$. Find necessary and sufficient conditions for the convergence $\lim_n e_n = 0$.

4. Let f be a function from $\mathbb{R}$ to $\mathbb{R}$ that satisfies the inequalities $f' > 0$ and $f'' > 0$. Prove that if f has a zero, then the zero is unique, and Newton's iteration, started at any point, converges to the zero.

5. How must the analysis in Theorem 1 be modified to accommodate functions from $\mathbb{C}$ to $\mathbb{C}$? (Remember that the Mean Value Theorem in its real-variable form is not valid.)

6. If r is a zero of a function f, then the corresponding "basin of attraction" is the set of all x such that the Newton sequence starting at x converges to r. For the function $f(z) = z^2 + 1$, $z \in \mathbb{C}$, and the zero $r = i$, prove that the basin of attraction contains the disk of radius $\frac{2}{3}$ about r.

3.4 Implicit Function Theorems

In this section we give several versions of the Implicit Function Theorem and prove its corollary, the Inverse Function Theorem. Theorems in this broad category are often used to establish the existence of solutions to nonlinear equations of the form $f(x) = y$. The conclusions are typically *local* in nature, and describe how the solution x depends on y in a neighborhood of a given solution (x_0, y_0). Usually, there will be a hypothesis involving invertibility of the derivative $f'(x_0)$.

The intuition gained from examining some simple cases proves to be completely reliable in attacking very general cases. Consider, then, a function $F : \mathbb{R}^2 \to \mathbb{R}$. We ask whether the equation $F(x, y) = 0$ defines y to be a unique function of x. For example, we can ask this question for the equation

$$x + y^2 - 1 = 0 \qquad (x, y \in \mathbb{R})$$

This can be "solved" to yield $y = \sqrt{1 - x}$. The graph of this is shown in the accompanying Figure 3.2. It is clear that we cannot let x be the point A in the figure, because there is no corresponding y for which $F(x, y) \equiv x + y^2 - 1 = 0$. One must start with a point (x_0, y_0) like B in the figure, where we already have $F(x_0, y_0) = 0$. Finally, observe that at the point C there will be a difficulty, for there are values of x near C to which no y's correspond. This is a point where $dy/dx = \infty$. Recall that if $y = y(x)$ and if $F(x, y(x)) = 0$, then y' can be obtained from the equation

$$0 = \frac{d}{dx}F(x, y(x)) = D_1F + (D_2F)y'$$

(In this equation, D_i is partial differentiation with respect to the ith argument.) Thus $y' = -D_1F/D_2F$, and the condition $y'(x_0) = \infty$ corresponds to $D_2F(x_0, y_0) = 0$. In this example, notice that another function arises from Equation (1), namely

$$y = -\sqrt{1-x}$$

In a neighborhood of $(1, 0)$, both functions solve Equation (1), and there is a failure of uniqueness.

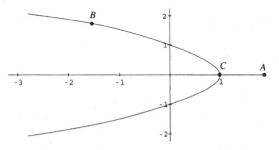

Figure 3.2

In the classical implicit function theorem we have a function F of two real variables in class C^1. That means simply that $\partial F/\partial x$ and $\partial F/\partial y$ exist and are continuous. It is convenient to denote these partial derivatives by F_1 and F_2.

Theorem 1. Classical Implicit Function Theorem.   *Let F be a C^1-function on the square*

$$\{(x, y) \; : \; |x - x_0| \leqslant \delta \, , \quad |y - y_0| \leqslant \delta\} \subset \mathbb{R}^2$$

If $F(x_0, y_0) = 0$ and $F_2(x_0, y_0) \neq 0$, then there is a continuously differentiable function f defined in a neighborhood of x_0 such that $y_0 = f(x_0)$ and $F(x, f(x)) = 0$ in that neighborhood. Furthermore, $f'(x) = -F_1(x, y)/F_2(x, y)$, where $y = f(x)$.

Proof.   Assume that $F_2(x_0, y_0) > 0$. Then by continuity, $F_2(x, y) > \alpha > 0$ in a neighborhood of (x_0, y_0), which we assume to be the original δ-neighborhood. The function $y \mapsto F(x_0, y)$ is strictly increasing for $y_0 - \delta \leqslant y \leqslant y_0 + \delta$. Hence

$$F(x_0, y_0 - \delta) < F(x_0, y_0) = 0 < F(x_0, y_0 + \delta)$$

By continuity there is an ε in $(0, \delta)$ such that $F(x, y_0 - \delta) < 0 < F(x, y_0 + \delta)$ if $|x - x_0| < \varepsilon$. By continuity, there corresponds to each such x a value of y such that $F(x, y) = 0$ and $y_0 - \delta < y < y_0 + \delta$. If there were two such y's, then by Rolle's theorem, $F_2(x, \overline{y}) = 0$ at some point, contrary to hypothesis. Hence y is unique, and we may put $y = f(x)$. Then we have $F(x, f(x)) = 0$ and $y_0 = f(x_0)$.

Now fix x_1 in the ε-neighborhood of x_0. Put $y_1 = f(x_1)$. Let Δx be a small number and let $y_1 + \Delta y = f(x_1 + \Delta x)$. Then

$$0 = F(x_1 + \Delta x, y_1 + \Delta y)$$
$$= F_1(x_1 + \theta \Delta x, y_1 + \theta \Delta y)\Delta x + F_2(x_1 + \theta \Delta x, y_1 + \theta \Delta y)\Delta y$$

for an appropriate θ satisfying $0 \leqslant \theta \leqslant 1$. (This is the Mean Value Theorem for a function from $\mathbb{R}^2$ to $\mathbb{R}$. See Problem 3.2.8, page 124.) This equation gives us

$$\frac{\Delta y}{\Delta x} = -\frac{F_1(x_1 + \theta \Delta x,\ y_1 + \theta \Delta y)}{F_2(x_1 + \theta \Delta x,\ y_1 + \theta \Delta y)}$$

As $\Delta x \to 0$ the right side remains bounded. Hence, so does the left side. This proves that as Δx converges to 0, Δy also converges to 0. Hence f is continuous at x_1. After all, $\Delta y = f(x_1 + \Delta x) - f(x_1)$. Furthermore,

$$f'(x_1) = \lim \frac{\Delta y}{\Delta x} = -\frac{F_1(x_1, y_1)}{F_2(x_1, y_1)}$$

Therefore, f is differentiable at x_1. The formula can be written

$$f'(x) = -F_1\big(x, f(x)\big)/F_2\big(x, f(x)\big)$$

and this shows that f' is continuous at x, provided that x is in the open interval $(x_0 - \varepsilon, x_0 + \varepsilon)$. ∎

Theorem 2. Implicit Function Theorem for Many Variables.
Let $F : \mathbb{R}^n \times \mathbb{R} \to \mathbb{R}$, and suppose that $F(x_0, y_0) = 0$ for some $x_0 \in \mathbb{R}^n$ and $y_0 \in \mathbb{R}$. If all $n+1$ partial derivatives $D_i F$ exist and are continuous in a neighborhood of (x_0, y_0) and if $D_{n+1}F(x_0, y_0) \neq 0$, then there is a continuously differentiable function f defined on a neighborhood of x_0 such that $F(x, f(x)) = 0$, $f(x_0) = y_0$, and

$$D_i f(x) = -D_i F\big(x, f(x)\big)/D_{n+1}F\big(x, f(x)\big) \quad (1 \leqslant i \leqslant n)$$

Proof. This is left as a problem (Problem 3.4.4). ∎

Example 1. $F(x, y) = x^2 + y^2 + 1$ or $x^2 + y^2$ or $x^2 + y^2 - 1$. (Three phenomena are illustrated.) ∎

If we expect to generalize the preceding theorems to normed linear spaces, there will be several difficulties. Of course, division by F_2 will become multiplication by F_2^{-1}, and the invertibility of the Fréchet derivative will have to be hypothesized. A more serious problem occurs in defining the value of y corresponding to x. The order properties of the real line were used in the preceding proofs; in the more general theorems, an appeal to a fixed point theorem will be substituted.

Definition. Let X, Y, Z be Banach spaces. Let $F : X \times Y \to Z$ be a mapping. The Cartesian product $X \times Y$ is also a Banach space if we give it the norm $\|(x, y)\| = \|x\| + \|y\|$. If they exist, the partial derivatives of F at (x_0, y_0) are bounded linear operators $D_1 F(x_0, y_0)$ and $D_2 F(x_0, y_0)$ such that

$$\lim \|F(x_0 + h, y_0) - F(x_0, y_0) - D_1 F(x_0, y_0)h\|/\|h\| = 0 \qquad (h \in X,\ h \to 0)$$

and

$$\lim \|F(x_0, y_0 + k) - F(x_0, y_0) - D_2 F(x_0, y_0)k\|/\|k\| = 0 \qquad (k \in Y,\ k \to 0)$$

Thus $D_1 F(x_0, y_0) \in \mathcal{L}(X, Z)$ and $D_2 F(x_0, y_0) \in \mathcal{L}(Y, Z)$. We often use the notation F_i in place of $D_i F$.

Theorem 3. General Implicit Function Theorem.   *Let* X, Y, *and* Z *be normed linear spaces,* Y *being assumed complete. Let* Ω *be an open set in* $X \times Y$. *Let* $F : \Omega \to Z$. *Let* $(x_0, y_0) \in \Omega$. *Assume that* F *is continuous at* (x_0, y_0), *that* $F(x_0, y_0) = 0$, *that* $D_2 F$ *exists in* Ω, *that* $D_2 F$ *is continuous at* (x_0, y_0), *and that* $D_2 F(x_0, y_0)$ *is invertible. Then there is a function* f *defined on a neighborhood of* x_0 *such that* $F\big(x, f(x)\big) = 0$, $f(x_0) = y_0$, f *is continuous at* x_0, *and* f *is unique in the sense that any other such function must agree with* f *on some neighborhood of* x_0.

Proof.   We can assume that $(x_0, y_0) = (0, 0)$. Select $\delta > 0$ so that

$$\{(x, y) \ : \ \|x\| \leqslant \delta, \ \|y\| \leqslant \delta\} \subset \Omega$$

Put $A = D_2 F(0, 0)$. Then $A \in \mathcal{L}(Y, Z)$ and $A^{-1} \in \mathcal{L}(Z, Y)$. For each x satisfying $\|x\| \leqslant \delta$ we define $G_x(y) = y - A^{-1} F(x, y)$. Here $\|y\| \leqslant \delta$. Observe that if G_x has a fixed point y^*, then

$$y^* = G_x(y^*) = y^* - A^{-1} F(x, y^*)$$

from which we conclude that $F(x, y^*) = 0$. Let us therefore set about proving that G_x has a fixed point. We shall employ the Contraction Mapping Theorem. (Chapter 4, Section 2, page 177). We have

$$G_x'(y) = I - A^{-1} D_2 F(x, y) = A^{-1}\{D_2 F(0, 0) - D_2 F(x, y)\}$$

By the continuity of $D_2 F$ at $(0, 0)$ we can reduce δ if necessary such that

$$\Big[\|x\| \leqslant \delta \ \text{ and } \ \|y\| \leqslant \delta\Big] \ \implies \ \|G_x'(y)\| \leqslant \tfrac{1}{2}$$

Now $G_x(0) = -A^{-1} F(x, 0) = -A^{-1}\{F(x, 0) - F(0, 0)\}$. Let $0 < \varepsilon < \delta$. By the continuity of F at $(0, 0)$ we can find $\delta_\varepsilon \in (0, \delta)$ so that

$$\|x\| \leqslant \delta_\varepsilon \ \implies \ \|G_x(0)\| < \tfrac{1}{2}\varepsilon$$

If $\|x\| \leqslant \delta_\varepsilon$ and $\|y\| \leqslant \varepsilon$, then by the Mean Value Theorem III of Section 2, page 123,

$$\begin{aligned}
\|G_x(y)\| &\leqslant \|G_x(0)\| + \|G_x(y) - G_x(0)\| \\
&\leqslant \tfrac{1}{2}\varepsilon + \sup_{0 \leqslant \lambda \leqslant 1} \|G_x'(\lambda y)\| \cdot \|y\| \\
&\leqslant \frac{\varepsilon}{2} + \frac{\varepsilon}{2} = \varepsilon
\end{aligned}$$

Define $U = \{y \in Y \ : \ \|y\| \leqslant \varepsilon\}$. We have shown that, for each x satisfying $\|x\| \leqslant \delta_\varepsilon$, the function G_x maps U into U. We also know that $\|G_x'(y)\| \leqslant \tfrac{1}{2}$. By Problem 1, G_x has a unique fixed point y in U. Since this fixed point depends on x, we write $y = f(x)$, thus defining f. From the observations above we infer that

$$F\big(x, f(x)\big) = 0$$

Since $F(0,0) = 0$, it follows that $G_0(0) = 0$. By the uniqueness of the fixed point, $0 = f(0)$. Since ε was arbitrary in $(0, \delta)$ we have this conclusion: For each ε in $(0, \delta)$ there is a δ_ε such that

$$\|x\| \leqslant \delta_\varepsilon \quad \Longrightarrow \quad \|G_x(0)\| < \tfrac{1}{2}\varepsilon$$

Our analysis then showed that $y = f(x) \in U$, or $\|f(x)\| \leqslant \varepsilon$. As a consequence, $\|x\| \leqslant \delta_\varepsilon \implies \|f(x)\| \leqslant \varepsilon$, showing continuity at 0. For the uniqueness, suppose that $\overline{f}$ is another function defined on a neighborhood of 0 such that $\overline{f}$ is continuous at 0, $\overline{f}(0) = 0$, and $F(x, \overline{f}(x)) = 0$. If $0 < \varepsilon < \delta$, find $\theta > 0$ such that $\theta < \delta_\varepsilon$ and

$$\|x\| \leqslant \theta \quad \Longrightarrow \quad \|\overline{f}(x)\| \leqslant \varepsilon$$

Then $\overline{f}(x) \in U$. So we have apparently two fixed points, $f(x)$ and $\overline{f}(x)$, for the function G_x. Since this is not possible, $f(x) = \overline{f}(x)$ whenever $\|x\| \leqslant \theta$. ∎

Theorem 4. Second Version of the Implicit Function Theorem.
In the preceding theorem, assume further that F is continuously differentiable in Ω and that $D_2F(x_0, y_0)$ is invertible. Then the function f will be continuously differentiable and

$$f'(x) = -D_2F(x, f(x))^{-1} \circ D_1F(x, f(x))$$

Furthermore, there will exist a neighborhood of x_0 in which f is unique.

This theorem can be found in [Dieu], page 265.

Theorem 5. Inverse Function Theorem I.   *Let f be a continuously differentiable map from an open set Ω in a Banach space into a normed linear space. If $x_0 \in \Omega$ and if $f'(x_0)$ is invertible, then there is a continuously differentiable function g defined on a neighborhood $\mathcal{N}$ of $f(x_0)$ such that $f(g(y)) = y$ for all $y \in \mathcal{N}$.*

Proof.   For x in Ω and y in the second space, define $F(x, y) = f(x) - y$. Put $y_0 = f(x_0)$ so that $F(x_0, y_0) = 0$. Note that $D_1F(x, y) = f'(x)$, and thus $D_1F(x_0, y_0)$ is invertible. By Theorem 4, there is a neighborhood $\mathcal{N}$ of y_0 and a continuously differentiable function g defined on $\mathcal{N}$ such that $F(g(y), y) = 0$, or $f(g(y)) - y = 0$ for all $y \in \mathcal{N}$. ∎

Theorem 6. Surjective Mapping Theorem I.   *Let X and Y be Banach spaces, Ω an open set in X. Let $f : \Omega \to Y$ be a continuously differentiable map. Let $x_0 \in \Omega$ and $y_0 = f(x_0)$. If $f'(x_0)$ is invertible, as an element of $\mathcal{L}(X, Y)$, then $f(\Omega)$ is a neighborhood of y_0.*

Proof.   Define $F : \Omega \times Y \to Y$ by putting $F(x, y) = f(x) - y$. Then $F(x_0, y_0) = 0$ and $D_1F(x_0, y_0) = f'(x_0)$. (D_1 is a partial derivative, as defined previously.) By hypothesis, $D_1F(x_0, y_0)$ is invertible. By the Implicit Function Theorem (with the rôles of x and y reversed!), there exist a neighborhood $\mathcal{N}$ of y_0 and a function $g : \mathcal{N} \to \Omega$ such that $g(y_0) = x_0$ and $F(g(y), y) = 0$ for all $y \in \mathcal{N}$. From the definition of F we have $f(g(y)) - y = 0$ for all $y \in \mathcal{N}$. In other words, each element y of $\mathcal{N}$ is the image under f of some point in Ω, namely, $g(y)$. ∎

Theorem 7. A Fixed Point Theorem.   *Let Ω be an open set in a Banach space X, and let G be a differentiable map from Ω to X. Suppose that there is a closed ball $B \equiv B(x_0, r)$ in Ω such that*

(i) $k \equiv \sup\limits_{x \in B} \left\| G'(x) \right\| < 1$

(ii) $\left\| G(x_0) - x_0 \right\| < r(1 - k)$

Then G has a unique fixed point in B.

Proof.   First, we show that $G|B$ is a contraction. If x_1 and x_2 are in B, then by the Mean Value Theorem (Theorem 4 in Section 3.2, page 123)

$$\left\| G(x_1) - G(x_2) \right\| \leqslant \sup\limits_{0 \leqslant \lambda \leqslant 1} \left\| G'(x_1 + \lambda(x_2 - x_1)) \right\| \left\| x_1 - x_2 \right\|$$
$$\leqslant k \left\| x_1 - x_2 \right\|$$

Second, we show that G maps B into B. If $x \in B$, then

$$\left\| G(x) - x_0 \right\| \leqslant \left\| G(x) - G(x_0) \right\| + \left\| G(x_0) - x_0 \right\|$$
$$\leqslant k \left\| x - x_0 \right\| + r(1 - k)$$
$$\leqslant kr + (1 - k)r = r$$

Since X is complete, B is a complete metric space. By the Contractive Mapping Theorem (page 177), G has a unique fixed point in B.   ∎

Theorem 8. Inverse Function Theorem II.   *Let Ω be an open set in a Banach space X. Let f be a differentiable map from Ω to a normed space Y. Assume that Ω contains a closed ball $B \equiv B(x_0, r)$ such that*

(i) *The linear transformation $A \equiv f'(x_0)$ is invertible.*

(ii) $k \equiv \sup_{x \in B} \left\| I - A^{-1} f'(x) \right\| < 1$

Then for each y in Y satisfying $\left\| y - f(x_0) \right\| < (1 - k)r \left\| A^{-1} \right\|^{-1}$ the equation $f(x) = y$ has a unique solution in B.

Proof.   Let y be as hypothesized, and define $G(x) = x - A^{-1}[f(x) - y]$. It is clear that $f(x) = y$ if and only if x is a fixed point of G. The map G is differentiable in Ω, and $G'(x) = I - A^{-1} f'(x)$. To verify the hypothesis (i) in the preceding theorem, write

$$\left\| G'(x) \right\| = \left\| I - A^{-1} f'(x) \right\| \leqslant k \qquad (x \in B)$$

By the assumptions made about y, we can verify hypothesis (ii) of the preceding theorem by writing

$$\left\| G(x_0) - x_0 \right\| = \left\| x_0 - A^{-1}[f(x_0) - y] - x_0 \right\|$$
$$= \left\| A^{-1} \right\| \left\| f(x_0) - y \right\|$$
$$\leqslant \left\| A^{-1} \right\| (1 - k)r \left\| A^{-1} \right\|^{-1} = (1 - k)r$$

By the preceding theorem, G has a unique fixed point in B, which is the unique solution of $f(x) = y$ in B.   ∎

Example 2.   Consider a nonlinear Volterra integral equation

$$x(t) - 2x(0) + \tfrac{1}{2}\int_0^t \cos(st)[x(s)]^2\, ds = y(t) \qquad (0 \leqslant t \leqslant 1)$$

in which $y \in C[0,1]$. Notice that when $y = 0$ the integral equation has the solution $x = 0$. We ask: Does the equation have solutions when $\|y\|$ is small? Here, we use the usual sup-norm on $C[0,1]$, as this makes the space complete. (Weighted sup-norms would have this property, too.) Write the integral equation as $f(x) = y$, where f has the obvious interpretation. Then $f'(x)$ is given by

$$[f'(x)h](t) = h(t) - 2h(0) + \int_0^t \cos(st)x(s)h(s)\, ds$$

Let $A = f'(0)$, so that $Ah = h - 2h(0)$. One verifies easily that $A^2 h = h$, from which it follows that $A^{-1} = A$. In order to use the preceding theorem, with $x_0 = 0$, we must verify its hypotheses. We have just seen that A is invertible. Let $\|x\| \leqslant r$, where r is to be chosen later so that $\|I - A^{-1}f'(x)\| \leqslant k < 1$. From an equation above,

$$\left| [f'(x)h](t) - (Ah)(t) \right| = \left| \int_0^t \cos(st)x(s)h(s)\, ds \right|$$
$$\leqslant \|h\|\,\|x\|$$

It follows that
$$\|f'(x)h - Ah\| \leqslant \|h\|\,\|x\|$$
and that
$$\|f'(x) - A\| \leqslant \|x\| \leqslant r$$
Since $\|A\| = \|A^{-1}\| = 3$, we have

$$\|I - A^{-1}f'(x)\| = \|A^{-1}(A - f'(x))\| \leqslant \|A^{-1}\|r = 3r$$

The hypothesis of the preceding theorem requires that $3r \leqslant k < 1$, where k is to be chosen later. By the preceding theorem, the equation $f(x) = y$ will have a unique solution if

$$\|y\| < (1-k)r\|A^{-1}\|^{-1} \leqslant \tfrac{1}{3}(1-k)\tfrac{k}{3}$$

In order for this bound to be as generous as possible, we let $k = \tfrac{1}{2}$, arriving at the restriction $\|y\| < \tfrac{1}{36}$.   ■

Lemma.   *Let X and Y be Banach spaces. Let Ω be an open set in X, and let $f : \Omega \to Y$ be a continuously differentiable mapping. If $x_0 \in \Omega$ and $\varepsilon > 0$, then there is a $\delta > 0$ such that*

$$\|x_1 - x_0\| < \delta,\ \|x_2 - x_0\| < \delta \implies \|f(x_1) - f(x_2) - f'(x_0)(x_1 - x_2)\| < \varepsilon\|x_1 - x_2\|$$

Proof.   The map $x \mapsto f'(x)$ is continuous from Ω to $\mathcal{L}(X,Y)$. Therefore, in correspondence with the given ε, there is a $\delta > 0$ such that

$$\|x - x_0\| < \delta \implies \|f'(x) - f'(x_0)\| < \varepsilon$$

(We may assume also that $B(x_0, \delta) \subset \Omega$.) If $\|x_1 - x_0\| < \delta$ and $\|x_2 - x_0\| < \delta$, then the line segment S joining x_1 to x_2 satisfies $S \subset B(x_0, S) \subset \Omega$. By Problem 2, page 145, we have

$$\|f(x_1) - f(x_2) - f'(x_0)(x_1 - x_2)\| \leqslant \|x_1 - x_2\| \cdot \sup_{x \in S} \|f'(x) - f'(x_0)\|$$

$$\leqslant \epsilon \|x_1 - x_2\| \qquad \blacksquare$$

Theorem 9. Surjective Mapping Theorem II. *Let X and Y be Banach spaces. Let Ω be an open set in X. Let $f : \Omega \to Y$ be continuously differentiable. If $x_0 \in \Omega$ and $f'(x_0)$ has a right inverse in $\mathcal{L}(Y, X)$, then $f(\Omega)$ is a neighborhood of $f(x_0)$.*

Proof. Put $A = f'(x_0)$ and let L be a member of $\mathcal{L}(Y, X)$ such that $AL = I$, where I denotes the identity map on Y. Let $c = \|L\|$. By the preceding lemma, there exists $\delta > 0$ such that $B(x_0, \delta) \subset \Omega$ and such that

$$\|u - x_0\| \leqslant \delta, \ \|v - x_0\| \leqslant \delta \Longrightarrow \|f(u) - f(v) - A(u - v)\| \leqslant \tfrac{1}{2c} \|u - v\|$$

Let $y_0 = f(x_0)$ and $y \in B(y_0, \ \delta/2c)$. We will find $x \in \Omega$ such that $f(x) = y$. The point x is constructed as the limit of a sequence $\{x_n\}$ defined inductively as follows. We start with the given x_0. Put $x_1 = x_0 + L(y - y_0)$. From then on we define

$$x_{n+1} = x_n - L\big[f(x_n) - f(x_{n-1}) - A(x_n - x_{n-1})\big]$$

By induction we establish that $\|x_n - x_{n-1}\| \leqslant \delta/2^n$ and $\|x_n - x_0\| \leqslant \delta$. Here are the details of the induction:

$$\|x_1 - x_0\| = \|L(y - y_0)\| \leqslant c\|y - y_0\| \leqslant c\delta/(2c) = \delta/2 \ .$$

$$\|x_{n+1} - x_n\| \leqslant c\|f(x_n) - f(x_{n-1}) - A(x_n - x_{n-1})\|$$

$$\leqslant c(1/2c)\|x_n - x_{n-1}\| \leqslant \delta/2^{n+1}$$

$$\|x_{n+1} - x_0\| \leqslant \|x_{n+1} - x_n\| + \|x_n - x_{n-1}\| + \cdots + \|x_1 - x_0\|$$

$$\leqslant \frac{\delta}{2^{n+1}} + \frac{\delta}{2^n} + \cdots + \frac{\delta}{2} \leqslant \delta$$

Next we observe that the sequence $[x_n]$ has the Cauchy property, since (for $m > n$)

$$\|x_n - x_m\| \leqslant \|x_n - x_{n+1}\| + \cdots + \|x_{m-1} - x_m\| \leqslant \delta\left(\frac{1}{2^{n+1}} + \frac{1}{2^{n+2}} + \cdots\right) \leqslant \delta/2^n$$

Since X is complete, we can define $x = \lim x_n$. All that remains is to prove $x \in \Omega$ and $f(x) = y$. Since $\|x_n - x_0\| \leqslant \delta$, we have $\|x - x_0\| \leqslant \delta$ and $x \in \Omega$. From the equation defining x_{n+1} we have

$$A(x_{n+1} - x_n) = -AL\{f(x_n) - f(x_{n-1}) - A(x_n - x_{n-1})\}$$

$$= A(x_n - x_{n-1}) - \{f(x_n) - f(x_{n-1})\}$$

By using this equation recursively we reach finally

$$A(x_{n+1} - x_n) = A(x_1 - x_0) - \{f(x_n) - f(x_{n-1})\}$$

$$- \{f(x_{n-1}) - f(x_{n-2})\} - \cdots - \{f(x_1) - f(x_0)\}$$

$$= AL(y - y_0) - f(x_n) + f(x_0)$$

$$= y - y_0 - f(x_n) + y_0 = y - f(x_n)$$

Let $n \to \infty$ in this equation to get $0 = y - f(x)$. $\qquad \blacksquare$

Corollary.   *Let f be a continuously differentiable function mapping an open set Ω in a Banach space into a finite-dimensional Banach space. If $x_0 \in \Omega$ and if $f'(x_0)$ is surjective, then $f(\Omega)$ is a neighborhood of $f(x_0)$.*

Proof. By comparing this assertion to the preceding theorem, we see that it will suffice to prove that $f'(x_0)$ has a right inverse. It suffices to give the proof in the case that f maps Ω into Euclidean n-space $\mathbb{R}^n$, because all finite-dimensional Banach spaces are topologically equivalent. Let $\{e_1, \ldots, e_n\}$ be the usual basis for $\mathbb{R}^n$. Let X be the Banach space containing Ω, and set $A = f'(x)$. Since A is surjective, there exist points $u_1, \ldots, u_n$ in X such that $Au_i = e_i$. Define $L : \mathbb{R}^n \to X$ by requiring $Le_i = u_i$ (and that L be linear, of course). Obviously, $ALe_i = e_i$, so $ALy = y$ for all $y \in \mathbb{R}^n$. Also, $\|L\| \leqslant (\sum \|u_i\|^2)^{1/2}$ because the Cauchy–Schwarz inequality yields (with $y = \sum c_i e_i$)

$$\|Ly\| = \left\| L \sum c_i e_i \right\| = \left\| \sum c_i Le_i \right\| = \left\| \sum c_i u_i \right\| \leqslant \sum |c_i| \, \|u_i\|$$

$$\leqslant \left(\sum c_i^2 \right)^{1/2} \left(\sum \|u_i\|^2 \right)^{1/2} = \|y\| \left(\sum \|u_i\|^2 \right)^{1/2} \qquad \blacksquare$$

Example 3.   Let $f : \mathbb{R}^3 \to \mathbb{R}^3$ be given by

$$f(x) = y \qquad x = (\xi_1, \xi_2, \xi_3) \qquad y = (\eta_1, \eta_2, \eta_3)$$
$$\eta_1 = 2\xi_1^4 + \xi_3 \cos \xi_2 - \xi_1 \xi_3$$
$$\eta_2 = (\xi_1 + \xi_3)^3 - 4 \sin \xi_2$$
$$\eta_3 = \log(\xi_2 + 1) + 5\xi_1 + \cos \xi_3 - 1$$

Notice that $f(0) = 0$. We ask: For y close to zero is there an x for which $f(x) = y$? To answer this, one can use the Inverse Function Theorem. We compute the Fréchet derivative or Jacobian:

$$f'(x) = \begin{bmatrix} 8\xi_1^3 - \xi_3 & -\xi_3 \sin \xi_2 & \cos \xi_2 - \xi_1 \\ 3(\xi_1 + \xi_3)^2 & -4 \cos \xi_2 & 3(\xi_1 + \xi_3)^2 \\ 5 & (\xi_2 + 1)^{-1} & -\sin \xi_3 \end{bmatrix}$$

At $x = 0$ we have

$$f'(0) = \begin{bmatrix} 0 & 0 & 1 \\ 0 & -4 & 0 \\ 5 & 1 & 0 \end{bmatrix}$$

Obviously, $f'(0)$ is invertible, and so we can conclude that in some neighborhood of $y = 0$ there is defined a function g such that $f(g(y)) = y$. $\blacksquare$

The **direct sum** of n normed linear spaces $X_1, X_2, \ldots, X_n$ is denoted by $\sum_{i=1}^{n} \oplus X_i$. Its elements are n-tuples $x = (x_1, x_2, \ldots, x_n)$, where $x_i \in X_i$ for $i = 1, 2, \ldots, n$. Although many definitions of the norm are suitable, we use $\|x\| = \sum_{i=1}^{n} \|x_i\|$.

If $X = \sum_{i=1}^{n} \oplus X_i$ and if f is a mapping from an open set of X into a normed space Y, then the *partial derivatives* $D_i f(x)$, if they exist, are continuous linear maps $(D_i f)(x) \in \mathcal{L}(X_i, Y)$ such that

$$\|f(x_1, \ldots, x_{i-1}, x_i + h, x_{i+1}, \ldots, x_n) - f(x) - (D_i f)(x)h\| = o(\|h\|)$$

The connection between partial derivatives and a "total" derivative is as one expects from multivariable calculus. That relationship is formalized next.

Theorem 10.   Let f be defined on an open set Ω in the direct-sum space $X = \sum_{i=1}^{n} \oplus X_i$ and take values in a normed space Y. Assume that all the partial derivatives $D_i f$ exist in Ω and are continuous at a point x in Ω. Then f is Fréchet differentiable at x, and its Fréchet derivative is given by

(1) $$f'(x)h = \sum_{i=1}^{n} D_i f(x)h_i \qquad (h \in X)$$

Proof.   Equation (1) defines a linear transformation from X to Y, and

$$\|f'(x)h\| \leqslant \sum_{i=1}^{n} \|D_i f(x)h_i\| \leqslant \sum_{i=1}^{n} \|D_i f(x)\| \, \|h_i\|$$

$$\leqslant \max_{1 \leqslant j \leqslant n} \|D_j f(x)\| \sum_{i=1}^{n} \|h_i\|$$

$$= \max_{1 \leqslant j \leqslant n} \|D_j f(x)\| \, \|h\|$$

Thus Equation (1) defines a bounded linear transformation. Let

$$G(h) = f(x + h) - f(x) - \sum_{i=1}^{n} D_i f(x)h_i$$

We want to prove that $\|G(h)\| = o(\|h\|)$. For sufficiently small h, $x + h$ is in Ω, and the partial derivatives of G exist at $x + h$. They are

$$D_i G(h) = D_i f(x + h) - D_i f(x)$$

If ε is a given positive number, we use the assumed continuity of $D_i f$ at x to find a positive δ such that for $\|h\| < \delta$ we have $\|D_i G(h)\| < \varepsilon$, for $1 \leqslant i \leqslant n$. Then, by the mean value theorem,

$$\|G(h)\| \leqslant \|G(h_1, h_2, \ldots, h_n) - G(0, h_2, \ldots, h_n)\|$$
$$+ \|G(0, h_2, \ldots, h_n) - G(0, 0, h_3, \ldots, h_n)\|$$
$$+ \cdots + \|G(0, 0, \ldots, h_n) - G(0, 0, \ldots, 0)\|$$
$$\leqslant \sum_{i=1}^{n} \varepsilon \|h_i\| = \varepsilon \|h\|$$

Since ε was arbitrary, this shows that $\|G(h)\| = o(\|h\|)$.   ∎

Problems 3.4

1. Let U be a closed ball in a Banach space. Let $F : U^+ \to U$, where U^+ is an open set containing U. Prove that if $\sup\{\|F'(x)\| : x \in U\} < 1$, then F has a unique fixed point in U.

2. Let X be a Banach space and $\mathcal{O}$ an open set in X containing a, b, x_0, and the line segment S joining a to b. Prove that

$$\|f(b) - f(a) - f'(x_0)(b - a)\| \leqslant \|b - a\| \sup_{x \in S} \|f'(x) - f'(x_0)\| .$$

[Suggestion: Use the function $g(x) = f(x) - f'(x_0)x$.] Determine whether the same inequality is true when $f'(x_0)$ is replaced by an arbitrary linear operator. In this problem, $f : X \to Y$, where Y is any normed space.

3. Suppose $F(x_0, y_0) = 0$. If x_1 is close to x_0, there should be a y_1 such that $F(x_1, y_1) = 0$. Show how Newton's method can be used to obtain y_1. (Here $F : X \times Y \to Z$, and X, Y, Z are Banach spaces.)

4. Prove Theorem 2.

5. Let $f : \Omega \to Y$ be a continuously differentiable map, where Ω is an open set in a Banach space, and Y is a normed linear space. Assume that $f'(x)$ is invertible for each $x \in \Omega$, and prove that $f(\Omega)$ is open.

6. Let α be the point in $[0, 1]$ where $\cos \alpha = \alpha$. Define X to be the vector space of all continuously differentiable functions on $[0, 1]$ that vanish at the point α. Define a norm on X by writing $\|x\| = \sup_{0 \leqslant t \leqslant 1} |x'(t)|$. Prove that there exists a positive number δ such that if $y \in X$ and $\|y\| < \delta$, then there exists an $x \in X$ satisfying

$$\sin \circ x + x \circ \cos = y$$

7. Let f be a continuous map from an open set Ω in a Banach space X into a Banach space Y. Suppose that for some x_0 in Ω, $f'(x_0)$ exists and is invertible. Prove that f is one-to-one in some neighborhood of x_0.

8. In Example 2, with the nonlinear integral equation, show that the mapping $x \mapsto f'(x)$ is continuous; indeed, it satisfies a Lipschitz condition.

9. Rework Example 2 when the term $2x(0)$ is replaced by $\alpha x(0)$, for an arbitrary constant α. In particular, treat the case when $\alpha = 0$.

3.5 Extremum Problems and Lagrange Multipliers

A *minimum* point of a real-valued function f defined on a set M is a point x_0 such that $f(x_0) \leqslant f(x)$ for all $x \in M$. If M has a topology, then the concept of *relative* minimum point is defined as a point $x_0 \in M$ such that for some neighborhood $\mathcal{N}$ of x_0 we have $f(x_0) \leqslant f(x)$, for all x in $\mathcal{N}$.

Theorem 1. Necessary Condition for Extremum.   Let Ω be *an open set in a normed linear space, and let $f : \Omega \to \mathbb{R}$. If x_0 is a minimum point of f and if $f'(x_0)$ exists, then $f'(x_0) = 0$.*

Proof. Let X be the Banach space, and assume $f'(x_0) \neq 0$. Then there exists $v \in X$ such that $f'(x_0)v = -1$. By the definition of $f'(x_0)$ we can take $\lambda > 0$ and so small that $x_0 + \lambda v$ is in Ω and

$$|f(x_0 + \lambda v) - f(x_0) - \lambda f'(x_0)v| \,/\, \lambda\|v\| < (2\|v\|)^{-1}$$

This means that $\frac{1}{\lambda}[f(x_0 + \lambda v) - f(x_0)]$ is within distance $\frac{1}{2}$ from -1, and so is negative. This implies $f(x_0 + \lambda v) < f(x_0)$. ∎

In this section we will be concerned mostly with *constrained* extremum problems. A simple illustrative case is the following. We have two nice functions, f and g, on $\mathbb{R}^2$ to $\mathbb{R}$. We put $M = \{(x, y) \,:\, g(x, y) = 0\}$ and seek an extremum of $f|M$. (That means f restricted to M.) If the equation $g(x, y) = 0$ defines y as a function of x, say $y = y(x)$, then we can look for an *unrestricted* extremum of $\phi(x) = f(x, y(x))$. Hence we try to solve the equation $\phi'(x) = 0$. This leads to

$$\begin{aligned} 0 &= f_1(x, y(x)) + f_2(x, y(x))y'(x) \\ &= f_1(x, y(x)) - f_2(x, y(x))g_1(x, y(x))/g_2(x, y(x)) \end{aligned}$$

Thus we must solve simultaneously

(1) $g(x, y) = 0$ and $f_1(x, y) - f_2(x, y)g_1(x, y)/g_2(x, y) = 0$

The method of Lagrange multipliers introduces the function

$$H(x, y, \lambda) = f(x, y) + \lambda g(x, y)$$

and solves simultaneously $H_1 = H_2 = H_3 = 0$. Thus

$$f_1(x, y) + \lambda g_1(x, y) = f_2(x, y) + \lambda g_2(x, y) = g(x, y) = 0$$

If $g_2(x, y) \neq 0$, then $\lambda = -f_2(x, y)/g_2(x, y)$, and we recover system (1). The method of Lagrange multipliers treats x and y symmetrically, and includes both cases of the implicit function theorem. Thus y can be a differentiable function of x, or x can be a differentiable function of y.

Example 1.   Let f and g be functions from $\mathbb{R}^2$ to $\mathbb{R}$ defined by $f(x, y) = x^2 + y^2$, $g(x, y) = x - y + 1$. The set $M = \{(x, y) : g(x, y) = 0\}$ is the straight line shown in Figure 3.3. Also shown are some level sets of f, i.e., sets of the type $\{(x, y) : f(x, y) = c\}$. At the solution, the gradient of f is parallel to the gradient of g. The function H is $H(x, y, \lambda) = x^2 + y^2 + \lambda(x - y + 1)$, and the three equations to be solved are $2x + \lambda = 2y - \lambda = x - y + 1 = 0$. The solution is $\left(-\frac{1}{2}, \frac{1}{2}\right)$. ∎

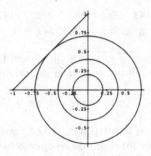

Figure 3.3

Example 2.   Let $f(x, y) = x^2 - y^2$ and $g(x, y) = x^2 + y^2 - 1$. Again we show M and some level sets of f, which are hyperbolas and straight lines. There are four extrema; some are maxima and some are minima. Which are which? The H-function is $H = x^2 - y^2 + \lambda(x^2 + y^2 - 1)$, and the three equations to solve are $2x + 2\lambda x = -2y + 2\lambda y = x^2 + y^2 - 1 = 0$. The (x, y, λ) solutions are $(0, 1, 1)$, $(0, -1, 1)$, $(1, 0, -1)$, $(-1, 0, -1)$. Figure 3.4 is pertinent.   ∎

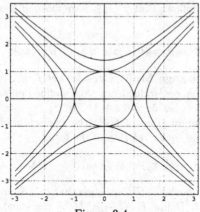

Figure 3.4

If there are several constraint functions, there will be several Lagrange multipliers, as in the next example.

Example 3.   Find the minimum distance from a point to a line in $\mathbb{R}^3$. Let the line be given as the intersection of two planes whose equations are $\langle a, x \rangle = k$ and $\langle b, x \rangle = \ell$. (Here, x, a, and b belong to $\mathbb{R}^3$.) Let the point be c. Then H should be

$$\|x - c\|^2 + \lambda[\langle a, x \rangle - k] + \mu[\langle b, x \rangle - \ell]$$

This H is a function of $(x_1, x_2, x_3, \lambda, \mu)$. The five equations to solve are

$$2(x_1 - c_1) + \lambda a_1 + \mu b_1 = 2(x_2 - c_2) + \lambda a_2 + \mu b_2 = 2(x_3 - c_3) + \lambda a_3 + \mu b_3 = 0$$
$$\langle a, x \rangle - k = \langle b, x \rangle - \ell = 0$$

We see that x is of the form $x = c + \alpha a + \beta b$. When this is substituted in the second set of equations, we obtain two linear equations for determining α and β:

$$\langle a, a \rangle \alpha + \langle a, b \rangle \beta = k - \langle a, c \rangle \quad \text{and} \quad \langle a, b \rangle \alpha + \langle b, b \rangle \beta = \ell - \langle b, c \rangle \qquad ∎$$

Theorem 2. Lagrange Multiplier. *Let f and g be continuously differentiable real-valued functions on an open set Ω in a Banach space. Let $M = \{x \in \Omega : g(x) = 0\}$. If x_0 is a local minimum point of $f|M$ and if $g'(x_0) \neq 0$, then $f'(x_0) = \lambda g'(x_0)$ for some $\lambda \in \mathbb{R}$.*

Proof. Let X be the Banach space in question. Select a neighborhood U of x_0 such that

$$x \in U \cap M \Longrightarrow f(x_0) \leqslant f(x)$$

We can assume $U \subset \Omega$. Define $F : U \to \mathbb{R}^2$ by $F(x) = (f(x), g(x))$. Then $F(x_0) = (f(x_0), 0)$ and $F'(x)v = (f'(x)v, g'(x)v)$ for all $v \in X$. Observe that if $r < f(x_0)$, then $(r, 0)$ is *not* in $F(U)$. Hence $F(U)$ is not a neighborhood of $F(x_0)$. By the Corollary in Section 4.4, $F'(x_0)$ is *not* surjective (as a linear map from X to $\mathbb{R}^2$). Hence $F'(x_0)v = \alpha(v)(\theta, \mu)$ for some continuous linear functional α. (Thus $\alpha \in X^*$.) It follows that $f'(x_0)v = \alpha(v)\theta$ and $g'(x_0)v = \alpha(v)\mu$. Since $g'(x_0) \neq 0$, $\mu \neq 0$. Therefore,

$$f'(x_0)v = (\theta/\mu)\alpha(v)\mu = (\theta/\mu)g'(x_0)v \qquad ∎$$

Theorem 3. Lagrange Multipliers. *Let $f, g_1, \ldots, g_n$ be continuously differentiable real-valued functions defined on an open set Ω in a Banach space X. Let $M = \{x \in \Omega : g_1(x) = \cdots = g_n(x) = 0\}$. If x_0 is a local minimum point of $f|M$ (the restriction of f to M), then there is a nontrivial linear relation of the form*

$$\mu f'(x_0) + \lambda_1 g_1'(x_0) + \lambda_2 g_2'(x_0) + \cdots + \lambda_n g_n'(x_0) = 0$$

Proof. Select a neighborhood U of x_0 such that $U \subset \Omega$ and such that $f(x_0) \leqslant f(x)$ for all $x \in U \cap M$. Define $F : U \to \mathbb{R}^{n+1}$ by the equation

$$F(x) = (f(x), g_1(x), g_2(x), \ldots, g_n(x))$$

If $r < f(x_0)$, then the point $(r, 0, 0, \ldots, 0)$ is *not* in $F(U)$. Thus $F(U)$ does *not* contain a neighborhood of the point $(f(x_0), g_1(x_0), \ldots, g_n(x_0)) \equiv (f(x_0), 0, 0, \ldots, 0)$. By the Corollary in Section 3.4, page 143, $F'(x_0)$ is *not* surjective. Since the range of $F'(x_0)$ is a linear subspace of $\mathbb{R}^{n+1}$, we now know that it is a *proper* subspace of $\mathbb{R}^{n+1}$. Hence it is contained in a hyperplane through the origin. This means that for some $\mu, \lambda_1, \ldots, \lambda_n$ (not all zero) we have

$$\mu f'(x_0)v + \lambda_1 g_1'(x_0)v + \cdots + \lambda_n g_n'(x_0)v = 0$$

for all $v \in X$. This implies the equation in the statement of the theorem. ∎

Example 4.   Let A be a compact Hermitian operator on a Hilbert space X. Then $\|A\| = \max\{|\lambda| : \lambda \in \Lambda(A)\}$, where $\Lambda(A)$ is the set of eigenvalues of A. This is proved by Lemma 2, page 92, together with Problem 22, page 101. Then by Lemma 2 in Section 2.3, page 85, we have $\|A\| = \sup\{|\langle Ax, x\rangle| : \|x\| = 1\}$. Hence we can find an eigenvalue of A by determining an extremum of $\langle Ax, x\rangle$ on the set defined by $\|x\| = 1$. An alternative is given by the next result. ∎

> **Lemma.**   If A is Hermitian, then the "Rayleigh Quotient" $f(x) = \langle Ax, x\rangle/\langle x, x\rangle$ has a stationary value at each eigenvector.

Proof.   Let $Ax = \lambda x$, $x \neq 0$. Then $f(x) = \langle Ax, x\rangle/\langle x, x\rangle = \lambda$. Recall that the eigenvalues of a Hermitian operator are real. Let us compute the derivative of f at x and show that it is 0.

$$\lim_{h \to 0} |f(x+h) - f(x)|/\|h\| = \lim \left| \frac{\langle Ax+Ah, x+h\rangle}{\langle x+h, x+h\rangle} - \lambda \right|/\|h\|$$
$$= \lim |\langle Ax,x\rangle + \langle Ah,x\rangle + \langle Ax,h\rangle + \langle Ah,h\rangle - \lambda\|x+h\|^2|/\|h\| \; \|x+h\|^2$$
$$= \lim |\langle h,Ax\rangle + \lambda\langle x,h\rangle + \langle Ah,h\rangle - 2\lambda\mathrm{Re}\langle x,h\rangle - \lambda\langle h,h\rangle|/\|h\| \; \|x+h\|^2$$
$$= \lim |\lambda\langle h,x\rangle + \lambda\langle x,h\rangle + \langle Ah,h\rangle - 2\lambda\mathrm{Re}\langle x,h\rangle - \lambda\langle h,h\rangle|/\|h\| \; \|x+h\|^2$$
$$= \lim |\langle Ah,h\rangle - \lambda\langle h,h\rangle|/\|h\| \; \|x+h\|^2$$
$$= \lim |\langle Ah - \lambda h, h\rangle|/\|h\| \; \|x+h\|^2$$
$$\leqslant \lim \|Ah - \lambda h\| \; \|h\|/\|h\| \; \|x+h\|^2$$
$$\leqslant \lim \|A - \lambda I\| \; \|h\|/\|x+h\|^2 = 0$$

Thus from the definition of $f'(x)$ as the operator that makes the equation

$$\lim_{h \to 0} |f(x+h) - f(x) - f'(x)h| \, / \, \|h\| = 0$$

true, we have $f'(x) = 0$. ∎

Since the Rayleigh quotient can be written as

$$\frac{\langle Ax, x\rangle}{\|x\|^2} = \left\langle A\left(\frac{x}{\|x\|}\right), \frac{x}{\|x\|}\right\rangle$$

it is possible to consider the simpler function $F(x) = \langle Ax, x\rangle$ restricted to the unit sphere.

> **Theorem 4.**   If A is a Hermitian operator on a Hilbert space, then each local constrained minimum or maximum point of $\langle Ax, x\rangle$ on the unit sphere is an eigenvector of A. The value of $\langle Ax, x\rangle$ is the corresponding eigenvalue.

Proof.   Use $F(x) = \langle Ax, x\rangle$ and $G(x) = \|x\|^2 - 1$. Then

$$F'(x)h = 2\langle Ax, h\rangle \qquad G'(x)h = 2\langle x, h\rangle$$

Our theorem about Lagrange multipliers gives a necessary condition in order that x be a local extremum, namely that $\mu F'(x) + \lambda G'(x) = 0$ in a nontrivial manner. Since $\|x\| = 1$, $G'(x) \neq 0$. Hence $\mu \neq 0$, and by the homogeneity we can set $\mu = -1$. This leads to

$$-2\langle Ax, h \rangle + 2\lambda \langle x, h \rangle = 0 \qquad (h \in X)$$

whence $Ax = \lambda x$.   ∎

Extremum problems with *inequality* constraints can also be discussed in a general setting free of dimensionality restrictions. This leads to the so-called Kuhn–Tucker Theory.

Inequalities in a vector space require some elucidation. An **ordered vector space** is a pair $(X, \geqslant)$ in which X is a real vector space and $\geqslant$ is a partial order in X that is consistent with the linear structure. This means simply that

$$x \geqslant y \quad \Longrightarrow \quad x + z \geqslant y + z$$
$$x \geqslant y \,, \; \lambda \geqslant 0 \quad \Longrightarrow \quad \lambda x \geqslant \lambda y$$

In an ordered vector space, the **positive cone** is

$$P = \{x : x \geqslant 0\}$$

A cone having vertex v is a set C such that $v + \lambda(x - v) \in C$ when $x \in C$ and $\lambda \geqslant 0$. It is elementary to prove that P is a convex cone having vertex at 0. Also, the partial order can be recovered from P by defining $x \geqslant y$ to mean $x - y \in P$.

If X is a normed space with an order as described, then X^* is ordered in a standard way; namely, we define $\phi \geqslant 0$ to mean $\phi(x) \geqslant 0$ for all $x \geqslant 0$. Here $\phi \in X^*$.

These matters are well illustrated by the space $C[a, b]$, in which the natural order $f \geqslant g$ is defined to mean $f(t) \geqslant g(t)$ for all $t \in [a, b]$. The conjugate space consists of signed measures.

In the next theorem, X and Y are normed linear spaces, and Y is an ordered vector space. Differentiable functions $f : X \to \mathbb{R}$ and $G : X \to Y$ are given. We seek necessary conditions for a point x_0 to maximize $f(x)$ subject to $G(x) \geqslant 0$.

Theorem 5.   *If x_0 is a local maximum point of f on the set $\{x : G(x) \geqslant 0\}$ and if there is an $h \in X$ such that $G(x_0) + G'(x_0)h$ is an interior point of the positive cone, then there is a nonnegative functional $\phi \in Y^*$ such that $\phi(G(x_0)) = 0$ and $f'(x_0) = -\phi \circ G'(x_0)$.*

Proof.   (Following Luenberger [Lue2]). Working in the space $\mathbb{R} \times Y$, we define two convex sets

$$H = \{ (t, y) : \text{for some } h, \; t \leqslant f'(x_0)h \text{ and } y \leqslant G(x_0) + G'(x_0)h \}$$
$$K = \{ (t, y) : t \geqslant 0 \,, \; y \geqslant 0 \} = [0, \infty) \times P$$

One of the hypotheses in the theorem shows that P has an interior point, and consequently K has an interior point. No interior point of K lies in H, however.

In order to prove this, suppose that (t, y) is an interior point of K and belongs to H. Then for some $h \in X$ we have

$$0 < t \leqslant f'(x_0)h$$
$$0 < y \leqslant G(x_0) + G'(x_0)h$$

Here the inequality $y > 0$ is interpreted to mean that y is an interior point of the positive cone P. For $\lambda \in (0, 1)$ we have

$$
\begin{aligned}
G(x_0 + \lambda h) &= G(x_0) + G'(x_0)\lambda h + o(\lambda) \\
&= (1 - \lambda)G(x_0) + \lambda\big[G(x_0) + G'(x_0)h\big] + o(\lambda) \\
&\geqslant \lambda\big[G(x_0) + G'(x_0)h\big] + o(\lambda) \\
&\geqslant \lambda y + o(\lambda)
\end{aligned}
$$

Since y is an interior point of P, there is an $\varepsilon > 0$ such that $B(y, \varepsilon) \subset P$. By Problem 1, $B(\lambda y, \lambda \varepsilon) \subset P$ for all $\lambda > 0$. Select λ small enough so that $\|o(\lambda)\| < \lambda \varepsilon$. Then

$$\lambda y + o(\lambda) \in B(\lambda y, \lambda \varepsilon) \subset P$$

Consequently, $G(x_0 + \lambda h) \geqslant 0$. Similarly, for small λ we have

$$
\begin{aligned}
f(x_0 + \lambda h) &= f(x_0) + f'(x_0)\lambda h + o(\lambda) \\
&\geqslant f(x_0) + \lambda t + o(\lambda) > f(x_0)
\end{aligned}
$$

Thus $x_0 + \lambda h$ lies in the constraint set and produces a larger value in f than $f(x_0)$. This contradiction shows that H is disjoint from the interior of K.

Now use the Separation Theorem (Theorem 2 in Section 7.3, page 343). It asserts the existence of a hyperplane separating K from H. Thus there exist $\mu \in \mathbb{R}$ and $\phi \in Y^*$ such that $|\mu| + \|\phi\| > 0$ and

$$\mu t + \phi(y) \leqslant c \quad \text{when} \quad (t, y) \in H$$
$$\mu t + \phi(y) \geqslant c \quad \text{when} \quad (t, y) \in K$$

Since $(0, 0) \in H \cap K$, we see that $c = 0$. From the definition of K, we see that $\mu \geqslant 0$ and $\phi \geqslant 0$. Actually, $\mu > 0$. To verify this, suppose $\mu = 0$. Then $\phi \neq 0$, and $\phi(y) \leqslant 0$ whenever $(t, y) \in H$. From the hypotheses of the theorem, there is an h such that $G(x_0) + G'(x_0)h \equiv z$ is interior to P. By the definition of H, $(f'(x_0)h, z) \in H$, and so $\phi(z) \leqslant 0$. Hence there are points y near z and in P where $\phi(y) < 0$. But this contradicts the fact that $\phi \geqslant 0$.

Since $\mu > 0$, we can take it to be 1. For any h, the point

$$(f'(x_0)h, \; G(x_0) + G'(x_0)h)$$

belongs to H. Consequently,

$$f'(x_0)h + \phi\big[G(x_0) + G'(x_0)h\big] \leqslant 0 \qquad (h \in X)$$

Taking $h = 0$ in this inequality gives us $\phi[G(x_0)] \leqslant 0$. Since $G(x_0) \geqslant 0$ and $\phi \geqslant 0$, we have $\phi[G(x_0)] \geqslant 0$. Thus $\phi[G(x_0)] = 0$.

Now we conclude that

$$f'(x_0)h + \phi[G'(x_0)h] \leqslant 0 \qquad (h \in X)$$

Since h can be replaced by $-h$ here, it follows that

$$f'(x_0)h + \phi\big[G'(x_0)h\big] = 0$$

In other words,

$$f'(x_0) = -\phi \circ G'(x_0) \qquad\qquad \blacksquare$$

Problems 3.5

1. (a) Use Lagrange multipliers to find the maximum of xy subject to $x + y = c$. (b) Find the shortest distance from the point $(1, 0)$ to the parabola given by $y^2 = 4x$.

2. Let the equations $f(x, y) = 0$ and $g(x, y) = 0$ define two non-intersecting curves in $\mathbb{R}^2$. What system of equations should be solved if we wish to find minimum or maximum distances between points on these two curves?

3. Show that in an ordered vector space with positive cone P, if $B(x, r) \subset P$, then $B(\lambda x, \lambda r) \subset P$ for $\lambda \geqslant 0$.

4. Prove that the positive cone P determines the vector order.

5. Let A be a Hermitian operator on a Hilbert space. Define $f(x) = \langle Ax, x\rangle$ and $g(x) = \langle x, x\rangle - 1$. What are $f'(x)$ and $g'(x)$? Find a necessary condition for the extrema of f on the set $M = \{x : g(x) = 0\}$. (Use the first theorem on Lagrange multipliers.) Prove that your necessary condition is fulfilled by any eigenvector of A in M.

6. What is $f'(x)$ in the lemma of this section if x is not an eigenvalue?

7. Use the method of Lagrange multipliers to find a point on the surface $(x - y)^2 - z^2 = 1$ as close as possible to the origin in $\mathbb{R}^3$.

8. Let A and B be Hermitian operators on a real Hilbert space. Prove that the stationary values of $\langle x, Ax\rangle$ on the manifold where $\langle x, Bx\rangle = 1$ are necessarily numbers λ for which $A - \lambda B$ is not invertible.

9. Find the dimensions of a rectangular box (whose edges are parallel to the coordinate axis) that is contained in the ellipsoid $a^2x^2 + b^2y^2 + c^2z^2 = 1$ and has maximum volume.

10. Find the least distance between two points, one on the parabola $y = x^2$ and the other on the parabola $y = -(x - 4)^2$.

11. Find the distance from the point $(3, 2)$ to the curve $xy = 2$.

12. In $\mathbb{R}^3$ the equation $x^2 + y^2 = 5$ describes a cylinder. The equation $6x + 3y + 2z = 6$ describes a plane. The intersection of the cylinder and the plane is an ellipse. Find the points on this ellipse that are nearest the origin and farthest from the origin.

13. Find the minimum and maximum values of $xy + yz + zx$ on the unit sphere in $\mathbb{R}^3$ $(x^2 + y^2 + z^2 = 1)$. See [Barb], page 21.

3.6 The Calculus of Variations

The "calculus of variations," interpreted broadly, deals with extremum problems involving *functions*. It is analogous to the theory of maxima and minima in elementary calculus, but with the added complication that the unknowns in the problems are not simple numbers but functions. We begin with some classical illustrations, posing the problems only, and postponing their solutions until after some techniques have been explained. Traditional notation is used, in which x and y are real variables, x being "independent" and y being "dependent." This harmonizes with most books on this subject.

Example 1.   Find the equation of an arc of minimal length joining two points in the plane. Let the points be (a, α) and (b, β), where $a < b$. Let the arc be given by a continuously differentiable function $y = y(x)$, where $y(a) = \alpha$ and $y(b) = \beta$. The arc length is given by the integral $\int_a^b \sqrt{1 + y'(x)^2}\, dx$. Here $y \in C^1[a, b]$. The solution, as we know, is a straight line, and this fact will be proved later.   ∎

Example 2.   Find a function y in $C^1[a, b]$, satisfying $y(a) = \alpha$ and $y(b) = \beta$, such that the surface of revolution obtained by rotating the graph of y about the x-axis has minimum area. To solve this, one starts by recalling from calculus that the area to be minimized is given by

$$(1) \qquad \int_a^b 2\pi y(x)\, ds = 2\pi \int_a^b y(x)\sqrt{1 + y'(x)^2}\, dx$$

The solution turns out to be (in many cases) a catenary, as shown later. Figure 3.5 shows one of these surfaces.   ∎

Example 3.   In a vertical plane, with gravity exerting a downward force, we imagine a particle sliding without friction along a curve joining two points, say $(0,0)$ and (b, β). There is no loss of generality in taking $b > 0$, and if the positive direction of the y-axis is downward, then $\beta > 0$ also. We ask for the curve along which the particle would fall in the least time. If the curve is the graph of a function y in $C^1[0, b]$, then the time of descent is

$$(2) \qquad \int_0^b \sqrt{\frac{1 + y'(x)^2}{2gy(x)}}\, dx$$

as is shown later. In the integral, g is the acceleration due to gravity. This problem is the "Brachistochrone Problem," posed as a challenge by John Bernoulli in 1696. Figure 3.6 shows two cases of such curves, corresponding to two choices of the terminal point (b, β). Both curves are cycloids, one being a subset of the other.

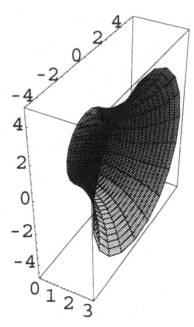

Figure 3.5

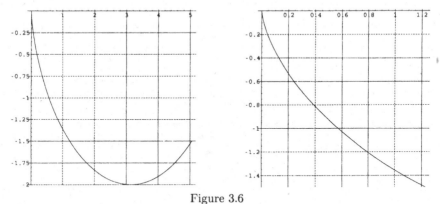

Figure 3.6

In 1696, Isaac Newton had just recently become Warden of the Mint and was in the midst of overseeing a massive recoinage. Nevertheless, when he heard of the problem, he found that he could not sleep until he had solved it, and having done so, he published the solution anonymously. Bernoulli, however, knew at once that the author of the solution was Newton, and in a famous remark asserted that he "recognized the Lion by the print of its paw". [West] ∎

The three examples given above have a common form; thus, in each one there is a nonlinear functional to be minimized, and it has the form

(3)
$$\int_a^b F\big(x, y(x), y'(x)\big)\, dx$$

The unknown function y is required to satisfy endpoint conditions $y(a) = \alpha$ and $y(b) = \beta$. In addition, some smoothness conditions must be imposed on y, since the functional is allowed to involve y'. The first theorem establishes a necessary condition for extrema, known as the Euler equation, or the Euler–Lagrange Equation.

> **Theorem 1.   The Euler Equation.**   *Let F be a mapping from $\mathbb{R}^3$ to $\mathbb{R}^1$, possessing piecewise continuous partial derivatives of the second order. In order that a function y in $C^1[a, b]$ minimize $\int_a^b F\big(x, y(x), y'(x)\big)\, dx$ subject to the constraints $y(a) = \alpha$, $y(b) = \beta$, it is necessary that Euler's equation hold:*

(4)
$$\frac{d}{dx} F_3\big(x, y(x), y'(x)\big) = F_2\big(x, y(x), y'(x)\big)$$

(Here F_2 and F_3 are partial derivatives.)

Proof.   Let $u \in C^1[a, b]$ and $u(a) = u(b) = 0$. Assume that y is a solution of the problem. For all real θ, $y + \theta u$ is a competing function. Hence

$$\frac{d}{d\theta} \int_a^b F\big(x, y(x) + \theta u(x), y'(x) + \theta u'(x)\big)\, dx \Big|_{\theta=0} = 0$$

This leads to $\int_a^b (F_2 u + F_3 u') = 0$. The second term can be integrated by parts. The result is

$$\int_a^b \left[F_2\big(x, y(x), y'(x)\big) - \frac{d}{dx} F_3\big(x, y(x), y'(x)\big) \right] u(x)\, dx = 0$$

By invoking the following lemma, we obtain the Euler equation.   ∎

> **Lemma.**   *If v is piecewise continuous on $[a, b]$ and if $\int_a^b u(x)v(x)\, dx = 0$ for every u in $C^1[a, b]$ that vanishes at the endpoints a and b, then $v = 0$.*

Proof.   Assume the hypotheses, and suppose that $v \neq 0$. Then there is a nonempty open interval (α, β) contained in $[a, b]$ in which v is continuous and has no zero. We may assume that $v(x) > 0$ on (α, β). There is a function u in $C^1[a, b]$ such that $u(x) > 0$ on (α, β) and $u(x) = 0$ elsewhere in $[a, b]$. Since $\int_a^b uv = \int_\alpha^\beta uv > 0$, we have a contradiction, and $v = 0$.   ∎

Example 1 *revisited.*   (Shortest distance between two points.) In this problem, $F(u, v, w) = \sqrt{1 + w^2}$. Hence $F_1 = F_2 = 0$ and $F_3 = w(1 + w^2)^{-1/2}$. Then

$$F_3\big(x, y(x), y'(x)\big) = y'(x) \left[1 + y'(x)^2 \right]^{-1/2}$$

The Euler equation is $\dfrac{d}{dx} F_3(x, y(x), y'(x)) = 0$. This can be integrated to yield $F_3(x, y(x), y'(x)) = $ constant. Then we find that y' must be constant and that $y(x) = \alpha + m(x - a)$, where $m = (\beta - \alpha)/(b - a)$.   ∎

Theorem 2.   In Theorem 1, if $F_1 = 0$, then the Euler equation implies that

(5) $y'(x)F_3\big(x, y(x), y'(x)\big) - F\big(x, y(x), y'(x)\big) = \text{constant}$

Proof.

$$\frac{d}{dx}\big[y'F_3 - F\big] = y''F_3 + y'\frac{d}{dx}F_3 - F_1 - F_2 y' - F_3 y'' = y'\left[\frac{d}{dx}F_3 - F_2\right] = 0 . \ \blacksquare$$

Example 2 *revisited.*   In Example 2, the function F in the general theory will be $F(u, v, w) = v(1 + w^2)^{1/2}$, where $u = x$, $v = y(x)$, and $w = y'(x)$. Then, by the preceding theorem, $wF_3 - F$ is constant. In the present case, it means that

$$w^2 v(1 + w^2)^{-1/2} - v(1 + w^2)^{1/2} = -c$$

In this equation, multiply by $(1 + w^2)^{1/2}$, obtaining

$$-c(1 + w^2)^{1/2} = w^2 v - v(1 + w^2) = -v$$

This gives us $c^2(1 + w^2) = v^2$, from which we get

$$\frac{dy}{dx} = w = \left[(v/c)^2 - 1\right]^{1/2} = \frac{1}{c}\left[v^2 - c^2\right]^{1/2} = \frac{1}{c}\left[y^2 - c^2\right]^{1/2}$$

Write this as

$$\frac{dy}{\sqrt{y^2 - c^2}} = \frac{dx}{c}$$

This can be integrated to give $\cosh^{-1}(y/c) = (x/c) + \lambda$. Without loss of generality, we take the left-hand endpoint to be $(0, \alpha)$. The curve $y = c\cosh((x/c) + \lambda)$ passes through this point if and only if $\alpha = c\cosh\lambda$. Hence c can be eliminated to give us a one-parameter family of catenaries:

(6) $$y = \frac{\alpha}{\cosh\lambda}\cosh\left(\frac{\cosh\lambda}{\alpha}x + \lambda\right)$$

Here λ is the parameter. If this catenary is to pass through the other given endpoint (b, β), then λ will have to satisfy the equation

$$\beta = (\alpha/\cosh\lambda)\cosh(b\cosh\lambda/\alpha + \lambda)$$

Here α, b, β are prescribed and λ is to be determined. In [Bl] (pages 85-119) you will find an exhaustive discussion. Here are the main conclusions, without proof:

I. The one-parameter family of catenaries in Equation (6) has the appearance shown in Figure 3.7. The "envelope" of the family is defined by $g(x) = \min_\lambda y_\lambda(x)$, where y_λ is the function in Equation 6.

II. If the terminal point (b, β) is below the envelope, no member of the family (6) passes through it. The problem is then solved by the "Goldschmidt solution" described below.

III. If the terminal point is above the envelope, two catenaries of the family (6) pass through it. One of these is a local minimum in the problem but not necessarily the absolute minimum. If it is not the absolute minimum, the Goldschmidt solution again solves the problem.

IV. For terminal points sufficiently far above the envelope, the upper catenary of the two passing through the point is the solution to the problem.

V. The Goldschmidt solution is a broken line from $(0, \alpha)$ to $(0, 0)$ to $(b, 0)$ and to (b, β). It generates a surface of revolution whose area is $\pi(\alpha^2 + \beta^2)$. ∎

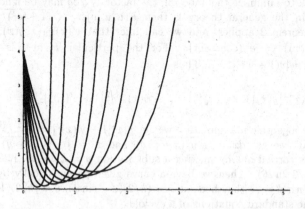

Figure 3.7

Example 3 *revisited.*   Consider again the Brachistochrone problem. We are using (x, y) to denote points in $\mathbb{R}^2$, t will denote time, and s will be arc length. The derivation of the integral given previously for the time of descent is as follows. At any point of the curve, the downward force of gravity is mg, where m is the mass of the particle and g is the constant acceleration due to gravity. The component of this force along the tangent to the curve is $mg \cos \theta = mg \, (dy/ds)$, where θ is the angle between the tangent and the vertical. (See Figure 3.8.) The velocity of the particle is ds/dt, and its acceleration is d^2s/dt^2.

Figure 3.8

By Newton's law of motion ($F = ma$) we have $mg(dy/ds) = m(d^2s/dt^2)$, or $d^2s/dt^2 = g(dy/ds)$. Multiply by $2(ds/dt)$ to get

$$2\frac{ds}{dt}\frac{d^2s}{dt^2} = 2g\frac{dy}{ds}\frac{ds}{dt}$$

whence

$$\frac{d\cdot}{dt}\left(\frac{ds}{dt}\right)^2 = 2g\frac{dy}{dt} \quad \text{and} \quad \left(\frac{ds}{dt}\right)^2 = 2gy + C$$

If the initial conditions are $t = 0$, $y = 0$, and $ds/dt = 0$, then $C = 0$. Hence

$$\frac{ds}{dt} = \sqrt{2gy} \qquad\qquad \frac{dt}{ds} = \frac{1}{\sqrt{2gy}}$$

$$T = \int dt = \int \frac{ds}{\sqrt{2gy}} = \int_0^b \sqrt{\frac{1 + y'(x)^2}{2gy(x)}}\, dx$$

Since we seek to minimize this integral, the factor $1/\sqrt{2g}$ may be ignored. The function F in the general theory is then $F(u,v,w) = \sqrt{(1+w^2)/v}$. Since $F_1 = 0$, Theorem 2 applies, and we can infer that $y'(x)F_3\big(x,y(x),y'(x)\big) - F\big(x,y(x),y'(x)\big) = c$ (constant). For the particular F in this example, $F_3(u,v,w) = w[v(1+w^2)]^{-1/2}$. Thus

$$y'(x)^2\big[y(x)\big(1 + y'(x)^2\big)\big]^{-1/2} - y(x)^{-1/2}\big[1 + y'(x)^2\big]^{1/2} = c$$

After a little algebraic manipulation we get $y(x)[1 + y'(x)^2] = c^{-2}$. When this is "separated" we get $dx = \sqrt{y/(k-y)}\, dy$ and $x = \int \sqrt{y/(k-y)}\, dy$. The integration is carried out by making a substitution $y = k\sin^2\theta$. The result is $x = k(\theta - 1/2\sin 2\theta)$. Then we have a curve given parametrically by the two formulas. With $\phi = 2\theta$ they become $x = (k/2)(\phi - \sin\phi)$, $y = (k/2)(1 - \cos\phi)$. These are the standard equations of a cycloid. ∎

Example 4.   This is the Brachistochrone Problem, except that the terminal point is allowed to be anywhere on a given vertical line. Following the previous discussion, we are led to minimize the expression

$$\int_0^b \sqrt{\frac{1 + y'(x)^2}{y(x)}}\, dx$$

subject to $y \in C^2[0,b]$ and $y(0) = 0$. Notice that the value $y(b)$ is **not** prescribed. To solve such a problem, we require a modification of Theorem 1, namely:

Theorem 3.   *Any function y in $C^2[a,b]$ that minimizes*

$$\int_a^b F\big(x,y(x),y'(x)\big)\, dx$$

subject to the constraint $y(a) = \alpha$ must satisfy the two conditions

(7) $\dfrac{d}{dx}F_3\big(x,y(x),y'(x)\big) = F_2\big(x,y(x)y'(x)\big)$ and $F_3\big(b,y(b),y'(b)\big) = 0$

Proof.   This is left as a problem. ∎

Returning now to Example 4, we conclude that $F_3\big(b,y(b),y'(b)\big) = 0$. This entails $y'(b)/\sqrt{y(b)[1 + y'(b)^2]} = 0$, or $y'(b) = 0$. Thus the slope of our cycloid must be zero at $x = b$. The cycloids going through the initial point are given parametrically by

$$\begin{cases} x = k(\phi - \sin \phi) \\ y = k(1 - \cos \phi) \end{cases}$$

The slope is

$$\frac{dy}{dx} = \frac{dy}{d\phi} \div \frac{dx}{d\phi} = \frac{\sin \phi}{1 - \cos \phi}$$

This is 0 at $\phi = \pi$. The value $x = b$ corresponds to $\phi = \pi$, and $k = b/\pi$. The solution is given by $x = (b/\pi)(\phi - \sin \phi)$, $y = (b/\pi)(1 - \cos \phi)$, $0 \leqslant \phi \leqslant \pi$.   ∎

Example 5.   (The Generalized Isoperimetric Problem.) Find the function y that minimizes an integral

$$\int_a^b F(x, y(x), y'(x))\, dx$$

subject to constraints that y belong to $C^1[a, b]$ and

$$\int_a^b G(x, y(x), y'(x))\, dx = 0 \qquad y(a) = \alpha \qquad y(b) = \beta \qquad ∎$$

The next theorem pertains to this problem.

Theorem 4.   *If F and G map $\mathbb{R}^3$ to $\mathbb{R}$ and have continuous partial derivatives of the second order, and if y is an element of $C^2[a, b]$ that minimizes $\int_a^b F(x, y(x), y'(x))\, dx$ subject to endpoint constraints and $\int_a^b G(x, y(x), y'(x))\, dx = 0$, then there is a nontrivial linear combination $H = \mu F + \lambda G$ such that*

$$(8) \qquad H_2(x, y(x), y'(x)) = \frac{d}{dx} H_3(x, y(x), y'(x))$$

Proof.   As in previous problems of this section, we try to obtain a necessary condition for a solution by perturbing the solution in such a way that the constraints are not violated. Suppose that y is a solution in $C^1[a, b]$. Let η_1 and η_2 be two functions in $C^1[a, b]$ that vanish at the endpoints. Consider the function $z = y + \theta_1 \eta_1 + \theta_2 \eta_2$. It belongs to $C^1[a, b]$ and takes correct values at the endpoints: $z(a) = \alpha$, $z(b) = \beta$. We require two perturbing functions, η_1 and η_2 because the constraint $\int_a^b G(x, z(x), z'(x))\, dx = 0$ will be true only if we allow a relationship between the two parameters θ_1 and θ_2. Let $I(\theta_1, \theta_2) = \int_a^b F(x, z(x), z'(x))\, dx$ and $J(\theta_1, \theta_2) = \int_a^b G(x, z(x), z'(x))\, dx$. The minimum of $I(\theta_1, \theta_2)$ under the constraint $J(\theta_1, \theta_2) = 0$ occurs at $\theta_1 = \theta_2 = 0$, because y is a solution of the original problem. By the Theorem on Lagrange Multipliers (Theorem 3, page 148), there is a nontrivial linear relation of the form $\mu I'(0, 0) + \lambda J'(0, 0) = 0$. Thus

$$\mu \frac{\partial I}{\partial \theta_1} + \lambda \frac{\partial J}{\partial \theta_1} = 0 \text{ at } (\theta_1, \theta_2) = (0, 0), \text{ and } \mu \frac{\partial I}{\partial \theta_2} + \lambda \frac{\partial J}{\partial \theta_2} = 0 \text{ at } (0, 0)$$

Following the usual procedure, including an integration by parts, we eventually obtain Equation (8).   ∎

Example 6.   It is required to find the curve of given length ℓ joining the point $(-1, 0)$ to the point $(1, 0)$ that, together with the interval $[-1, 1]$ on the horizontal axis, encloses the greatest possible area.

We assume that $2 < \ell < \pi$. Let the curve be given by $y = y(x)$, where y belongs to $C^1[-1, 1]$. The area to be maximized is then

$$\int_{-1}^{1} y(x)\, dx$$

and the constraints are

$$\int_{-1}^{1} \sqrt{1 + y'(x)^2}\, dx = \ell \qquad y(-1) = y(1) = 0$$

This problem can be treated with Theorem 4, taking

$$F(x, y, y') = y \qquad \text{and} \qquad G(x, y, y') = \sqrt{1 + (y')^2} - \ell/2$$

The necessary condition of Theorem 4 is that for a suitable nontrivial pair of coefficients μ and λ

$$\left[(\mu F + \lambda G)_2 - \frac{d}{dx}(\mu F + \lambda G)_3 \right](x, y(x), y'(x)) = 0$$

(In these equations, subscript 2 means a partial derivative with respect to the second argument of the function, and so on.) In the case being considered, we have $F_2 = 1$, $F_3 = 0$, $G_2 = 0$, and $G_3 = y'(x)([1 + y'(x)^2]^{-1/2}$. The necessary condition then reads

$$\mu - \lambda \frac{d}{dx} \frac{y'(x)}{\sqrt{1 + y'(x)^2}} = 0$$

If $\mu = 0$, then λ must be 0 as well. Hence we are free to set $\mu = 1$ and integrate the previous equation, arriving at

$$x - \frac{\lambda y'(x)}{\sqrt{1 + y'(x)^2}} = c_1$$

This can be solved for $y'(x)$:

$$y'(x) = \frac{x - c_1}{\sqrt{\lambda^2 - (x - c_1)^2}}$$

Another integration leads to

$$y(x) = -\sqrt{\lambda^2 - (x - c_1)^2} + c_2$$

We see that the curve is a circle by writing this last equation in the form

$$(x - c_1)^2 + (y - c_2)^2 = \lambda^2$$

Since the circle must pass through the points $(-1, 0)$ and $(1, 0)$, we find that $c_1 = 0$ and that $1 + c_2^2 = \lambda^2$. When the condition on the length of the arc is imposed, we obtain $\ell = 2\lambda \operatorname{Arcsin} \frac{1}{\lambda}$, from which λ can be computed.   ∎

Example 7.   (The Classical Isoperimetric Problem.)   Among all the plane curves having a prescribed length, find one enclosing the greatest area. We assume a parametric representation $x = x(t)$ and $y = y(t)$ with continuously differentiable functions. We can also assume that $0 \leqslant t \leqslant b$ and that $x(0) = x(b)$, $y(0) = y(b)$ so the curve is closed. Let us assume further that as t increases from 0 to b, the curve is described in the counterclockwise direction. The region enclosed is then always on the left. Recall Green's Theorem, [Wid1], page 223:

$$\int_\Gamma (P\, dx + Q\, dy) = \int\!\!\int_R [Q_1(x, y) - P_2(x, y)]\, dx\, dy$$

where R is the region enclosed by the curve Γ and the subscripts denote partial derivatives. A special case of Green's Theorem is

$$\frac{1}{2} \int_\Gamma (-y\, dx + x\, dy) = \frac{1}{2} \int\!\!\int_R (1 + 1)\, dx\, dy = \text{Area of } R$$

Thus our isoperimetric problem is to maximize the integral

$$\int_0^b \left(-y\, \frac{dx}{dt} + x\, \frac{dy}{dt} \right) dt$$

subject to

$$\int_0^b \sqrt{ \left(\frac{dx}{dt} \right)^2 + \left(\frac{dy}{dt} \right)^2 }\, dt = \text{constant}$$

This isoperimetric problem involves again the minimization of an integral subject to a constraint expressed as an integral. But now we have two unknown functions to be determined. A straightforward extension of Theorem 4 applies in this situation. Suppose that we wish to minimize

$$\int_a^b F\big(t, x(t), x'(t), y(t), y'(t)\big)\, dt$$

subject to the usual endpoint constraints and a constraint

$$\int_a^b G\big(t, x(t), x'(t), y(t), y'(t)\big)\, dt = 0$$

The Euler necessary condition is that for a suitable nontrivial linear combination $H = \mu F + \lambda G$,

$$H_2(t, x, x', y, y') = \frac{d}{dt} H_3(t, x, x', y, y')$$

$$H_4(t, x, x'y, y') = \frac{d}{dt} H_5(t, x, x', y, y')$$

If we apply this result to Example 7, we will use

$$H(t, x, x'y, y') = \mu(xy' - yx') + \lambda\sqrt{x'^2 + y'^2}$$

The Euler equations are

$$\begin{cases} \mu y' = \dfrac{d}{dt}\left[-\mu y + \lambda x'\left(x'^2 + y'^2\right)^{-1/2}\right] \\[2mm] -\mu x' = \dfrac{d}{dt}\left[\mu x + \lambda y'\left(x'^2 + y'^2\right)^{-1/2}\right] \end{cases}$$

Upon integrating these with respect to t, we obtain

$$2\mu y = \lambda x'\left(x'^2 + y'^2\right)^{-1/2} + A$$
$$-2\mu x = \lambda y'\left(x'^2 + y'^2\right)^{-1/2} - B$$

If $\mu = 0$, we infer that $x' = y' = 0$, and then the "curve" is a straight line. Hence $\mu \neq 0$, and by homogeneity we can assume $\mu = \frac{1}{2}$. Then $y - A = \lambda x'(x'^2 + y'^2)^{-1/2}$ and $x - B = -\lambda y'(x'^2 + y'^2)^{-1/2}$. Square these two equations and add to obtain the equation of a circle: $(x - B)^2 + (y - A)^2 = \lambda^2$. ∎

Applications to Geometrical Optics. Fermat's Principle states that a ray of light passing between two points will follow a path that minimizes the elapsed time. In a homogeneous medium, the velocity of light is constant, and the least elapsed time will occur for the shortest path, which is a straight line. Consider now two homogeneous media separated by a plane. Let the velocities of light in the two media be c_1 and c_2. What is the path of a ray of light from a point in the first medium to a point in the second? We assume that the path lies in a plane. By the remarks made above, the path consists of two lines meeting at the plane that separates the two media.

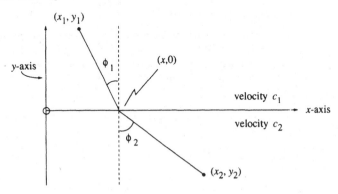

Figure 3.9

If the coordinate system is as shown in Figure 3.9, and if the unknown point on the x-axis is $(x, 0)$ then the time of passage is

$$T = \frac{1}{c_1}\sqrt{(x - x_1)^2 + y_1^2} + \frac{1}{c_2}\sqrt{(x - x_2)^2 + y_2^2} = c_1^{-1}p_1 + c_2^{-1}p_2$$

For an extremum we want $dT/dx = 0$. Thus

$$c_1^{-1}\frac{dp_1}{dx} + c_2^{-1}\frac{dp_2}{dx} = 0 \; ,$$

$$c_1^{-1}\frac{1}{p_1}(x - x_1) + c_2^{-1}\frac{1}{p_2}(x - x_2) = 0$$

$$c_1^{-1}\sin\phi_1 = c_2^{-1}\sin\phi_2$$

This last equation is known as Snell's Law.

Now consider a medium in which the velocity of light is a function of y; let us say $c = c(y)$. This would be the case in the Earth's atmosphere or in the ocean. Think of the medium as being composed of many thin layers, in each of which the velocity of light is constant. See Figure 3.10, in which three layers are shown.

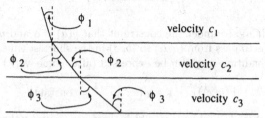

Figure 3.10

Snell's Law yields

$$\frac{\sin\phi_1}{c_1} = \frac{\sin\phi_2}{c_2} = \frac{\sin\phi_3}{c_3} = \cdots = k \text{ constant}$$

For a continuously varying speed $c(y)$, the path of a ray of light should satisfy $\sin\phi(y)/c(y) = k$. Notice that the slope of the curve is

$$y'(x) = \tan\left(\frac{\pi}{2} - \phi\right) = \cot\phi = \cos\phi/\sin\phi$$

$$= \sqrt{1 - \sin^2\phi}\Big/\sin\phi = \sqrt{1 - k^2c^2}/kc$$

Hence

$$\frac{kc}{\sqrt{1 - k^2c^2}}\,dy = dx \quad \text{and} \quad x = \int \frac{kc(y)\,dy}{\sqrt{1 - k^2c(y)^2}}$$

Example 8.   What is the path of a light beam if the velocity of light in the medium is $c = \alpha y$ (where α is a constant)?

Solution.   The path is the graph of a function y such that

$$x = \int \frac{k\alpha y}{\sqrt{1 - k^2\alpha^2 y^2}}\,dy$$

The integration produces

$$x = \frac{-1}{k\alpha}\sqrt{1 - k^2\alpha^2 y^2} + A$$

Here A and k are constants that can be adjusted so that the path passes through two given points. The equation can be written in the form

$$(x - A)^2 = \frac{1}{k^2\alpha^2}(1 - k^2\alpha^2 y^2)$$

or in the standard form of a circle:

$$(x - A)^2 + y^2 = \frac{1}{k^2\alpha^2}$$

The analysis above can also be based directly on Fermat's Principle. The time taken to traverse a small piece of the path having length Δs is $\Delta s/c(y)$. The total time elapsed is then

$$\int_a^b \frac{\sqrt{1 + y'(x)^2}}{c(y)} \, dx$$

This is to be minimized under the constraint that $y(a) = \alpha$ and $y(b) = \beta$. Here the ray of light is to pass from (a, α) to (b, β) in the shortest time. By Theorem 2, a necessary condition on y can be expressed (after some work) as

$$\left[c\big(y(x)\big)\right]^{-1}\left[1 + y'(x)^2\right]^{-1/2} = \text{constant} \qquad \blacksquare$$

In order to handle problems in which there are several unknown functions to be determined, one needs the following theorem.

Theorem 5.   *Suppose that $y_1, \ldots, y_n$ are functions (of t) in $C^2[a, b]$ that minimize the integral*

$$\int_a^b F(y_1, \ldots, y_n, y_1', \ldots, y_n') \, dt$$

subject to endpoint constraints that prescribe values for all $y_i(a)$, $y_i(b)$. Then the Euler Equations hold:

(9) $$\frac{d}{dt}\frac{\partial F}{\partial y_i'} = \frac{\partial F}{\partial y_i} \qquad (1 \leqslant i \leqslant n)$$

Proof.   Take functions $\eta_1, \ldots, \eta_n$ in $C^2[a, b]$ that vanish at the endpoints. The expression

$$\int_a^b F(y_1 + \theta_1\eta_1, \ldots, y_n + \theta_n\eta_n) \, dt$$

will have a minimum when $(\theta_1, \ldots, \theta_n) = (0, 0, \ldots, 0)$. Proceeding as in previous proofs, one arrives at the given equations. $\blacksquare$

Geodesic Problems.   Find the shortest arc lying on a given surface and joining two points on the surface. Let the surface be defined by $z = z(x, y)$. Let the two points be (x_0, y_0, z_0) and (x_1, y_1, z_1). Arc length is

$$\int_{(x_0 y_0 z_0)}^{(x_1 y_1 z_1)} ds = \int_{(x_0 y_0 z_0)}^{(x_1 y_1 z_1)} \sqrt{dx^2 + dy^2 + dz^2}$$

If the curve is given parametrically as $x = x(t)$, $y = y(t)$, $z = z(x(t), y(t))$, $0 \leqslant t \leqslant 1$, then our problem is to minimize

(10 $$\int_0^1 \sqrt{x'^2 + y'^2 + (z_x x' + z_y y')^2}\, dt$$

subject to $x \in C^2[0,1]$, $y \in C^2[0,1]$, $x(0) = x_0$, $x(1) = x_1$, $y(0) = y_0$, $y(1) = y_1$.

Example 9.   We search for geodesics on a cylinder. Let the surface be the cylinder $x^2 + z^2 = 1$, or $z = (1 - x^2)^{1/2}$ (upper-half cylinder). In the general theory, $F(x, y, x', y') = \sqrt{x'^2 + y'^2 + (z_x x' + z_y y')^2}$. In this particular case this is

$$F = \left[x'^2 + y'^2 + z_x^2 x'^2\right]^{1/2} = \left[(1 - x^2)^{-1} x'^2 + y'^2\right]^{1/2}$$

Then computations show that

$$\frac{\partial F}{\partial x} = \frac{x x'^2}{(1 - x^2)^2 F} \qquad \frac{\partial F}{\partial x'} = \frac{x'}{(1 - x^2) F} \qquad \frac{\partial F}{\partial y} = 0 \qquad \frac{\partial F}{\partial y'} = \frac{y'}{F}$$

To simplify the work we take t to be arc length and drop the requirement that $0 \leqslant t \leqslant 1$. Since $dt = ds = \sqrt{x'^2 + y'^2 + z'^2}\, dt$, we have $x'^2 + y'^2 + z'^2 = 1$ and $F(x, y, x', y') = 1$ along the minimizing curve. The Euler equations yield

$$\frac{(1 - x^2) x'' + 2 x x'^2}{(1 - x^2)^2} = \frac{x x'^2}{(1 - x^2)^2} \qquad \text{and} \qquad y'' = 0$$

The first of these can be written $x'' = x x'^2/(x^2 - 1)$. The second one gives $y = at + b$, for appropriate constants a and b that depend on the boundary conditions. The condition $1 = F^2$ leads to $x'^2/(1 - x^2) + y'^2 = 1$ and then to $x'^2/(1 - x^2) = 1 - a^2$. The Euler equation for x then simplifies to $x'' = (a^2 - 1)x$. There are three interesting cases:

Case 1: $a = 1$.   Then $x'' = 0$, and thus both $x(t)$ and $y(t)$ are linear expressions in t. The path is a straight line on the surface (necessarily parallel to the y-axis).

Case 2: $a = 0$.   Then $x'' = -x$, and $x = c \cos(t + d)$ for suitable constants c and d. The condition $x'^2/(1 - x^2) = 1$ gives us $c = 1$. It follows that $x = \cos(t + d)$, $y = b$, and $z = \sqrt{1 - x^2} = \sin(t + d)$. The curve is a circle parallel to the xz-plane.

Case 3: $0 < a < 1$.   Then $x = c \cos(\sqrt{1 - a^2}\, t + d)$, and as before, $c = 1$. Again $z = \sin(\sqrt{1 - a^2}\, t + d)$, and $y = at + b$. The curve is a spiral.   ∎

Examples of Problems in the Calculus of Variations with No Solutions.
Some interesting examples are given in [CH], Vol. 1.

I. Minimize the integral $\int_0^1 \sqrt{1 + y'^2}\, dx$ subject to constraints $y(0) = y(1) = 0$, $y'(0) = y'(1) = 1$. An admissible curve is shown in Figure 3.11, but there is none of least length, since the infimum of the admissible lengths is 1, but is not attained by an admissible y.

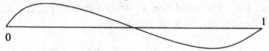

0

Figure 3.11

II. Minimize $\int_{-1}^{1} x^2 \, y'(x)^2 \, dx$ subject to constraints that y be piecewise continuously differentiable, continuous, and satisfy $y(-1) = -1$, $y(1) = 1$. An admissible y is shown in Figure 3.12, and the value of the integral for this function is $2\varepsilon/3$. The infimum is 0 but is not attained by an admissible y. This example was given by Weierstrass himself!

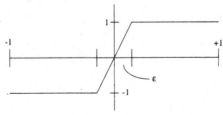

Figure 3.12

Direct Methods in the Calculus of Variations. These are methods that proceed directly to the minimization of the given functional without first looking at necessary conditions. Such methods sometimes yield a constructive proof of existence of the solution. (Methods based solely on the use of necessary conditions never establish existence of the solution.)

The Rayleigh–Ritz Method. (We shall consider this again in Chapter 4.) Suppose that U is a set of "admissible" functions, and Φ is a functional on U that we desire to minimize. Put $\rho = \inf\{\Phi(u) : u \in U\}$. We assume $\rho > -\infty$, and seek a $u \in U$ such that $\Phi(u) = \rho$. The problem, of course, is that the infimum defining ρ need not be *attained*. In the Rayleigh–Ritz method, we start with a sequence of functions $w_1, w_2, \ldots$ such that every linear combination $c_1 w_1 + \cdots + c_n w_n$ is admissible. Also, we must assume that for each $u \in U$ and for each $\epsilon > 0$ there is a linear combination v of the w_i such that $\Phi(v) \leqslant \Phi(u) + \epsilon$. For each n we select v_n in the linear span of $w_1, \ldots, w_n$ to minimize $\Phi(v_n)$. This is an ordinary minimization problem for n real parameters $c_1, \ldots, c_n$. It can be attacked with the ordinary techniques of calculus.

Example 10.   We wish to minimize the expression $\int_0^b \int_0^a (\phi_x^2 + \phi_y^2) \, dx \, dy$ subject to the constraints that ϕ be a continuously differentiable function on the rectangle $R = \{(x, y) : 0 \leqslant x \leqslant a, \ 0 \leqslant y \leqslant b\}$, that $\phi = 0$ on the perimeter of R, and that $\iint_R \phi^2 \, dx \, dy = 1$. A suitable set of base functions for this problem is the doubly indexed sequence

$$u_{nm}(x, y) = \frac{2}{\sqrt{ab}} \sin \frac{n\pi x}{a} \sin \frac{m\pi y}{b} \qquad (n, m \geqslant 1)$$

It turns out that this is an orthonormal set with respect to the inner product $\langle u, v \rangle = \iint_R u(x, y) v(x, y) \, dx \, dy$. We are looking for a function $\phi = \sum_{n,m=1}^{\infty} c_{nm} u_{nm}$ that will solve the problem. Clearly, the function ϕ vanishes on the perimeter of R. The condition $\iint \phi^2 = 1$ means $\sum_{n,m=1}^{\infty} c_{nm}^2 = 1$ by the Parseval identity (page 73). Now we compute

$$\frac{\partial}{\partial x} u_{nm}(x, y) = \frac{2}{\sqrt{ab}} \frac{n\pi}{a} \cos \frac{n\pi x}{a} \sin \frac{m\pi y}{b}$$

The system of functions

$$\frac{2}{\sqrt{ab}} \cos \frac{n\pi x}{a} \sin \frac{m\pi y}{b}$$

is also orthonormal. Thus $\iint_R \phi_x^2 = \sum_{n,m} (\frac{n\pi}{a} c_{nm})^2$. Similarly,

$$\iint_R \phi_y^2 = \sum_{n,m} (\frac{m\pi}{b} c_{nm})^2$$

Hence we are trying to minimize the expression

$$\iint_R (\phi_x^2 + \phi_y^2) = \sum_{n,m} \left[\left(\frac{n\pi}{a} c_{nm}\right)^2 + \left(\frac{m\pi}{b} c_{nm}\right)^2 \right] = \pi^2 \sum \left(\frac{n^2}{a^2} + \frac{m^2}{b^2}\right) c_{nm}^2$$

subject to the constraint $\sum c_{nm}^2 = 1$. Because of the coefficients n^2 and m^2, we obtain a solution by letting $c_{11} = 1$ and the remaining coefficients all be zero. Hence $\phi = u_{11} = \frac{2}{\sqrt{ab}} \sin \frac{\pi x}{a} \sin \frac{\pi y}{b}$. (This example is taken from [CH], page 178.) ∎

Example 11. (Dirichlet Problem for the Circle) The 2-dimensional Dirichlet problem is to find a function that is harmonic on the interior of a given 2-dimensional region and takes prescribed values on the boundary of that region. "Harmonic" means that the function u satisfies Laplace's equation:

$$u_{xx} + u_{yy} = 0 \quad \text{or} \quad \left(\frac{\partial^2}{\partial x^2} + \frac{\partial^2}{\partial y^2} \right) u(x, y) = 0$$

Laplace's equation arises as the Euler equation in the calculus of variations when we seek to minimize the integral $\iint_R (u_x^2 + u_y^2)\, dx\, dy$ subject to the constraint that u be twice continuously differentiable and take prescribed values on the boundary of the region R. To illustrate the Rayleigh–Ritz method, we take R to be the circle $\{(x, y) : x^2 + y^2 \leqslant 1\}$. Then polar coordinates are appropriate. Here are the formulas that are useful. (They are easy but tedious to derive.)

$$x = r \cos \theta \qquad y = r \sin \theta \qquad\qquad r = \sqrt{x^2 + y^2} \qquad \theta = \tan^{-1}(y/x)$$

$$u_x = u_r \frac{x}{r} - u_\theta \frac{y}{r^2} \qquad\qquad u_y = u_r \frac{y}{r} + u_\theta \frac{x}{r^2}$$

$$u_x^2 + u_y^2 = u_r^2 + r^{-2} u_\theta^2 \qquad\qquad dx\, dy = r\, dr\, d\theta$$

The integral to be minimized is now

$$I \equiv \int_0^{2\pi} \int_0^1 (u_r^2 + r^{-2} u_\theta^2)\, r\, dr\, d\theta$$

The boundary points of the domain are characterized by their value of θ. Let the prescribed boundary values of u be given by $f(\theta)$. Let f'' be continuous. Then (by classical theorems) f is represented by its Fourier series:

$$f(\theta) = \sum_{n=0}^{\infty}{}' (a_n \cos n\theta + b_n \sin n\theta) \quad (' \text{ means that the first term is halved})$$

These are the values that $u(r, \theta)$ must assume when $r = 1$. We therefore postulate that $u(r, \theta)$ have the form

$$u(r, \theta) = \sum_{n=0}^{\infty}{}' \Big[f_n(r) \cos n\theta + g_n(r) \sin n\theta \Big] \qquad g_0 = 0$$

The integral I consists of two parts, of which the first is $I_1 = \int_0^1 r \int_0^{2\pi} u_r^2 \, d\theta \, dr$. Now

$$u_r = \sum{}' [f_n'(r) \cos n\theta + g_n'(r) \sin n\theta]$$

and consequently,

$$\int_0^{2\pi} u_r^2 \, d\theta = \|u_r\|^2 = \pi \sum_{n=0}^{\infty} [f_n'(r)^2 + g_n'(r)^2]$$

Thus $I_1 = \pi \sum \int_0^1 r [f_n'(r)^2 + g_n'(r)^2] \, dr$.   Similarly, the other part of I is $I_2 = \int_0^1 r^{-1} \int_0^{2\pi} u_\theta^2 \, d\theta$. We have

$$u_\theta = \sum_{n=0}^{\infty}{}' \Big[-n f_n(r) \sin n\theta + n g_n(r) \cos n\theta \Big]$$

and consequently,

$$\int_0^{2\pi} u_\theta^2 \, d\theta = \pi \sum_{n=1}^{\infty} [n^2 f_n(r)^2 + n^2 g_n(r)^2]$$

Thus $I_2 = \pi \sum_{n=1}^{\infty} n^2 \int_0^1 r^{-1} [f_n(r)^2 + g_n(r)^2] \, dr$. Hence

$$I = \pi \sum_{n=0}^{\infty} \int_0^1 \Big[r f_n'(r)^2 + r g_n'(r)^2 + n^2 r^{-1} f_n(r)^2 + n^2 r^{-1} g_n(r)^2 \Big] \, dr$$

We therefore must solve these minimization problems individually:

$$\text{minimize} \int_0^1 \Big[r f_n'(r)^2 + n^2 r^{-1} f_n(r)^2 \Big] \, dr \text{ subject to } f_n(1) = a_n$$

$$\text{minimize} \int_0^1 \Big[r g_n'(r)^2 + n^2 r^{-1} g_n(r)^2 \Big] \, dr \text{ subject to } g_n(1) = b_n$$

Fixing n and concentrating on the function f_n, we suppose it is a polynomial of high degree, $m \geqslant n$. Thus $f_n(r) = c_0 + c_1 r + c_2 r^2 + \cdots + c_m r^m$. The integral to be minimized becomes a function of $(c_0, c_1, c_2, \ldots, c_m)$, and the constraint is $c_0 + c_1 + c_2 + \cdots + c_m = a_n$. The solution of this minimization problem is

$c_n = a_n$, all other c_j being $= 0$. Hence $f_n(r) = a_n r^n$. Similarly, $g_n(r) = b_n r^n$. Thus the u-function is

$$u(r,\theta) = \frac{a_0}{2} + \sum_{n=1}^{\infty} r^n (a_n \cos n\theta + b_n \sin n\theta) \qquad \blacksquare$$

Problems 3.6

1. Among all the functions x in $C^1[a,b]$ that satisfy $x(a) = \alpha$, $x(b) = \beta$, find the one for which $\int_a^b u(t)^2 x'(t)^2\, dt$ is a minimum. Here u is given as an element of $C[a,b]$.

2. Prove Theorem 3, assuming that F has continuous second derivatives.

3. Find a function $y \in C^2[0,1]$ that minimizes the integral $\int_0^1 [\frac{1}{2} y'(x)^2 + y(x)y'(x) + y'(x) + y(x)]\, dx$. Note that $y(0)$ and $y(1)$ are *not* specified.

4. Prove this theorem: If $\{u_n\}$ is an orthonormal system in $L^2(S,\mu)$ and $\{v_n\}$ is an orthonormal system in $L^2(T,\nu)$, then $\{u_n \otimes v_m : 1 \leqslant n < \infty, 1 \leqslant m < \infty\}$ is orthonormal in $L^2(S \times T)$. Here $u_n \otimes v_m$ is the function whose value at (s,t) is $u_n(s)v_m(t)$. Explain how this theorem is pertinent to Example 10.

5. Determine the path of a light beam in the xy-plane if the velocity of light is $1/y$.

6. Find a function u in $C^1[0,1]$ that minimizes the integral $\int_0^1 [u'(t)^2 + u'(t)]\, dt$ subject to the constraints $u(0) = 0$ and $u(1) = 1$.

7. Repeat the preceding problem when the integrand is $u(t)^2 + u'(t)^2$.

8. Explain what happens in Example 6 if $\ell > \pi$. Try to solve the problem using polar coordinates: $r = r(\theta)$, where $0 \leqslant \theta \leqslant \pi$.

9. Verify that the family of catenaries in Example 2 passing through the point $(0,4)$ is given by $y = f(c,x)$, where c is a parameter and

$$f(c,x) = \frac{4}{\cosh c} \cosh\left[\frac{\cosh c}{4} x - c\right]$$

10. Suppose that the path of a ray of light in the xy-plane is along the parabola described by $2x = y^2$. What function describes the speed of light in the medium?

11. Find the function u in $C^2[0,1]$ that minimizes the integral $\int_0^1 \{[u(t)]^2 + [u'(t)]^2\}\, dt$ subject to the constraints $u(0) = u(1) = 1$.

Chapter 4

Basic Approximate Methods in Analysis

4.1   Discretization   170
4.2   The Method of Iteration   176
4.3   Methods Based on the Neumann Series   186
4.4   Projections and Projection Methods   191
4.5   The Galerkin Method   198
4.6   The Rayleigh–Ritz Method   205
4.7   Collocation Methods   213
4.8   Descent Methods   225
4.9   Conjugate Direction Methods   232
4.10   Methods Based on Homotopy and Continuation   237

In this chapter we will explain and illustrate some important strategies that can often be used to solve operator equations such as differential equations, integral equations, and two-point boundary value problems. The methods to be discussed are as indicated in the table above.

There is often not a clear-cut distinction between these methods, and numerical procedures may combine several different methods in the solution of a problem. Thus, for example, the Galerkin technique can be interpreted as a projection method, and iterative procedures can be combined with a discretization of a problem to effect its solution.

4.1 Discretization

The term "discretize" has come to mean the replacement of a continuum by a finite subset (or at least a discrete subset) of it. A discrete set is characterized by the property that each of its points has a neighborhood that contains no other points of the set. A function defined on the continuum can be restricted to that discrete set, and the restricted function is a simpler object to determine. For example, in the numerical solution of a differential equation on an interval $[a, b]$, it is usual to determine an approximate solution only on a discrete subset of that interval, say at points $a = t_0 < t_1 < \cdots < t_{n+1} = b$. From values of a function u at these points, one can create a function $\bar{u}$ on $[a, b]$ by some interpolation process. It is important to recognize that the problem itself is usually changed by the passage to a discrete set.

Let us consider an idealized situation, and enumerate the steps involved in a solution by "discretization."

1. At the beginning, a problem $\mathbf{P}$ is posed that has as its solution a function u defined on a domain D. Our objective is to determine u, or an approximation to it.

2. The domain D is replaced by a discrete subset D_h, where h is a parameter that ideally will be allowed to approach zero in order to get finer discrete sets. The problem $\mathbf{P}$ is replaced by a "discrete version" $\mathbf{P}_h$.

3. Problem $\mathbf{P}_h$ is solved, yielding a function v_h defined on D_h.

4. By means of an interpolation process, a function $\bar{v}_h$ is obtained whose domain is D and whose values agree with v_h on D_h.

5. The function $\bar{v}_h$ is regarded as an *approximate* solution of the original problem $\mathbf{P}$. Error estimates are made to justify this. In particular, as $h \to 0$, $\bar{v}_h$ should converge to a solution of $\mathbf{P}$.

Example 1. This strategy will now be illustrated by a two-point boundary-value problem:

$$(1) \qquad \begin{cases} u'' + au' + bu = c & 0 < t < 1 \\ u(0) = 0 \qquad u(1) = 0 \end{cases}$$

The coefficients a, b, and c are allowed to be functions, assumed to be continuous in the independent variable t. Notice that the differential equation is linear. That is, the unknown function u occurs in a linear fashion. The boundary-value problem (1) is problem $\mathbf{P}$ in the previous discussion. The domain D is the interval $[0, 1]$.

For a discretization, let us choose equally spaced points in the interval as follows:

$$0 = t_0 < t_1 < \cdots < t_n < t_{n+1} = 1 \qquad t_i = ih\,, \quad h = 1/(n+1)$$

The parameter h in the previous discussion is just the step size in the boundary-value problem.

Directly from Equations (1), we have

$$u''(t_i) + a(t_i)u'(t_i) + b(t_i)u(t_i) = c(t_i) \qquad (1 \leqslant i \leqslant n)$$

That equation is written in abbreviated form as

$$u_i'' + a_i u_i' + b_i u_i = c_i \qquad (1 \leqslant i \leqslant n)$$

A "discrete version" of the original problem is obtained by replacing derivatives in Equation (1) by approximations. Two commonly used formulas (discussed later, in Lemma 2) are these:

(2) $$f''(t) = \frac{f(t+h) - 2f(t) + f(t-h)}{h^2} - \frac{1}{12}h^2 f^{(4)}(\tau)$$

(3) $$f'(t) = \frac{f(t+h) - f(t-h)}{2h} - \frac{1}{6}h^2 f'''(\tau)$$

We use these (without the error terms involving τ) to approximate u' and u'' at the points t_i. Since we wish to use u as the solution of the original problem, we use a different letter to denote the solution of the discretized problem. Thus v will be a vector of $n+2$ components, and v_i is expected to be an approximate value of $u(t_i)$. The problem $\mathbf{P}_h$ is

(4) $$\begin{cases} \dfrac{v_{i+1} - 2v_i + v_{i-1}}{h^2} + a_i \dfrac{v_{i+1} - v_{i-1}}{2h} + b_i v_i = c_i \quad (1 \leqslant i \leqslant n) \\ v_0 = v_{n+1} = 0 \end{cases}$$

Here we have written $a_i = a(t_i)$, and so on. Problem (4) is a system of n linear equations in n unknowns v_i. It is solved by standard methods of linear algebra, such as Gaussian elimination. The ith equation in the system can be written in the form

(5) $$v_{i-1}\left(h^{-2} - \frac{1}{2}h^{-1}a_i\right) + v_i(-2h^{-2} + b_i) + v_{i+1}\left(h^{-2} + \frac{1}{2}h^{-1}a_i\right) = c_i$$

It is clear that the coefficient matrix for this system is tridiagonal, because the ith equation contains only the three unknowns v_{i-1}, v_i, and v_{i+1}. Furthermore, if h is small enough and if $b(t) < 0$, the matrix will be diagonally dominant. Indeed, assume that $h|a_i| \leqslant 2$. Then $h^{-2} \pm \frac{1}{2}h^{-1}a_i$ is nonnegative and $-2h^{-2} + b_i$ is nonpositive. The condition for diagonal dominance in a generic $n \times n$ matrix $A = (A_{ij})$ is

$$|A_{ii}| - \sum_{\substack{j=1 \\ j \neq i}}^{n} |A_{ij}| > 0 \qquad (i = 1, \ldots, n)$$

In this particular case, the condition becomes

(6) $$2h^{-2} - b_i - \left(h^{-2} - \frac{1}{2}h^{-1}a_i\right) - \left(h^{-2} + \frac{1}{2}h^{-1}a_i\right) \equiv -b_i > 0$$

We write System (5) in the form $Av = c$, where A is the tridiagonal matrix and c now denotes the vector having components c_i. The vectors v and c should be "column vectors."

Let us assume that the linear system has been solved to produce the vector v. The next step is to "fill in" the values of a continuous function $\bar{v}$ such that $\bar{v}(t_i) = v_i$ $(1 \leqslant i \leqslant n)$. This can be done in many ways, such as by means of a cubic spline interpolant. Another way of interpreting this step is to say that we have extended the function v to the function $\bar{v}$.

In order to investigate how close the approximate solution may be to the true solution, we begin by recording three equations that are satisfied by the true solution:

$$u'' + au' + bu = c$$

$$u''(t_i) + a(t_i)u'(t_i) + b(t_i)u(t_i) = c(t_i)$$

(7) $\dfrac{u_{i+1} - 2u_i + u_{i-1}}{h^2} - \dfrac{h^2 u^{(4)}(\tau_i)}{12} + a_i\left[\dfrac{u_{i+1} - u_{i-1}}{2h} - \dfrac{h^2 u^{(3)}(\xi_i)}{6}\right] + b_i u_i = c_i$

On the other hand, the solution to the discrete problem satisfies the equation

(8) $\qquad h^{-2}(v_{i+1} - 2v_i + v_{i-1}) + a_i(2h)^{-1}(v_{i+1} - v_{i-1}) + b_i v_i = c_i$

By subtracting one equation from the other, we arrive at an equation for the "error" $e = u - v$:

(9) $\qquad h^{-2}(e_{i+1} - 2e_i + e_{i-1}) + a_i(2h)^{-1}(e_{i+1} - e_{i-1}) + b_i e_i = h^2 d_i$

Here $d_i = \frac{1}{12}u^{(4)}(\tau_i) + \frac{1}{6}a_i u^{(3)}(\xi_i)$. Equation (9) has the same coefficient matrix as Equation (8). If we denote that matrix by A_h, Equation (9) has the form

$$A_h e = h^2 d$$

Now by the lemma that follows, and by Equation (6),

$$\left\|A_h^{-1}\right\|_\infty \leqslant \max_i\left\{|a_{ii}| - \sum_{\substack{j=1 \\ j \neq i}}^{n} |a_{ij}|\right\}^{-1} = \max_i(-b_i)^{-1}$$

Since b is continuous and negative on $[0, 1]$, there is a positive δ, independent of i and h, such that $-b_i \geqslant \delta$. Thus $\left\|A_h^{-1}\right\|_\infty \leqslant 1/\delta$, and we have

$$\|e\|_\infty \leqslant h^2\left\|A_h^{-1}\right\|_\infty \|d\|_\infty \leqslant h^2 \delta^{-1}\left(\frac{1}{12}\left\|u^{(4)}\right\|_\infty + \frac{1}{6}\left\|u^{(3)}\right\|_\infty\right)$$

Thus as $h \to \infty$, the discrete solution converges to the true solution at the speed $\mathcal{O}(h^2)$. ($\mathcal{O}(h)$ is a generic function such that $|\mathcal{O}(h)| \leqslant ch$.) ∎

Lemma 1. If an $n \times n$ matrix A is diagonally dominant, then it is nonsingular, and

$$\left\|A^{-1}\right\|_\infty \leqslant \max_i\left\{|a_{ii}| - \sum_{\substack{j=1 \\ j \neq i}}^{n} |a_{ij}|\right\}^{-1}$$

Proof. Let x be any nonzero vector, and let $y = Ax$. Select i so that $|x_i| = \|x\|_\infty$. Then

$$a_{ii}x_i + \sum_{\substack{j=1 \\ j\neq i}}^{n} a_{ij}x_j = y_i$$

$$|a_{ii}x_i| \leqslant |y_i| + \sum_{\substack{j=1 \\ j\neq i}}^{n} |a_{ij}|\,|x_j|$$

$$|a_{ii}|\,\|x\|_\infty \leqslant |y_i| + |x_i| \sum_{\substack{j=1 \\ j\neq i}}^{n} |a_{ij}| \leqslant \|y\|_\infty + \|x\|_\infty \sum_{\substack{j=1 \\ j\neq i}}^{n} |a_{ij}|$$

Hence

$$\|x\|_\infty \left(|a_{ii}| - \sum_{\substack{j=1 \\ j\neq i}}^{n} |a_{ij}| \right) \leqslant \|y\|_\infty$$

This shows that $y \neq 0$. Thus A maps no nonzero vector into 0, and A is nonsingular. If we write $x = A^{-1}y$ in the above inequality, we obtain

$$\|A^{-1}y\|_\infty \leqslant \|y\|_\infty \left(|a_{ii}| - \sum_{\substack{j=1 \\ j\neq i}}^{n} |a_{ij}| \right)^{-1}$$

and this implies the upper bound in the lemma.   ∎

Lemma 2.   If $f^{(4)}$ is continuous on $(t-h, t+h)$, then

$$f''(t) = h^{-2}\big[f(t+h) - 2f(t) + f(t-h)\big] - \frac{1}{12}h^2 f^{(4)}(\xi)$$

$$f'(t) = (2h)^{-1}\big[f(t+h) - f(t-h)\big] - \frac{1}{6}h^2 f'''(\eta)$$

Proof. We derive the first formula and leave the second as a problem. By Taylor's Theorem we have

$$f(t+h) = f(t) + hf'(t) + \frac{1}{2}h^2 f''(t) + \frac{1}{6}h^3 f'''(t) + \frac{1}{24}h^4 f^{(4)}(\xi_1)$$

$$f(t-h) = f(t) - hf'(t) + \frac{1}{2}h^2 f''(t) - \frac{1}{6}h^3 f'''(t) + \frac{1}{24}h^4 f^{(4)}(\xi_2)$$

Upon adding these two equations, we get

$$f(t+h) + f(t-h) = 2f(t) + h^2 f''(t) + \frac{1}{24}h^4 \big[f^{(4)}(\xi_1) + f^{(4)}(\xi_2)\big]$$

Upon rearranging this, we obtain

$$f''(t) = h^{-2}[f(t+h) - 2f(t) + f(t-h)] - \frac{1}{24}h^2[f^{(4)}(\xi_1) + f^{(4)}(\xi_2)]$$

Observe that the expression $\frac{1}{2}[f^{(4)}(\xi_1) + f^{(4)}(\xi_2)]$ is the average of two values of $f^{(4)}$ on the interval $[t-h, t+h]$. Its value therefore lies between the maximum and minimum of $f^{(4)}$ on this interval. If $f^{(4)}$ is continuous, this value is assumed at some point ξ in the same interval. Hence the error term can be written as $-h^2 f^{(4)}(\xi)/12$. ∎

Example 2.   We give another illustration of the discretization strategy. Consider a linear integral equation, such as

$$\int_a^b k(s,t)x(s)\,ds = v(t) \qquad (a \leqslant t \leqslant b)$$

In this equation, the kernel k and the function v are prescribed. We seek the unknown function x.

Suppose that a quadrature formula of the type

$$\int_a^b f(s)\,ds \approx \sum_{j=1}^{n} c_j f(s_j)$$

is available. (The points s_j need not be equally spaced.) Taking $t = s_i$ in the integral equation, we have

$$\int_a^b k(s, s_i)x(s)\,ds = v(s_i) \qquad (1 \leqslant i \leqslant n)$$

Applying the quadrature formula leads to a discrete version of the integral equation:

$$\sum_{j=1}^{n} c_j k(s_j, s_i)x(s_j) = v(s_i) \qquad (1 \leqslant i \leqslant n)$$

This is a system of n linear equations in the unknowns $x(s_j)$; it can be solved by standard methods. Then an interpolation method can be used to reconstruct $x(t)$ on the interval $a \leqslant t \leqslant b$. Approximations have been made at two stages, and the resulting function x is *not* the solution of the original problem. This strategy is considered later in more detail (Section 4.7). ∎

References for this chapter are [AR], [AY], [AlG], [At], [Au], [Bak], [Bar], [Brez], [BrS], [CCC], [CMY], [Cia], [CH], [Dav], [DMo], [Det], [Dzy], [Eav], [Fic], [GG], [GZ1], [GZ2], [GZ3], [Gold], [Gre], [Gri], [Hen], [HS], [IK], [KA], [KK], [Kee], [KC], [Kras], [Kr], [LSU], [Leis], [Li], [LM], [Lo], [Lue1], [Lue2], [Mey], [Mil], [Moo], [Mor1], [Mor2], [Naz1], [Naz2], [OD], [OR], [Ped], [Pry], [Red], [Rh1], [Rh2], [Ros], [Sa], [Sm], [Tod], [Wac], [Was], [Wat], [Wilf], and [Zien].

Problems 4.1

1. Establish the second formula in the second lemma.

2. Derive the following formula and its error term:

$$f'''(x) \approx [f(x + 2h) - 2f(x + h) + 2f(x - h) - f(x - 2h)]/(2h^3)$$

3. To change a boundary value problem

$$u'' + au' + bu = c \qquad u(\alpha) = 0 \qquad u(\beta) = 0$$

   into an equivalent one on the interval $[0, 1]$, we change the independent variable from t to s using the equation $t = \beta s + \alpha(1 - s)$. What is the new boundary value problem?

4. To change a boundary value problem

$$u'' + au' + bu = c \qquad u(0) = \alpha \qquad u(1) = \beta$$

   into an equivalent one having homogeneous boundary conditions, we make a change in the dependent variable $v = u - \ell$, where $\ell(t) = \alpha + (\beta - \alpha)t$. What is the new boundary value problem?

5. A kernel of the form $K(s, t) = \sum_{i=1}^{n} u_i(s)v_i(t)$ is said to be "separable" or "degenerate." If such a kernel occurs in the integral equation

$$x(t) = \int_0^1 K(s, t)x(s)\, ds + w(t)$$

   then a solution can be found in the form $x = w + \sum_{i=1}^{n} \alpha_i v_i$. Carry out the solution based on this idea. Illustrate, by using the separable kernel e^{s-t}.

4.2 The Method of Iteration

The term **iteration** can be applied to any repetitive process, but traditionally it refers to an algorithm of the following nature:

(1) x_0 given, $x_{n+1} = Fx_n$, $n = 0, 1, 2, \ldots$

We can also write $x_n = F^n x_0$, where F^0 is the identity map and $F^{n+1} = F \circ F^n$. In such a procedure, the entities $x_0, x_1, \ldots$ are usually elements in some topological space X, and the map $F : X \to X$ should be continuous. If $\lim_{n \to \infty} x_n$ exists, then it is a fixed point of F, because

(2) $F(\lim x_n) = \lim F x_n = \lim x_{n+1} = \lim x_n$

The method of iteration can be considered as one technique for finding fixed points of operators.

   The Contraction Mapping Theorem (due to Banach, 1922) is an elegant and powerful tool for establishing that the sequence defined in Equation (1) converges. We require the notion of a **contraction,** or **contractive mapping**.

Such a mapping is defined from a metric space X into itself and satisfies an inequality

(3) $$d(Fx, Fy) \leqslant \theta d(x, y) \qquad (x, y \in X)$$

in which θ is a positive constant less than 1. Complete metric spaces were the subject of Problem 48 in Section 1.2, page 15. Every Banach space is necessarily a complete metric space, it being assumed that the distance function is $d(x, y) = \|x - y\|$. A closed set in a Banach space is also a complete metric space. Since most of our examples occur in this setting, the reader will lose very little generality by letting X be a closed subset of a Banach space in the Contraction Mapping Theorem.

Theorem 1. Contraction Mapping Theorem.   *If F is a contraction on a complete metric space X, then F has a unique fixed point ξ. The point ξ is the limit of every sequence generated from an arbitrary point x by iteration*

$$[x, Fx, F^2 x, \dots] \qquad (x \in X)$$

Proof.   Reverting to the previous notation, we select x_0 arbitrarily in X and define $x_{n+1} = F x_n$ for $n = 0, 1, 2, \dots$. We have

$$d(x_n, x_{n-1}) = d(F x_{n-1}, F x_{n-2}) \leqslant \theta d(x_{n-1}, x_{n-2})$$

This argument can be repeated, and we conclude that

(4) $$d(x_n, x_{n-1}) \leqslant \theta^{n-1} d(x_1, x_0)$$

In order to establish the Cauchy property of the sequence $[x_n]$, let $n > N$ and $m > N$. There is no loss of generality in supposing that $m \geqslant n$. Then from Equation (4),

$$
\begin{aligned}
d(x_m, x_n) &\leqslant d(x_m, x_{m-1}) + d(x_{m-1}, x_{m-2}) + \cdots + d(x_{n+1}, x_n) \\
&\leqslant \left[\theta^{m-1} + \theta^{m-2} + \cdots + \theta^n\right] d(x_1, x_0) \\
&\leqslant \left[\theta^N + \theta^{N+1} + \cdots\right] d(x_1, x_0) \\
&= \theta^N (1 - \theta)^{-1} d(x_1, x_0)
\end{aligned}
$$

Since $0 \leqslant \theta < 1$, $\lim_{N \to \infty} \theta^N = 0$. This proves the Cauchy property. Since the space X is complete, the sequence converges to a point ξ. Since the contractive property implies directly that F is continuous, the argument in Equation (2) shows that ξ is a fixed point of F.

If η is also a fixed point of F, then we have

(5) $$d(\xi, \eta) = d(F\xi, F\eta) \leqslant \theta d(\xi, \eta)$$

If $\xi \neq \eta$, then $d(\xi, \eta) > 0$, and Inequality (5) leads to the contradiction $\theta \geqslant 1$. This proves the uniqueness of the fixed point.   ∎

The iterative procedure is well illustrated by a Fredholm integral equation, $x = Fx$, where

(6) $$(Fx)(t) = \int_0^1 K\big(s,t,x(s)\big)\,ds + w(t)   \qquad (0 \leqslant t \leqslant 1)$$

It is assumed that w is continuous and that $K(s,t,r)$ is continuous on the domain in $\mathbb{R}^3$ defined by the inequalities

$$0 \leqslant s \leqslant 1 \qquad 0 \leqslant t \leqslant 1 \qquad -\infty < r < \infty$$

We will seek a solution x in the space $C[0,1]$. This space is complete if it is given the standard norm

$$\|x\|_\infty = \sup_t |x(t)|$$

In order to see whether F is a contraction, we estimate $\|Fu - Fv\|$:

(7) $$|(Fu)(t) - (Fv)(t)| \leqslant \int_0^1 |K\big(s,t,u(s)\big) - K\big(s,t,v(s)\big)|\,ds$$

If K satisfies a Lipschitz condition of the type

(8) $$|K(s,t,\xi) - K(s,t,\eta)| \leqslant \theta|\xi - \eta| \qquad (\theta < 1)$$

then from Equation (7) we get

$$|(Fu)(t) - (Fv)(t)| \leqslant \theta\|u - v\|$$

and the contraction condition

(9) $$\|Fu - Fv\| \leqslant \theta\|u - v\|$$

By Banach's theorem, the iteration $x_{n+1} = Fx_n$ leads to a solution, starting from any function x_0 in $C[0,1]$. This proves the following result.

Theorem 2.   *If K satisfies the hypotheses in the preceding paragraph, then the integral equation (6) has a unique solution in the space $C[0,1]$.*

Example 1.   Consider the nonlinear Fredholm equation

(10) $$x(t) = \frac{1}{2} \int_0^1 \cos\big(stx(s)\big)\,ds$$

By the mean value theorem,

$$|\cos(st\xi) - \cos(st\eta)| = |\sin(st\zeta)|\,|st\xi - st\eta| \leqslant |\xi - \eta|$$

Thus the preceding theory is applicable with $\theta = \frac{1}{2}$. If the iteration is begun with $x_0 = 0$, the next two steps are $x_1(t) = \frac{1}{2}$ and $x_2(t) = t^{-1}\sin(t/2)$. The next element in the sequence is given by

$$x_3(t) = \frac{1}{2}\int_0^1 \cos\left(t\sin\frac{s}{2}\right)ds$$

This integration cannot be effected in elementary functions. In fact, it is analogous to the Bessel function J_0, whose definition is

$$J_0(z) = \frac{1}{\pi}\int_0^\pi \cos(z\sin\theta)\,d\theta$$

If the iteration method is to be continued in this example, numerical procedures for indefinite integration will be needed. ∎

The method of iteration can also be applied to differential equations, usually by first turning them into equivalent integral equations. The procedure is of great theoretical importance, as it is capable of yielding existence theorems with very little effort. We present some important theorems to illustrate this topic.

Theorem 3.   *Let S be an interval of the form $S = [0, b]$. Let f be a continuous map of $S \times \mathbb{R}$ to $\mathbb{R}$. Assume a Lipschitz condition in the second argument:*

$$|f(s, t_1) - f(s, t_2)| \leqslant \lambda|t_1 - t_2|$$

where λ is a constant depending only on f. Then the initial-value problem

$$x'(s) = f(s, x(s)) \qquad x(0) = \beta$$

has a unique solution in $C(S)$.

Proof.   We introduce a new norm in $C(S)$ by defining

$$\|x\|_w = \sup_{s\in S} |x(s)|\, e^{-2\lambda s}$$

The space $C(S)$, accompanied by this norm, is complete. Since the initial-value problem is equivalent to the integral equation

$$x = Ax \qquad (Ax)(s) = \beta + \int_0^s f(t, x(t))\,dt \qquad x \in C(S)$$

all we have to do is show that the mapping A has a fixed point. In order for the Contraction Mapping Theorem to be used, it suffices to establish that A is a contraction. Let $u, v \in C(S)$. Then we have, for $0 \leqslant s \leqslant b$,

$$|(Au - Av)(s)| \leqslant \int_0^s |f(t, u(t)) - f(t, v(t))|\,dt$$

$$\leqslant \int_0^s \lambda|u(t) - v(t)|\,dt$$

$$= \lambda \int_0^s e^{2\lambda t}\, e^{-2\lambda t}|u(t) - v(t)|\,dt$$

$$\leqslant \lambda\|u - v\|_w \int_0^s e^{2\lambda t}\,dt$$

$$\leqslant \lambda\|u - v\|_w (2\lambda)^{-1} e^{2\lambda s}$$

From this we conclude that

$$e^{-2\lambda s}\left|(Au - Av)(s)\right| \leqslant \tfrac{1}{2}\left\|u - v\right\|_w$$

and that

$$\left\|Au - Av\right\|_w \leqslant \tfrac{1}{2}\left\|u - v\right\|_w \qquad\blacksquare$$

Example 2.   Does the following initial value problem have a solution in the space $C[0, 10]$?

$$x' = \cos(xe^s) \qquad x(0) = 0$$

This is an illustration of the general theory in which $f(s, t) = \cos(te^s)$. By the mean value theorem,

$$|f(s, t_1) - f(s, t_2)| = \left|\frac{\partial f}{\partial t}(s, \tau)\right| \, |t_1 - t_2|$$

For $0 \leqslant s \leqslant 10$ and $t \in \mathbb{R}$,

$$\left|\frac{\partial f}{\partial t}\right| = |-\sin(te^s)e^s| \leqslant e^{10}$$

Hence, the hypothesis of Theorem 3 is satisfied, and our problem has a unique solution in $C[0, 10]$. $\qquad\blacksquare$

Example 3.   If f is continuous but does not satisfy the Lipschitz condition in Theorem 3, the conclusions of the theorem may fail. For example, the problem $x' = x^{2/3}$, $x(0) = 0$ has two solutions, $x(s) = 0$ and $x(s) = s^3/27$. There is no Lipschitz condition of the form

$$\left|t_1^{2/3} - t_2^{2/3}\right| \leqslant \lambda|t_1 - t_2|$$

(Consider the implications of this inequality when $t_2 = 0$.) $\qquad\blacksquare$

Example 4.   The problem $x' = x^2$, $x(0) = 1$ does not conform to the hypotheses of Theorem 3 because there is no Lipschitz condition. By appealing to other theorems in the theory of differential equations, one can conclude that there is a solution in some interval about the initial point $s = 0$. $\qquad\blacksquare$

In order for us to handle *systems* of differential equations, the preceding theorem must be extended. A system of n differential equations accompanied by initial values has this form:

$$
\begin{aligned}
x_1'(s) &= f_1(s, x_1(s), x_2(s), \ldots, x_n(s)) & x_1(0) &= 0 \\
x_2'(s) &= f_2(s, x_1(s), x_2(s), \ldots, x_n(s)) & x_2(0) &= 0 \\
&\;\cdots\cdots\cdots & &\cdots\cdots \\
x_n'(s) &= f_n(s, x_1(s), x_2(s), \ldots, x_n(s)) & x_n(0) &= 0
\end{aligned}
$$

The right way of viewing this is as a *single* equation involving a function $\mathbf{x} : S \to \mathbb{R}^n$, where S is an interval of the form $[0, b]$. Likewise, we must have a function $\mathbf{f} : S \times \mathbb{R}^n \to \mathbb{R}^n$. We then adopt any convenient norm on $\mathbb{R}^n$, and define the norm of $\mathbf{x}$ to be

$$\|\mathbf{x}\|_w = \sup_{s \in S} e^{-2\lambda s} \|\mathbf{x}(s)\|$$

The Lipschitz condition on $\mathbf{f}$ is

$$\|\mathbf{f}(s, \mathbf{u}) - \mathbf{f}(s, \mathbf{v})\| \leqslant \lambda \|\mathbf{u} - \mathbf{v}\| \qquad \mathbf{u}, \mathbf{v} \in \mathbb{R}^n$$

The setting for the theorem is now $C(S, \mathbb{R}^n)$, which is the space of all continuous maps $\mathbf{x} : S \to \mathbb{R}^n$, normed with $\| \ \|_w$. The equation $\mathbf{x}'(s) = \mathbf{f}(s, \mathbf{x}(s))$ now represents the system of differential equations referred to earlier. For further discussion see the book by Edwards [Edw], pp. 153–155.

The use of iteration to solve differential equations predates Banach's result by many years. Ince [4] says that it was probably known to Cauchy, but was apparently first published by Liouville in 1838. Picard described it in its general form in 1893. It is often referred to as *Picard* iteration. It is rarely used directly in the numerical solution of initial value problems because the step-by-step methods of numerical integration are superior. Here is an artificial example to show how it works.

Example 5.

$$x' = 2t(1 + x) \qquad x(0) = 0$$

The formula for the Picard iteration in this example is

$$x_{n+1}(t) = \int_0^t 2s\big(1 + x_n(s)\big)\, ds = t^2 + \int_0^t 2s x_n(s)\, ds$$

If $x_0 = 0$, then successive computations yield

$$x_1(t) = t^2 \qquad x_2(t) = t^2 + \frac{1}{2}t^4 \qquad x_3(t) = t^2 + \frac{1}{2}t^4 + \frac{1}{6}t^6$$

It appears that we are producing the partial sums in the Taylor series for $e^{t^2} - 1$, and one verifies readily that this is indeed the solution.   ∎

In some applications it is useful to have the following extension of Banach's Theorem:

Theorem 4.   *Let F be a mapping of a complete metric space into itself such that for some m, F^m is contractive. Then F has a unique fixed point. It is the limit of every sequence $[F^k x]$, for arbitrary x.*

Proof.   Since F^m is contractive, it has a unique fixed point ξ by Theorem 1. Then

$$F\xi = F(F^m \xi) = F^{m+1}\xi = F^m(F\xi)$$

This shows that $F\xi$ is also a fixed point of F^m. By the uniqueness of ξ, $F\xi = \xi$. Thus F has at least one fixed point (namely ξ), and ξ can be obtained by iteration using the function F^m. If x is any fixed point of F, then

$$Fx = x \qquad F^2x = x \ldots \qquad F^mx = x$$

Thus x is a fixed point of F^m, and $x = \xi$.

It remains to be proved that the sequence F^nx converges to ξ as $n \to \infty$. Observe that for $i \in \{1, 2, \ldots, m\}$ we have

$$F^{nm+i}x = F^{nm}(F^ix) \to \xi \quad \text{as} \quad n \to \infty$$

by the first part of the proof. If $\varepsilon > 0$, we can select an integer N having the property

$$n \geqslant N \quad \Longrightarrow \quad d(F^{nm+i}x, \xi) < \varepsilon \qquad (1 \leqslant i \leqslant m)$$

Since each integer j greater than Nm can be written as $j = nm+i$, where $n \geqslant N$ and $1 \leqslant i \leqslant m$, we have

$$j > Nm \quad \Longrightarrow \quad d(F^jx, \xi) < \varepsilon$$

This proves that $\lim F^jx = \xi$.   ∎

To illustrate the application of this theorem, we consider a linear Volterra integral equation, which typically would have the form

$$(11) \qquad x(t) = \int_a^t K(t, s)x(s)\, ds + v(t) \qquad x \in C[a, b]$$

(The presence of an *indefinite* integral classifies this as a Volterra equation.) Equation (11) can be written as

$$x = Ax + v$$

in which A is a linear operator defined by writing

$$(Ax)(t) = \int_a^t K(t, s)x(s)\, ds$$

It is clear that
$$|(Ax)(t)| \leqslant \|K\|_\infty \|x\|_\infty (t - a)$$

From this it follows that

$$|(A^2x)(t)| \leqslant \int_a^t |K(t, s)|\, |(Ax)(s)|\, ds$$

$$\leqslant \int_a^t |K(t, s)|\, \|K\|_\infty \|x\|_\infty (s - a)\, ds$$

$$\leqslant \|K\|_\infty^2 \|x\|_\infty \frac{(t - a)^2}{2}$$

Repetition of this argument leads to the estimate

$$|(A^n x)(t)| \leqslant \|K\|_\infty^n \|x\|_\infty \frac{(t-a)^n}{n!} \leqslant \|K\|_\infty^n \|x\|_\infty \frac{(b-a)^n}{n!}$$

This tells us that

$$\|A^n\| \leqslant c^n/n! \qquad c = \|K\|_\infty (b-a)$$

Now select m so that $\|A^m\| < 1$. Denote the right side of Equation (11) by $(Fx)(t)$, so that $Fx = Ax + v$. Then

$$F^2 x = A^2 x + Av + v$$
$$F^3 x = A^3 x + A^2 v + Av + v$$

and so on. Thus

$$\|F^m x - F^m y\| = \|A^m x - A^m y\| \leqslant \|A^m\| \, \|x - y\|$$

This shows that F^m is a contraction. By Theorem 4, F has a unique fixed point, which can be obtained by iteration of the map F. Observe that this conclusion is reached without making strong assumptions about the kernel K. Our work establishes the following existence theorem.

Theorem 5.   *Let v be continuous on $[a, b]$ and let K be continuous on the square $[a, b] \times [a, b]$. Then the integral equation*

$$x(t) = \int_a^t K(s, t)x(s)\, ds + v(t)$$

has a unique solution in $C[a, b]$.

The next theorem concerns the solvability of a nonlinear equation $F(x) = b$ in a Hilbert space. The result is due to Zarantonello [Za].

Theorem 6.   *Let F be a mapping of a Hilbert space into itself such that*

   (a) $\langle Fx - Fy, x - y \rangle \geqslant \alpha \|x - y\|^2 \qquad (\alpha > 0)$
   (b) $\|Fx - Fy\| \leqslant \beta \|x - y\|$

Then F is surjective and injective. Consequently, F^{-1} exists.

Proof.   The injectivity follows at once from (a): If $x \neq y$, then $Fx \neq Fy$. For the surjectivity, let w be any point in the Hilbert space. It is to be proved that, for some x, $Fx = w$. It is equivalent to prove, for any $\lambda > 0$, that an x exists satisfying $x - \lambda(Fx - w) = x$. Define $Gx = x - \lambda(Fx - w)$, so that our task is to prove the existence of a fixed point for G. The Contraction Mapping

Theorem will be applied. To prove that G is a contraction, we let $\lambda = \alpha/\beta^2$ in the following calculation:

$$
\begin{aligned}
\left\|Gx - Gy\right\|^2 &= \left\|x - \lambda(Fx - w) - y + \lambda(Fy - w)\right\|^2 \\
&= \left\|x - y - \lambda(Fx - Fy)\right\|^2 \\
&= \left\|x - y\right\|^2 - 2\lambda\langle Fx - Fy, x - y\rangle + \lambda^2\left\|Fx - Fy\right\|^2 \\
&\leqslant \left\|x - y\right\|^2 - 2\lambda\alpha\left\|x - y\right\|^2 + \lambda^2\beta^2\left\|x - y\right\|^2 \\
&= \left\|x - y\right\|^2(1 - 2\lambda\alpha + \lambda^2\beta^2) \\
&= \left\|x - y\right\|^2(1 - 2\alpha^2/\beta^2 + \alpha^2/\beta^2) \\
&= \left\|x - y\right\|^2(1 - \alpha^2/\beta^2) \qquad\qquad \blacksquare
\end{aligned}
$$

Problems 4.2

1. From the existence theorem proved in the text deduce a similar theorem for the initial value problem
$$
x'(s) = f(s, x(s)) \qquad x(a) = c \qquad a \leqslant s \leqslant b
$$

2. Let F be a mapping of a Banach space X into itself. Let $x_0 \in X$, $r > 0$, and $0 \leqslant \lambda < 1$. Assume that on the closed ball $B(x_0, r)$ we have
$$
\|Fx - Fy\| \leqslant \lambda\|x - y\|
$$
Assume also that $\|x_0 - Fx_0\| < (1 - \lambda)r$. Prove that $F^n x_0 \in B(x_0, r)$ and that $x^* = \lim F^n x_0$ exists. Prove that x^* is a fixed point of F and that $\|F^n x_0 - x^*\| \leqslant \lambda^n r$.

3. For what values of λ can we be sure that the integral equation
$$
x(t) = \lambda \int_0^1 e^{st} \cos x(s)\, ds + \tan t
$$
has a continuous solution on $[0,1]$?

4. Prove that there is no contraction of X onto X if X is a compact metric space having at least two points.

5. Give an example of a Banach space X and a map $F : X \to X$ having both of the following properties:
   (a) $\|Fx - Fy\| < \|x - y\|$ whenever $x \neq y$
   (b) $Fx \neq x$ for all x

6. Prove that the integral equation
$$
x(t) = \int_0^t [x(s) + s]\sin s\, ds
$$
has a unique solution in $C[0, \pi/2]$, and give an iterative process whose limit is the solution.

7. Prove that if X is a compact metric space and F is a mapping from X to X such that $d(Fx, Fy) < d(x, y)$ when $x \neq y$, then F has a unique fixed point.

8. Let F be a contraction defined on a metric space that is not assumed to be complete. Prove that
$$
\inf_x d(x, Fx) = 0
$$

9. Let F be a mapping on a metric space such that $d(Fx, Fy) < d(x, y)$ when $x \neq y$. Let x be a point such that the sequence $F^n x$ has a cluster point. Show that this cluster point is a fixed point of F (Edelstein).

10. Carry out 4 steps of Picard iteration in the initial value problem $x' = x + 1$, $x(0) = 0$.

11. Give an example of a discontinuous map $F : \mathbb{R} \to \mathbb{R}$ such that $F \circ F$ is a contraction. Find the fixed point of F.

12. Extend the theorem in Problem 7 by showing that the fixed point is the limit of $F^n x$, for arbitrary x.

13. The **diameter** of a metric space X is

$$\mathrm{diam}(X) = \sup\{d(x, y) : x \in X , y \in X\}$$

This is allowed to be $+\infty$. Show that there cannot exist a surjective contraction on a metric space of finite nonzero diameter. (Cf. Problem 4.)

14. Let X be a Banach space and f a mapping of X into X. Are these two properties of f equivalent?
   (i) f has a fixed point.
   (ii) There is a nonempty closed set E in X such that $f(E) \subset E$ and such that $\|f(x) - f(y)\| < \frac{1}{2}\|x - y\|$ for all x, y in E.

15. Let T be a contraction on a metric space:

$$d(Tx, Ty) \leqslant \lambda d(x, y) \qquad (\lambda < 1)$$

Prove that the set $\{x : d(x, Tx) \leqslant \varepsilon\}$ is nonempty, closed, and of diameter at most $\varepsilon(1 - \lambda)^{-1}$.

16. The Volterra integral equation

$$x(t) = w(t) + \int_0^t (t + s)x(s)\, ds$$

is equivalent to an initial-value problem involving a second-order linear differential equation. Find it.

17. Let F be a mapping of a complete metric space into itself. If ξ is a fixed point of F^m (for some m), does it follow that ξ is a fixed point of F?

18. Let $f : [0, b] \times \mathbb{R} \to \mathbb{R}$. Prove that if f and $\partial f/\partial t$ are continuous (t being the second argument of f), then the initial-value problem $x'(s) = f(s, x(s))$, $x(0) = \beta$ has a unique solution on $[0, b]$.

19. Prove that if $F : \mathbb{R}^n \to \mathbb{R}^n$ is a contraction, then $I + F$ is a homeomorphism of $\mathbb{R}^n$ onto $\mathbb{R}^n$.

20. Prove that if $g \in C[0, b]$, then the differential equation $x'(s) = \cos(x(s)g(s))$ with prescribed initial value $x(0) = \beta$ has a unique solution.

21. A map $F : X \to X$, where X is a metric space, is said to be **nonexpansive** if $d(Fx, Fy) \leqslant d(x, y)$ for all x and y. Prove that if F is nonexpansive and if the sequence $[F^n x]$ converges for each x, then the map $x \mapsto \lim_n F^n x$ is continuous.

22. Let c_0 be the Banach space of all real sequences $x = [x_1, x_2, \ldots]$ that tend to zero. The norm is defined by $\|x\| = \max_i |x_i|$. Find all the fixed points of the mapping $f : c_0 \to c_0$ defined by $f(x) = [1, x_1, x_2, \ldots]$. Show that f is nonexpansive, according to the definition in the preceding problem. Explain the significance of this example vis-à-vis the Contraction Mapping Theorem.

23. Let U be a closed set in a complete metric space X. Let f be a contraction defined on U and taking values in X. Let λ be the contraction constant. Suppose that $x_0 \in U$,

$x_1 = f(x_0)$, $r = \lambda(1-\lambda)^{-1}d(x_0, x_1)$, and $B(x_1, r) \subset U$. Prove that f has a fixed point in $B(x_1, r)$.

24. Prove that the following integral equation (in which u is the unknown function) has a continuous solution if $|\lambda| < \frac{1}{4}$.

$$u(t) - \lambda \int_0^1 \sqrt{\frac{t+3}{s}} \, \sin[tu(s)] \, ds = e^t \qquad (0 \leqslant t \leqslant 1)$$

25. Let F be a contraction defined on a Banach space. Prove that $I - F$ is invertible and that $(I - F)^{-1} = \lim_n H_n$, where $H_0 = I$ and $H_{n+1} = I + FH_n$.

26. Prove that the following integral equation has a solution in the space $C[0, 1]$.

$$x(t) = e^t + \int_0^1 \cos[t^2 - s^2 + x(s)\sin(s)] \, ds$$

27. In the study of radiative transfer, one encounters integral equations of the form

$$u(t) = a(t) + \int_0^1 u(s)k(t - s) \, ds$$

in which u represents the flux density of the radiation at a specified wave length. Prove that this equation has a solution in the special case $k(t) = \sin t$.

4.3 Methods Based on the Neumann Series

Recall the Neumann Theorem in Section 1.5, page 28, which asserts that if a linear operator on a Banach space, $A : X \to X$, has operator norm less than 1, then $I - A$ is invertible, and

$$(1) \qquad\qquad (I - A)^{-1} = \sum_{n=0}^{\infty} A^n$$

The series in this equation is known as the **Neumann series** for $(I - A)^{-1}$.

   This theorem is easy to remember because it is the analogue of the familiar geometric series for complex numbers:

$$(2) \qquad\qquad \frac{1}{1 - z} = 1 + z + z^2 + z^3 + \cdots \qquad\qquad |z| < 1$$

   In using the Neumann series, one can generate the partial sums $x_n = \sum_{k=0}^n A^k v$ by setting $x_0 = v_0 = v$ and computing inductively

$$(3) \qquad\qquad v_n = Av_{n-1} \qquad x_n = x_{n-1} + v_n \qquad (n = 1, 2, \ldots)$$

Another iteration is suggested in Problem 25.

   For linear operator equations in a Banach space, one should not overlook the possibility of a solution by the Neumann series. This remark will be illustrated with some examples.

Example 1. First, consider an integral equation of the form

(4)
$$x(t) - \lambda \int_0^1 K(s,t)x(s)\, ds = v(t) \qquad (0 \leqslant t \leqslant 1)$$

In this equation, K and v are prescribed, and x is to be determined. For certain values of λ, solutions will exist. Write the equation in the form

(5)
$$(I - \lambda A)x = v$$

in which A is the integral operator in Equation (4). If we have chosen a suitable Banach space and if $\|\lambda A\| < 1$, then the Neumann series gives a formula for the solution:

(6)
$$x = (I - \lambda A)^{-1}v = \sum_{n=0}^{\infty} (\lambda A)^n v = v + \lambda Av + \lambda^2 A^2 v + \cdots \qquad \blacksquare$$

Example 2. For a concrete example of this, consider

(7)
$$x(t) = \lambda \int_0^1 e^{t-s}x(s)\, ds + v(t)$$

Here, we use an operator A defined by

$$(Ax)(t) = \int_0^1 e^{t-s}x(s)\, ds$$

If we compute A^2x, the result is

$$(A^2x)(t) = \int_0^1 e^{t-s}(Ax)(s)\, ds$$

$$= \int_0^1 e^{t-s} \int_0^1 e^{s-\sigma}x(\sigma)\, d\sigma\, ds$$

$$= \int_0^1 \left[\int_0^1 e^{t-s}e^{s-\sigma}\, ds \right] x(\sigma)\, d\sigma$$

$$= \int_0^1 e^{t-\sigma}x(\sigma)\, d\sigma = (Ax)(t)$$

This shows that $A^2 = A$. Consequently, the solution by the Neumann series in Equation (6) simplifies to

$$x = v + \lambda Av + \lambda^2 Av + \lambda^3 Av + \cdots$$

$$= v + \left(\frac{1}{1-\lambda} - 1 \right) Av = v + \frac{\lambda}{1-\lambda} Av \qquad \blacksquare$$

Example 3. Another important application of the Neumann series occurs in a process called **iterative refinement**. Suppose that we wish to solve an operator equation $Ax = v$. If A is invertible, the solution is $x = A^{-1}v$. Suppose now that we are in possession of an **approximate right inverse** of A. We mean by that an operator B such that $\|I - AB\| < 1$. Can we use B to solve the problem? It is amazing that the answer is "Yes." Obviously, $x_0 = Bv$ is a first approximation to x. By the Neumann theorem, we know that AB is invertible and that

$$(AB)^{-1} = \sum_{k=0}^{\infty}(I - AB)^k$$

It is clear that the vector $x = B(AB)^{-1}v$ is a solution, because $Ax = AB(AB)^{-1}v = v$. Hence $x =$

(8) $B(AB)^{-1}v = B\sum_{k=0}^{\infty}(I - AB)^k v = Bv + B(I - AB)v + B(I - AB)^2 v + \cdots$ ∎

Theorem.   *The partial sums in the series of Equation (8) can be computed by the algorithm $x_0 = Bv$, $x_{n+1} = x_n + B(v - Ax_n)$.*

Proof.   Let the sequence $[x_n]$ be defined by the algorithm, and let

$$y_n = B\sum_{k=0}^{n}(I - AB)^k v$$

We wish to prove that $x_n = y_n$ for all n. For $n = 0$, we have $y_0 = Bv = x_0$. Now assume that for some n, $x_n = y_n$. We shall prove that $x_{n+1} = y_{n+1}$. Put $S_n = \sum_{k=0}^{n}(I - AB)^k$, so that $y_n = BS_n v$. Then

$$
\begin{aligned}
x_{n+1} &= x_n + B(v - Ax_n) = y_n + Bv - BAy_n \\
&= BS_n v + Bv - BABS_n v = B(S_n + I - ABS_n)v \\
&= B[(I - AB)S_n + I]v = B\left[\sum_{k=0}^{n}(I - AB)^{k+1} + I\right]v \\
&= BS_{n+1}v = y_{n+1}
\end{aligned}
$$
∎

The algorithm in the preceding theorem is known as **iterative refinement**. The vector $v - Ax_n$ is called the **residual vector** associated with x_n. If the hypothesis $\|I - AB\| < 1$ is fulfilled, the Neumann series converges, the partial sums in Equation (8) converge, and by the theorem, the sequence $[x_n]$ converges to the solution of the problem. The residuals therefore converge to zero.

The method of iterative refinement is commonly applied in numerically solving systems of linear equations. Let such a system be written in the form $Ax = v$, in which A is an $N \times N$ matrix, assumed invertible. The numerical solution of such a system on a computer involves a finite sequence of linear operations being applied to v to produce the numerical solution x_0. Thus (since every linear operator on $\mathbb{R}^n$ is effected by a matrix) we have $x_0 = Bv$, in which B is a certain $N \times N$ matrix. Ideally, B would equal A^{-1} and x_0 would be the correct solution. Because of roundoff errors (which are inevitable), B is only an approximate inverse of A. If, in fact, $\|I - AB\| < 1$, then the theory outlined previously applies. Thus the initial approximation $x_0 = Bv$ can be "refined" by adding Br_0 to it, where r_0 is the residual corresponding to x_0. Thus $r_0 = v - Ax_0$. From $x_1 = x_0 + Br_0$ further refinement is possible by adding Br_1, and so on. The numerical success of this process depends upon the residuals being computed with higher precision than is used in the remaining calculations.

Problems 4.3

1. The problem $Ax = v$ is equivalent to the fixed-point problem $Fx = x$, if we define $Fx = x - Ax + v$. Suppose that A is a linear operator and that $\|I - A\| < 1$. Show that F is a contraction. Let $x_0 = v$ and $x_{n+1} = Fx_n$. Show that x_n is the partial sum of the Neumann series appropriate to the problem $Ax = v$.

2. Prove that if A is invertible and if the operator B satisfies $\|A - B\| < \|A^{-1}\|^{-1}$, then B is invertible. What does this imply about the set of invertible elements in $\mathcal{L}(X, X)$? (Here X is a Banach space.)

3. Prove that if $\inf_\lambda \|I - \lambda A\| < 1$, then A is invertible.

4. If $\|A\|$ is small, then $(I - A)^{-1} \approx I + A$. Find $\varepsilon > 0$ such that the condition $\|A\| < \varepsilon$ implies
$$\|(I - A)^{-1} - (I + A)\| \leqslant 3\|A\|^2$$

5. Make this statement precise: If $\|AB - I\| < 1$, then $2B - BAB$ is superior to B as an approximate inverse of A.

6. Prove that if X is a Banach space, if $A \in \mathcal{L}(X, X)$, and if $\|A\| < 1$, then the iteration
$$x_{n+1} = Ax_n + b$$
converges to a solution of the equation $x = Ax + b$ from any starting point x_0.

7. Let X and Y be Banach spaces. Show that the set Ω of invertible elements in $\mathcal{L}(X, Y)$ is an open set and that the map $f : \Omega \to \mathcal{L}(Y, X)$ defined by $f(A) = A^{-1}$ is continuously differentiable.

8. Give an example of an operator A that has a right inverse but is not invertible. Observe that in the theory of iterative refinement, A need not be invertible.

9. Prove that if the equation $Ax = v$ has a solution x_0 and if $\|I - BA\| < 1$, then $x_0 = (BA)^{-1}Bv$, and a suitable modification of iterative refinement will work.

10. In Example 2, prove that the solution given there is correct for all λ satisfying $\lambda \neq 1$. In particular, it is not necessary to assume that $\|\lambda A\| < 1$ in this example.

11. In Example 2, compute $\|A\|$.

12. Show how to solve the equation $(I - \lambda A)x = v$ when A is idempotent (i.e., $A^2 = A$).

13. Let A be a bounded linear operator on a normed linear space. Prove that if A is nilpotent (i.e., $A^m = 0$ for some $m \geqslant 0$), then $I - A$ is invertible. Give a formula for $(I - A)^{-1}$.

14. Prove this generalization of the Neumann Theorem. If A is a bounded linear transformation from a Banach space X into X such that the sequence $S_n = \sum_{k=0}^{n} A^k$ has the Cauchy property, then $(I - A)^{-1}$ exists and equals $\lim_{n \to \infty} S_n$. Give an example to show that this is a generalization.

15. Prove or disprove: If A is a bounded linear operator on a Banach space and if $\|A^m\| < 1$ for some m, then $(I - A)^{-1} = \sum_{k=0}^{\infty} A^k$.

16. A **Volterra** integral operator is one of the form
$$(Ax)(t) = \int_a^t K(s, t) x(s) \, ds$$

Assume that A maps $C[a,b]$ into $C[a,b]$. Prove that $(I-A)^{-1}$ exists and is given by the usual Neumann series. Refer to Section 4.2 for further information about Volterra integral equations.

17. Define $A : C[0,1] \to C[0,1]$ by the following equation, and prove that A is surjective.

$$(Ax)(t) = \int_0^1 \cos(st)x(s)\,ds + 2x(t)$$

18. Prove that the set of nonsingular $n \times n$ matrices is open and dense in the set of all $n \times n$ matrices.

19. Let $\phi \in C[0,1]$ and satisfy $\phi(t) > 0$ on $[0,1]$. Put $K(s,t) = \phi(s)/\phi(t)$ and

$$(Ax)(t) = \int_0^1 K(s,t)x(s)\,ds$$

Prove that $A^2 = A$. What are the implications for the integral equations $\lambda Ax = x + w$?

20. Suppose that the operator A satisfies a polynomial equation $\sum_{j=0}^n c_j A^j = 0$ in which $c_0 \neq 0$. Prove that A is invertible and give a formula for its inverse.

21. Prove that if $\|A\| < 1$, then

$$\frac{1}{1 + \|A\|} \leqslant \|(I-A)^{-1}\| \leqslant \frac{1}{1 - \|A\|}$$

22. Assume that $A^{m+1} = A$ for some integer $m \geqslant 1$, and show how to solve the equation $x - \lambda Ax = b$.

23. Investigate the nature of the solutions to the Fredholm integral equation $x(t) = 1 + \int_0^1 x(st)\,ds$.

24. Define $F : C[0,1] \to C[0,1]$ by the equation

$$(Fx)(t) = x(t) - \int_0^1 K\big(t, x(s)\big)\,ds$$

Make reasonable assumptions about K and compute the Fréchet derivative $F'(x)$. Make further assumptions about K and prove that $F'(x)$ is invertible.

25. In Example 3, do we have $A^{-1} = B(AB)^{-1}$?

26. Find the connection between the Neumann Theorem and Lemma 1 in Section 4.1. (That lemma concerns diagonally dominant matrices.)

27. Let a and b be elements of a (possibly noncommutative) ring with unit 1. Show that the partial sums of the series $\sum_{k=0}^\infty b(1-ab)^k$ can be computed by the formulas $x_0 = b$ and $x_{n+1} = x_n + b(1 - ax_n)$.

28. Define an operator A from $C[0,1]$ into $C[0,1]$ by the equation

$$(Ax)(t) = x(t) - \int_0^1 x(s)\left[s^2 + \frac{t^2}{2}\right]ds$$

Prove that A is invertible and give a series for its inverse.

29. Consider the integral operator $(Kx)(t) = \int_0^t k(t,s)x(s)\,ds$, where $x \in C[0,1]$ and $k \in C([0,1] \times [0,1])$. Prove that $I - K$ is invertible. Prove that this assertion may be false if the upper limit in the integral is replaced by 1.

4.4 Projections and Projection Methods

Consider a normed linear space X. An element P of $\mathcal{L}(X,X)$ is called a **projection** if P is *idempotent*: $P^2 = P$. Notice particularly that linearity and continuity are incorporated in the definition. Obvious examples are $P = 0$ and $P = I$. In Hilbert space, if $\{v_1, v_2, \ldots\}$ is a finite or infinite orthonormal system, the equation

$$(1) \qquad Px = \sum_j \langle x, v_j \rangle v_j$$

defines a projection. To prove this, notice first that

$$Pv_i = \sum_j \langle v_i, v_j \rangle v_j = \sum_j \delta_{ij} v_j = v_i$$

Thus P leaves undisturbed each v_i. Consequently,

$$P^2 x = P(Px) = \sum_j \langle x, v_j \rangle Pv_j = \sum_j \langle x, v_j \rangle v_j = Px$$

Here are some elementary results about projections.

> **Theorem 1.**   *The range of a projection is identical with the set of its fixed points.*

Proof.   Let $P : X \to X$ be a projection and V its range. If $v \in V$, then $v = Px$ for some x, and consequently,

$$Pv = P^2 x = Px = v$$

Thus v is a fixed point of P. The reverse inclusion is obvious: If $x = Px$, then x is in the range of P.   ∎

> **Theorem 2.**   *If P is a projection, then so is $I - P$. The range of each is the null space of the other.*

Proof.
$$(I - P)^2 = (I - P)(I - P) = I - 2P + P^2 = I - P$$

Use $\mathcal{R}$ and $\mathcal{N}$ for "range" and "null space." Then the preceding theorem shows that

$$(2) \qquad \mathcal{R}(P) = \mathcal{N}(I - P)$$

Applying this to $I - P$, we get

$$(3) \qquad \mathcal{R}(I - P) = \mathcal{N}(P)$$ ∎

It should be noted that the range of a projection P is necessarily closed, because the continuity of P implies that the set $\{\, x \,:\, (I - P)x = 0 \,\}$ is closed. This is a special property not possessed by all elements of $\mathcal{L}(X,X)$. Notice also that P acts like the identity on its range. Thus, every projection can be regarded as a continuous linear extension of the identity operator defined initially on a subspace of the given space. If P is a projection of X onto V, then V is closed, and we say that "P is a projection of X onto V," writing $P : X \twoheadrightarrow V$, where the double arrow signifies a surjection.

Theorem 3.   *The adjoint of a projection is also a projection.*

Proof. Let P be a projection defined on the normed space X. Recall that its adjoint P^* maps X^* to X^* and is defined by the equation

$$P^*\phi = \phi \circ P \qquad \phi \in X^*$$

Thus it follows that

$$(P^*)^2\phi = P^*(P^*\phi) = (\phi \circ P) \circ P = \phi \circ P^2 = \phi \circ P = P^*\phi \qquad \blacksquare$$

In the next theorem, the structure of a projection having a finite-dimensional range is revealed.

Theorem 4.   *Let P be a projection of a normed space X onto a finite-dimensional subspace V. If $\{v_1, \ldots, v_n\}$ is any basis for V then there exist functionals ϕ_i in X^* such that*

(4) $$\phi_i(v_j) = \delta_{ij} \qquad (1 \leqslant i, j \leqslant n)$$

(5) $$Px = \sum_{i=1}^{n} \phi_i(x)v_i \qquad (x \in X)$$

Proof.   Select $\psi_i \in V^*$ such that for any $v \in V$,

$$v = \sum_{i=1}^{n} \psi_i(v)v_i$$

The functionals ψ_i are linear and continuous, by Corollary 1 in Section 1.5, page 26. (See also the proof of Corollary 2 in the same section.) For each $x \in X$, $Px \in V$, and so $Px = \sum_{i=1}^{n} \psi_i(Px)v_i$. Hence we can let $\phi_i(x) = \psi_i(Px)$. Being a composition of continuous linear maps, ϕ_i is also linear and continuous. The equation $Pv_j = v_j$ implies that $\phi_i(v_j) = \delta_{ij}$ by the uniqueness of the representation of Px as a linear combination of the basis vectors v_i. $\blacksquare$

A set of vectors $v_1, v_2, \ldots$ and a set of functionals $\phi_1, \phi_2, \ldots$ is said to form a **biorthogonal system** if $\phi_i(v_j) = \delta_{ij}$ for all i and j. The book [Brez] is devoted to this topic.

In practical problems, a projection is often used to provide approximations. Thus, if x is an element of a normed space X and if V is a subspace of X, it may be desired to approximate x by an element of V. If $v \in V$, the **error** or **deviation** of x from v is $\|x - v\|$, and the **minimum deviation** or **distance** from x to V is

$$\text{dist}(x, V) = \inf_{v \in V} \|x - v\|$$

This quantity represents the best that can be done in approximating x by an element of V. In most normed spaces it is quite difficult to determine a **best approximation** to x. That would be an element $v \in V$ such that

(6) $$\|x - v\| = \text{dist}(x, V)$$

Such an element need not exist. It will exist if V is finite dimensional and in certain other special cases. Usually, we find a convenient projection $P : X \twoheadrightarrow V$ and accept Px as an approximation to x. In Hilbert space, we can use the **orthogonal** projection of X onto V and thereby obtain the **best** approximation. In most spaces, best approximations—even if they exist—cannot be obtained by using linear maps.

Theorem 5.   *If P is a projection of X onto a subspace V, then for all $x \in X$,*

(7) $$\|x - Px\| \leqslant \|I - P\| \operatorname{dist}(x, V)$$

Proof.   For any $v \in V$, $Pv = v$. Hence

$$\|x - Px\| = \|(x - v) - P(x - v)\| = \|(I - P)(x - v)\| \leqslant \|I - P\|\, \|x - v\|$$

Now take an infimum as v ranges over V.   ∎

As remarked above, the Hilbert space case is especially favorable, since the *orthogonal* projection onto a subspace does yield best approximations. If X is a Hilbert space and $P : X \twoheadrightarrow V$ is a projection, we call it the **orthogonal** projection onto V if

(8) $$x - Px \perp V \qquad (x \in X)$$

If this is the case, then by the Pythagorean Law,

(9) $$\|x\|^2 = \|Px\|^2 + \|x - Px\|^2$$

Hence P and $I - P$ have operator norms at most 1. Inequality (7) now shows that Px is the best approximation to x in the subspace V. Furthermore, $x - Px$ is the best approximation of x in $V^\perp$. (This last space is the orthogonal complement of V in the Hilbert space X.)

Example 1.   Consider the familiar space $C[a, b]$. In it we single out for special attention the subspace Π_{n-1} consisting of all polynomials of degree at most $n-1$. This has dimension n. Now select points $t_1 < t_2 < \cdots < t_n$ in $[a, b]$, and define polynomials

$$\ell_i(s) = \prod_{\substack{j=1 \\ j \neq i}}^{n} \frac{s - t_j}{t_i - t_j} \qquad (1 \leqslant i \leqslant n)$$

These polynomials have degree $n - 1$ and satisfy the equation

$$\ell_i(t_j) = \delta_{ij} \qquad (1 \leqslant i, j \leqslant n)$$

This is a special case of Equation (4) above. The operator L, defined for $x \in C[a, b]$ by the equation

$$Lx = \sum_{i=1}^{n} x(t_i)\ell_i$$

is the Lagrange Interpolation Operator; it is a projection.   ∎

For any $t \in [a, b]$, we can define a functional t^* on the space $C[a, b]$ by writing $t^*(x) = x(t)$, where x runs over $C[a, b]$. This functional is called a **point evaluation functional**. Notice that in Example 1, the functions $\ell_1, \ell_2, \ldots, \ell_n$ and the functionals $t_1^*, t_2^*, \ldots, t_n^*$ form together a biorthogonal system, as defined just after Theorem 4.

Besides being used directly to provide approximations to elements in a normed linear space, projections are used to solve operator equations. Let us illustrate this in a Hilbert space X. Suppose that it is desired to solve an operator equation $Ax = b$. Here, $A \in \mathcal{L}(X, X)$; it could be an integral operator, for example. Let $[u_j : j \in \mathbb{N}]$ be an orthonormal basis for X. (In such a case, X must be separable.) The equation $Ax = b$ is equivalent to an infinite system of equations:

$$(10) \qquad \langle Ax, u_j \rangle = \langle b, u_j \rangle \qquad (j \in \mathbb{N})$$

In attempting to find an approximate solution to this problem, we solve the finite system

$$\langle Ax_n, u_j \rangle = \langle b, u_j \rangle \qquad (1 \leqslant j \leqslant n)$$

This is the same as

$$(11) \qquad P_n(Ax_n - b) = 0$$

where P_n is the orthogonal projection defined by

$$(12) \qquad P_n y = \sum_{j=1}^{n} \langle y, u_j \rangle u_j$$

Does this strategy have any chance of success? It depends on whether the sequence $[x_n]$, arising as outlined above, converges. Assume that $x_n \to x$. Let us verify that x is a solution: $Ax = b$. By the continuity of A, $Ax_n \to Ax$. Since $\|P_n\| = 1$, $P_n(Ax_n - Ax) \to 0$. But $P_n Ax_n = P_n b$ by our choice of x_n. Hence $P_n b - P_n Ax \to 0$. In the limit, this gives us $b = Ax$.

Notice that this proof uses the essential fact that $P_n y \to y$ for all y. For our general theorem (in any Banach space) this assumption is needed.

> **Theorem 6.**   *Let $[P_n]$ be a sequence of projections on a Banach space X, and assume that $P_n y \to y$ for each y in X. Let $b \in X$ and $A \in \mathcal{L}(X, X)$. For each n let x_n be a point such that $P_n(Ax_n - b) = 0$. If $x_n \to x$, then $Ax = b$.*

Proof.   Since $P_n y \to y$, we also have $\|P_n y\| \to \|y\|$, and therefore $\sup_n \|P_n y\| < \infty$ for each y. Since X is complete, we may apply the Uniform Boundedness Principle (Section 1.7, page 42) and conclude that $\sup_n \|P_n\| < \infty$. By the continuity of A, $Ax_n \to Ax$. By the boundedness of $\|P_n\|$, we have $P_n(AX_n - Ax) \to 0$. By the choice of x_n, $P_n Ax_n = P_n^{\circ} b$. Hence $P_n b - P_n Ax \to 0$. In the limit, this yields $b = Ax$. ∎

A **projection method** for solving an equation of the form $Ax = b$, where $A \in \mathcal{L}(X, X)$, begins by selecting a sequence of subspaces

$$(13) \qquad V_1 \subset V_2 \subset V_3 \subset \cdots \subset X$$

and associated projections $P_n : X \twoheadrightarrow V_n$. For each n, we find an x_n that satisfies

$$(14) \qquad\qquad P_n(Ax_n - b) = 0$$

Often we insist that $x_n \in V_n$, but this is not essential. One hopes that the sequence $[x_n]$ will converge to a solution of the original problem. We shall give some positive results in this direction. These apply to a problem of the form

$$(15) \qquad\qquad x - Ax = b$$

> **Theorem 7.**   Let P be a projection of the normed space X onto a subspace V. Suppose that $x \in X$, $x - Ax - b = 0$, $\widetilde{x} \in V$, and $P(\widetilde{x} - A\widetilde{x} - b) = 0$. If $I - PA$ is invertible, then

$$(16) \qquad\qquad \left\| x - \widetilde{x} \right\| \leqslant \left\| (I - PA)^{-1} \right\| \, \left\| x - Px \right\|$$

Proof.   From Equation (15), $PAx = Px - Pb$. Hence

$$(17) \qquad\qquad x - PAx = x - Px + Pb$$

Since $\widetilde{x} \in V$, it follows that $P\widetilde{x} = \widetilde{x}$. Consequently,

$$(18) \qquad\qquad \widetilde{x} - PA\widetilde{x} = Pb$$

Subtraction between Equations (17) and (18) gives

$$x - \widetilde{x} - PA(x - \widetilde{x}) = x - Px$$

or

$$(I - PA)(x - \widetilde{x}) = x - Px$$

Thus we have

$$x - \widetilde{x} = (I - PA)^{-1}(x - Px)$$

This leads to Inequality (16).   ∎

Let us see how this theorem can be applied in the case where X is a Hilbert space. Let $[v_1, v_2, \ldots]$ be an orthonormal basis for X, and suppose that we wish to solve

$$(19) \qquad\qquad x - Ax = b$$

in which $A \in \mathcal{L}(X, X)$ and $\|A\| < 1$. Let V_n be the linear subspace of dimension n generated by $\{v_1, \ldots, v_n\}$, and let P_n be the orthogonal projection of X onto V_n. The familiar formula for P_n is

$$(20) \qquad\qquad P_n x = \sum_{j=1}^{n} \langle x, v_j \rangle v_j$$

In applying the projection method to Equation (19), let us select elements x_n in V_n for which

$$(21) \qquad P_n(I - A)x_n = P_n b$$

By the preceding theorem, the actual solution x is related to the approximate solution x_n by the inequality

$$(22) \qquad \|x - x_n\| \leqslant \|(I - P_nA)^{-1}\| \, \|x - P_n x\|$$

Now $\|P_n\| = 1$, and so $\|P_n A\| \leqslant \|A\| < 1$. Hence

$$\|(I - P_nA)^{-1}\| \leqslant (1 - \|A\|)^{-1}$$

(This estimate comes from the proof of the Neumann Theorem.) Also, we know that

$$\|x - P_n x\|^2 = \left\| \sum_{i=n+1}^{\infty} \langle x, v_i \rangle v_i \right\|^2 = \sum_{i=n+1}^{\infty} |\langle x, v_i \rangle|^2 \to 0$$

We conclude therefore from Inequality (22) that the approximate solutions x_n converge to x as $n \to \infty$. We summarize this discussion in the next theorem.

Theorem 8.   Let $[v_1, v_2, \dots]$ be an orthonormal basis in a Hilbert space X. Let $A \in \mathcal{L}(X, X)$ and $\|A\| < 1$. For each n, let x_n be a linear combination of $v_1, \dots, v_n$ chosen so that

$$(23) \qquad \langle x_n - Ax_n, v_i \rangle = \langle b, v_i \rangle \qquad (1 \leqslant i \leqslant n)$$

Then $[x_n]$ converges to the solution of the equation $x - Ax = b$.

Of course, the Neumann Theorem can be used to solve the equation $(I - A)x = b$. It gives $x = (I - A)^{-1}b = \sum_{n=0}^{\infty} A^n b$. There seems to be no obvious connection between this solution and the one provided by Theorem 8.

In the general projection method, in solving the equation

$$(24) \qquad P_n(Ax - b) = 0$$

we need not confine ourselves to the case where x is chosen in the range of P_n. Instead, we can let x be a linear combination of other prescribed elements, say $x = \sum_{i=1}^{n} c_i u_i$. In this case, we attempt to choose c_i so that

$$(25) \qquad P_n \left(\sum_{j=1}^{n} c_j A u_j - b \right) = 0$$

Suppose that P_n is a projection of rank n having the explicit formula

$$(26) \qquad P_n x = \sum_{i=1}^{n} \phi_i(x) v_i$$

Then Equation (25) gives us

$$(27) \qquad \phi_i\left(\sum_{j=1}^n c_j Au_j - b\right) = 0 \qquad (1 \leqslant i \leqslant n)$$

or

$$(28) \qquad \sum_{j=1}^n c_j \phi_i(Au_j) = \phi_i(b) \qquad (1 \leqslant i \leqslant n)$$

This is a system of linear equations having coefficient matrix $(\phi_i(Au_j))$. In the next two sections we shall see examples of this procedure.

Problems 4.4

1. Let $\{u_1,\ldots,u_n\}$ be a linearly independent set in a inner-product space. Prove that the Gram matrix, whose elements are $\langle u_i, u_j \rangle$, is nonsingular.

2. Let P be a projection of a normed space X onto a subspace V. Prove that $Px = 0$ if and only if $\phi(x) = 0$ for each ϕ in the range of P^*.

3. Let $P_1, P_2, \ldots$ be a sequence of projections on a normed space X. Suppose that $P_{n+1}P_n = P_n$ for all n and that the union of the ranges of these projections is dense in X. Suppose further that $\sup_n \|P_n\| < \infty$. Prove that $P_n x \to x$ for all $x \in X$.

4. Let $P_1, P_2, \ldots$ be a sequence of projections on a Banach space X. Prove that if $P_n x \to x$ for all $x \in X$, then $\sup_n \|P_n\| < \infty$ and the union of the ranges of the projections is dense in X. Hint: The Uniform Boundedness Theorem is useful.

5. Let X be a Banach space and $P_1, P_2, \ldots$ projections on X such that $P_n x \to x$ for every x. Suppose that A is an invertible element of $\mathcal{L}(X, X)$. For each n, let x_n be a point such that $P_n x_n = x_n$ and $P_n(Ax_n - b) = 0$. Prove or disprove that the sequence $[x_n]$ necessarily converges to the solution of the equation $Ax = b$.

6. Let $\{\phi_1, \ldots, \phi_n\}$ be a linearly independent set in X^*. Is there a projection $P : X \to X$ having rank n of the form $Px = \sum_{i=1}^n \phi_i(x)v_i$? (The **rank** of a linear operator is the dimension of its range.)

7. Adopt the notation of Theorem 4, and prove that $\phi_i(Px) = \phi_i(x)$ for all i in $\{1, 2, \ldots, n\}$ and for all x in X.

8. Prove that the operator L in Example 1 is a projection. Prove that $\|L\| = \| \sum |\ell_i| \|_\infty$.

9. Let A and P be elements of $\mathcal{L}(X, X)$, where $P^2 = P$. Let V denote the range of P. Show that $PA|V \in \mathcal{L}(V, V)$. Is $PA|V$ invertible?

10. Prove a variant of Theorem 6 in which P is an arbitrary linear operator and $\tilde{x}$ satisfies $P\tilde{x} = \tilde{x}$.

11. In the setting of Theorem 8, prove that the solution to the problem is given by $x = \sum_{n=0}^\infty A^n b$. How is this solution related to the one given in the theorem, namely, $x = \lim x_n$?

12. Consider the familiar sequence space c_0. (It was described in Problem 1.2.16, page 12.) We define a projection $P : c_0 \to c_0$ by selecting any set of integers J and setting

$$(Px)(n) = \begin{cases} x(n) & (n \in \mathbb{N} \smallsetminus J) \\ 0 & (n \in J) \end{cases}$$

Prove that P is a projection. Identify the null space and range of P. Give the formula for $I - P$. Compute $\|P\|$ and $\|I - P\|$. How many projections of this type are there? What is the distance between any two different such projections?

13. Let $\{u_1, u_2, \ldots, u_n\}$ and $\{v_1, v_2, \ldots, v_n\}$ be sets in a Hilbert space, the second set being assumed to be linearly independent. Define $Ax = \sum_{i=1}^{n} \langle x, u_i \rangle v_i$. Determine the necessary and sufficient conditions on $\{u_i\}$ in order that A be a projection, i.e., $A^2 = A$.

14. (Variation on Problem 5.) Let X be a Banach space, and let $P_1, P_2, \ldots$ be projections on X such that $P_n x \to x$ for each x in X. Assume that $\|P_n\| = 1$ for all n. Let A be a linear operator such that $\|I - A\| < 1$. If the points x_n satisfy $P_n x_n = x_n$ and $P_n(Ax_n - b) = 0$, then the sequence $[x_n]$ converges to a solution of the equation $Ax = b$.

15. In $\mathbb{R}^2$, let $u = (1,1)$, $v = (1,0)$, and $Px = \langle x, u \rangle v$. Prove that P is a projection. Using the Euclidean norm in $\mathbb{R}^2$, compute $\|P\|$. This problem illustrates the fact that projections on Hilbert spaces need not have norm 1.

16. Is a norm-1 projection defined on a Hilbert space necessarily an orthogonal projection?

17. Explain why point-evaluation functionals, as defined on the space $C[a,b]$, cannot be defined on any of the spaces $L^p[a,b]$.

4.5 The Galerkin Method

The procedure that goes by the name of the mathematician Galerkin is one of the projection methods, in fact, the one described at length in the preceding section. We review the method briefly, and then discuss concrete examples of its use.

We wish to solve an equation of the form

$$(1) \qquad\qquad Au = b$$

in which A is an operator acting on a Hilbert space U. A finite-dimensional subspace V is chosen in U, and we let P denote the orthogonal projection of U onto V. Then we find $\tilde{u} \in V$ such that

$$(2) \qquad\qquad P(A\tilde{u} - b) = 0$$

If $v_1, v_2, \ldots, v_n$ is a basis for V, and if we set $\tilde{u} = \sum_{j=1}^{n} c_j v_j$, then Equation (2) leads to

$$(3) \qquad\qquad \sum_{j=1}^{n} c_j \langle Av_j, v_i \rangle = \langle b, v_i \rangle \qquad (1 \leqslant i \leqslant n)$$

These are the "Galerkin Equations" for $\tilde{u}$.

Here is an example of the Galerkin method in the subject of partial differential equations. We recall the definition of the **Laplacian** operator ∇^2 (also denoted by Δ):

$$\nabla^2 u = \frac{\partial^2 u}{\partial x^2} + \frac{\partial^2 u}{\partial y^2}$$

An important problem, known as the **Dirichlet** problem, is to find a function $u = u(x,y)$ that obeys Laplace's equation $\nabla^2 u = 0$ in a bounded open set ("region") Ω in the plane and takes prescribed values on the boundary of the region (denoted by $\partial\Omega$). Thus there are two conditions on the unknown function u:

$$(4) \qquad\qquad \begin{cases} \nabla^2 u = 0 \ \text{ in } \ \Omega \\ u(x,y) = g(x,y) \ \text{ on } \ \partial\Omega \end{cases}$$

A function u that has continuous second–order partial derivatives and satisfies $\nabla^2 u = 0$ in a region Ω is said to be **harmonic** in Ω. In the Dirichlet problem, the function g is defined only on $\partial\Omega$ and should be continuous. Thus the Dirichlet problem has the goal of reconstructing a harmonic function in a region Ω from a knowledge of its values on the boundary of Ω. It furnishes a nice example of the **recovery** of a function from incomplete information. (The general topic of optimal recovery of functions has many other specific examples, such as in computed tomography, where the density function of a solid object is to be recovered from information produced by X-ray scanning.)

In applying Galerkin's method to the Dirichlet problem, it is advantageous to select base functions $u_1, \ldots, u_n$ that are harmonic in Ω. Then an arbitrary linear combination $\sum_{j=1}^{n} c_j u_j$ will also be harmonic, and we need only to adjust the coefficients so that the boundary conditions are approximately satisfied. That could mean that $\left\| \sum_{j=1}^{n} c_j u_j - g \right\|$ is to be minimized, where the norm is one that involves only function values on $\partial\Omega$. In Galerkin's method, however, we select c_j so that

$$(5) \qquad \left\langle \sum_{j=1}^{n} c_j u_j - g, \ u_i \right\rangle = 0 \qquad i = 1, \ldots, n$$

where the inner product could be a line integral around the boundary of Ω.

A plenitude of harmonic functions can be obtained from the fact that the real and imaginary parts of a holomorphic function of a complex variable are harmonic. To prove this, let w be holomorphic, and let $w = u + iv$, where u and v are the real and imaginary parts of w. By the Cauchy–Riemann Equations, we have

$$\frac{\partial^2 u}{\partial x^2} + \frac{\partial^2 u}{\partial y^2} = \frac{\partial}{\partial x}\frac{\partial u}{\partial x} + \frac{\partial}{\partial y}\frac{\partial u}{\partial y} = \frac{\partial}{\partial x}\frac{\partial v}{\partial y} + \frac{\partial}{\partial y}\left(-\frac{\partial v}{\partial x}\right) = 0$$

The proof for v comes from observing that it is the real part of $-iw$.

To illustrate this, consider the function $z \mapsto z^2$. We have

$$w = z^2 = (x + iy)^2 = x^2 - y^2 + 2ixy = u + iv$$

Thus the functions $u = x^2 - y^2$ and $v = 2xy$ are harmonic. (See Problem 10.)

The Dirichlet problem is frequently encountered with Poisson's Equation:

$$(6) \qquad \begin{cases} \nabla^2 u = f & \text{in } \Omega \\ u = g & \text{on } \partial\Omega \end{cases}$$

One way of solving (6) is to solve two related but easier problems:

$$(7) \qquad \begin{cases} \nabla^2 v = f & \text{on } \Omega \\ v = 0 & \text{on } \partial\Omega \end{cases} \qquad \begin{cases} \nabla^2 w = 0 & \text{on } \Omega \\ w = g & \text{on } \partial\Omega \end{cases}$$

Clearly, the function $u = v + w$ will then solve (6). The problem involving w was discussed previously. The Galerkin procedure for approximating v begins with the selection of base functions $v_1, v_2, \ldots$ that vanish on $\partial\Omega$. Then an approximate solution v is sought having the form $v = \sum_{j=1}^{n} c_j v_j$. The usual Galerkin criterion is applied, so we have to solve the linear equations

$$(8) \qquad \sum_{j=1}^{n} c_j \langle \nabla^2 v_j, v_i \rangle = \langle f, v_i \rangle \qquad i = 1, \ldots, n$$

The inner product here should be defined by the equation

$$(9) \qquad \langle u, v \rangle = \int_{\Omega} u(x, y) v(x, y) \, dx \, dy$$

Theorem 1.   *If Ω is a region to which Green's Theorem applies, then the Laplacian is self-adjoint with respect to the inner product (9) when applied to functions vanishing on $\partial\Omega$.*

Proof.   Using subscripts to denote partial derivatives, we write Green's Theorem in the form

$$\int_{\Omega} (Q_x - P_y) = \int_{\partial\Omega} (P \, dx + Q \, dy)$$

Applying this to the functions $Q = uv_x - vu_x$ and $P = vu_y - uv_y$, we obtain

$$\int_{\Omega} (u\nabla^2 v - v\nabla^2 u) = \int_{\partial\Omega} \left[(vu_y - uv_y) \, dx + (uv_x - vu_x) \, dy \right]$$

Since v and u vanish on $\partial\Omega$, we conclude that $\langle u, \nabla^2 v \rangle = \langle \nabla^2 u, v \rangle$.   ∎

A remark about Equation (8) is in order. Some authors argue that the coefficients c_j should be chosen to minimize the expression

$$\left\| \nabla^2 \left(\sum_{j=1}^{n} c_j v_j \right) - f \right\|$$

where the Hilbert space norm is being used, corresponding to the inner product in Equation (9). This is a problem of approximating f as well as possible by a linear combination of the functions $\nabla^2 v_i$ ($1 \leqslant i \leqslant n$). The solution is obtained via the *normal* equations

$$\sum_{j=1}^{n} c_j \nabla^2 v_j - f \perp \nabla^2 v_i \qquad (1 \leqslant i \leqslant n)$$

This leads to the system

$$(10) \qquad \sum_{j=1}^{n} c_j \langle \nabla^2 v_j, \nabla^2 v_i \rangle = \langle f, \nabla^2 v_i \rangle \qquad (1 \leqslant i \leqslant n)$$

This is *not* the classical Galerkin method, although it *is* an example of the general theory presented in Section 6.4.

An existence theorem for solutions of the Dirichlet problem is quoted here, from [Gar], page 288. It concerns domains with smooth boundary. Other such theorems can be found in [Kello].

Theorem 2.   Let Ω be a bounded open set in $\mathbb{R}^2$ whose boundary $\partial\Omega$ consists of a finite number of simple closed curves. Assume the existence of a number $r > 0$ such that at each point of $\partial\Omega$ there are two circles of radius r tangent to $\partial\Omega$ at that point, one circle in $\overline{\Omega}$ and the other in $\mathbb{R}^2 \setminus \Omega$. Let g be a twice continuously differentiable function on Ω. Then the Dirichlet problem (4) has a unique solution.

For boundary-value problems involving differential equations, the Galerkin strategy can be applied after first turning the boundary-value problem into a "variational form." This typically leads to particular cases of a general problem that we now describe.

Two Hilbert spaces U and V are prescribed, and there is given a bilinear functional on $U \times V$. ("Bilinear" means linear in each variable.) Calling this functional B, let us make further assumptions as follows:

(a) $|B(u,v)| \leqslant \alpha\|u\|\,\|v\|$
(b) $\inf_{\|u\|=1} \sup_{\|v\|=1} |B(u,v)| = \beta > 0$
(c) If $v \neq 0$, then $\sup_u B(u,v) > 0$

With this setting established, there is a standard problem to be solved, namely, given a specific point z in V, to find w in U such that, for all v in V, $B(w,v) = \langle z,v\rangle$. The following theorem concerning this problem is Babuška's generalization of a theorem proved first by Lax and Milgram. The proof given is adapted from [OdR].

Theorem 3. Babuška–Lax–Milgram   *Under the hypotheses listed above we have the following: For each z in V there is a unique w in U such that*

$$B(w,v) = \langle z,v\rangle \quad \text{for all } v \text{ in } V$$

Furthermore, w depends linearly and continuously on z.

Proof.   As usual, we define the u-sections of B by $B_u(v) = B(u,v)$. Then each B_u is a continuous linear functional on V. Indeed,

$$\|B_u\| = \sup\{|B(u,v)| \,:\, v \in V, \|v\| = 1\} \leqslant \alpha\|u\|$$

By the Riesz Representation Theorem (Section 2.3, Theorem 1, page 81), there corresponds to each u in U a unique point Au in V such that $B_u(v) = \langle Au,v\rangle$. Elementary arguments show that A is a linear map of U into V. Thus,

$$\langle Au,v\rangle = B_u(v) = B(u,v)$$

The continuity of A follows from the inequality $\|Au\| = \|B_u\| \leqslant \alpha\|u\|$. The operator A is also bounded from below:

$$\|Au\| = \sup_{\|v\|=1} \langle Au,v\rangle = \sup_{\|v\|=1} |B(u,v)| \geqslant \beta\|u\|$$

In order to prove that the range of A is closed, let $[v_n]$ be a convergent sequence in the range of A. Write $v_n = Au_n$, and note that by the Cauchy property

$$0 = \lim_{n,m \to \infty} \left\| v_n - v_m \right\| = \lim_{n,m} \left\| Au_n - Au_m \right\| \geqslant \beta \lim_{n,m} \left\| u_n - u_m \right\|$$

Consequently, $[u_n]$ is a Cauchy sequence. Let $u = \lim_n u_n$. By the continuity of A, $v_n = Au_n \to Au$, showing that $\lim_n v_n$ is in the range of A. Next, we wish to establish that the range of A is dense in V. If it is not, then the closure of the range is a proper subspace of V. Select a nonzero vector p orthogonal to the range of A. Then $\langle Au, p \rangle = 0$ for all u. Equivalently, $B(u,p) = 0$, contrary to the hypothesis (3) on B. At this juncture, we know that A^{-1} exists as a linear map. Its continuity follows from the fact that A is bounded below: If $u = A^{-1}v$, then the inequality $\|Au\| \geqslant \beta\|u\|$ implies $\|v\| \geqslant \beta \|A^{-1}v\|$. The equation we seek to solve is $B(w,v) = \langle z, v \rangle$ for all v. Equivalently, $\langle Aw, v \rangle = \langle z, v \rangle$. Hence $Aw = z$ and $w = A^{-1}z$. Since there is no other choice for w, we conclude that it is unique and depends continuously and linearly on z.   ∎

If a problem has been recast into the form of finding a vector w for which $B(w,v) = \langle z, v \rangle$, as described above, then the Galerkin procedure can be used to solve this problem on a succession of finite-dimensional subspaces $U_n \subset U$ and $V_n \subset V$.

Reviewing the details of this strategy, we start by assuming that $\dim(U_n) = \dim(V_n) = n$. Select bases $\{u_i\}$ for U_n and $\{v_i\}$ for V_n. A solution w_n to the "partial problem" is sought:

$$B(w_n, v_i) = \langle z, v_i \rangle \qquad (1 \leqslant i \leqslant n)$$

A "trial solution" is hypothesized: $w_n = \sum_1^n c_j u_j$. We must now solve the following system of n linear equations in the n unknown quantities c_j:

$$\sum_{j=1}^{n} c_j B(u_j, v_i) = \langle z, v_i \rangle \qquad (1 \leqslant i \leqslant n)$$

In order to have at this stage a nonsingular $n \times n$ matrix $B(u_j, v_i)$, we would have to make an assumption like hypothesis (b) for the two spaces U_n and V_n. For example, we could assume

(b*) There is a positive β_n such that

$$\sup_{v \in V_n, \; \|v\|=1} |B(u, v)| \geqslant \beta_n \|u\| \qquad (u \in U_n)$$

Problem 14 asks for a proof that this hypothesis will guarantee the nonsingularity of the matrix described above.

Example 1.   Consider the two-point boundary-value problem

$$(pu')' - qu = f \qquad u(a) = 0 \qquad u(b) = 0$$

This is a Sturm–Liouville problem, the subject of Theorem 1 in the next section (page 206) as well as Section 2.5, pages 105ff. In order to apply Theorem 3, one requires the bilinear form and linear functional appropriate to the problem. They are revealed by a standard procedure: Multiply the differential equation by a function v that vanishes at the endpoints a and b, and then use integration by parts:

$$\int_a^b [v(pu')' - vqu] = \int_a^b vf$$

$$vpu'\Big|_a^b - \int_a^b v'pu' - \int_a^b vqu = \int_a^b vf$$

$$\int_a^b [pu'v' + quv] = \int_a^b -fv$$

$$B(u,v) = \int_a^b (pu'v' + quv) \qquad \langle f, v \rangle = - \int_a^b fv$$

There is much more to be said about this problem, but here we wish to emphasize only the formal construction of the maps that enter into Theorem 3.

Example 2.   The steady-state distribution of heat in a two-dimensional domain Ω is governed by Poisson's Equation:

$$\nabla^2 u = f \qquad \text{in } \Omega$$

Here, $u(x,y)$ is the temperature at the location (x,y) in $\mathbb{R}^2$, and f is the heat-source function. If the temperature on the boundary $\partial\Omega$ is held constant, then, with suitable units for the measurement of temperature, we may take $u(x,y) = 0$ on $\partial\Omega$. This simple case leads to the problem of discovering u such that $B(u,v) = \langle f,v \rangle$ for all v, where

$$B(u,v) = - \int_\Omega (u_x v_x + u_y v_y)$$

To arrive at this form of the problem, first write the equivalent equation

$$\langle \nabla^2 u, v \rangle = \langle f, v \rangle \qquad \text{for all } v$$

The integral form of this is

$$\int_\Omega v\nabla^2 u = \int_\Omega vf \qquad \text{for all } v$$

The integral on the left is treated by using Green's Theorem (also known as Gauss's Theorem). (This theorem plays the role of integration by parts for multivariate functions.) It states that

$$\int_\Omega (P_x + Q_y) = \int_{\partial\Omega} (P\,dy - Q\,dx)$$

This equation holds true under mild assumptions on P, Q, Ω, and $\partial\Omega$. (See [Wid].) Exploiting the hypothesis of zero boundary values, we have

$$
\int_\Omega v\nabla^2 u = \int_\Omega (u_{xx} + u_{yy})v
$$
$$
= \int_\Omega \left[(u_x v)_x + (u_y v)_y - u_x v_x - u_y v_y \right]
$$
$$
= \int_{\partial\Omega} (u_x v - u_y v) - \int_\Omega (u_x v_x + u_y v_y)
$$
$$
= -\int_\Omega (u_x v_x + u_y v_y) = B(u,v) \qquad \blacksquare
$$

References.   For the classical theory of harmonic functions consult [Kello]. For the existence theory for solutions to the Dirichlet Problem, see [Gar]; in particular, Theorem 2 above can be found in that reference. For the Galerkin method, consult [KK], [KA], [OdR], [Kee], and [Gre], [Gri].

Problems 4.5

1. ([Mil], page 115) Find an approximate solution of the two-point boundary-value problem

$$
x'' + tx' + x = 2t \qquad x(0) = 1 \qquad x(1) = 0
$$

by using Galerkin's method with trial functions

$$
x(t) = (1-t)(1 + c_1 t + c_2 t^2 + c_3 t^3)
$$

2. Find an approximate solution of the problem

$$
(tx')' + x = t \qquad x(0) = 0 \qquad x(1) = 1
$$

by using Galerkin's method and the trial solution

$$
x(t) = t + t(1-t)(c_1 + c_2 t)
$$

3. Invent an efficient algorithm for generating the sequences of harmonic functions $[u_n], [v_n]$, where $z^n = u_n + iv_n$ and $n = 0, 1, 2, \ldots$

4. Prove that a differentiable function $f : \mathbb{R} \to \mathbb{R}$ such that $\inf_x f'(x) > 0$ is necessarily surjective. Show by an example that the simpler condition $f'(x) > 0$ is not adequate.

5. A sequence $v_1, v_2, \ldots$ in a Banach space U is called a **basis** (or more exactly a **Schauder basis**) if each $x \in U$ has a unique representation as a convergent series in U of the form $x = \sum_{n=1}^\infty a_n(x)v_n$. The a_n depend on x in a continuous and linear manner, i.e., $a_n \in U^*$. If U has a Schauder basis, then U is separable. Prove that $a_n(v_m) = \delta_{nm}$. Prove that no loss of generality occurs in assuming that $\|v_n\| = 1$ for all n. See problems 24–26 in Section 1.6, pages 38 and 39.

6. (Continuation) Prove that for each n, the map P_n defined by the equation $P_n x = \sum_{k=1}^n a_k(x)v_k$ is a (bounded, linear) projection.

7. (Continuation) Prove that $\sup_n \|P_n\| < \infty$.

8. (Continuation) What are the Galerkin equations for the problem $Ax = b$ when the projections in the preceding problems are employed?

9. (Continuation) Use Galerkin's method to solve $Ax = b$ when A is defined by $Av_n = \sum_{j=n}^{\infty} j^{-2} v_j$ and $b = \sum_{n=1}^{\infty} 2^{-n} v_n$.

10. Use the computer system Mathematica to find the real and imaginary parts of z^n for $n = 1$ to $n = 10$. Version 2 (or later) of Mathematica will be necessary. The input to do this all at once is

    ```
    n=1; While[(n=n+1)<11, Print[ComplexExpand[(x+I y)∧n]]]
    ```

11. Use Green's Theorem to show that the operator A defined by

    $$Au = au_{xx} + bu_{xy} + cu_{yy}$$

    is Hermitian on the space of functions having continuous partial derivatives of orders 0, 1, 2 in Ω and vanishing on $\partial\Omega$. In the definition of A, the coefficients are constants.

12. Give an elementary proof of Green's Theorem for any rectangle in $\mathbb{R}^2$ whose sides are parallel to the coordinate axes.

13. Solve the Dirichlet problem on the unit disk in $\mathbb{R}^2$ when the boundary values are given by the expression $8x^4 - 8y^4 + 1$.

14. Prove that the matrix $(B(u_j, v_i))$ described following Theorem 3 of this section is nonsingular if the hypothesis (b*) is fulfilled by the spaces U_n and V_n.

4.6 The Rayleigh–Ritz Method

The basic strategy of this method is to recast a problem such as

$$(1) \qquad\qquad F(x) = 0 \qquad (x \in X)$$

into an *extremal* problem, i.e., a problem of finding the maximum or minimum of some functional Φ. This extremum problem is then solved on an increasing family of subspaces in the normed space X:

$$U_1 \subset U_2 \subset U_3 \subset \cdots \subset X$$

It is obvious that there are many ways to create an extremum problem equivalent to the problem in (1). For example, we can put

$$(2) \qquad\qquad \Phi(x) = \left\| F(x) \right\|$$

If this choice is made with a linear problem in a Hilbert space, we are led to a procedure very much like Galerkin's Method. Suppose that $F(x) = Ax - v$, where A is a linear operator on a Hilbert space. The minimization of $\left\| Ax - v \right\|$, where $x \in U_n$, reduces to a standard least-squares calculation. Suppose that U_n has a basis $[u_1, u_2, \ldots, u_n]$. Then let $x = \sum_{j=1}^{n} c_j u_j$. The minimum of

$$\left\| \sum_{j=1}^{n} c_j Au_j - v \right\|$$

is obtained when the coefficient vector $(c_1, c_2, \ldots, c_n)$ satisfies the "normal" equations

(3)
$$\sum_{j=1}^{n} c_j A u_j - v \perp A(U_n)$$

This means that

(4)
$$\sum_{j=1}^{n} c_j \langle A u_j, A u_i \rangle = \langle v, A u_i \rangle \qquad (1 \leqslant i \leqslant n)$$

These are not the Galerkin equations (Equation (3) in Section 4.5, page 198).

The Rayleigh–Ritz method (in its classical formulation) applies to differential equations, and the functional Φ is directly related to the differential equation. We illustrate with a two-point boundary-value problem, in which all the functions are assumed to be sufficiently smooth. (They are functions of t.)

(5)
$$\begin{cases} (px')' - qx = f & a \leqslant t \leqslant b \\ x(a) = \alpha \qquad x(b) = \beta \end{cases}$$

In correspondence with this problem, a functional Φ is defined by

(6)
$$\Phi(x) = \int_a^b \left[(x')^2 p + x^2 q + 2xf \right] dt$$

Theorem 1.   *If x is a function in $C^2[a, b]$ that minimizes $\Phi(x)$ locally subject to the constraints $x(a) = \alpha$ and $x(b) = \beta$, then x solves the problem in (5).*

Proof.   Assume the hypotheses, and let y be any element of $C^2[a, b]$ such that $y(a) = y(b) = 0$. We use what is known as a variational argument. For each real number λ, $x + \lambda y$ is a competitor of x in the minimization of Φ. Hence the function $\lambda \mapsto \Phi(x + \lambda y)$ has a local minimum at $\lambda = 0$. We compute

$$\frac{d}{d\lambda} \Phi(x + \lambda y) = \frac{d}{d\lambda} \int_a^b \left[(x' + \lambda y')^2 p + (x + \lambda y)^2 q + 2(x + \lambda y)f \right] dt$$

$$= 2 \int_a^b \left[(x' + \lambda y')y'p + (x + \lambda y)yq + yf \right] dt$$

Evaluating this derivative at $\lambda = 0$ and setting the result equal to 0 yields the necessary condition

(7)
$$\int_a^b (px'y' + qxy + fy)\, dt = 0$$

We use integration by parts on the first term, like this:

$$\int_a^b px'y' = px'y \Big|_a^b - \int_a^b (px')'y = -\int_a^b (px')'y$$

Here the fact that $y(a) = y(b) = 0$ has been exploited. Equation (7) now reads

$$(8) \qquad \int_a^b \left[-(px')' + qx + f \right] y \, dt = 0$$

(The steps just described are the same as those in Example 1, page 203.) Since y is an arbitrary element of $C^2[a, b]$ vanishing at a and b, we conclude from Equation (8) that

$$-(px')' + qx + f = 0$$

The details of this last argument are as follows. Let $z = -(px')' + qx + f$. Then $\int_a^b z(t)y(t) \, dt = 0$ for all functions y of the type described above. Suppose that $z \neq 0$. Then for some τ, $z(\tau) \neq 0$. For definiteness, let us assume that $z(\tau) \equiv \varepsilon > 0$. Then there is a closed interval $J \subset (a, b)$ in which $z(t) \geqslant \varepsilon/2$. There is an open interval I containing J in which $z(t) > 0$. Let y be a C^2 function that is constantly equal to 1 on J and constantly equal to 0 on the complement of I. Then $\int_a^b z(t)y(t) \, dt > 0$. ∎

> **Theorem 2.**   Assume that $p(t) > 0$ and $q(t) \geqslant 0$ on $[a, b]$. If x is a function in $C^2[a, b]$ that solves the boundary-value problem (5) then x is the unique local minimizer of Φ subject to the boundary conditions as constraints.

Proof.   Let $z \in C^2[a, b]$, $z \neq x$, $z(a) = \alpha$, and $z(b) = \beta$. Then the function $y = z - x$ satisfies 0-boundary conditions but is not 0. By calculations like those in the preceding proof,

$$(9) \qquad \Phi(x + y) = \Phi(x) + 2 \int_a^b (x'y'p + xyq + yf) + \int_a^b \left[(y')^2 p + y^2 q \right]$$

Using integration by parts on the middle term, we find that it is zero. Then Equation (9) shows that $\Phi(z) > \Phi(x)$. ∎

Returning now to the two-point boundary-value problem in Equations (5), we make the simplifying assumption that the boundary conditions are $x(0) = x(1) = 0$. (In Problems 4.1.3 – 4, page 176, it was noted that simple changes of variable can be employed to arrange this state of affairs.) The subspaces U_n should now be composed of functions that satisfy $x(0) = x(1) = 0$. For example, we could use linear combinations of terms $t^i(1 - t)^j$ where $i, j \geqslant 1$. If $x = \sum_{j=1}^n c_j u_j$ is substituted in the functional Φ of Equation (6), the result is a real–valued function of $(c_1, c_2, \ldots, c_n)$ to be minimized. The minimum occurs when

$$\frac{\partial}{\partial c_i} \Phi \left(\sum_{j=1}^n c_j u_j \right) = 0 \qquad (i = 1, 2, \ldots, n)$$

When the calculations indicated here are carried out, the system of equations to be solved emerges as

$$\sum_{j=1}^n a_{ij} c_j = b_i \qquad (1 \leqslant i \leqslant n)$$

where

$$a_{ij} = \int_0^1 \left[p(t)u_i'(t)u_j'(t) + q(t)u_i(t)u_j(t) \right] dt$$

$$b_i = -\int_0^1 f(t)u_i(t)\, dt$$

This completes the description of the Rayleigh–Ritz method for this problem.

In order to prove theorems about the *convergence* of the method, some preliminaries must be dealt with. The following lemma is formulated for an arbitrary topological space. Refer to Chapter 7, Section 6, pages 361ff, for basic topology.

Lemma.   *Let Ψ be an upper semicontinuous nonlinear functional defined on a topological space X. Let $X_1 \subset X_2 \subset \cdots$ be subsets of X such that $\bigcup_{n=1}^{\infty} X_n$ is dense in X. Then as $n \uparrow \infty$, we have*

(10)
$$\inf_{x \in X_n} \Psi(x) \downarrow \inf_{x \in X} \Psi(x)$$

Proof.   Let $\rho = \inf_{x \in X} \Psi(x)$. (We permit $\rho = -\infty$.) For any $r > \rho$, the set

$$\mathcal{O} = \left\{ \, x \in X : \Psi(x) < r \, \right\}$$

is nonempty. It is open, because Ψ is upper semicontinuous. Since $\bigcup X_n$ is dense, it intersects $\mathcal{O}$. Select m such that $X_m \cap \mathcal{O}$ contains a point, say ξ. Then for $n \geqslant m$

$$\rho \leqslant \inf_{x \in X_n} \Psi(x) \leqslant \inf_{x \in X_m} \Psi(x) \leqslant \Psi(\xi) < r \qquad \blacksquare$$

In applying the Rayleigh–Ritz method to the two-point boundary-value problem

(11)
$$(px')' - qx = f$$

(12)
$$x(0) = x(1) = 0$$

we take X to be the space $\{x \in C^2[0,1] : x(0) = x(1) = 0\}$, normed by defining $\|x\| = \|x\|_\infty + \|x'\|_\infty$. Next, we assume that the finite-dimensional subspaces U_n in X are nested ($U_n \subset U_{n+1}$) and that $\bigcup_{n=1}^{\infty} U_n$ is dense in X. Thus, the elements of U_n satisfy (12). Notice that X has codimension 2 in $C^2[0,1]$. In fact, $X \oplus \Pi_1 = C^2[0,1]$, where Π_1 denotes the subspace of all first-degree polynomials. The functional Φ is

$$\Phi(x) = \int_0^1 \left[p(x')^2 + qx^2 + 2fx \right] dt$$

The functional Φ attains its infimum on each finite-dimensional subspace of X. A proof of this is outlined in Problems 6, 7, and 8.

Theorem 3.   *Let x denote the solution of the boundary-value problem (11, 12), and let x_n be a point in U_n that minimizes Φ on U_n. Assume $q \geqslant 0$ and $p > 0$ on $[0, 1]$. Then $x_n(t) \to x(t)$ uniformly.*

Proof.   Since we have assumed that $\bigcup_{n=1}^{\infty} U_n$ is dense in X, the preceding lemma and Theorem 2 imply that $\Phi(x_n) \downarrow \Phi(x)$. Notice that our choice of norm on X guarantees the continuity of Φ. In the following, we use the standard inner-product notation

$$\langle u, v \rangle = \int_0^1 u(t)v(t)\, dt$$

and $\|\ \|_2$ is the accompanying quadratic norm. From Equation (9) in the proof of Theorem 2, we have

$$\Phi(x_n) - \Phi(x) = \int_0^1 \left[p \cdot (x_n' - x')^2 + q \cdot (x_n - x)^2 \right] dt$$

$$\geqslant \int_0^1 p \cdot (x_n' - x')^2\, dt \geqslant \min_{0 \leqslant t \leqslant 1} p(t) \|x_n' - x'\|_2^2$$

This shows that $\|x_n' - x'\|_2 \to 0$. By obvious manipulations, including the Cauchy–Schwarz inequality, we have now

$$|x_n(s) - x(s)| = \left| \int_0^s \left[x_n'(t) - x'(t) \right] dt \right| \leqslant \int_0^s |x_n'(t) - x'(t)|\, dt$$

$$\leqslant \int_0^1 |x_n'(t) - x'(t)|\, dt = \langle |x_n' - x'|, 1 \rangle \leqslant \|x_n' - x'\|_2 \cdot \|1\|_2$$

This shows that

$$\|x_n - x\|_\infty \leqslant \|x_n' - x'\|_2 \to 0 \qquad\qquad \blacksquare$$

As an illustration of the preceding theorem, consider the subspaces

$$U_n = \{\, wv \; : \; v \in \Pi_n \,\} \qquad w(t) = t(1 - t)$$

The union of these subspaces is the space of all functions having the form

$$t \longmapsto t(1 - t) \sum_{k=0}^n a_k t^k$$

Is this subspace dense in the space

$$X = \{\, x \in C^2[0, 1] \; : \; x(0) = x(1) = 0 \,\} \; ?$$

The norm on X is defined to be $\|x\| = \|x\|_\infty + \|x'\|_\infty$. To prove density, let x be any element of X, and let $\varepsilon > 0$. By the Weierstrass Approximation Theorem, there is a polynomial p such that $\|p - x'\|_\infty < \infty$. Define $u(s) =$

$\int_0^s p(t)\,dt$, so that $u' = p$, $u(0) = 0$, and $\|u' - x'\|_\infty < \varepsilon$. Then $|u(s) - x(s)| = |\int_0^s [u'(t) - x'(t)]\,dt| < \varepsilon$. Since $x(1) = 0$, $|u(1)| < \varepsilon$. Put $v(t) = tu(1)$. Then $|v'(t)| = |u(1)| < \varepsilon$ and $|v(t)| \leqslant |u(1)| < \varepsilon$. Notice that $u - v$ is a polynomial that takes the value 0 at 0 and 1. Hence $u - v$ contains $w(t) = t(1 - t)$ as a factor, and belongs to one of the spaces U_n. Also,

$$\|u - v - x\|_\infty < \|u - x\|_\infty + \|v\|_\infty < 2\varepsilon$$
$$\|u' - v' - x'\|_\infty < \|u' - x'\|_\infty + \|v'\|_\infty < 2\varepsilon$$

Thus x can be approximated with arbitrary precision by elements of $\bigcup_{n=1}^\infty U_n$. To summarize, we state the following theorem.

> **Theorem 4.**   *In the two-point boundary-value problem described in Equations* $(11, 12)$, *assume that* p', q, *and* f *are continuous. Assume further that* $p > 0$ *and* $q \geqslant 0$. *If, for each* n, x_n *is the polynomial of degree* n *that minimizes* Φ *subject to the constraint* $x_n(0) = x_n(1) = 0$, *then* $[x_n]$ *converges uniformly to a solution of Equations* $(11, 12)$.

Another illustration of the Rayleigh–Ritz method is provided by a boundary-value problem involving Poisson's equation in two variables:

(13)
$$\begin{cases} \nabla^2 u = f \ \text{ in } \ \Omega \\ u = g \ \text{ on } \ \partial\Omega \end{cases}$$

In correspondence with this problem, we set up the functional

$$\Phi(u) = \int_\Omega (u_x^2 + u_y^2 + 2uf)\,dx\,dy$$

We shall now show that any function u that minimizes $\Phi(u)$ under the constraint that $u = g$ on $\partial\Omega$ must solve the boundary-value problem (13).

Proceeding as before, take a function w that vanishes on $\partial\Omega$, and consider $\Phi(u + \lambda w)$. Since u is a minimum point of Φ,

$$\frac{d}{d\lambda}\Phi(u + \lambda w)\Big|_{\lambda=0} = 0$$

This leads, by straightforward calculation, to the equation

(14)
$$\int_\Omega (u_x w_x + u_y w_y + fw)\,dx\,dy = 0$$

In order to proceed, we require Green's Theorem, which asserts (under reasonable hypotheses concerning Ω) that

$$\int_\Omega (P_x + Q_y)\,dx\,dy = \int_{\partial\Omega} (P\,dy - Q\,dx)$$

Using this, we have

$$\int_\Omega (u_x w_x + u_y w_y) = \int_\Omega \left[(wu_x)_x + (wu_y)_y - w\nabla^2 u \right]$$

$$= \int_{\partial\Omega} (-wu_y \, dx + wu_x \, dy) - \int_\Omega w\nabla^2 u$$

Since w vanishes on $\partial\Omega$, we can now write (14) in the form

$$\iint_\Omega (-w\nabla^2 u + fw) \, dx \, dy = 0$$

Since w is almost arbitrary, this equation implies that

$$\nabla^2 u = f$$

There are many problems in applied mathematics that arise naturally as minimization problems. This occurs, for example when a configuration involving minimum energy is sought in a mechanical system. One inclusive setting for such problems is described here, and an elegant, easy, theorem is proved concerning it.

Let X be a Banach space, and suppose that a continuous, symmetric bilinear functional B is given. Let ϕ be a continuous linear functional on X, and let a closed convex set K be prescribed. We seek the minimum of $B(x,x) + \phi(x)$ on K.

Theorem 5.   *In addition to the hypotheses in the preceding paragraph, assume that B is "elliptic," in the sense that $B(x,x) \geqslant \beta\|x\|^2$ for some $\beta > 0$. Then the minimum of $B(x,x) + \phi(x)$ on K is attained at a unique point of K.*

Proof.   The bilinear form B defines an inner product on X. The norm arising from the inner product is written $\|x\|_B = \sqrt{B(x,x)}$. Since B is continuous, there is a positive constant α such that $|B(x,y)| \leqslant \alpha\|x\|\|y\|$. Consequently, $\|x\|_B \leqslant \alpha\|x\|$. On the other hand, from the condition of ellipticity, we have $\|x\|_B \geqslant \sqrt{\beta}\|x\|$. Thus the two norms on X are equivalent. Hence, $(X, \|\cdot\|_B)$ is complete and therefore a Hilbert space. Also, K is a closed convex set in this Hilbert space. By the Riesz Representation Theorem, $\phi(x) = -2B(v,x)$ for some v in X. Write

$$B(x,x) + \phi(x) = B(x - v, x - v) - B(v,v)$$

This shows that our minimization problem is the standard one of finding a point of K closest to v, in the Hilbert space setting. Theorem 2 in Section 2.1 (page 64) applies and establishes the existence of a unique point x in K solving the problem.   ∎

Historical note.   A biographical article about George Green is [Cann1], and a book by the same author is [Cann2]. When you are next in England, you would

enjoy visiting Nottingham and seeing the well-restored mill, of which George
Green was the proprietor. His collected works have been published in [Green].

Problems 4.6

1. Let u be an element of $C[0,1]$ such that

$$\int_0^1 u(t)x(t)\, dt = 0$$

   whenever x is an element of $C[0,1]$ satisfying the equation $\sum_{n=1}^{\infty}[x(1/n)]^2 = 0$. Prove
   or disprove that u is necessarily 0.

2. Let $t_1, t_2, \ldots, t_m$ be specified points in $[0,1]$, and let x be an element of $C[0,1]$ that
   vanishes at the points $t_1, \ldots, t_m$. Can we approximate x with arbitrary precision by a
   polynomial that also vanishes at these points?

3. Solve the two-point boundary-value problem

$$x'' = x^2 \qquad\qquad x(0) = 6 \qquad\qquad x(1) = \tfrac{3}{2}$$

   Suggestion: Multiply by x' and integrate, or try some likely functions containing parameters.

4. Let x be an element of $C^2[a,b]$ that minimizes the functional

$$\Phi(x) = \int_a^b \left[p(x')^2 + qx^2 + 2fx\right] dt$$

   subject to the constraints $x(a) = \alpha$ and $x'(b) = 0$. Find the two-point boundary-value
   problem that x solves.

5. Use an elementary change of variable in the problem

$$\begin{cases} (px')' - qx = f \\ x(a) = \alpha \quad x(b) = \beta \end{cases}$$

   to find an equivalent problem having homogeneous boundary conditions. Thus, $y(a) =
   y(b) = 0$ in the new variable. Is the new problem also of Sturm–Liouville form?

6. Define $X = \{x \in C^2[a,b] : x(a) = x(b) = 0\}$. Prove that if $x \in X$ then $\|x'\|_\infty \geqslant
   \|x\|_\infty (b-a)^{-1}$. Try to improve the bound.

7. (Continuation) In the Sturm–Liouville problem described by Equations (11) and (12),
   assume that $p(t) \geqslant \delta > 0$ and that $q \geqslant 0$. Prove that for $x \in X$,

$$\Phi(x) \geqslant \delta\|x'\|_2^2 - \|f\|_2\|x\|_2$$

8. (Continuation) Use the two preceding problems to show that on any finite-dimensional
   subspace in X, the infimum of Φ is attained.

9. Solve the two-point boundary-value problem

$$x'' = x^2 \qquad\qquad x(0) = 0 \qquad\qquad x(1) = 1$$

   This is deceptively similar to Problem 3, but harder. Look for a solution of the form
   $x(t) = \sum_{n=0}^{\infty} a_n t^n$. You should find a "general" solution of the differential equation
   containing two arbitrary constants a_0 and a_1. All remaining coefficients can then be

obtained by a recurrence relation. After imposing the condition $x(0) = 0$, your solution will contain only the powers $t, t^4, t^7, t^{10}, \ldots$ The parameter a_1 will be available to secure the remaining boundary condition, $x(1) = 1$. Reference: [Dav].

10. Let $[a, b]$ be a compact interval in $\mathbb{R}$, and define

$$\|x\|_\infty = \sup_{a \leqslant t \leqslant b} |x(t)| \qquad \|x\|_2 = \left\{ \int_a^b |x(t)|^2 \, dt \right\}^{1/2}$$

Prove that if $\|x'_n\|_2 \to 0$ and if $\inf_{a \leqslant t \leqslant b} |x_n(t)| \to 0$, then $\|x_n\|_\infty \to 0$. Is the result true when the interval is replaced by $[a, \infty)$? Assume x'_n continuous.

11. In $C^1[a, b]$ consider the two norms in the preceding problem. Prove that

$$\|x\|_\infty \leqslant |x(a)| + k\|x'\|_2 \qquad \text{where} \quad k = (b - a)^{1/2}$$

12. Consider the two-point boundary-value problem

$$x'' = f(t, x) \qquad x(0) = \alpha \qquad x(1) = \beta$$

and the functional

$$\Phi(x) = \int_0^1 [(x')^2 + 2g(t, x)] \, dt$$

Assume that $f = \partial g / \partial x$. Prove that any C^2-function that minimizes $\Phi(x)$ subject to the constraints $x(0) = \alpha$ and $x(1) = \beta$ is a solution of the boundary-value problem. Show by example that the converse is not necessarily true.

13. Prove that there exists a polynomial of degree 5 such that $p(0) = p'(0) = p''(0) = p'(1) = p''(1) = 0$ and $p(1) = 1$.

14. (Continuation) Let $a < b$. Using the polynomial in the preceding problem, show that there exists a polynomial q of degree 5 such that $q(a) = q'(a) = q''(a) = q'(b) = q''(b) = 0$ and $q(b) = 1$. With the help of q construct a nondecreasing C^2-function f such that $f(t) = 0$ on $(-\infty, a)$ and $f(t) = 1$ on (b, ∞).

15. Solve the integral equation

$$x(t) = \sin t + \int_0^\pi (t^2 + s^2) x(s) \, ds$$

16. This two-point boundary value problem is easily solved:

$$x''(t) = \cos t \qquad x(0) = x(\pi) = 0 \qquad (0 \leqslant t \leqslant \pi)$$

Use the Ritz method to solve it, employing trial functions of the form $x(t) = \sum a_n \sin nt$. Orthogonality relations among the trigonometric functions will be useful in minimizing the Ritz functional.

4.7 Collocation Methods

The collocation method for solving a linear operator equation

(1) $$Ax = b$$

is one of the projection methods, as described in Section 4.4. It begins by selecting base functions $u_1, \ldots, u_n$ and linear functionals $\phi_1, \ldots, \phi_n$. We try to satisfy Equation (1) with an x of form $x = \sum_{j=1}^{n} c_j u_j$. This leads to the equation

$$(2) \qquad \sum_{j=1}^{n} c_j A u_j = b$$

In general, this system is inconsistent, because b is usually not in the linear space generated by $Au_1, Au_2, \ldots, Au_n$. We apply the functionals ϕ_i to Equation (2), however, and arrive at a set of n linear equations for the n unknowns $c_1, c_2, \ldots, c_n$:

$$\sum_{j=1}^{n} c_j \phi_i(A u_j) = \phi_i(b) \qquad 1 \leqslant i \leqslant n$$

Of course, care must be taken to ensure that the $n \times n$ matrix whose elements are $\phi_i(Au_j)$ is nonsingular. In the classical collocation method, the problem (1) involves a function space; i.e., the unknown x is a function. Then the functionals are chosen to be **point-evaluation functionals**:

$$\phi_i(x) = x(t_i)$$

Here, the points t_i have been specified in the domain of x.

Let us see how a two-point boundary-value problem can be solved approximately by the method of collocation. We take a linear problem with zero boundary conditions:

$$(3) \qquad \begin{cases} x'' + px' + qx = f \\ x(0) = x(1) = 0 \end{cases}$$

As usual, it makes matters easier to select base functions that satisfy the homogeneous part of the problem. Suppose that we let $u_j(t) = (1-t)t^j$ for $j = 1, 2, \ldots, n$. As the functionals, we use $\phi_i(x) = x(t_i)$, where the points $t_1, \ldots, t_n$ can be chosen in the interval $[0, 1]$. For example, we can let $t_i = (i-1)h$, where $h = 1/(n-1)$. The operator A is defined by

$$Ax = x'' + px' + qx$$

and by computing we find that

$$(Au_j)(t) = j(j-1)t^{j-2} - j(j+1)t^{j-1} + p(t)\big[jt^{j-1} - (j+1)t^j\big] + q(t)[t^j - t^{j+1}]$$

The matrix whose elements are $(Au_j)(t_i)$ is easily written down, but it is not instructive to do so. It will probably be an ill-conditioned matrix, because the base functions we have chosen are not suited to numerical work. Better choices for the base functions u_i would be the Chebyshev polynomials (suitable to the interval in question) or a set of B-splines.

More examples of collocation techniques will be given later in this section, but first we shall discuss the important technique of turning a two-point boundary-value problem into an equivalent integral equation. This continues a theme introduced in Section 2.5. The integral equation can then be shown to have a solution by applying a fixed-point theorem, and in this way we obtain an existence theorem for the original two-point boundary-value problem.

We consider a two-point problem of the form

$$
(4) \qquad \begin{cases} x'' = f(t,x) & 0 \leqslant t \leqslant 1 \\ x(0) = 0 \quad x(1) = 0 \end{cases}
$$

There is no loss of generality in assuming that the interval of interest is $[0,1]$ and that the boundary conditions are homogeneous, because if these hypotheses are not fulfilled at the beginning, they can be brought about by suitable changes in the variables. (In this connection, refer to Problems 3 and 4 in Section 4.1, page 176.) Observe that Equation (1) is, in general, *nonlinear*. We assume that f is continuous on $[0,1] \times \mathbb{R}$.

The **Green's function** for the boundary value problem (4) is defined to be the function

$$
(5) \qquad G(t,s) = \begin{cases} t(1-s) & 0 \leqslant t \leqslant s \leqslant 1 \\ s(1-t) & 0 \leqslant s \leqslant t \leqslant 1 \end{cases}
$$

Notice that G is defined on the unit square in the st-plane, and vanishes on the boundary of the square. Although G is continuous, its partial derivatives have jump discontinuities along the line $s = t$. Using the Green's function as kernel, we define an integral equation

$$
(6) \qquad x(t) = -\int_0^1 G(t,s)f\big(s,x(s)\big)\,ds
$$

Theorem 1.   *Each solution of the boundary-value problem (4) solves the integral equation (6) and conversely.*

Proof.   Let x be any function in $C[0,1]$, and define y by writing

$$
y(t) = -\int_0^1 G(t,s)f\big(s,x(s)\big)\,ds
$$

$$
= -\int_0^t G(t,s)f\big(s,x(s)\big)\,ds + \int_1^t G(t,s)f\big(s,x(s)\big)\,ds
$$

We intend to differentiate in this equation, and the reader should recall the general rule:

$$
\frac{d}{dt}\int_a^t h(s)\,ds = h(t)
$$

Then the chain rule gives us

$$
\frac{d}{dt}\int_a^{k(t)} h(s)\,ds = h(k(t))k'(t)
$$

Now, for $y'(t)$ we have

$$y'(t) = -G(t,t)f\big(t,x(t)\big) - \int_0^t G_t(t,s)f\big(s,x(s)\big)\,ds$$

$$+ G(t,t)f\big(t,x(t)\big) + \int_1^t G_t(t,s)f\big(s,x(s)\big)\,ds$$

$$= \int_0^t s\,f\big(s,x(s)\big)\,ds + \int_1^t (1-s)f\big(s,x(s)\big)\,ds$$

A second differentiation yields

(7) $$y''(t) = tf\big(t,x(t)\big) + (1-t)f\big(t,x(t)\big) = f\big(t,x(t)\big)$$

If x is a solution of the integral equation, then $y = x$, and our calculation (7) shows that $x'' = y'' = f(t,x)$. Since $G(t,s) = 0$ on the boundary of the square, $y(0) = y(1) = 0$. Hence $x(0) = x(1) = 0$. This proves half of the theorem.

   Now suppose that x is a solution of the boundary-value problem. The above calculation (7) shows that

$$y''(t) = f\big(t,x(t)\big) = x''(t)$$

It follows that the two functions x and y can differ only by a *linear* function of t (because $x'' - y'' = 0$). Since $x(t)$ and $y(t)$ take the same values at $t = 0$ and $t = 1$, we conclude that $x = y$. Thus x solves the integral equation.   ∎

   Theorem 2.   *Let $f(s,t)$ be continuous in the domain defined by the inequalities $0 \leqslant s \leqslant 1$, $-\infty < t < \infty$. Assume also that f satisfies a Lipschitz condition in this domain:*

(8) $$\big|f(s,t_1) - f(s,t_2)\big| \leqslant k\big|t_1 - t_2\big| (k < 8)$$

Then the integral equation (6) has a unique solution in $C[0,1]$.

Proof.   Consider the nonlinear mapping $F : C[0,1] \to C[0,1]$ defined by

$$(Fx)(t) = -\int_0^1 G(t,s)f\big(s,x(s)\big)\,ds   x \in C[0,1]$$

We shall prove that F is a contraction. We have

$$\big|(Fu)(t) - (Fv)(t)\big| \leqslant \int_0^1 G(t,s)\big|f\big(s,u(s)\big) - f\big(s,v(s)\big)\big|\,ds$$

$$\leqslant k \int_0^1 G(t,s)\big|u(s) - v(s)\big|\,ds$$

$$\leqslant k\|u - v\|_\infty \int_0^1 G(t,s)\,ds$$

$$= (k/8)\|u - v\|_\infty$$

It follows that

$$\big\|Fu - Fv\big\|_\infty \leqslant (k/8)\|u - v\|_\infty$$

and that F is a contraction. Now apply Banach's Theorem, page 177, taking note of the fact that $C[0,1]$, with the supremum norm, is complete.   ∎

Corollary.   *If the function f satisfies the hypotheses of Theorem 2, then the boundary-value problem* (1) *has a unique solution in* $C[0,1]$.

Example 1.   Consider the two-point boundary-value problem

(9)
$$\begin{cases} x''(t) = \tfrac{1}{2}\exp\left\{\tfrac{1}{2}(t+1)\cos\left[x(t)+7-3t\right]\right\} & -1 \leqslant t \leqslant 1 \\ x(-1) = -10 \quad x(1) = -4 \end{cases}$$

Our existence theorem does not apply to this directly, and some changes of variables are called for. We set

$$z(t) = x(t) - 3t + 7$$

and find that z should solve this problem:

(10)
$$\begin{cases} z''(t) = \tfrac{1}{2}\exp\left\{\tfrac{1}{2}(t+1)\cos z(t)\right\} \\ z(-1) = z(+1) = 0 \end{cases}$$

Next we set

$$t = -1 + 2s \qquad y(s) = z(t)$$

and find that y should solve this problem:

(11)
$$\begin{cases} y''(s) = 2\exp\left\{s\cos y(s)\right\} \\ y(0) = y(1) = 0 \end{cases}$$

To this problem we can apply the preceding corollary. The function $f(s,r) = 2e^{s\cos r}$ satisfies a Lipschitz condition, as we see by applying the mean value theorem:

$$|f(s,r_1) - f(s,r_2)| = \left|\frac{\partial f}{\partial r}(s,r_3)\right| |r_1 - r_2|$$

The derivative here is bounded as follows

$$|2e^{s\cos r}(-s\sin r)| \leqslant 2e \approx 5.436$$

Since the Lipschitz constant $2e$ is less than 8, the boundary-value problem (11) has a solution y. Hence (9) has a solution x, and it is given by

$$x(t) = y((t+1)/2) + 3t - 7 \qquad\qquad\blacksquare$$

For the practical solution of boundary-value problems, one usually relies on numerical methods (some of which have already been discussed) such as discretization, Galerkin's method, and collocation. For the problem considered above, namely

(12)
$$\begin{cases} x'' = f(t,x) & 0 \leqslant t \leqslant 1 \\ x(0) = x(1) = 0 \end{cases}$$

there is now an additional method of proceeding. One can set up an equivalent integral equation,

$$(13) \qquad x(t) = \int_0^1 -G(t,s) f\big(s, x(s)\big)\, ds$$

and solve it instead. If we discretize both problems (12) and (13) in a certain uniform way, the two new problems will be equivalent, a result to which we now turn our attention.

The standard discretization of the boundary value problem (12) is done by introducing a formula for numerical differentiation, as in Section 4.1. For the integral equation, we require a formula for numerical integration, and choose for this purpose a simple Riemann sum. Thus the discretized problems are

$$(14) \qquad \begin{cases} y_{i-1} - 2y_i + y_{i+1} = h^2 f(t_i, y_i) & 1 \leqslant i \leqslant n \\[2mm] y_0 = y_{n+1} = 0 \end{cases}$$

$$(15) \qquad y_i = -h \sum_{j=1}^n G(t_i, t_j) f(t_j, y_j) \qquad 0 \leqslant i \leqslant n+1$$

In both of these we have set $h = 1/(n+1)$ and $t_i = ih$. Of course, $y \in \mathbb{R}^{n+2}$. Notice that we have used the fact that G vanishes on the boundary of the square.

Theorem 3.   *Problems (14) and (15) are equivalent.*

Proof.   We proceed as in the proof of Theorem 1, which concerns the "undiscretized" problems. The rôle of the second derivative is now played by a set of linear functionals L_i defined on $\mathbb{R}^{n+2}$ by the equation

$$(16) \qquad L_i(z) = h^{-2}[z_{i-1} - 2z_i + z_{i+1}] \qquad 1 \leqslant i \leqslant n$$

Here z is an arbitrary element of $\mathbb{R}^{n+2}$ written in the form $z = (z_0, z_1, \ldots, z_{n+1})$. Now let $(y_0, \ldots, y_{n+1})$ be arbitrary, and let z be defined by

$$(17) \qquad z_i = -h \sum_{j=1}^n G(t_i, t_j) f(t_j, y_j) \qquad 0 \leqslant i \leqslant n+1$$

We assert that $L_i z = f(t_i, y_i)$ for $i = 1, 2, \ldots, n$. In order to prove this, apply L_i to z, using the linearity of L_i. The result is

$$(18) \qquad L_i z = -h \sum_{j=1}^n f(t_j, y_j) L_i G(\cdot, t_j)$$

Now $G(t, s)$ is a linear function of t in each of the two intervals $0 \leqslant t \leqslant s$ and $s \leqslant t \leqslant 1$. Thus $L_i G(\cdot, t_j) = 0$ unless $i = j$. In the case $i = j$ we have

$$\begin{aligned} L_i G(\cdot, t_i) &= h^{-2}\big[G(t_{i-1}, t_i) - 2G(t_i, t_i) + G(t_{i+1}, t_i)\big] \\ &= h^{-2}\big[t_{i-1}(1 - t_i) - 2t_i(1 - t_i) + t_i(1 - t_{i+1})\big] \\ &= h^{-2}\big[(1 - t_i)(t_{i-1} - t_i) + t_i(1 - t_{i+1} - 1 + t_i)\big] \\ &= h^{-2}\big[-h(1 - t_i) - ht_i\big] = -h^{-1} \end{aligned}$$

Thus from Equation (18) we have, as asserted,

$$L_i z = -hf(t_i, y_i)(-h^{-1}) = f(t_i, y_i)$$

Now suppose that $(y_0, \ldots, y_{n+1})$ solves the equations in (15). Then $z_i = y_i$ for $0 \leqslant i \leqslant n+1$. Consequently, $L_i y = L_i z = f(t_i, y_i)$. Since $z_0 = z_{n+1} = 0$, from (17), we have also $y_0 = y_{n+1} = 0$. Thus y solves the equations in (14).

Conversely, if y solves the equations in (14), then $L_i y = f(t_i, y_i) = L_i z$. Since the second divided differences of y and z are equal, these two vectors can differ only by an arithmetic progression. But $y_{n+1} = z_{n+1}$ and $y_0 = z_0$, so the vectors are in fact identical. Thus y satisfies the equations in (15). ∎

Now reconsider the integral equation (13), which is equivalent to the boundary-value problem (12). One advantage of the integral equation is that many different numerical quadrature formulas can be applied to it. The most accurate of these formulas do *not* employ equally spaced nodes. The idea of using unequally spaced points in the discretized problem of (14) would not normally be entertained, as that would only complicate matters without producing any obvious advantage in precision. The quadrature formulas of maximal accuracy are well known, however, and are certainly to be recommended in the numerical solution of integral equations, in spite of their involving unequally spaced nodes.

A quadrature formula of the type needed here will have the form

$$(19) \qquad \int_0^1 g(s)\,ds \approx \sum_{j=1}^{n} A_j g(s_j)$$

in which $g \in C[0,1]$, $0 \leqslant s_j \leqslant 1$, and the A_j are coefficients, often called *weights*. The Riemann sums employed in Equation (15) are obviously obtained by a formula of the type in (19). However, there are other formulas that are markedly superior. The result of using formula (19) to discretize the integral equation (13) is

$$y(t) = -\sum A_j G(t, s_j) f\big(s_j, y(s_j)\big)$$

Notice that this equation can be used used in a practical way in functional iteration. We can start with any y_0 in $C[0,1]$ and define inductively

$$y_{m+1}(t) = -\sum_{j=1}^{n} A_j G(t, s_j) f(s_j, y_m(s_j))$$

The right–hand side of this equation is certainly computable. It is a linear combination of sections G^{s_j}. One can also use collocation at the points s_i to proceed. This will lead to

$$(20) \qquad y_i = -\sum_{j=1}^{n} A_j G(s_i, s_j) f(s_j, y_j) \qquad (1 \leqslant i \leqslant n)$$

In this equation, y_i is an approximation to $x(s_i)$. In general, Equation (20) will represent a system of n nonlinear equations in the n unknowns $(y_1, \ldots, y_n)$. The solution of such a system may be a difficult matter, but for the moment we shall suppose that a solution $y = (y_1, \ldots, y_n)$ has been obtained. Let us use x to denote the solution function for the integral equation (13). It is to be hoped that $|y_i - x(s_i)|$ will be small. Here the nodes of the quadrature formula are $s_1, \ldots, s_n$. Two functions that enter the theorem are

$$(21) \qquad u(t) = \int_0^1 G(t, s) f\big(s, x(s)\big)\, ds - \sum_{j=1}^n A_j G(t, s_j) f\big(s_j, x(s_j)\big)$$

$$(22) \qquad v(t) = \int_0^1 G(t, s)\, ds - \sum_{j=1}^n A_j G(t, s_j)$$

If the quadrature formula is a good one, these functions will be small in norm. We continue to assume the Lipschitz inequality (8) on f.

Theorem 4. *If $k(1 + 8\|v\|_\infty) < 8$, and if the weights A_j in Equation (19) are all positive, then for $i = 1, 2, \ldots, n$,*

$$|x(s_i) - y_i| \leqslant \lambda \|u\|_\infty \qquad \text{where } \lambda = \left[1 - k\left(\frac{1}{8} + \|v\|_\infty\right)\right]^{-1}$$

Proof. Let $\varepsilon_i = |x(s_i) - y_i|$ and $\varepsilon = \max \varepsilon_i$. Then for each i we have

$$\varepsilon_i = \left| \int_0^1 G(s_i, s) f\big(s, x(s)\big)\, ds - \sum_{j=1}^n A_j G(s_i, s_j) f(s_j, y_j) \right|$$

$$= \left| u(s_i) + \sum_{j=1}^n A_j G(s_i, s_j) f\big(s_j, x(s_j)\big) - \sum_{j=1}^n A_j G(s_i, s_j) f(s_j, y_j) \right|$$

$$\leqslant \|u\|_\infty + \sum_{j=1}^n A_j G(s_i, s_j) \big| f\big(s_j, x(s_j)\big) - f(s_j, y_j) \big|$$

$$\leqslant \|u\|_\infty + k\varepsilon \sum_{j=1}^n A_j G(s_i, s_j) = \|u\|_\infty + k\varepsilon \left[\int_0^1 G(s_i, s)\, ds - v(s_i) \right]$$

$$\leqslant \|u\|_\infty + k\varepsilon \left(\frac{1}{8} + \|v\|_\infty \right)$$

It follows that $\varepsilon \leqslant \|u\|_\infty + k\varepsilon \left(\frac{1}{8} + \|v\|_\infty \right)$. When this inequality is solved for ε, the result is the one stated in the theorem. ∎

Theorem 5. *The "discretized" integral equation (20) can be solved by the method of iteration if $k \left(\frac{1}{8} + \|v\|_\infty \right) < 1$ and $A_j > 0$. Note that k, v, and A_j are as in Equations (8), (22), and (19).*

Proof. Interpret Equation (20) as posing a fixed-point problem for a mapping $U : \mathbb{R}^n \to \mathbb{R}^n$ whose definition is

$$(Uy)_i = -\sum_{j=1}^n A_j G(s_i, s_j) f(s_j, y_j) \qquad (1 \leqslant i \leqslant n)$$

A short calculation will show that this is a contraction:

$$\left|(Uy - Uz)_i\right| \leqslant \sum_{j=1}^{n} A_j G(s_i, s_j)\left|f(s_j, y_j) - f(s_j, z_j)\right|$$

$$\leqslant k \max_j \left|y_j - z_j\right| \sum_{\nu=1}^{n} A_\nu G(s_i, s_\nu)$$

As in the preceding proof, the sum in this last inequality has the upper bound $\frac{1}{8} + \|v\|_\infty$. Hence we have

$$\|Uy - Uz\|_\infty \leqslant k\left(\frac{1}{8} + \|v\|_\infty\right)\|y - z\|_\infty \qquad\blacksquare$$

The Lipschitz condition in (8) is usually established by estimating the partial derivative $f_2 \equiv \partial f(t, s)/\partial s$ and using the mean value theorem. If $|f_2| \leqslant k < 8$ on the domain where $0 \leqslant t \leqslant 1$ and $-\infty < s < \infty$, then we can also use Newton's method to solve the discretized integral equation in (20). The equations that govern the procedure can be derived in the following way. Suppose that an approximate solution $(y_1, y_2, \ldots, y_n)$ for system (20) is available. We seek to calculate corrections h_i so that the vector $(y_1 + h_1, \ldots, y_n + h_n)$ will be an exact solution of (20). Thus we desire that

$$(23) \qquad\qquad y_i + h_i = -\sum_{j=1}^{n} A_j G(s_i, s_j) f(s_j, y_j + h_j)$$

Of course, we take just the linear terms in the Taylor expansion of the nonlinear expression $f(s_j, y_j + h_j)$ and use the resulting *linear* equations to solve for the h_i. These linear equations are

$$(24) \qquad\qquad y_i + h_i = -\sum_{j=1}^{n} A_j G(s_i, s_j)\left[f(s_j, y_j) + h_j f_2(s_j, y_j)\right]$$

(Having made this approximation, we can no longer expect the corrections to produce the exact solution; hence the need for iteration.) When Equation (24) is rearranged we have

$$(25) \qquad\qquad h_i + \sum_{j=1}^{n} E_{ij} h_j = d_i \qquad (1 \leqslant i \leqslant n)$$

in which

$$E_{ij} = A_j G(s_i, s_j) f_2(s_j, y_j)$$

and

$$d_i = -y_i - \sum_{j=1}^{n} A_j G(s_i, s_j) f(s_j, y_j)$$

Equation (25) has the form $(I + E)h = d$. We can see that $I + E$ is invertible (nonsingular) by verifying that $\|E\|_\infty < 1$:

$$\|E\|_\infty = \max_i \sum_{j=1}^{n} |E_{ij}| = \max_i \sum_{j=1}^{n} A_j G(s_i, s_j) |f_2(s_j, y_j)|$$

$$\leqslant k \max_i \sum_{j=1}^{n} A_j G(s_i, s_j) \leqslant k\left(\frac{1}{8} + \|v\|_\infty\right)$$

If our numerical integration formula is sufficiently accurate, then $\|v\|_\infty$ will be small enough to yield $\|E\|_\infty < 1$, since $k < 8$.

This section is concluded with a few remarks about the quadrature formulas mentioned above. Such a formula is of the type

$$(26) \qquad \int_a^b x(t)w(t)\,dt \approx \sum_{i=1}^{n} A_i x(t_i) \qquad x \in C[a, b]$$

The function w is assumed to be positive on $[a, b]$ and remains fixed in the discussion. The points t_i are called "nodes." They are fixed in $[a, b]$. The coefficients A_i are termed "weights." The formula is expected to be used on arbitrary functions x in $C[a, b]$.

Theorem 6.   *If the nodes and the function w are prescribed, then there exists a formula of the type displayed in Equation (26) that is exact for all polynomials of degree at most $n - 1$.*

Proof.   Recall the Lagrange interpolation operator described in Example 1 of Section 4.4, page 193. Its formula is

$$(27) \qquad Lx = \sum_{i=1}^{n} x(t_i)\ell_i$$

Since L is a projection of $C[a, b]$ onto $\prod_{n-1}$, we have $Lx = x$ for all $x \in \prod_{n-1}$, and consequently, for such an x,

$$(28) \quad \int_a^b x(t)w(t)\,dt = \int_a^b \sum_{i=1}^{n} x(t_i)\ell_i(t)w(t)\,dt = \sum_{i=1}^{n} x(t_i) \int_a^b \ell_i(t)w(t)\,dt \quad \blacksquare$$

Example 2.   If $(t_1, t_2, t_3) = (-1, 0, +1)$ and $[a, b] = [-1, 1]$, what is the quadrature formula produced by the preceding method when $w(t) = 1$? We follow the prescription and begin with the functions ℓ_i:

$$\ell_1(t) = (t - t_2)(t - t_3)(t_1 - t_2)^{-1}(t_1 - t_3)^{-1} = t(t-1)/2$$
$$\ell_2(t) = (t - t_1)(t - t_3)(t_2 - t_1)^{-1}(t_2 - t_3)^{-1} = 1 - t^2$$
$$\ell_3(t) = t(t + 1)/2 \quad \text{(by symmetry)}$$

The integrals $\int_{-1}^{1} \ell_i(t)\,dt$ are $\frac{1}{3}$, $\frac{4}{3}$, and $\frac{1}{3}$, and the quadrature formula is therefore

$$(29) \qquad \int_{-1}^{1} x(t)\,dt \approx \tfrac{1}{3}x(-1) + \tfrac{4}{3}x(0) + \tfrac{1}{3}x(1)$$

The formula gives correct values for each $x \in \prod_2$, by the analysis in the proof of Theorem 6. As a matter of fact, the formula is correct on $\prod_3$ because of symmetries in Equation (29). This formula is **Simpson's Rule.**   ∎

Theorem 7. Gaussian Quadrature.   *For appropriate nodes and weights, the quadrature formula* (26) *is exact on* $\prod_{2n-1}$.

Proof.   Define an inner product on $C[a, b]$ by putting

$$(30) \qquad \langle x, y \rangle = \int_a^b x(t)y(t)w(t)\,dt$$

Let p be the unique monic polynomial in $\prod_n$ that is orthogonal to $\prod_{n-1}$, orthogonality being defined by the inner product (30). Let the nodes $t_1, \ldots, t_n$ be the zeros of p. These are known to be simple zeros and lie in (a, b), although we do not stop to prove this. (See [Ch], page 111.) By Theorem 6, there is a set of weights A_i for which the quadrature formula (26) is exact on $\prod_{n-1}$. We now show that it is exact on $\prod_{2n-1}$. Let $x \in \prod_{2n-1}$. By the division algorithm, we can write $x = qp + r$, where q (the *quotient*) and r (the *remainder*) belong to $\prod_{n-1}$. Now write

$$\int_a^b xw = \int_a^b qpw + \int_a^b rw$$

Since $p \perp \prod_{n-1}$ and $q \in \prod_{n-1}$ the integral $\int qpw$ is zero. Since $p(t_i) = 0$, we have $x(t_i) = r(t_i)$. Finally, since $r \in \prod_{n-1}$, the quadrature formula (26) is exact for r. Putting these facts together yields

$$\int_a^b x(t)w(t)\,dt = \int_a^b r(t)w(t)\,dt = \sum_{i=1}^{n} A_i r(t_i) = \sum_{i=1}^{n} A_i x(t_i) \qquad ∎$$

Formulas that conform to Theorem 7 are known as **Gaussian quadrature formulas.**

Theorem 8.   *The weights in a Gaussian quadrature formula are positive.*

Proof.   Suppose that Formula (26) is exact on $\prod_{2n-1}$. Then it will integrate ℓ_j^2 exactly:

$$0 < \int_a^b \ell_j^2(t)w(t)\,dt = \sum_{i=1}^{n} A_i \ell_j^2(t_i) = A_j \qquad ∎$$

Problems 4.7

1. Refer to the proof of Theorem 3 and show that if z is a vector in $\mathbb{R}^{n+2}$ for which $L_i z = 0$ $(1 \leqslant i \leqslant n)$, then z is an arithmetic progression.

2. Prove that if, in Equation (28), $a = -b$, w is even, and the nodes are symmetrically placed about the origin, then the formula will give correct results on $\prod_n$ when n is odd.

3. Prove that if the formula (26) is exact on $\prod_{2n-1}$, then the nodes *must* be the zeros of a polynomial orthogonal to $\prod_{n-1}$.

4. Let $\langle x, y \rangle = \int_{-1}^{1} x(t)y(t)\, dt$. Verify that the polynomial $p(t) = t^3 - \frac{3}{5}t$ is orthogonal to $\prod_2$. Find the Gaussian quadrature formula for this case, i.e., $n = 3$, $w(t) \equiv 1$, $a = -1$, $b = +1$.

5. Define

$$\langle x, y \rangle = \int_{-1}^{1} x(t)y(t)(1 - t^2)^{-1/2}\, dt$$

   Verify that this improper integral converges whenever x and y are continuous functions on the interval $[-1, 1]$. Accepting the fact that the Chebyshev polynomial $T_3(t) = 4t^3 - 3t$ is orthogonal to $\prod_2$, find the Gaussian quadrature formula in this case. Hint: $T_3(\cos \theta) = \cos 3\theta$. Use the change of variable $t = \cos \theta$ to facilitate the work.

6. Consider this 2-point boundary value problem:

$$x'' = (x^2 + 1)^{-1/2} \qquad x(0) = 0 \qquad x(1) = 1$$

   By using Theorem 2, show that the problem has a unique solution in the space in $C[0, 1]$.

7. Prove that the general second-order linear differential equation

$$ux'' + vx' + wx = f$$

   can be put into Sturm–Liouville form, assuming that $u > 0$, by applying an integrating factor $\exp \int (v - u')/u$.

8. Prove that $\int_0^1 |G(t, s)|\, ds = \frac{1}{8}$.

9. Find the Green's function for the problem

$$\begin{cases} x' = f(t, x) & 0 \leqslant t \leqslant 1 \\ x(0) = 0 \end{cases}$$

   Prove that it is correct.

10. Prove that if $x \in C[0, 1]$ and if x satisfies the integral relation (6), in which f is continuous, then $x \in C^2[0, 1]$.

11. Prove that this two-point boundary value problem has no solution:

$$x'' + x = 0 \qquad x(0) = 3 \qquad x(\pi) = 7$$

12. Convert the two-point boundary value problem in Problem 11 to an equivalent homogeneous problem on the interval $[0, 1]$, and explain why Theorem 2 and its corollary do not apply.

13. An integral equation of the form

$$x(t) + \int_a^b K(t, s)f(s, x(s))\, ds = v(t)$$

is called a **Hammerstein** equation. Show that it can be written in the form $x + AFx = v$, where A and F are respectively a linear and a nonlinear operator defined by

$$(Ax)(t) = \int_a^b K(t,s)x(s)\,ds \qquad (F(x))(t) = f(t,x(t))$$

14. (Continuation) Show that the boundary-value problem

$$x''(t) + g(x(t)) = v(t) \qquad x(0) = x(1) = 0$$

    is equivalent to a Hammerstein integral equation.

15. Consider the initial-value problem

$$x'' + ux' + vx = f \qquad x(0) = \alpha, \quad x'(0) = \beta \quad (t \geqslant 0)$$

    Show that this is equivalent to the Volterra integral equation

$$x(t) + \int_0^t K(t,s)x(s)\,ds = f(t) - \beta u(t) - \alpha v(t) - \beta t v(t)$$

    in which $K(t,s) = u(t) + v(t)(t-s)$.

16. For what two-point boundary-value problem is this the Green's Function?

$$g(s,t) = \min\{s,t\} - \tfrac{1}{2}st \qquad 0 \leqslant s,t \leqslant 1$$

17. Prove that if $u_0 = 0$ and $L_i u = 0$ for $i = 1, 2, \ldots$, then $u_i = ia$ for a suitable constant a. (Refer to the proof of Theorem 3, page 218, for definitions.)

18. Write down the fixed-point problem that is equivalent to the boundary-value problem in Equation (11), page 217. Take one step in the iteration, starting with $y_0(t) = 0$. Check your answer against ours: $y_1(t) = 2[e^t + (1-e)t - 1]$.

19. Consider a numerical integration formula

$$\int_a^b x(t)w(t)\,dt \approx \sum_{i=1}^n A_i x(t_i)$$

    Assume that w is positive and continuous on $[a, b]$. Assume also that t_i are n distinct points in $[a, b]$. Prove that the formula gives correct results for "most" functions in $C[a,b]$. Interpret the word "most" in terms of dimension of certain subspaces.

20. Prove that the following two-point boundary-value problem has a continuous solution and that the solution satisfies $x(t) = x(1-t)$:

$$x''(t) = \sin(x-t)\sin(x+t-1) \qquad x(0) = x(1) = 0 \qquad (0 < t < 1)$$

4.8 Descent Methods

Here, as in the Rayleigh–Ritz method, we assume that a problem confronting us has been somehow recast as a minimization problem. Assume therefore that a functional $\Phi : K \to \mathbb{R}$ is given, where K (the domain of Φ) is a subset of some Banach space X. Usually Φ is nonlinear. Let

(1) $$\rho = \inf_{x \in K} \Phi(x)$$

We admit the possibility that $\rho = -\infty$. The objective is to find a point x_0 in K that yields

$$\Phi(x_0) = \rho$$

It is obvious that many problems of best approximation are of this nature; in such problems $\Phi(x)$ would be the distance between x and some element z that was to be approximated. The domain of Φ would typically be a linear subspace consisting of all the approximants.

Another familiar problem that can be recast as a minimization problem is the two-point boundary value problem for a Sturm–Liouville equation. When the Rayleigh–Ritz method is applied, we seek a minimum of the functional

$$\Phi(x) = \int_a^b \left[p \cdot (x')^2 + q \cdot x^2 + 2f \cdot x \right]$$

In the calculus of variations, similar functionals are encountered. For example, in the "brachistochrone" problem (page 153),

$$\Phi(x) = \int_0^a \left[1 + (x')^2 \right]^{1/2} (2gx)^{-1/2}$$

A goal somewhat more modest than finding the minimum point is to generate a **minimizing sequence** for Φ. That means a sequence $x_1, x_2, \ldots$ in K such that

(2) $$\lim_{k \to \infty} \Phi(x_k) = \rho$$

The sequence itself may or may not converge.

Theorem 1.   *A lower semicontinuous functional on a compact set attains its infimum.*

Proof.   Recall that lower semicontinuity of Φ means that each set of the form

$$K_\lambda = \left\{ x \in K : \Phi(x) \leqslant \lambda \right\}$$

is closed. If $\lambda > \rho$, then K_λ is nonempty. The family of closed sets $\{K_\lambda : \lambda > \rho\}$ has the finite-intersection property (i.e., the intersection of any finite subcollection is nonempty). Since the space is compact,

$$\bigcap \{K_\lambda : \lambda > \rho\} \neq \varnothing$$

(see [Kel]). Any point in this intersection satisfies the inequality $\Phi(x) \leqslant \rho$.   ∎

The preceding theorem can be proved also for a space that is only **countably compact**. This term signifies that any *countable* open cover of the space has a finite subcover. ([Kel] page 162). A consequence is that each sequence in the space has a cluster point. Let x_n be chosen so that $\Phi(x_n) < \rho + 1/n$. Let x^* be a cluster point of the sequence $[x_n]$. Then $\Phi(x^*) \leqslant \rho$. Indeed, if this inequality is false, then for some m, $\Phi(x^*) > \rho + 1/m$. Since Φ is lower semicontinuous, the set

$$\mathcal{O} = \{x : \Phi(x) > \rho + 1/m\}$$

is a neighborhood of x^*. Since x^* is a cluster point, the sequence $[x_n]$ is frequently in $\mathcal{O}$. But this is absurd, since $x_m, x_{m+1}, \ldots$ lie outside of $\mathcal{O}$.

A standard strategy for proving existence theorems consists of the following three steps:

I. Formulate the existence question in terms of minimizing a functional Φ on a set K.

II. Find a topology τ for K such that K is τ-compact and Φ is τ-lower-semicontinuous.

III. Apply the preceding theorem.

(The two requirements on the topology are in opposition to each other. The bigger the topology, the more difficult it is for K to be compact, but the easier it is for Φ to be lower semicontinuous.) Examples of this strategy can be given in spectral theory, approximation theory, and other fields. Here we wish to concentrate on methods for constructing a minimizing sequence for a functional Φ.

If Φ is a functional on a normed space X, and $x \in X$, then the Fréchet derivative of Φ at x may or may not exist. If it exists, it is an element $\Phi'(x)$ of X^* and has the property

$$(3) \qquad \Phi(x + h) - \Phi(x) - \Phi'(x)h = o(h) \qquad (h \in X)$$

These matters are discussed fully in Chapter 3. The linear functional $\Phi'(x)$ is usually called the **gradient** of Φ at x. In the special case when $X = \mathbb{R}^n$, and $x = (\xi_1, \xi_2, \ldots, \xi_n)$, $h = (\eta_1, \eta_2, \ldots, \eta_n)$, it has the form

$$(4) \qquad \Phi'(x)h = \sum_{i=1}^{n} \frac{\partial \Phi}{\partial \xi_i}(x)\eta_i \qquad h \in \mathbb{R}^n$$

If $\Phi'(x)$ exists at a specific point x, then for any $h \in X$, we have

$$(5) \qquad \frac{d}{dt}\Phi(x + th)\Big|_{t=0} = \Phi'(x)h$$

Indeed, by the chain rule (Section 3.2, page 121),

$$\frac{d}{dt}\Phi(x + th) = \Phi'(x + th)h$$

The left-hand side of Equation (5) is called the **directional derivative** of Φ at x in the direction h. The existence of the Fréchet derivative is *sufficient* for the existence of the directional derivative, but not *necessary*. (An example occurs in Problem 2.) The mapping

$$h \longmapsto \frac{d}{dt}\Phi(x+th)\Big|_{t=0}$$

is called the **Gâteaux** derivative. (The mathematician R. Gâteaux was killed while serving as a soldier in the First World War, September 1914.)

If, among all h of norm 1 in X, there is a vector for which $\Phi'(x)h$ is a maximum, this vector is said to point in the direction of **steepest ascent**. Its negative gives the direction of **steepest descent**. These matters are most easily understood when X is a Hilbert space. Suppose, then, that Φ is a functional on a Hilbert space X and that $\Phi'(x)$ exists for some point x. By the Riesz representation theorem for functionals on a Hilbert space, the functional $\Phi'(x)$ is represented by a vector v in X, so that $\Phi'(x)h = \langle h, v \rangle$. If $\|h\| = 1$, then by the Cauchy–Schwarz inequality,

$$\langle h, v \rangle \leqslant |\langle h, v \rangle| \leqslant \|h\| \, \|v\| = \|v\|$$

We have equality here if and only if h is taken to be $v/\|v\|$. Thus the (unnormalized) direction of steepest ascent is v.

An iterative procedure called the **method of steepest descent** can now be described. If any point x is given, the direction of steepest descent at x is computed. Let this be v. The functional Φ is now minimized along the "ray" consisting of points $x + tv$, $t \in \mathbb{R}$. This is done by a familiar technique from elementary calculus; namely, we solve for t in the equation

$$\frac{d}{dt}\Phi(x+tv) = 0$$

If the appropriate t is denoted by $\tilde{t}$, then the process is repeated with x replaced by $x + \tilde{t}v$.

These matters will now be illustrated by a special but important problem, namely, the problem of solving the equation

(6) $$Ax = b$$

in which A is a positive definite self-adjoint operator on a real Hilbert space X, and $b \in X$. In symbols, the hypotheses on A are that

(7) $$\langle Ax, y \rangle = \langle x, Ay \rangle$$

(8) $$\langle Ax, x \rangle > 0 \text{ if } x \neq 0$$

For this problem, we define the functional

(9) $$\Phi(x) = \langle Ax - 2b, x \rangle$$

Theorem 2.   *Under the hypotheses above, a point y satisfies the equation $Ay = b$ if and only if y is a global minimum point of Φ.*

Proof.   Let x be an arbitrary point and v any nonzero vector. Then

$$\Phi(x+tv) = \langle Ax + tAv - 2b, x + tv \rangle$$

(10)
$$= \langle Ax - 2b, x \rangle + t\langle Ax - 2b, v \rangle + t\langle Av, x \rangle + t^2\langle Av, v \rangle$$

$$= \Phi(x) + 2t\langle Ax - b, v \rangle + t^2\langle Av, v \rangle$$

The derivative of this expression, as a function of t, is

$$(11) \qquad \frac{d}{dt}\Phi(x + tv) = 2\langle Ax - b, v\rangle + 2t\langle Av, v\rangle$$

The minimum of $\Phi(x + tv)$ occurs when this derivative is zero. The value of t for which this happens is

$$(12) \qquad \tilde{t} = \langle b - Ax, v\rangle \langle Av, v\rangle^{-1}$$

When this value is substituted in Equation (10) the result is

$$(13) \qquad \Phi(x + \tilde{t}v) = \Phi(x) - \langle b - Ax, v\rangle^2 \langle Av, v\rangle^{-1}$$

This shows that we can cause $\Phi(x)$ to decrease by passing to the point $x + \tilde{t}v$, except when $b - Ax \perp v$. If $b - Ax \neq 0$, then many directions v can be chosen for our purpose, but if $Ax = b$, we cannot decrease $\Phi(x)$. ∎

In the problem under consideration, the directional derivative of Φ is obtained by putting $t = 0$ in Equation (11):

$$(14) \qquad \frac{d}{dt}\Phi(x + tv)\bigg|_{t=0} = 2\langle Ax - b, v\rangle$$

It follows that the direction of steepest descent is the **residual** vector $r = b - Ax$. (Positive scalar factors can be ignored in specifying a direction vector.) The algorithm for steepest descent in this problem is therefore described by these formulas:

$$(15) \qquad r_n = b - Ax_n \qquad t_n = \langle r_n, r_n\rangle / \langle Ar_n, r_n\rangle \qquad x_{n+1} = x_n + t_n r_n$$

Since the method of steepest descent is not competitive with the conjugate direction methods on this problem, we will not go into further detail, but simply state without proof the following theorem. See [KA], pages 606–608.

Theorem 3.   *If A is self-adjoint and satisfies*

$$\inf_{||x||=1} \langle Ax, x\rangle > 0$$

then the steepest-descent sequence in Equation (15) converges to the solution of the equation $Ax = b$.

There is more to this theorem than meets the eye, because the hypotheses on A imply its invertibility, and consequently the equation $Ax = b$ has a unique solution for each b in the Hilbert space. See the lemma in Section 4.9, page 234, for the appropriate formal result.

Example.   We consider the problem $Ax = b$ when

$$A = \begin{bmatrix} 1 & 2 \\ 2 & 5 \end{bmatrix} \qquad b = \begin{bmatrix} 3 \\ 1 \end{bmatrix}$$

How does the method of steepest descent perform on this example? We prefer to let Mathematica do the work, and give it these inputs:

```
A={{1.,2.},{2.,5.}}
b={3.,1.}
Inverse[A]
%.b
```

The output is $A^{-1} = \begin{bmatrix} 5 & -2 \\ -2 & 1 \end{bmatrix}$ and the solution, $x = (13, -5)^T$. Next, we program Mathematica to compute 10 steps of steepest descent, starting at $x = (0,0)$. The following input accomplishes this.

```
x={0.,0.}
Do[r=b-A.x;Print[r];phi=-x.(r+b);Print[phi];
    t=(r.r)/(r.A.r);y=x+t r;x=y;Print[x],{10}]
```

After 10 steps, the output is $x = (5.7, -1.7)$ and $\Phi = -22.4587$. Since the solution is $x^* = (13, -5)$ and $\Phi = -34$, the algorithm works very slowly. Of course, with some starting points, the solution will be obtained in one step. Such starting points are $x^* + sv$, for any eigenvector v. Here are Mathematica commands to compute eigenvectors of A:

```
A={{1.,2.},{2.,5.}}
Eigenvectors[N[A]]
```

If we start the steepest descent process at a remote point such as $x^* + 100v$, the first step (carried out numerically) gives a point very close to x^*. The contours (level sets) of Φ for this example are shown in Figure 4.1.

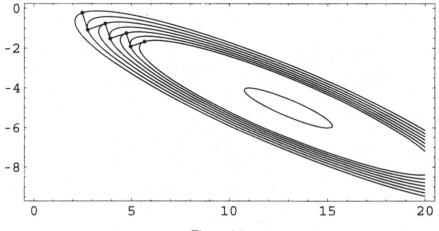

Figure 4.1

Problems 4.8

1. Refer to Equation (13) and discuss the problem of determining v so as to maximize

$$\langle b - Ax, v \rangle^2 \langle Av, v \rangle^{-1}$$

   The solution should be, of course, $v = x_0 - x$ (or a multiple of it), where $x_0 = A^{-1}b$.

2. Define $f : \mathbb{R}^2 \to \mathbb{R}$ by putting $f(x, y) = 0$ if $x = 0$ and $f(x, y) = xy^2(x^2 + y^4)^{-1}$ otherwise. Prove that f has a Gâteaux derivative at 0 in every direction, but that $f'(0)$ does not exist. Show, in fact, that f is discontinuous at 0.

3. Denote by C_2 the linear space of all continuous real-valued functions on $[0,1]$ with L_2-norm. Prove that point–evaluation functionals are discontinuous on C_2. A point-evaluation functional is of the form t^* for some $t \in [0,1]$, where $t^*(x) = x(t)$ for all $x \in C_2$. Does t^* have a directional derivative?

4. What is the direction of steepest descent for the function

$$\Phi(x) = \xi_1^2 + \sin(\xi_1 \xi_2) + \exp \xi_3 \qquad x = (\xi_1, \xi_2, \xi_3)$$

   at the point $(1, 2, 3)$? Does this function attain its minimum on $\mathbb{R}^3$?

5. Let A be a bounded linear operator on a real Hilbert space X. How does the functional

$$\Phi(x) = \| Ax - b \|^2$$

   behave on a ray $x + tv$? Where is the minimum point on this ray, and what is the minimum value of Φ on this ray? What is the direction of steepest descent? What are the answers if A is self-adjoint?

6. A functional Φ is said to be **convex** if the condition $0 < \lambda < 1$ implies

$$\Phi(\lambda x + (1 - \lambda)y) \leqslant \lambda \Phi(x) + (1 - \lambda)\Phi(y)$$

   for any two points x and y. Is the functional $x \mapsto \langle Ax - 2b, x \rangle$ convex when A is Hermitian and positive definite?

7. Let A be any bounded linear operator on a Hilbert space, and let H be a positive definite Hermitian operator. Put

$$\Phi(x) = \langle b - Ax, H(b - Ax) \rangle$$

   Discuss methods for solving $Ax = b$ based upon the minimization of Φ. Investigate the equivalence of the two problems, give the Gâteaux derivative of Φ, and derive the formula for steepest descent. In the latter, the method of Lagrange multipliers would be helpful. Determine the amount by which $\Phi(x)$ decreases in each step.

8. What happens to the theory if the coefficient 2 is replaced by 1 in Equation (9)?

9. Prove that when the method of steepest descent is applied to the problem $Ax = b$ the minimum value of Φ is $-\langle x, b \rangle$, where x is the solution of the problem.

10. Let the method of steepest descent be applied to solve the equation $Ax = b$, as described in the text. Show that

$$x_{n+2} = x_n + (t_n + t_{n+1})r_n - t_n t_{n+1} Ar_n$$
$$r_{n+1} = (I - t_n A)r_n$$

11. Prove that if v is an eigenvector of A and if $Ax_0 = b$, then the method of steepest descent will produce the solution in one step if started at $x_0 + sv$.

12. Let A be a bounded linear operator on a complex Hilbert space. We assume that A is self-adjoint and positive definite. Prove that if $Ax_0 = b$, then x_0 minimizes the functional

$$\Phi(x) = \langle Ax, x \rangle - \langle x, b \rangle - \langle b, x \rangle$$

13. Let $f : \mathbb{R} \to \mathbb{R}$ be continuous. Let $g(t) = f(t)$ everywhere except at $t = 0$, where we define $g(0) = f(0) - 1$. Is g lower semicontinuous? Generalize.

14. Prove that the method of steepest descent, as given in Equation (15), has this property:

$$\Phi(x_n) - \Phi(x_{n+1}) \geqslant \|r_n\|^2/\|A\|$$

Show that if b is not in the closure of the range of A, then $\Phi(x_n) \to -\infty$.

15. Prove that if $\inf_{\|x\|=1}\langle Ax, x\rangle = m > 0$, then the method of steepest descent (described in Equation (15)) has this property:

$$\Phi(x_n) - \Phi(x_{n+1}) \leqslant \|r_n\|^2/m$$

16. In the method of steepest descent, we expect successive direction vectors to be orthogonal to each other. Why? Prove that this actually occurs in the example described by Equation (15).

17. In the method of steepest descent applied to the equation $Ax = b$, explain how it is possible for Φ to be bounded below on each line yet not bounded below on the whole Hilbert space.

18. Use the definition of Gâteaux derivative given on page 228 or in Problem 3.1.21 (page 120) to verify that Equation (5) gives the Gâteaux derivative of Φ.

4.9 Conjugate Direction Methods

In this section we continue our study of algorithms for solving the equation

$$(1) \qquad\qquad\qquad Ax = b$$

assuming throughout that A is an operator on a real Hilbert space X. Later, the application of these methods to general optimization problems will be considered.

Recall (from Theorem 2 in the preceding section, page 228) that when A is self-adjoint and positive definite, solving Equation (1) is equivalent to minimizing the functional

$$(2) \qquad\qquad\qquad \Phi(x) = \langle Ax - 2b, x\rangle$$

A general descent algorithm goes as follows. At the nth step, a vector x_n is available from prior computations. By means of some strategy, a "search direction" is determined. This is a vector v_n. Then we let

$$(3) \qquad x_{n+1} = x_n + \alpha_n v_n \qquad\qquad \alpha_n = \langle b - Ax_n, v_n\rangle/\langle Av_n, v_n\rangle$$

Formula (3) ensures that $\Phi(x_{n+1})$ will be as small as possible when x_{n+1} is restricted to the ray $x_n + tv_n$.

In this algorithm, considerable freedom is present in choosing the search direction v_n. For example, in the method of steepest descent, $v_n = b - Ax_n$. We shall discuss an alternative that has many advantages over steepest descent. One advantage is that the idea of searching for a minimum value of a functional is abandoned, and we retain only the algorithm in Equations (3). The operator (or matrix in the finite-dimensional case) need not be self-adjoint or positive definite. Finally, the direction vectors v_n are subject to weaker hypotheses.

First some definitions are needed. For an operator A, a sequence of vectors $v_1, v_2, \ldots$ in X is said to be **A-orthogonal** if

$$(4) \qquad \langle v_i, Av_j \rangle = 0 \quad \text{when} \quad i \neq j$$

This new concept reduces to the familiar type of orthogonality if A is the identity operator. The descent algorithm (3) is called a **conjugate direction** method if the search directions $v_1, v_2, \ldots$ are nonzero and form an A-orthogonal set.

A slightly stronger hypothesis is that our set of vectors v_i is **A-orthonormal**, meaning that the condition $\langle v_i, Av_j \rangle = \delta_{ij}$ is fulfilled. The formula for α_n in Equation (3) is then simpler.

Theorem 1.   *In the conjugate direction algorithm (3), using an A-orthonormal set of direction vectors, each residual $r_n = b - Ax_n$ is orthogonal to all the previous search directions $v_1, \ldots, v_{n-1}$.*

Proof.   Let $r_n = b - Ax_n$. We wish to prove that

$$(5) \qquad r_n \perp v_1, v_2, \ldots, v_{n-1} \qquad (n = 2, 3, \ldots)$$

First we observe that

$$(6) \qquad r_{n+1} = b - Ax_{n+1} = b - A(x_n + \alpha_n v_n) = r_n - \alpha_n Av_n$$

Consequently, by the definition of α_1, we have

$$(7) \qquad \langle r_2, v_1 \rangle = \langle r_1, v_1 \rangle - \alpha_1 \langle Av_1, v_1 \rangle = \langle r_1, v_1 \rangle - \langle r_1, v_1 \rangle = 0$$

Now assume that Equation (5) is true for a certain index n. In order to prove Equation (5) for $n + 1$, let $1 \leqslant i \leqslant n$ and use Equation (6) to write

$$(8) \qquad \langle r_{n+1}, v_i \rangle = \langle r_n, v_i \rangle - \alpha_n \langle Av_n, v_i \rangle$$

For $i < n$, both terms on the right side of Equation (8) are zero. For $i = n$ the definition of α_n shows that the right side is zero, as in Equation (7).   ∎

Corollary.   *Let A be an $m \times m$ matrix. Let $\{v_1, \ldots, v_m\}$ be an A-orthonormal set of vectors. Then the conjugate direction algorithm (3) produces a solution to the problem in (1) no later than the $(m + 1)$st step. Thus $Ax_{m+1} = b$.*

Proof.   By the preceding theorem, $r_{m+1} \perp \{v_1, \ldots, v_m\}$. But this set of m vectors is linearly independent, because

$$\left\langle \sum_{j=1}^{m} c_j v_j, Av_i \right\rangle = c_i \langle v_i, Av_i \rangle = c_i$$

Hence $r_{m+1} = 0$. Thus $b - Ax_{m+1} = 0$.   ∎

Lemma.   *If A is an Hermitian operator on a Hilbert space and satisfies the inequality*

(9) $$\|Ax\| \geqslant m\|x\| (m > 0)$$

then A is invertible, and $\|A^{-1}\| \leqslant 1/m$.

Proof. (The techniques of this proof were used previously, in Theorem 3 of Section 4.5, page 201.) Recall from Theorem 3 of Section 2.3 (page 84) that the Hermitian property implies continuity and self-adjointness. In order to prove that the range of A, $\mathcal{R}(A)$, is closed, let $[y_n]$ be a convergent sequence in $\mathcal{R}(A)$. Write $y_n = Ax_n$ and $y_n \to y$. Then $[y_n]$ has the Cauchy property. By using (9) we get

$$m\|x_n - x_m\| \leqslant \|A(x_n - x_m)\| = \|y_n - y_m\|$$

This shows that $[x_n]$ is a Cauchy sequence. Hence $x_n \to x$ for some x. By the continuity of A, $y_n = Ax_n \to Ax$, and $y = Ax \in \mathcal{R}(A)$.

Next we observe that $\mathcal{R}(A)^\perp = 0$. Indeed, if $y \in \mathcal{R}(A)^\perp$ then for every x,

$$\langle Ay, x \rangle = \langle y, Ax \rangle = 0$$

This implies that $Ay = 0$, and then $y = 0$ by Inequality (9). Since $\mathcal{R}(A)$ is closed and $\mathcal{R}(A)^\perp = 0$, we infer that $\mathcal{R}(A) = X$. Since the null space of A is 0, A^{-1} exists as a (possibly unbounded) operator. But if $y = Ax$, then

$$\|y\| = \|Ax\| \geqslant m\|x\| = m\|A^{-1}y\|$$

whence $\|A^{-1}y\| \leqslant \frac{1}{m}\|y\|$.   ∎

Theorem 2.   *Let A be a self-adjoint operator on a Hilbert space X. Assume that*

(10) $$m\|x\|^2 \leqslant \langle x, Ax \rangle \leqslant M\|x\|^2   (m > 0)$$

Let $v_1, v_2, \ldots$ be an A-orthonormal sequence whose linear span is dense in X. Then the conjugate direction algorithm

(11) $$x_{n+1} = x_n + \langle v_n, b - Ax_n \rangle v_n$$

produces a sequence that converges to $A^{-1}b$ from any starting point x_1.

Proof. (After [Lue2]) Putting $\alpha_n = \langle v_n, b - Ax_n \rangle$, we have, from Equation (11),

$$x_2 = x_1 + \alpha_1 v_1$$
$$x_3 = x_2 + \alpha_2 v_2 = x_1 + \alpha_1 v_1 + \alpha_2 v_2$$

and so on. Thus, in general,

(12) $$x_n - x_1 = \alpha_1 v_1 + \alpha_o 2 v_2 + \cdots + \alpha_{n-1} v_{n-1}$$

From Equation (12) and the A-orthogonal property,

(13) $$\langle x_n - x_1, Av_n \rangle = 0 = \langle Ax_n - Ax_1, v_n \rangle$$

From the definition of α_n and Equation (13), we get

$$\alpha_n = \langle v_n, b - Ax_n \rangle = \langle v_n, b - Ax_1 - Ax_n + Ax_1 \rangle$$
$$= \langle v_n, b - Ax_1 \rangle = \langle v_n, A(A^{-1}b - x_1) \rangle$$

This shows that the right side of Equation (12) represents the partial sum of the Fourier series of $A^{-1}b - x_1$, if we use for this expansion the inner product

$$[x, y] = \langle x, Ay \rangle$$

These two inner products lead to the same topology on X because of Equation (10). Hence $x_n - x_1 \to A^{-1}b - x_1$. ∎

In the conjugate direction algorithm, there is still some freedom in the choice of the direction vectors v_i. In the **conjugate gradient** method, these vectors are generated in such a way that (for each n) x_n minimizes Φ on a certain linear variety of dimension $n - 1$. The conjugate gradient algorithm appears in a number of different versions. For a theoretical analysis of the method, this version seems to be the best:

   I. To start, let x_1 be arbitrary, and define $v_1 = b - Ax_1$.
   II. Given x_n and v_n, we set

(14) $x_{n+1} = x_n + \alpha_n v_n \qquad \alpha_n = \langle b - Ax_n, v_n \rangle \langle v_n, Av_n \rangle^{-1}$

(15) $v_{n+1} = b - Ax_{n+1} - \beta_n v_n \qquad \beta_n = \langle b - Ax_{n+1}, Av_n \rangle \langle v_n, Av_n \rangle^{-1}$

Theorem 3.   *Let A be a self-adjoint operator on a Hilbert space. Assume that for some positive m and M,*

(16) $$m\|x\|^2 \leqslant \langle x, Ax \rangle \leqslant M\|x\|^2$$

Then the sequence $[x_n]$ generated by the conjugate gradient algorithm converges to $A^{-1}b$.

Proof.   Throughout the proof, the nth residual is defined to be

(17) $$r_n = b - Ax_n \qquad n = 1, 2, 3, \ldots$$

It follows that

(18) $$r_{n+1} = b - Ax_{n+1} = b - A(x_n + \alpha_n v_n) = r_n - \alpha_n Av_n$$

First we establish two orthogonality relations:

(19) $\langle r_{n+1}, \ v_n \rangle = 0 \qquad n = 1, 2, \ldots$

(20) $\langle v_{n+1}, Av_n \rangle = 0 \qquad n = 1, 2, \ldots$

These are consequences of formulas (18), (14), and (15), as follows:

$$\langle r_{n+1}, v_n \rangle = \langle r_n - \alpha_n A v_n, v_n \rangle = \langle r_n, v_n \rangle - \alpha_n \langle A v_n, v_n \rangle$$
$$= \langle r_n, v_n \rangle - \langle r_n, v_n \rangle = 0$$
$$\langle v_{n+1}, A v_n \rangle = \langle r_{n+1} - \beta_n v_n, A v_n \rangle = \langle r_{n+1}, A v_n \rangle - \beta_n \langle v_n, A v_n \rangle$$
$$= \langle r_{n+1}, A v_n \rangle - \langle r_{n+1}, A v_n \rangle = 0$$

Define a sequence e_n by the equation

$$e_n = \langle r_n, A^{-1} r_n \rangle$$

From this we have

(21) $$e_n = \langle A A^{-1} r_n, A^{-1} r_n \rangle \geqslant m \| A^{-1} r_n \|^2 \geqslant 0$$

Using Equation (18) we can express e_{n+1} as follows:

$$e_{n+1} = \langle r_{n+1}, A^{-1} r_{n+1} \rangle$$
$$= \langle r_n - \alpha_n A v_n, A^{-1}(r_n - \alpha_n A v_n) \rangle$$
$$= \langle r_n - \alpha_n A v_n, A^{-1} r_n - \alpha_n v_n \rangle$$
$$= \langle r_n, A^{-1} r_n \rangle - \alpha_n \langle A v_n, A^{-1} r_n \rangle - \alpha_n \langle r_n, v_n \rangle + \alpha_n^2 \langle A v_n, v_n \rangle$$
$$= e_n - \alpha_n \langle v_n, r_n \rangle - \alpha_n \langle v_n, r_n \rangle + \alpha_n \langle v_n, r_n \rangle$$
$$= e_n - \alpha_n \langle v_n, r_n \rangle$$
$$= e_n \big[1 - \alpha_n \langle v_n, r_n \rangle / e_n \big]$$

In order to show that e_n converges geometrically to zero, it suffices to prove that the bracketed expression in the previous equation is less than $1 - m/M$. We will prove two inequalities that accomplish this objective, namely

(22) $$\alpha_n \geqslant 1/M$$
(23) $$\langle v_n, r_n \rangle / e_n \geqslant m$$

From Equations (15) and (19) we have

(24) $$\langle r_n, v_n \rangle = \langle r_n, r_n - \beta_{n-1} v_{n-1} \rangle = \langle r_n, r_n \rangle$$

Equation (21) and the Cauchy–Schwarz inequality imply that

$$m \| A^{-1} r_n \|^2 \leqslant \langle A^{-1} r_n, r_n \rangle \leqslant \| A^{-1} r_n \| \, \| r_n \|$$

This leads to $m \| A^{-1} r_n \| \leqslant \| r_n \|$. Inequality (23) now follows from

$$m e_n = m \langle r_n, A^{-1} r_n \rangle \leqslant m \| r_n \| \, \| A^{-1} r_n \| \leqslant \| r_n \|^2 = \langle r_n, r_n \rangle = \langle v_n, r_n \rangle$$

To prove Inequality (22), we start with Equation (15), written in the form

$$r_n = v_n + \beta_{n-1} v_{n-1}$$

From this we conclude that

$$\langle r_n, Ar_n \rangle = \langle v_n + \beta_{n-1}v_{n-1},\ A(v_n + \beta_{n-1}v_{n-1}) \rangle$$

Since $\langle v_{n-1}, Av_n \rangle = \langle Av_{n-1}, v_n \rangle = 0$ by Equation (20), we obtain

$$\langle r_n, Ar_n \rangle = \langle v_n, Av_n \rangle + \beta_{n-1}^2 \langle v_{n-1}, Av_{n-1} \rangle \geqslant \langle v_n, Av_n \rangle$$

Thus, using (16) and (24) we have

$$\langle v_n, Av_n \rangle \leqslant \langle r_n, Ar_n \rangle \leqslant M\langle r_n, r_n \rangle = M\langle r_n, v_n \rangle$$

Hence

$$\alpha_n = \langle r_n, v_n \rangle \langle v_n, Au_n \rangle^{-1} \geqslant 1/M$$

At this stage, we have established that

$$e_{n+1} \leqslant \left(1 - \frac{m}{M}\right) e_n$$

Consequently, $e_n \to 0$. From Inequality (21) we conclude that $A^{-1}r_n \to 0$, or $A^{-1}b - x_n \to 0$. ∎

Problems 4.9

1. Let A be an $m \times m$ matrix that is symmetric and positive definite. Let U be an $m \times m$ matrix whose columns form an A-orthonormal set. Prove that $U^T A U = I$.

2. Let A be an $m \times m$ symmetric matrix such that $\langle x, Ax \rangle \neq 0$ when $x \neq 0$. Let $\{u_1, \ldots, u_m\}$ be a basis for $\mathbb{R}^m$. Define $v_1 = u_1$ and

$$v_{k+1} = u_{k+1} - \sum_{i=1}^{k} \frac{\langle u_{k+1}, Av_i \rangle}{\langle v_i, Av_i \rangle} v_i \qquad (k = 1, 2, \ldots, m-1)$$

Prove that $\{v_1, v_2, \ldots, v_m\}$ is an A-orthogonal basis for $\mathbb{R}^m$.

3. Show that if A is an $m \times m$ symmetric positive definite matrix and if $\{v_1, \ldots, v_m\}$ is an A-orthonormal set, then the solution of $Ax = b$ is $x = \sum_{i=1}^{m} \langle b, v_i \rangle v_i$.

4.10 Methods Based on Homotopy and Continuation

In this section we address the problem of finding the roots of an equation or the zeros of a mapping

$$(1) \qquad\qquad\qquad f(x) = 0$$

Here f can be a mapping from one Banach space to another, say $f : X \to Y$. This problem is so general that it includes systems of algebraic equations, integral equations, differential equations, and so on. We will describe a tactic called the

continuation method for attacking this problem. The discussion is adapted from that in [KC].

The fundamental idea of the continuation method is to embed the given problem in a one-parameter family of problems, using a parameter t that runs over the interval $[0,1]$. The original problem will correspond to $t = 1$, and another problem whose solution is known will correspond to $t = 0$. For example, we can define

$$(2) \qquad\qquad h(t,x) = tf(x) + (1-t)g(x)$$

The equation $g(x) = 0$ should have a known solution. The next step is to select points $t_0, t_1, \ldots$ such that

$$0 = t_0 < t_1 < t_2 < \cdots < t_m = 1$$

One then attempts to solve each equation $h(t_i, x) = 0$, $(0 \leqslant i \leqslant m)$. Assuming that some iterative method will be used (such as Newton's method), it makes sense to use the solution at the ith step as the starting point in computing a solution at the $(i+1)$st step.

This whole procedure is designed to cure the difficulty that plagues Newton's method, viz., the need for a good starting point.

The relationship (2), which embeds the original problem (1) in a family of problems, is an example of a **homotopy** that connects the two functions f and g. In general, a homotopy can be *any* continuous connection between f and g. Formally, a homotopy between two functions $f, g : X \to Y$ is a continuous map

$$(3) \qquad\qquad h : [0,1] \times X \to Y$$

such that $h(0,x) = g(x)$ and $h(1,x) = f(x)$. If such a map exists, we say that f is **homotopic** to g. This is an equivalence relation among the continuous maps from X to Y, where X and Y can be any two topological spaces.

An elementary homotopy that is often used in the continuation method is

$$(4) \qquad \begin{aligned} h(t,x) &= tf(x) + (1-t)\big[f(x) - f(x_0)\big] \\ &= f(x) + (t-1)f(x_0) \end{aligned}$$

Here x_0 can be any point in X, and it is clear that x_0 will be a solution of the problem when $t = 0$.

If the equation $h(t,x) = 0$ has a unique root for each $t \in [0,1]$, then that root is a function of t, and we can write $x(t)$ as the unique member of X that makes the equation $h(t, x(t)) = 0$ true. The set

$$(5) \qquad\qquad \{\, x(t) \;:\; 0 \leqslant t \leqslant 1 \,\}$$

can be interpreted as an arc or curve in X, parametrized by t. This arc starts at the known point $x(0)$ and proceeds to the solution of our problem, $x(1)$. The continuation method determines this curve by computing points on it, $x(t_0), x(t_1), \ldots, x(t_m)$.

If the function $t \mapsto x(t)$ is differentiable and if h is differentiable, then the Implicit Function Theorems of Section 3.4, pages 136ff, enable us to compute $x'(t)$. By following this idea, we can describe the curve in Equation (5) by a differential equation. Assuming an arbitrary homotopy, we have

$$(6) \qquad\qquad 0 = h\big(t, x(t)\big)$$

Differentiating with respect to t, we obtain

$$(7) \qquad\qquad 0 = h_1\big(t, x(t)\big) + h_2\big(t, x(t)\big)x'(t)$$

in which subscripts denote partial derivatives. Thus

$$(8) \qquad\qquad x'(t) = -\Big[h_2\big(t, x(t)\big)\Big]^{-1} h_1\big(t, x(t)\big)$$

This is a differential equation for x. Its initial value is known, because $x(0)$ has been assumed to be known. Upon integrating this differential equation across the interval $0 \leqslant t \leqslant 1$ (usually by numerical procedures), one reaches the value $x(1)$, which is the solution to Equation (1).

Example 1.   Let $X = Y = \mathbb{R}^2$, and define

$$f(x) = \begin{bmatrix} \sin \xi_1 + e^{\xi_2} - 3 \\ (\xi_2 + 3)^2 - \xi_1 - 4 \end{bmatrix} \qquad\qquad x = (\xi_1, \xi_2) \in \mathbb{R}^2$$

A convenient homotopy is defined by Equation (4), and we select the starting point $x_0 = (5, 3)$. The derivatives on the right side of Equation (8) are computed to be

$$h_2 = f'(x) = \begin{bmatrix} \cos \xi_1 & e^{\xi_2} \\ -1 & 2\xi_2 + 6 \end{bmatrix}$$

$$h_1 = f(x_0) = \begin{bmatrix} a \\ b \end{bmatrix}$$

where $a = \sin 5 + e^3 - 3$ and $b = 27$. The inverse of $f'(x)$ is

$$[f'(x)]^{-1} = \frac{1}{\Delta} \begin{bmatrix} 2\xi_2 + 6 & -e^{\xi_2} \\ 1 & \cos \xi_1 \end{bmatrix} \qquad\qquad \Delta = 2(\xi_2 + 3)\cos \xi_1 + e^{\xi_2}$$

The differential equation that controls the path leading away from the point x_0 is Equation (8). In this concrete case it is a pair of ordinary differential equations:

$$\begin{bmatrix} \xi_1' \\ \xi_2' \end{bmatrix} = \frac{-1}{\Delta} \begin{bmatrix} 2\xi_2 + 6 & -e^{\xi_2} \\ 1 & \cos \xi_1 \end{bmatrix} \begin{bmatrix} a \\ b \end{bmatrix} = \frac{-1}{\Delta} \begin{bmatrix} 2a(\xi_2 + 3) - be^{\xi_2} \\ a + b\cos \xi_1 \end{bmatrix}$$

When this system was integrated numerically on the interval $0 \leqslant t \leqslant 1$, the terminal value of x (at $t = 1$) was close to $(12, 1)$. In order to find a more accurate solution, we can use Newton's iteration starting at the point produced by the homotopy method. The Newton iteration replaces any approximate root x by $x - \delta$, the correction δ being defined by

$$\delta = \left[f'(x)\right]^{-1}f(x)$$

(These matters are the subject of Section 3.3, beginning at page 125.) In the current example, the vector δ is

$$\begin{bmatrix} \delta_1 \\ \delta_2 \end{bmatrix} = \frac{1}{\Delta}\begin{bmatrix} 2\xi_2 + 6 & -e^{\xi_2} \\ 1 & \cos\xi_1 \end{bmatrix}\begin{bmatrix} \sin\xi_1 + e^{\xi_2} - 3 \\ (\xi_2 + 3)^2 - \xi_1 - 4 \end{bmatrix}$$

Five steps of the Newton iteration produced these results:

	ξ_1	ξ_2
$k = 0$	12.000000000000000000	1.0000000000000000000
$k = 1$	12.691334908752890571	1.0864168635941113213
$k = 2$	12.628177397290770959	1.0777753827891591357
$k = 3$	12.628268254380085321	1.0777773669468545670
$k = 4$	12.628268254564651450	1.0777773669690025700
$k = 5$	12.628268254564651450	1.0777773669690025700

The curve $\{x(t) \;:\; 0 \leqslant t \leqslant 1\}$ is shown in Figure 4.2

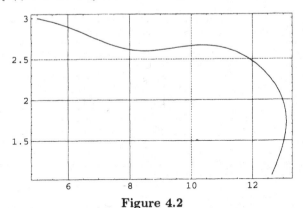

Figure 4.2

In an example such as this one, the differential equation need not be solved numerically with high precision, because the objective is to end at a point near the solution—in fact, near enough so that the classical Newton method will succeed if started at that point.

A formal result that gives some conditions under which the homotopy method will succeed is as follows. This result is from [OR].

Theorem.   *If $f : \mathbb{R}^n \to \mathbb{R}^n$ is continuously differentiable and if $\|f'(x)^{-1}\|$ is bounded on $\mathbb{R}^n$, then for any $x_0 \in \mathbb{R}^n$ there is a unique curve $\{x(t) : 0 \leqslant t \leqslant 1\}$ in $\mathbb{R}^n$ such that $f(x(t)) + (t-1)f(x_0) = 0$,*

$0 \leqslant t \leqslant 1$. *The function* $t \mapsto x(t)$ *is a continuously-differentiable solution of the initial value problem* $x' = -f'(x)^{-1} f(x_0)$, $x(0) = x_0$.

Another way of describing the path $x(t)$ has been given by Garcia and Zangwill [GZ]. We start with the equation $h(t, x) = 0$, assuming now that $x \in \mathbb{R}^n$ and $t \in [0, 1]$. A vector $y \in \mathbb{R}^{n+1}$ is defined by

$$y = (t, \xi_1, \xi_2, \ldots, \xi_n)$$

where $\xi_1, \xi_2, \ldots$ are the components of x. Thus our equation is simply $h(y) = 0$. Each component of y, including t, is now allowed to be a function of an independent variable s, and we write $h(y(s)) = 0$. Differentiation with respect to s leads to the basic differential equation

(9) $$h'(y) y'(s) = 0$$

The variables s and t start at 0. The initial value of x is $x(0) = x_0$. Thus suitable starting values are available for the differential equation (9).

Since f and g are maps of $\mathbb{R}^n$ into $\mathbb{R}^n$, h is a map of $\mathbb{R}^{n+1}$ into $\mathbb{R}^n$. The Fréchet derivative $h'(y)$ is therefore represented by an $n \times (n + 1)$ matrix, A. The vector $y'(s)$ has $n + 1$ components, which we denote by $\eta_1', \eta_2', \ldots, \eta_{n+1}'$. By appealing to the lemma below, we can obtain another form for Equation (9), namely

(10) $$\eta_j' = (-1)^j \det(A_j) \qquad (1 \leqslant j \leqslant n + 1)$$

where A_j is the $n \times n$ matrix that results from A by deleting its jth column. Let us illustrate this formalism with a problem similar to the one in Example 1.

Example 2. Let f be the mapping

$$f(x) = \begin{bmatrix} \xi_1^2 - 3\xi_2^2 + 3 \\ \xi_1 \xi_2 + 6 \end{bmatrix} \qquad x = (\xi_1, \xi_2) \in \mathbb{R}^2$$

We take the starting point $x_0 = (1, 1)$ and use the homotopy of Equation (4). Then

$$h(t, x) = \begin{bmatrix} \xi_1^2 - 3\xi_2^2 + 2 + t \\ \xi_1 \xi_2 - 1 + 7t \end{bmatrix}$$

The differential equation (9) is given by

(11) $$\begin{bmatrix} 1 & 2\xi_1 & -6\xi_2 \\ 7 & \xi_2 & \xi_1 \end{bmatrix} \begin{bmatrix} t' \\ \xi_1' \\ \xi_2' \end{bmatrix} = \begin{bmatrix} 0 \\ 0 \end{bmatrix}$$

It is preferable to use Equation (10), however, and to write the differential equations in the form

(12) $$\begin{cases} t' = -(2\xi_1^2 + 6\xi_2^2) & t(0) = 0 \\ \xi_1' = \xi_1 + 42\xi_2 & \xi_1(0) = 1 \\ \xi_2' = -(\xi_2 - 14\xi_1) & \xi_2(0) = 1 \end{cases}$$

The derivatives in this system are with respect to s. Since we want t to run from 0 to 1, it is clear (from the equation governing t) that we must let s proceed to the left. Alternatively, we can appeal to the homogeneity in the system, and simply change the signs on the right side of (12). Following the latter course, and performing a numerical integration, we arrive at these two points:

$$s = .087 \,, \qquad t = .969 \,, \qquad \xi_1 = -2.94 \,, \qquad \xi_2 = 1.97$$
$$s = .088 \,, \qquad t = 1.010 \,, \qquad \xi_1 = -3.02 \,, \qquad \xi_2 = 2.01$$

Either of these can be used to start a Newton iteration, as was done in Example 1. The path generated by this homotopy is shown in Figure 4.3.   ∎

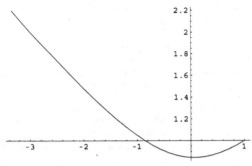

Figure 4.3

A drawback to the method used in Example 2 is that one has no *a priori* knowledge of the value of s corresponding to $t = 1$. In practice, this may necessitate several computer runs.

Lemma. *Let A be an $n \times (n+1)$ matrix. A solution of the equation $Ax = 0$ is given by $x_j = (-1)^j \det(A_j)$, where A_j is the matrix A without its column j.*

Proof. Select any row (for example the ith row) in A and adjoin a copy of it as a new row at the top of A. This creates an $(n+1) \times (n+1)$ matrix B that is obviously singular, because row i of A occurs twice in B. In expanding the determinant of B by the elements in its top row we obtain

$$0 = \det B = \sum_{j=1}^{n+1} (-1)^j a_{ij} \det(A_j) = \sum_{j=1}^{n+1} a_{ij} x_j$$

Since this is true for $i = 1, 2, \ldots, n$, we have $Ax = 0$.   ∎

The connection between the homotopy methods and Newton's method is deeper than may be seen at first glance. Let us start with the homotopy

$$h(t, x) = f(x) - e^{-t} f(x_0)$$

In this equation t will run from 0 to ∞. We seek a curve or path, $x = x(t)$, on which

$$0 = h\big(t, x(t)\big) = f\big(x(t)\big) - e^{-t} f(x_0)$$

As usual, differentiation with respect to t will lead to a differential equation describing the path:

$$(13) \qquad 0 = f'\big(x(t)\big)x'(t) + e^{-t}f(x_0) = f'\big(x(t)\big)x'(t) + f\big(x(t)\big)$$

$$(14) \qquad\qquad x'(t) = -f'\big(x(t)\big)^{-1}f\big(x(t)\big)$$

If this differential equation is integrated using Euler's method and step size 1, the result is the formula

$$x_{n+1} = x_n - f'(x_n)^{-1}f(x_n)$$

This is, of course, the formula for Newton's method. It is clear that one can expect to obtain better results by solving the differential equation (14) with a more accurate numerical method (incorporating a variable step size). These matters have been thoroughly explored by Smale and others. See, for example, [Sm].

Application to Linear Programming. The homotopy method can be used to solve linear programming problems. This approach leads naturally to the algorithm proposed in 1984 by Karmarkar [Kar]. In explaining the homotopy method in this context, we follow closely the description in [BroS].

Consider the standard linear programming problem

$$(15) \qquad \begin{cases} \text{maximize } c^T x \\ \text{subject to } Ax = b \text{ and } x \geqslant 0 \end{cases}$$

Here, $c \in \mathbb{R}^n$, $x \in \mathbb{R}^n$, $b \in \mathbb{R}^m$, and A is an $m \times n$ matrix. We start with a **feasible point**, i.e., a point x^0 that satisfies the constraints. The **feasible set** is

$$\mathcal{F} = \big\{\, x \in \mathbb{R}^n \ : \ Ax = b \text{ and } x \geqslant 0 \,\big\}$$

Our intention is to move from x^0 to a succession of other points, remaining always in $\mathcal{F}$, and increasing the value of the **objective function**, $c^T x$. It is clear that if we move from x^0 to x^1, the difference $x^1 - x^0$ must lie in the null space of A. We shall try to find a curve $t \mapsto x(t)$ in the feasible set, starting at x^0 and leading to a solution of the extremal problem. Our requirements are

(i) $x(t) \geqslant 0$ for $t \geqslant 0$
(ii) $Ax(t) = b$ for $t \geqslant 0$
(iii) $c^T x(t)$ is increasing for $t \geqslant 0$.

The curve will be defined by an initial-value problem:

$$(16) \qquad\qquad x' = F(x) \qquad x(0) = x^0$$

The task facing us is to determine a suitable F. In order to satisfy condition (i), we shall arrange that whenever a component x_i approaches 0, its velocity $x'_i(t)$ shall also approach 0. This can be accomplished by letting $D(x)$ be the diagonal matrix

$$D(x) = \begin{bmatrix} x_1 & & & 0 \\ & x_2 & & \\ & & \ddots & \\ 0 & & & x_n \end{bmatrix}$$

and assuming that for some bounded function G,

(17) $$F(x) = D(x)G(x)$$

If this is the case, then from Equations (15) and (17) we shall have

$$x_i' = x_i G_i(x)$$

and clearly $x_i' \to 0$ if $x_i \to 0$.

In order to satisfy requirement (ii), it suffices to require $Ax' = 0$. Indeed, if $Ax' = 0$ then $Ax(t)$ is constant as a function of t. Since $Ax(0) = b$, we have $Ax(t) = b$ for all t. Since $x' = F = DG$, we must require $ADG = 0$. This is most conveniently arranged by letting $G = PH$, where H is any function, and P is the orthogonal projection onto the null space of AD.

Finally, in order to secure property (iii), we should select H so that $c^T x(t)$ is increasing. Thus, we want

$$0 < \frac{d}{dt}\big(c^T x(t)\big) = c^T x' = c^T F(x) = c^T DG = c^T DPH$$

A convenient choice for H is Dc, for then we have, (using $v = Dc$),

$$c^T DPH = c^T DPDc = v^T Pv = \langle v, Pv \rangle$$
$$= \langle v - Pv + Pv, Pv \rangle = \langle Pv, Pv \rangle \geqslant 0$$

Notice that $v - Pv$ is orthogonal to the range of P, and $\langle v - Pv, Pv \rangle = 0$.

The final version of our initial-value problem is

(18) $$x' = D(x)P(x)D(x)c \qquad x(0) = x^0$$

The theoretical formula for P is

(19) $$P = I - (AD)^T \big[(AD)(AD)^T\big]^{-1} AD$$

The validity of this depends upon $B \equiv AD$ having full rank, so that BB^T will be nonsingular. This will, in turn, require $x_i > 0$ for each component. Thus the points $x(t)$ should remain in the interior of the set

$$\{ x \ : \ x \geqslant 0 \}$$

In particular, x^0 should be so chosen. In practice, Pv is computed not by (19) but by solving the equation $BB^T z = Bv$ and noting that

$$Pv = v - B^T z$$

As mentioned earlier, the initial-value problem (18) need not be solved very accurately. A variation of the Euler Method can be used. Recall that the Euler Method for Equation (16) advances the solution by

$$x(t + \delta) = x(t) + \delta x'(t) = x(t) + \delta F(x)$$

Using this type of formula, we generate a sequence of vectors $x^0, x^1, \ldots$ by the equation

$$x^{k+1} = x^k + \delta_k F(x^k)$$

Although it is tempting to take δ_k as large as possible subject to the requirement $x^{k+1} \in \mathcal{F}$, that will lead to a point x^{k+1} having at least one zero component. As pointed out previously, that will introduce other difficulties. What seems to work well in practice is to take δ_k approximately 9/10 of the maximum possible step. This maximum step is easily computed; it is the maximum λ for which $x^{k+1} \geqslant 0$. (The constraint $Ax = b$ is maintained automatically.)

Problems 4.10

1. Solve the system of equations

$$x - 2y + y^2 + y^3 - 4 = -x - y + 2y^2 - 1 = 0$$

   by the homotopy method used in Example 2, starting with the point $(0,0)$. (All the calculations can be performed without recourse to numerical methods.)

2. Consider the homotopy $h(t, x) = t f(x) + (1 - t)g(x)$, in which

$$f(x) = x^2 - 5x + 6 \qquad g(x) = x^2 - 1$$

   Show that there is no path connecting a zero of g to a zero of f.

3. Let $y = y(s)$ be a differentiable function from $\mathbb{R}$ to $\mathbb{R}^n$ satisfying the differential equation (9). Assume that $h(y(0)) = 0$. Prove that $h(y(s)) = 0$.

4. If the homotopy method of Example 2 is to be used on the system

$$\sin x + \cos y + e^{xy} = \tan^{-1}(x + y) - xy = 0$$

   starting at $(0,0)$, what is the system of differential equations that will govern the path?

5. Prove that homotopy is an equivalence relation among the continuous maps from one topological space to another.

6. Are the functions $f(x) = \sin x$ and $g(x) = \cos x$ homotopic?

7. Consider these maps of $[0, 1]$ into $[0, 1] \cup [2, 3]$:

$$f(t) = 0 \qquad g(t) = 2$$

   Are they homotopic?

8. To find $\sqrt{2}$ we can solve the equation $f(x) = x^2 - 2 = 0$. Let $x_0 = 1$ and $h(t, x) = t f(x) + (1-t)[f(x) - f(x_0)]$. Determine the initial value problem that arises from Equation (8). Solve it in closed form and verify that $x(1) = \sqrt{2}$.

9. In Example 1 are the hypotheses of the Ortega–Rheinboldt theorem fulfilled?

10. Prove that any two continuous maps from a topological space into a normed linear space are homotopic.

Chapter 5

Distributions

5.1 Definition and Examples   246
5.2 Derivatives of Distributions   253
5.3 Convergence of Distributions   257
5.4 Multiplication of Distributions by Functions   260
5.5 Convolutions   268
5.6 Differential Operators   273
5.7 Distributions with Compact Support   280

5.1 Definition and Examples

The theory of distributions originated in the work of Laurent Schwartz in the era 1945–1952 [Schl2]. Earlier work by Sobolev was along similar lines. The objective was to treat functions as functionals, and to notice that when so interpreted, differentiation was always possible. This opened the way to the study of partial differential equations by new methods that bypassed the classical restrictions on functions. The functionals that now become the focus of study are called "distributions"—not to be confused with distributions in probability theory! The term "generalized functions" is also used, especially by Russian authors.

Since this exposition of distribution theory is addressed to readers who are seeing these matters for the first time, we have used notation that maintains distinctions between entities that, in other literature, are often denoted by a single symbol. For example, we use $\tilde{f}$ to denote the distribution arising from a locally integrable function f. We use $\phi_j \twoheadrightarrow \phi$ to signify the special convergence defined in a space of test functions, and we use ∂^α for a distributional derivative, in contrast to D^α for a classical derivative.

In these first sections of Chapter 5 we consider all test functions to be defined on $\mathbb{R}^n$, not on a prescribed open set Ω. This frees the exposition from some additional complication.

We begin with the notion of a **multi-index**. This is any n-tuple of nonnegative integers

$$\alpha = (\alpha_1, \alpha_2, \ldots, \alpha_n)$$

The **order** of a multi-index α is the quantity

$$|\alpha| = \sum_{i=1}^{n} \alpha_i$$

If α is a multi-index, there is a partial differential operator D^α corresponding to it. Its definition is

$$D^\alpha = \left(\frac{\partial}{\partial x_1}\right)^{\alpha_1} \left(\frac{\partial}{\partial x_2}\right)^{\alpha_2} \cdots \left(\frac{\partial}{\partial x_n}\right)^{\alpha_n} = \frac{\partial^{|\alpha|}}{\partial x_1^{\alpha_1} \cdots \partial x_n^{\alpha_n}}$$

This operates on functions of n real variables $x_1, \ldots, x_n$. Thus, for example, if $n = 3$ and $\alpha = (3, 0, 4)$, then

$$D^\alpha \phi = \frac{\partial^7 \phi}{\partial x_1^3 \partial x_3^4}$$

The space $C^\infty(\mathbb{R}^n)$ consists of all functions $\phi : \mathbb{R}^n \to \mathbb{R}$ such that $D^\alpha \phi \in C(\mathbb{R}^n)$ for each multi-index α. Thus, the partial derivatives of ϕ of all orders exist and are continuous.

A vector space $\mathcal{D}$, called the **space of test functions**, is now introduced. Its elements are all the functions in $C^\infty(\mathbb{R}^n)$ having compact support. The **support** of a function ϕ is the closure of $\{x : \phi(x) \neq 0\}$. Another notation for $\mathcal{D}$ is $C_c^\infty(\mathbb{R}^n)$. The value of n is usually fixed in our discussion. If we want to show n in the notation, we can write $\mathcal{D}(\mathbb{R}^n)$.

At first glance, it may seem that $\mathcal{D}$ is empty! After all, an analytic function that vanishes on an open nonempty set must be 0 everywhere. But that is a theorem about complex-valued functions of complex variables, whereas we are here considering real-valued functions of real variables.

An important example of a function in $\mathcal{D}$ is given by the formula

(1) $$\rho(x) = \begin{cases} c \cdot \exp(|x|^2 - 1)^{-1} & \text{if } x \in \mathbb{R}^n \text{ and } |x| < 1 \\ 0 & \text{if } x \in \mathbb{R}^n \text{ and } |x| \geqslant 1 \end{cases}$$

where c is chosen so that $\int \rho(x)\,dx = 1$. Here and elsewhere we use $|x|$ for the Euclidean norm:

$$|x| = \left(\sum_{i=1}^{n} x_i^2\right)^{1/2}$$

The graph of ρ in the case $n = 1$ is shown in Figure 5.1.

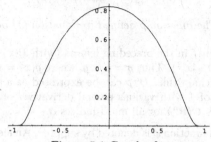

Figure 5.1 Graph of ρ

The fact that $\rho \in \mathcal{D}$ is not at all obvious, and the next two lemmas are inserted solely to establish this fact.

Lemma 1.   For any polynomial P, the function $f : \mathbb{R} \to \mathbb{R}$ defined by

$$f(x) = \begin{cases} P(1/x)e^{-1/x} & x > 0 \\ 0 & x \leqslant 0 \end{cases}$$

is in $C^\infty(\mathbb{R})$.

Proof.   First we show that f is continuous. The only questionable point is $x = 0$. We have

$$\lim_{x \downarrow 0} f(x) = \lim_{x \downarrow 0} \frac{P(1/x)}{\exp(1/x)} = \lim_{t \uparrow \infty} \frac{P(t)}{\exp(t)}$$

By using L'Hôpital's rule repeatedly on this last limit, we see that its value is 0. Hence f is continuous. Differentiation of f gives

$$f'(x) = \begin{cases} Q(1/x)e^{-1/x} & x > 0 \\ 0 & x < 0 \end{cases}$$

where $Q(x) = x^2[P(x) - P'(x)]$. By the first part of the proof, $\lim_{x \downarrow 0} f'(x) = 0$. It remains only to be proved that $f'(0) = 0$. We have, by the mean value theorem,

$$f'(0) = \lim_{h \to 0} \frac{f(h) - f(0)}{h} = \lim_{h \to 0} f'(\xi(h)) = 0$$

where $\xi(h)$ is strictly between 0 and h. (Note that h can be positive or negative in this argument.) We have shown that

$$f'(x) = \begin{cases} Q(1/x)e^{-1/x} & x > 0 \\ 0 & x \leqslant 0 \end{cases}$$

This has the same form as f, and therefore f' is continuous. The argument can be repeated indefinitely. The reader should observe that our argument requires the following version of the mean value theorem: If g is continuous on $[a, b]$ and differentiable on (a, b), then for some ξ in (a, b)

$$g'(\xi) = [g(b) - g(a)]/(b - a) \qquad\qquad \blacksquare$$

Lemma 2.   The function ρ defined in Equation (1) belongs to $\mathcal{D}$.

Proof.   The function f in the preceding lemma (with $P(x) = 1$) has the property that $\rho(x) = c\,f(1 - |x|^2)$. Thus $\rho = c\,f \circ g$, where $g(x) = 1 - |x|^2$ and belongs to $C^\infty(\mathbb{R}^n)$. By the chain rule, $D^\alpha \rho$ can be expressed as a sum of products of ordinary derivatives of f with various partial derivatives of g. Since these are all continuous, $D^\alpha \rho \in C(\mathbb{R}^n)$ for all multi-indices α. $\blacksquare$

The support of a function ϕ is denoted by $\mathrm{supp}(\phi)$. An element ϕ of $C^\infty(\mathbb{R}^n)$ such that

$$\phi \geqslant 0 \qquad\qquad \int \phi = 1 \qquad\qquad \mathrm{supp}(\phi) \subset \{x : |x| \leqslant 1\}$$

is called a **mollifier**. The function ρ defined above is thus a mollifier. If ϕ is a mollifier, then the scaled versions of ϕ, defined by

$$(2) \qquad\qquad \phi_j(x) = j^n \phi(jx) \qquad (x \in \mathbb{R}^n, \ j \in \mathbb{N})$$

play a role in certain arguments, such as in Sections 5.5 and 6.8. They, too, are mollifiers.

The linear space $\mathcal{D}$ is now furnished with a notion of sequential convergence. A sequence $[\phi_j]$ in $\mathcal{D}$ **converges to** 0 if there is a single compact set K containing the supports of all ϕ_j, and if for each multi-index α,

$$D^\alpha \phi_j \to 0 \ \text{ uniformly on } \ K$$

We write $\phi_j \twoheadrightarrow 0$ if these two conditions are fulfilled. Further, we write $\phi_j \twoheadrightarrow \phi$ if and only if $\phi_j - \phi \twoheadrightarrow 0$. The use of the symbol $\twoheadrightarrow$ is to remind the reader of the special nature of convergence in $\mathcal{D}$. Uniform convergence to 0 on K of the sequence $D^\alpha \phi_j$ means that

$$\sup_{x \in K} \left| (D^\alpha \phi_j)(x) \right| \to 0 \ \text{ as } \ j \to \infty$$

Since all ϕ_j vanish outside of K, we also have

$$\sup_{x \in \mathbb{R}^n} \left| (D^\alpha \phi_j)(x) \right| \to 0$$

Continuity and other topological notions will be based upon the convergence of sequences as just defined. In particular, a map F from $\mathcal{D}$ into a topological space is continuous if the condition $\phi_j \twoheadrightarrow \phi$ implies the condition $F(\phi_j) \to F(\phi)$. The legitimacy of defining topological notions by means of sequential convergence is a matter that would require an excursus into the theory of locally convex linear topological spaces. We refer the reader to [Ru1] for these matters. The next result gives an example of this type of continuity.

Theorem 1.   *For every multi-index α, D^α is a continuous linear transformation of $\mathcal{D}$ into $\mathcal{D}$.*

Proof.   The linearity is a familiar feature of differentiation. For the continuity, it suffices to prove continuity at 0 because D^α is linear. Thus, suppose that $\phi_j \in \mathcal{D}$ and $\phi_j \twoheadrightarrow 0$. Let K be a compact set containing the supports of all the functions ϕ_j. Then $D^\beta \phi_j(x) \to 0$ uniformly on K for every multi-index β. Consequently, $D^\beta D^\alpha \phi_j(x) \to 0$ uniformly for each β, and so $D^\alpha \phi_j \twoheadrightarrow 0$, by the definition of convergence in $\mathcal{D}$.   ∎

A **distribution** is a continuous linear functional on $\mathcal{D}$. Continuity of such a linear function T is defined by this implication:

$$\left[\phi_j \in \mathcal{D} \ \& \ \phi_j \twoheadrightarrow 0 \right] \implies T(\phi_j) \to 0$$

The space of all distributions is denoted by $\mathcal{D}'$, or by $\mathcal{D}'(\mathbb{R}^n)$.

Example 1.   A **Dirac** distribution δ_ξ is defined by selecting $\xi \in \mathbb{R}^n$ and writing

(3) $$\delta_\xi(\phi) = \phi(\xi) \qquad (\phi \in \mathcal{D})$$

It is a distribution, because firstly, it is linear:

$$\delta_\xi(\lambda_1\phi_1 + \lambda_2\phi_2) = (\lambda_1\phi_1 + \lambda_2\phi_2)(\xi) = \lambda_1\phi_1(\xi) + \lambda_2\phi_2(\xi) = \lambda_1\delta_\xi(\phi_1) + \lambda_2\delta_\xi(\phi_2)$$

Secondly, it is continuous because the condition $\phi_j \twoheadrightarrow 0$ implies that $\phi_j(\xi) \to 0$. If we write δ without a subscript it refers to evaluation at 0; i.e., $\delta = \delta_0$.   ∎

Example 2.   The **Heaviside** distribution is defined, when $n = 1$, by

(4) $$\widetilde{H}(\phi) = \int_0^\infty \phi(x)\,dx \qquad (\phi \in \mathcal{D}) \qquad\qquad ∎$$

Example 3.   Let $f : \mathbb{R}^n \to \mathbb{R}$ be continuous. With f we associate a distribution $\widetilde{f}$ by means of the definition

(5) $$\widetilde{f}(\phi) = \int f(x)\phi(x)\,dx \qquad (\phi \in \mathcal{D})$$

The linearity of $\widetilde{f}$ is obvious. For the continuity, we observe that if $\phi_j \twoheadrightarrow 0$, then there is a compact K containing the supports of the ϕ_j. Then we have

$$\left|\widetilde{f}(\phi_j)\right| = \left|\int_K f(x)\phi_j(x)\,dx\right| \leqslant \sup_x \left|\phi_j(x)\right| \int_K \left|f(y)\right|\,dy \to 0$$

because $\phi_j \twoheadrightarrow 0$ entails $\sup_x \left|\phi_j(x)\right| \to 0$.   ∎

Example 4.   Fix a multi-index α and define

$$T(\phi) = \int_{\mathbb{R}^n} D^\alpha\phi \qquad (\phi \in \mathcal{D})$$

This is a distribution. (The proof involves the use of Theorem 1.)   ∎

Example 5.   If H is the Heaviside function, defined by the equation

$$H(x) = \begin{cases} 1 & \text{if } x \geqslant 0 \\ 0 & \text{if } x < 0 \end{cases}$$

then Example 2 above illustrates the principle in Example 3, although H is obviously not continuous.   ∎

The distributions $\widetilde{f}$ described in Example 3 are the subject of the next theorem.

Theorem 2.　*If $f \in C(\mathbb{R}^n)$, then $\widetilde{f}$, as defined in Equation (5), is a distribution. The map $f \mapsto \widetilde{f}$ is linear and injective from $C(\mathbb{R}^n)$ into $\mathcal{D}'$.*

Proof. We have already seen that $\widetilde{f}$ is a distribution. The linearity of the mapping $f \mapsto \widetilde{f}$ follows from the equation

$$(\alpha_1 f_1 + \alpha_2 f_2)^{\sim}(\phi) = \int (\alpha_1 f_1 + \alpha_2 f_2)\phi = \alpha_1 \int f_1 \phi + \alpha_2 \int f_2 \phi$$

$$= \alpha_1 \widetilde{f_1}(\phi) + \alpha_2 \widetilde{f_2}(\phi) = (\alpha_1 \widetilde{f_1} + \alpha_2 \widetilde{f})(\phi)$$

For the injective property it suffices to prove that if $f \neq 0$, then $\widetilde{f} \neq 0$. Supposing that $f \neq 0$, let ξ be a point where $f(\xi) \neq 0$. Select j such that $f(x)$ is of one sign in the ball around ξ of radius $1/j$. Then $\rho_j(x - \xi)$, as defined in Equation 2, is positive in this same ball about ξ and vanishes elsewhere. Hence $\int f(x)\rho_j(x - \xi)\,dx \neq 0$. This means that $\widetilde{f}(\phi) \neq 0$ if $\phi(x) = \rho_j(x - \xi)$.　∎

Example 3 shows that in a certain natural way, each continuous function $f : \mathbb{R}^n \to \mathbb{R}$ "is" a distribution. That is, we can associate a distribution $\widetilde{f}$ with f. In fact, the same is true for some functions that are not continuous. The appropriate family of functions is described now.

A Lebesgue-measurable function $f : \mathbb{R}^n \to \mathbb{R}$ is said to be **locally integrable** if for every compact set $K \subset \mathbb{R}^n$, $\int_K |f(x)|\,dx < \infty$. As is usual when dealing with measurable functions, we define two functions to be **equivalent** if they differ only on a set of measure zero. The equivalence classes of locally integrable functions make up the space $L^1_{\text{loc}}(\mathbb{R}^n)$.

We mention, without proof, the result corresponding to the preceding theorem for the case of locally integrable functions. See [Ru1] page 142, or [Lan1] page 277.

Theorem 3.　*If f is locally integrable, then the equation $\widetilde{f}(\phi) = \int f\phi$ defines a distribution $\widetilde{f}$ that does not depend on the representative selected from the equivalence class of f. The mapping $f \mapsto \widetilde{f}$ is linear and injective from $L^1_{\text{loc}}(\mathbb{R}^n)$ into $\mathcal{D}'$.*

Theorem 4.　*Let μ be a positive Borel measure on $\mathbb{R}^n$ such that $\mu(K) < \infty$ for each compact set K in $\mathbb{R}^n$. Then μ induces a distribution T by the formula*

$$T(\phi) = \int_{\mathbb{R}^n} \phi(x)\,d\mu(x) \qquad (\phi \in \mathcal{D})$$

Proof. The linearity is obvious. For the continuity of T, let $\phi_j \in \mathcal{D}$ and $\phi_j \twoheadrightarrow 0$. Then there is a compact set K containing the supports of all ϕ_j. Consequently,

$$|T(\phi_j)| \leqslant \int_K |\phi_j(x)|\,d\mu(x) \leqslant \sup_{y \in K} |\phi_j(y)| \int_K d\mu(x)$$

$$= \mu(K) \sup_{y \in K} |\phi_j(y)| \to 0 \qquad\qquad ∎$$

The distributions described in Theorem 3 are said to be **regular**.

Suggested references for this chapter are [Ad], [Con], [Dono], [Edw], [Fol], [Fri], [Frie1], [Fried], [GV], [Gri], [Ho], [Horv], [Hu], [Jon], [Maz], [OD], [RS], [Ru1], [Schl], [Schl2], [So], [Yo], [Ze], [Zem], and [Zie].

Problems 5.1

1. Describe the null space of D^α in the case $n = 2$. Do this first when the domain is $C^\infty(\mathbb{R}^n)$ and second when it is $\mathcal{D}$.

2. Let $f : \mathbb{R} \to \mathbb{R}$. Suppose that f' exists and is continuous in the two intervals $(-\infty, 0)$, $(0, \infty)$. Assume further that $\lim_{x\downarrow 0} f'(x) = \lim_{x\uparrow 0} f'(x)$. Does it follow that f' is continuous on $\mathbb{R}$? Examples and theorems are wanted.

3. Prove that for each $x_0 \in \mathbb{R}^n$ and for each $r > 0$ there is an element ϕ of $\mathcal{D}$ such that the set $\{x : \phi(x) \neq 0\}$ is the open ball $B(x_0, r)$ having center x_0 and radius r.

4. Prove that if $\mathcal{O}$ is any bounded open set in $\mathbb{R}^n$, then there exists an element ϕ of $\mathcal{D}$ such that $\{x : \phi(x) \neq 0\} = \mathcal{O}$. Hints: Use the functions in Problem 3. Maybe a series of such functions $\sum 2^{-k}\phi_k$ will be useful. Don't forget that the points of $\mathbb{R}^n$ whose coordinates are rational form a dense set.

5. For each $v \in \mathbb{R}^n$ there is a **translation operator** E_v on $\mathcal{D}$. Its definition is $(E_v\phi)(x) = \phi(x - v)$. Prove that E_v is linear, continuous, injective, surjective, and invertible from $\mathcal{D}$ to $\mathcal{D}$.

6. For each ϕ in $C^\infty(\mathbb{R}^n)$ there is a **multiplication operator** M_ϕ defined on $\mathcal{D}$ by the equation $M_\phi\psi = \phi\psi$. Prove that M_ϕ is linear and continuous from $\mathcal{D}$ into $\mathcal{D}$. Under what conditions will M_ϕ be injective? surjective? invertible?

7. For suitable ϕ there is a **composition operator** C_ϕ defined on $\mathcal{D}$ by the equation $C_\phi\psi = \psi \circ \phi$. What must be assumed about ϕ in order that C_ϕ map $\mathcal{D}$ into $\mathcal{D}$? Prove that C_ϕ is linear and continuous from $\mathcal{D}$ to $\mathcal{D}$. Find conditions for C_ϕ to be injective, surjective, or invertible.

8. Prove that E_v, as defined in Problem 5, has this property for all test functions ϕ and ψ:

$$\int \phi E_v \psi = \int \psi E_{-v}\phi$$

What is the analogous property for M_ϕ in Problem 6?

9. Prove that if T is a distribution and if A is a continuous linear map of $\mathcal{D}$ into $\mathcal{D}$, then $T \circ A$ is a distribution. Use the notation of the preceding problems and identify $\delta_\xi \circ E_v$, $\delta_\xi \circ M_\phi$, and $\delta_\xi \circ C_\phi$ in elementary terms. What is $(\delta_\xi \circ E_v \circ M_\theta \circ C_\psi)(\phi)$?

10. Show that $D^\alpha D^\beta = D^{\alpha+\beta}$, and that consequently $D^\alpha D^\beta = D^\beta D^\alpha$.

11. Let $\phi \in \mathcal{D}$. Prove that if there exists a multi-index α for which $D^\alpha\phi = 0$, then $\phi = 0$. Suggestion: Do the cases $|\alpha| = 0$ and $|\alpha| = 1$ first. Proceed by induction on $|\alpha|$.

12. Prove (in detail) that each test function is uniformly continuous.

13. Prove that $\mathcal{D}$ is a *ring* without unit under pointwise multiplication. Prove that $\mathcal{D}$ is an *ideal* in the ring $C^\infty(\mathbb{R}^n)$. This means that $f\phi \in \mathcal{D}$ when $f \in C^\infty$ and $\phi \in \mathcal{D}$.

14. For $\phi \in \mathcal{D}(\mathbb{R})$, define $T(\phi) = \sum_{k=0}^\infty (D^k\phi)(k)$. Prove that T is a distribution.

15. Give an example of a sequence $[\phi_j]$ in $\mathcal{D}$ such that $[D^\alpha\phi_j]$ converges uniformly to 0 for each multi-index α, yet $[\phi_j]$ does not converge to 0 in the topology of $\mathcal{D}$.

16. Show that $\operatorname{supp}(\phi)$ is not always the same as $\{x : \phi(x) \neq 0\}$. Which of these sets contains the other? When are these sets identical?

17. A distribution T is said to be of **order** 0 if there is a constant C such that $|T(\phi)| \leqslant C\|\phi\|_\infty$ (for all test functions ϕ). Which regular distributions are of order 0?

18. Prove that the Dirac distributions in Example 1 are not regular.

19. Give a rough estimate of c in Equation 1. (Start with $n = 1$.)

20. Show that the notion of convergence in $\mathcal{D}$ is consistent with the linear structure in $\mathcal{D}$.

5.2 Derivatives of Distributions

We have seen that the space $\mathcal{D}'$ of distributions is very large; it contains (images of) all continuous functions on $\mathbb{R}^n$ and even all locally integrable functions. Then, too, it contains functionals on $\mathcal{D}$ that are not readily identified with functions. Such, for example, is the Dirac distribution, which is a "point-evaluation" functional. We now will define derivatives of distributions, taking care that the new notion of derivative will coincide with the classical one when both are meaningful.

Definition. If T is a distribution and α is a multi-index, then $\partial^\alpha T$ is the distribution defined by

$$(1) \qquad \partial^\alpha T = (-1)^{|\alpha|} \, T \circ D^\alpha$$

Notice that it is a little simpler to write $\partial^\alpha T = T \circ (-D)^\alpha$. The first question is whether $\partial^\alpha T$ is a distribution. Its linearity is clear, since T and D^α are linear. Its continuity follows by the same reasoning. (Here Theorem 1 from the preceding section is needed.)

The next question is whether this new definition is consistent with the old. Let f be a function on $\mathbb{R}^n$ such that $D^\alpha f$ exists and is continuous whenever $|\alpha| \leqslant k$. Then $\tilde{f}$ is a distribution, and when $|\alpha| \leqslant k$,

$$(2) \qquad \partial^\alpha \tilde{f} = (D^\alpha f)^\sim$$

To verify this, we write (for any test function ϕ)

$$(3) \qquad (D^\alpha f)^\sim(\phi) = \int (D^\alpha f)\phi = (-1)^{|\alpha|} \int f D^\alpha \phi = (-1)^{|\alpha|} \, \tilde{f}(D^\alpha \phi)$$

$$= (\partial^\alpha \tilde{f})(\phi)$$

In this calculation integration by parts was used repeatedly. Here is how a single integration by parts works:

$$\int_{-\infty}^{\infty} \frac{\partial f}{\partial x_i} \phi \, dx_i = f\phi \Big|_{-\infty}^{\infty} - \int_{-\infty}^{\infty} f \, \frac{\partial \phi}{\partial x_i} \, dx_i$$

Since $\phi \in \mathcal{D}$, ϕ vanishes outside some compact set, and the first term on the right–hand side of the equation is zero. Each application of integration by parts transfers one derivative from f to ϕ and changes the sign of the integral. The number of these steps is $|\alpha| = \sum_{i=1}^n \alpha_i$.

Now, it *can* happen that $\partial^\alpha \tilde{f} \neq (D^\alpha f)^\sim$ for a function f that does not have continuous partial derivatives. For an example, the reader should consult [Ru1], page 144.

Example 1. Let $\tilde{H}$ be the Heaviside distribution (Example 2, page 250), and
let δ be the Dirac distribution at 0 (Example 1, page 250). Then with $n = 1$
and $\alpha = (1)$, we have $\partial\tilde{H} = \delta$. Indeed, for any test function ϕ,

$$(\partial\tilde{H})(\phi) = -\tilde{H}(D\phi) = -\int_0^\infty \phi' = \phi(0) - \phi(\infty) = \phi(0) = \delta(\phi) \qquad \blacksquare$$

Example 2. Again let $n = 1$ and $\alpha = (1)$, so that D is an ordinary derivative.
Let

$$f(x) = \begin{cases} x & \text{if } x \geqslant 0 \\ 0 & \text{if } x \leqslant 0 \end{cases}$$

One is tempted to say that f' is the Heaviside function H. But this is not true,
since $f'(0)$ is undefined in the classical sense. However, $\partial\tilde{f} = \tilde{H}$, and so in the
sense of distributions the equation $f' = H$ becomes correct. $\qquad \blacksquare$

The nomenclature that is often used in these matters is as follows: A "distribution derivative" (or a "distributional derivative") of a function f is a distribution T such that $(\tilde{f})' = T$. In the general case of an operator D^α, we require
$\partial^\alpha \tilde{f} = T$. If T is a regular distribution, say $T = \tilde{g}$, then the defining equation is

$$\int g\phi = (-1)^{|\alpha|} \int f D^\alpha \phi \qquad (\phi \in \mathcal{D})$$

Example 3. What is the distribution derivative of the function $f(x) = |x|$?
It is a distribution $\tilde{g}$, where g is a function such that for all test functions ϕ,

$$\int g\phi = -\int f\phi' = -\int_{-\infty}^0 (-x)\phi'(x)\,dx - \int_0^\infty x\phi'(x)\,dx$$

$$= x\phi(x)\Big|_{-\infty}^0 - \int_{-\infty}^0 \phi(x)\,dx - x\phi(x)\Big|_0^\infty + \int_0^\infty \phi(x)\,dx$$

$$= \int_{-\infty}^0 (-1)\phi(x)\,dx + \int_0^\infty (+1)\phi(x)\,dx$$

Thus

$$g(x) = \begin{cases} -1 & x < 0 \\ +1 & x \geqslant 0 \end{cases} = 2H(x) - 1$$

We say that $f' = g$ in the sense of distributions, or $\partial\tilde{f} = \tilde{g}$. Note particularly
that f does not have a "classical" derivative. $\qquad \blacksquare$

Example 4. What is the distribution derivative f'' when $f(x) = |x|$? If we
blindly use the techniques of classical calculus, we have from Examples 1 and
3, $f' = 2H - 1$ and $f'' = 2\delta$. This procedure is justified by the next theorem.
$\qquad \blacksquare$

Theorem 1.   *The operators ∂^α are linear from $\mathcal{D}'$ into $\mathcal{D}'$. Furthermore, $\partial^\alpha \partial^\beta = \partial^\beta \partial^\alpha = \partial^{\alpha+\beta}$ for any pair of multi-indices.*

Proof.   The linearity of ∂^α is obvious from the definition, Equation (1). The commutative property rests upon a theorem of classical calculus that states that for any function f of two variables, if $\dfrac{\partial^2 f}{\partial x \partial y}$ and $\dfrac{\partial^2 f}{\partial y \partial x}$ exist and are continuous, then they are equal. Therefore, for any $\phi \in \mathcal{D}$, we have $D^\alpha D^\beta \phi = D^\beta D^\alpha \phi$. Consequently, for an arbitrary distribution T we have

$$\partial^\alpha(\partial^\beta T) = (-1)^{|\alpha|}(\partial^\beta T) \circ D^\alpha = (-1)^{|\alpha|}(-1)^{|\beta|}\, T \circ D^\beta \circ D^\alpha$$
$$= (-1)^{|\beta|}(-1)^{|\alpha|}\, T \circ D^\alpha \circ D^\beta$$
$$= (-1)^{|\beta|}(\partial^\alpha T) \circ D^\beta$$
$$= \partial^\beta \partial^\alpha T \qquad\qquad\blacksquare$$

Theorem 2.   *For $n = 1$ (i.e., for functions of one variable), every distribution is the derivative of another distribution.*

Proof.   Prior to beginning the proof we define some linear maps. Let $\tilde{1}$ be the distribution defined by the constant 1:

$$\tilde{1}(\phi) = \int_{-\infty}^{\infty} \phi(x)\, dx \qquad (\phi \in \mathcal{D})$$

Let M be the kernel (null space) of $\tilde{1}$. Then M is a closed hyperplane in $\mathcal{D}$. Select a test function ψ such that $\tilde{1}(\psi) = 1$, and define

$$A\phi = \phi - \tilde{1}(\phi)\psi \qquad (\phi \in \mathcal{D})$$

$$(B\phi)(x) = \int_{-\infty}^{x} \phi(y)\, dy \qquad (\phi \in M)$$

We observe that if $\phi \in M$, then $B\phi \in \mathcal{D}$.

Now let T be any distribution, and set $S = -T \circ B \circ A$. It is to be shown that S is a distribution and that $\partial S = T$. Because $A\phi \in M$ for every test function, $BA\phi \in \mathcal{D}$. Since $B \circ A$ is continuous from $\mathcal{D}$ into $\mathcal{D}$, one concludes that S is a distribution. Finally, we compute

$$(\partial S)(\phi) = -S(\phi') = T(BA\phi') = T(B\phi') = T\phi$$

Here we used the elementary facts that $\tilde{1}(\phi') = 0$ and that $B\phi' = \phi$. $\blacksquare$

Theorem 3.   Let $n = 1$, and let T be a distribution for which $\partial T = 0$. Then T is $\tilde{c}$ for some constant c.

Proof.   Adopt the notation of the preceding proof. The familiar equation

$$\phi(x) = \frac{d}{dx} \int_{-\infty}^{x} \phi(y)\, dy$$

says that $\phi = DB\phi$, and this is valid for all $\phi \in M$. Since $A\phi \in M$ for all $\phi \in \mathcal{D}$, we have $A\phi = DBA\phi$ for all $\phi \in \mathcal{D}$. Consequently, if $\partial T = 0$, then for all test functions

$$T(\phi) = T\big(A\phi + \tilde{1}(\phi)\psi\big) = T(DBA\phi) + \tilde{1}(\phi)T(\psi)$$
$$= -(\partial T)(BA\phi) + T(\psi)\tilde{1}(\phi)$$
$$= T(\psi)\tilde{1}(\phi)$$

Thus $T = \tilde{c}$, with $c = T(\psi)$.   ∎

We state without proof a generalization of Theorem 2.

Theorem 4.   If T is a distribution and K is a compact set in $\mathbb{R}^n$, then there exists an $f \in C(\mathbb{R}^n)$ and a multi-index α such that for all $\phi \in \mathcal{D}$ whose supports are in K,

$$T(\phi) = (\partial^\alpha \tilde{f})(\phi)$$

Proof.   For the proof, consult [Ru1], page 152.   ∎

Problems 5.2

1. Let f be a C^1-function on $(-\infty, 0]$ and on $[0, \infty)$. Let $\alpha = \lim_{x \downarrow 0} f(x) - \lim_{x \uparrow 0} f(x)$. Express the distribution derivative of f in terms of α and familiar distributions.

2. Let δ and $\tilde{H}$ be the Dirac and Heaviside distributions. What are $\partial^n \delta$ and $\partial^n \tilde{H}$?

3. Find all the distributions T for which $\partial^\alpha T = 0$ whenever $|\alpha| = 1$.

4. Use notation introduced in the proof of Theorem 2. Prove that $\tilde{1} \circ A = 0$, that $DBA = A$, and that $A \circ D = D$. Prove that $B \circ A$ is continuous on $\mathcal{D}$.

5. The characteristic function of a set A is the function $\mathcal{X}_A$ defined by

$$\mathcal{X}_A(s) = \begin{cases} 1 & \text{if } s \in A \\ 0 & \text{if } s \notin A \end{cases}$$

   If $A = (a, b) \subset \mathbb{R}$, what is the distributional derivative of $\mathcal{X}_A$?

6. For what functions f on $\mathbb{R}$ is the equation $\partial \tilde{f} = \tilde{f'}$ true?

7. Let $n = 1$. Prove that $D : \mathcal{D} \to \mathcal{D}$ is injective. Prove that $\partial : \mathcal{D}' \to \mathcal{D}'$ is not injective.

8. Work in $\mathcal{D}(\mathbb{R}^n)$. Let α be a multi-index such that $|\alpha| = 1$. Prove or disprove that $D^\alpha : \mathcal{D} \to \mathcal{D}$ is injective. Prove or disprove that $\partial^\alpha : \mathcal{D}' \to \mathcal{D}'$ is injective.

9. Let $n = 1$. Is every test function the derivative of a test function?

10. Let $f \in C^\infty(\mathbb{R})$ and let H be the Heaviside function. Compute the distributional derivatives of fH. Show by induction that $\partial^m(fH) = HD^m f + \sum_{k=0}^{m-1} D^k f(0)\partial^{m-k-1}\delta$.

11. Prove that the hyperplane M defined in the proof of Theorem 2 is the range of the operator $\dfrac{d}{dx}$ when the latter is interpreted as acting from $\mathcal{D}(\mathbb{R}^1)$ into $\mathcal{D}(\mathbb{R}^1)$.

12. If two locally integrable functions are the same except on a set of measure 0, then the corresponding distributions are the same. If H is the Heaviside function, then $H'(x) = 0$ except on a set of measure 0. Therefore, the distributional derivative of H should be 0. Explain the fallacy in this argument.

13. Find the distributional derivative of this function:

$$f(x) = \begin{cases} \cos x & x > 0 \\ \sin x & x \leqslant 0 \end{cases}$$

14. Define A on $\mathcal{D}(\mathbb{R}^2)$ by putting

$$A(\phi) = \int_{-\infty}^{\infty} \phi(x_1, x_2)\, dx_1$$

Prove that A maps $\mathcal{D}(\mathbb{R}^2)$ into $\mathcal{D}(\mathbb{R}^1)$.

15. Refer to the proof of Theorem 2 and show that BA is a surjective map of $\mathcal{D}$ to $\mathcal{D}$.

16. Refer to Problem 5. Does the characteristic function of every measurable set have a distributional derivative?

17. Consider the maps $f \mapsto \tilde{f}$, D^α, and ∂^α. Draw a commutative diagram expressing the consistency of D^α and ∂^α in Equation (1).

18. Find a distribution T on $\mathbb{R}$ such that $\partial^2 T + T = \delta$.

5.3 Convergence of Distributions

If $[T_j]$ is a sequence of distributions, we will write $T_j \to 0$ if and only if $T_j(\phi) \to 0$ for each test function ϕ. If T is a distribution, $T_j \to T$ means that $T_j - T \to 0$. The reader will recognize this as weak* convergence of a sequence of linear functionals. It is also "pointwise" convergence, meaning convergence at each point in the domain. Topological notions in $\mathcal{D}'$, such as continuity, will be based on this notion of convergence (which we refer to simply as "convergence of distributions"). For example, we have the following theorem.

Theorem 1.   *For every multi-index α, ∂^α is a continuous linear map of $\mathcal{D}'$ into $\mathcal{D}'$.*

Proof.   Let $T_j \to 0$. In order to prove that $\partial^\alpha T_j \to 0$, we select an arbitrary test function ϕ, and attempt to prove that $(\partial^\alpha T_j)(\phi) \to 0$. This means that $(-1)^{|\alpha|} T_j(D^\alpha \phi) \to 0$, which is certainly true, because $D^\alpha \phi$ is a test function. See Theorem 1 in Section 5.2.   ∎

An important result, whose proof can be found, for example, in [Ru1] page 146, or [Ho] page 38, is the following.

Theorem 2.   *If a sequence of distributions $[T_j]$ has the property that $[T_j(\phi)]$ is convergent for each test function ϕ, then the equation $T(\phi) = \lim_j T_j(\phi)$ defines a distribution T, and $T_j \to T$.*

The theorem asserts that for a sequence of distributions T_j if $\lim_j T_j(\phi)$ exists in $\mathbb{R}$ for every test function ϕ, then the equation

$$T(\phi) = \lim_j T_j(\phi)$$

defines a distribution. There is no question about T being well-defined. Its linearity is also trivial, since we have

$$T(\phi_1 + \phi_2) = \lim_j T_j(\phi_1 + \phi_2) = \lim_j T_j(\phi_1) + \lim_j T_j(\phi_2) = T(\phi_1) + T(\phi_2)$$

The only real issue is whether T is continuous, and the proof of this requires some topological vector space theory beyond the scope of this chapter.

Corollary 1.   *A series of distributions, $\sum_{j=1}^{\infty} T_j$, converges to a distribution if and only if for each test function ϕ the series $\sum_{j=1}^{\infty} T_j(\phi)$ is convergent in $\mathbb{R}$.*

Corollary 2.   *If $\sum T_j$ is a convergent series of distributions, then for any multi-index α, $\partial^\alpha \sum T_j = \sum \partial^\alpha T_j$.*

Proof.   By Theorem 1, ∂^α is continuous. Hence

$$\partial^\alpha \left(\sum_{j=1}^{\infty} T_j \right) = \partial^\alpha \left(\lim_{m \to \infty} \sum_{j=1}^{m} T_j \right) = \lim_{m \to \infty} \left(\partial^\alpha \sum_{j=1}^{m} T_j \right)$$

$$= \lim_{m \to \infty} \sum_{j=1}^{m} \partial^\alpha T_j = \sum_{j=1}^{\infty} \partial^\alpha T_j \qquad \blacksquare$$

The previous theorem and its corollaries stand in sharp contrast to the situation that prevails for classical derivatives and functions. Thus one can construct a pointwise convergent sequence of continuous functions whose limit is discontinuous. For example, consider the functions f_k shown in Figure 5.2.

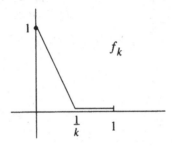

Figure 5.2

Similarly, even a *uniformly* convergent series of continuously differentiable functions can fail to satisfy the equation

$$\frac{d}{dx} \sum f_k = \sum \frac{d}{dx} f_k$$

A famous example of this phenomenon is provided by the Weierstrass nondifferentiable function

$$f(x) = \sum_{k=1}^{\infty} 2^{-k} \cos 3^k x$$

This function is continuous but not differentiable at any point! (This example is treated in [Ti2] and [Ch]. See also Section 7.8 in this book, pages 374ff, where some graphics are displayed.)

Example. Let $f_n(x) = \cos nx$. This sequence of functions does not converge. Is the same true for the accompanying distributions $\tilde{f}_n$? To answer this, we take any test function ϕ and contemplate the effect of $\tilde{f}_n$ on it:

$$\tilde{f}_n(\phi) = \int_{-\infty}^{\infty} \phi(x) \cos nx \, dx = \int_a^b \phi(x) \cos nx \, dx$$

Here the interval $[a, b]$ is chosen to contain the support of ϕ. For large values of n the C^∞ function is being integrated with the highly oscillatory function f_n. This produces very small values because of a cancellation of positive areas and negative areas. The limit will be zero, and hence $\tilde{f}_n \to 0$. This conclusion can also be justified by writing $f_n = g'_n$, where $g_n(x) = \sin nx / n$. We see that $g_n \to 0$ uniformly, and that the equations $\tilde{g}_n \to 0$ and $\tilde{f}_n \to 0$ follow, in $\mathcal{D}'$. ∎

Theorem 3. Let $f, f_1, f_2, \ldots$ belong to $L^1_{\text{loc}}(\mathbb{R}^n)$, and suppose that $f_j \to f$ pointwise almost everywhere. If there is an element $g \in L^1_{\text{loc}}(\mathbb{R}^n)$ such that $|f_j| \leqslant g$, then $\tilde{f}_j \to \tilde{f}$ in $\mathcal{D}'$.

Proof. The question is whether

$$(1) \qquad \int f_j \phi \to \int f \phi$$

for all test functions ϕ. We have $f_j \phi \in L^1(K)$ if K is the support of ϕ. Furthermore, $|f_j \phi| \leqslant g|\phi|$ and $(f_j \phi)(x) \to (f\phi)(x)$ almost everywhere. Hence by the Lebesgue Dominated Convergence Theorem (Section 8.6, page 406), Equation (1) is valid. ∎

Theorem 4.   Let $[f_j]$ be a sequence of nonnegative functions in $L^1_{loc}(\mathbb{R}^n)$ such that $\int f_j = 1$ for each j and such that

$$\lim_{j\to\infty} \int_{|x|\geqslant r} f_j = 0$$

for all positive r. Then $\tilde{f}_j \to \delta$ (the Dirac distribution).

Proof.   Let $\phi \in \mathcal{D}$ and put $\psi = \phi - \phi(0)$. Let $\varepsilon > 0$, and select $r > 0$ so that $|\psi(x)| < \varepsilon$ when $|x| < r$. Then

$$\left|\int f_j\phi - \phi(0)\right| = \left|\int f_j[\phi - \phi(0)]\right| = \left|\int f_j\psi\right| \leqslant \int |f_j\psi|$$

$$\leqslant \int_{|x|<r} |f_j\psi| + \int_{|x|\geqslant r} |f_j\psi|$$

$$\leqslant \varepsilon \int |f_j| + \max|\psi(x)| \int_{|x|\geqslant r} f_j$$

Taking the limit as $j \to \infty$, we obtain

$$\left|\lim_j \tilde{f}_j(\phi) - \delta(\phi)\right| \leqslant \varepsilon$$

Since ε was arbitrary, $\lim_j \tilde{f}_j(\phi) = \delta(\phi)$.   ∎

Problems 5.3

1. What is the distributional derivative of the Weierstrass nondifferentiable function mentioned in this section?

2. Let $\{T_\theta : \theta \in A\}$ be a set of distributions indexed with a real parameter θ; i.e. $A \subset \mathbb{R}$. Make a suitable definition for $\lim_{\theta\to r} T_\theta$.

3. (Continuation) If $\phi \in \mathcal{D}$, if $n = 1$, and if $\theta \in \mathbb{R}$, let $\phi_\theta(x) = \phi(x - \theta)$. If T is a distribution, let T_θ be the distribution defined by $T_\theta(\phi) = T(\phi_\theta)$. Prove that $\lim_{\theta\to 0} \theta^{-1}(T_\theta - T) = \partial T$.

4. Let $[\xi_j]$ be a sequence of distinct points in $\mathbb{R}^n$, and let δ_j be the corresponding Dirac distributions. Under what conditions does $\sum_{j=1}^\infty c_j\delta_j$ represent a distribution? (Here $c_j \in \mathbb{R}$.) A necessary and sufficient condition on $[c_i]$ would be ideal.

5. Use Theorem 4 to prove that if f is a nonnegative element of $L^1_{loc}(\mathbb{R}^n)$ such that $\int f = 1$ and if $f_j(x) = j^n f(jx)$ for $j = 1, 2, \ldots$, then $\tilde{f}_j \to \delta$.

6. Do these sequences have the properties described in Theorem 4?

   (a) $f_j(x) = j/[\pi(1 + j^2x^2)]$

   (b) $f_j(x) = j\pi^{-1/2}\exp(-j^2x^2)$

7. Let the real line be partitioned by points $-\infty = x_0 < x_1 < \cdots < x_{n+1} = \infty$, and suppose that f is a piecewise continuously differentiable function, with breaks at $x_1, \ldots, x_n$. Prove that the distributional derivative of f is $(f')^\sim + \sum_{i=1}^n c_i\delta_{x_i}$, where c_i is the magnitude of the jump in f at x_i; i.e., $c_i = f(x_i + 0) - f(x_i - 0)$. Notice that this problem emphasizes

the difference between $(f')^\sim$ and $(f^\sim)'$. The prime symbol has different meanings in different contexts.

5.4 Multiplication of Distributions by Functions

Before getting to the main topic of this section, let us record some results from multivariate algebra and calculus.

Recall the definition of the classical **binomial coefficients**:

(1)
$$\binom{m}{k} = \begin{cases} \dfrac{m!}{k!(m-k)!} & \text{if } 0 \leqslant k \leqslant m \\ 0 & \text{otherwise} \end{cases}$$

These are the coefficients that make the Binomial Theorem true:

(2)
$$(a+b)^m = \sum_{k=0}^{m} \binom{m}{k} a^k \, b^{m-k}$$

The multivariate version of this theorem is presented below.

Definitions. For two multi-indices α and β in $\mathbb{Z}^n$, we write $\beta \leqslant \alpha$ if $\beta_i \leqslant \alpha_i$ for each $i = 1, 2, \ldots, n$. We denote by $\alpha + \beta$ the multi-index having components

$$(\alpha + \beta)_i = \alpha_i + \beta_i \qquad (1 \leqslant i \leqslant n)$$

If $\beta \leqslant \alpha$, then $\alpha - \beta$ is the multi-index whose components are $\alpha_i - \beta_i$. Finally, if $\beta \leqslant \alpha$, we define

(3)
$$\binom{\alpha}{\beta} = \binom{\alpha_1}{\beta_1}\binom{\alpha_2}{\beta_2}\cdots\binom{\alpha_n}{\beta_n}$$

If $x = (x_1, x_2, \ldots, x_n)$ and $\alpha = (\alpha_1, \alpha_2, \ldots, \alpha_n)$, then by definition,

$$x^\alpha := \prod_{i=1}^{n} x_i^{\alpha_i} = x_1^{\alpha_1} \cdot x_2^{\alpha_2} \cdots x_n^{\alpha_n}$$

The function $x \mapsto x^\alpha$ is a **monomial**. For $n = 3$, here are seven typical monomials:

$$x_1 x_2^7 x_3^2 \qquad x_1^4 x_3^5 \qquad x_2^3 \qquad 1 \qquad x_1 \qquad x_2 \qquad x_3$$

These are the building blocks for **polynomials**. The **degree** of a monomial x^α is defined to be $|\alpha|$. Thus, in the examples, the degrees are 10, 9, 3, 0, 1, 1, and 1. A polynomial in n variables is a function

$$p(x) = \sum_\alpha c_\alpha x^\alpha \qquad (x \in \mathbb{R}^n)$$

in which the sum is finite, the c_α are real numbers, and $\alpha \in \mathbb{Z}_+^n$. The **degree** of p is

$$\max\{|\alpha| : c_\alpha \neq 0\}$$

If all c_α are 0, then $p(x) = 0$, and we assign the degree $-\infty$ in this case. A polynomial of degree 0 is a constant function. Here are some examples, again with $n = 3$:

$$p_1(x) = 3 + 2x_1 - 7x_2^2 x_3 + 2x_1 x_2^4 x_3^7$$
$$p_2(x) = \sqrt{2}\, x_1 x_3 - \pi x_1^3 x_2^5 x_3$$

These have degrees 12 and 9, respectively.

The completely general polynomial of degree at most k in n variables can be written as

$$x \mapsto \sum_{|\alpha| \leqslant k} c_\alpha x^\alpha \qquad (x \in \mathbb{R}^n)$$

This sum has $\binom{k+n}{n}$ terms, as established later in Theorem 2.

We have seen in Section 5.1 that multi-indices are also useful in defining differential operators. If we set $D = \left(\dfrac{\partial}{\partial x_1}, \dfrac{\partial}{\partial x_2}, \cdots, \dfrac{\partial}{\partial x_n}\right)$. Then in a natural way we define

$$D^\alpha = \prod_{i=1}^n \frac{\partial^{\alpha_i}}{\partial x_i^{\alpha_i}} = \frac{\partial^{|\alpha|}}{\partial x_1^{\alpha_1} \partial x_2^{\alpha_2} \cdots \partial x_n^{\alpha_n}}$$

A further definition is

$$\alpha! = \alpha_1! \, \alpha_2! \cdots \alpha_n! = \prod_{i=1}^n \alpha_i!$$

The n−dimensional binomial coefficients are then expressible in the form

$$\binom{\alpha}{\beta} = \begin{cases} \dfrac{\alpha!}{\beta!(\alpha - \beta)!} & \text{if } 0 \leqslant \beta \leqslant \alpha \\ 0 & \text{otherwise} \end{cases}$$

Multivariate Binomial Theorem.   For all x and y in $\mathbb{R}^n$ and all α in $\mathbb{Z}_+^n$,

$$(x + y)^\alpha = \sum_{0 \leqslant \beta \leqslant \alpha} \binom{\alpha}{\beta} x^\beta y^{\alpha - \beta}$$

Proof.

$$(x + y)^\alpha = \prod_{i=1}^n (x_i + y_i)^{\alpha_i} = \prod_{i=1}^n \sum_{\beta_i=0}^{\alpha_i} \binom{\alpha_i}{\beta_i} x_i^{\beta_i} y_i^{\alpha_i - \beta_i}$$

$$= \left[\sum_{\beta_1=0}^{\alpha_1} \binom{\alpha_1}{\beta_1} x_1^{\beta_1} y_1^{\alpha_1 - \beta_1}\right] \cdots \left[\sum_{\beta_n=0}^{\alpha_n} \binom{\alpha_n}{\beta_n} x_n^{\beta_n} y_n^{\alpha_n - \beta_n}\right]$$

$$= \sum_{\beta_1=0}^{\alpha_1} \cdots \sum_{\beta_n=0}^{\alpha_n} \prod_{i=1}^{n} \binom{\alpha_i}{\beta_i} x_i^{\beta_i} y_i^{\alpha_i - \beta_i} = \sum_{0 \leqslant \beta \leqslant \alpha} \prod_{i=1}^{n} \binom{\alpha_i}{\beta_i} \prod_{j=1}^{n} x_j^{\beta_j} y_j^{\alpha_j - \beta_j}$$

$$= \sum_{0 \leqslant \beta \leqslant \alpha} \binom{\alpha}{\beta} x^\beta y^{\alpha - \beta} \qquad \blacksquare$$

We will usually abbreviate the inner product $\langle x, y \rangle$ of two vectors $x, y \in \mathbb{R}^n$ by the simpler notation xy.

Multinomial Theorem   *Let $x, y \in \mathbb{R}^n$ and $m \in \mathbb{N}$. Then*

$$(xy)^m = \sum_{|\alpha|=m} \frac{m!}{\alpha!} x^\alpha y^\alpha$$

Proof.   It suffices to consider only the special case $y = (1, 1, \dots, 1)$, because

$$xy = \langle (x_1 y_1, \dots, x_n y_n), (1, \dots, 1) \rangle$$

For this special case we proceed by induction on n. The case $n = 1$ is trivially true, and the case $n = 2$ is the usual binomial formula:

$$(x_1 + x_2)^m = \sum_{\alpha_1 + \alpha_2 = m} \frac{m!}{\alpha_1! \alpha_2!} x_1^{\alpha_1} x_2^{\alpha_2} = \sum_{j=0}^{m} \frac{m!}{j!(m-j)!} x_1^j x_2^{m-j}$$

Suppose that the multinomial formula is true for a particular value of n. The proof for the next case goes as follows. Let $x = (x_1, \dots, x_n)$, $w = (x_1, \dots, x_{n+1})$, $\alpha = (\alpha_1, \dots, \alpha_n)$, and $\beta = (\alpha_1, \dots, \alpha_{n+1})$. Then

$$(x_1 + \cdots + x_{n+1})^m = \left[(x_1 + \cdots + x_n) + x_{n+1} \right]^m$$

$$= \sum_{j=0}^{m} \binom{m}{j} (x_1 + \cdots + x_n)^j x_{n+1}^{m-j}$$

$$= \sum_{j=0}^{m} \binom{m}{j} \sum_{|\alpha|=j} \frac{j!}{\alpha!} x^\alpha x_{n+1}^{m-j}$$

$$= \sum_{j=0}^{m} \sum_{|\alpha|=j} \frac{m!}{j!(m-j)!} \frac{j!}{\alpha!} x^\alpha x_{n+1}^{m-j}$$

$$= \sum_{|\beta|=m} \frac{m!}{(m-j)!\alpha!} w^\beta = \sum_{|\beta|=m} \frac{m!}{\beta!} w^\beta$$

In this calculation, we let $\beta = (\alpha_1, \dots, \alpha_n, m - j)$, where $|\alpha| = j$. $\blacksquare$

The linear space of all polynomials of degree at most m in n real variables is denoted by $\Pi_m(\mathbb{R}^n)$. Thus each element of this space can be written as

$$p(x) = \sum_{|\alpha| \leqslant m} c_\alpha x^\alpha$$

Consequently, the set of monomials

$$\{ x \mapsto x^\alpha : |\alpha| \leqslant m \}$$

spans $\Pi_m(\mathbb{R}^n)$. Is this set in fact a basis?

Theorem 1   *The set of monomials $x \mapsto x^\alpha$ on $\mathbb{R}^n$ is linearly independent.*

Proof. If $n = 1$, the monomials are the elementary functions $x_1 \mapsto x_1^j$ for $j = 0, 1, 2, \ldots$ They form a linearly independent set, because a nontrivial linear combination $\sum_{j=0}^m c_j x_1^j$ cannot vanish as a function. (Indeed, it can have at most m zeros.)

Suppose now that our assertion has been proved for dimension $n - 1$. Let $x = (x_1, \ldots, x_n)$ and $\alpha = (\alpha_1, \ldots, \alpha_n)$. Suppose that $\sum_{\alpha \in J} c_\alpha x^\alpha = 0$, where the sum is over a finite set $J \subset \mathbb{Z}_+^n$. Put

$$J_k = \{\alpha \in J : \alpha_1 = k\}$$

Then for some m, $J = J_0 \cup \cdots \cup J_m$, and we can write

$$0 = \sum_{k=0}^m \sum_{\alpha \in J_k} c_\alpha x_1^{\alpha_1} \cdots x_n^{\alpha_n} = \sum_{k=0}^m x_1^k \sum_{\alpha \in J_k} c_\alpha x_2^{\alpha_2} \cdots x_n^{\alpha_n}$$

By the one-variable case, we infer that for $k = 0, \ldots, m$,

$$\sum_{\alpha \in J_k} c_\alpha x_2^{\alpha_2} \cdots x_n^{\alpha_n} = 0$$

Note that as α runs over J_k, the multi-indices $(\alpha_2, \ldots, \alpha_n)$ are all distinct. By the induction hypothesis, we then infer that for all $\alpha \in J_k$, $c_\alpha = 0$. Since k runs from 0 to m, all c_α are 0. ∎

We want to calculate the dimension of $\Pi_m(\mathbb{R}^n)$. The following lemma is needed before this can be done. Its proof is left as a problem.

Lemma 1   *For $n = 1, 2, 3, \ldots$ and $m = 0, 1, 2, \ldots$ we have*

$$\sum_{k=0}^m \binom{k+n}{n} = \binom{m+1+n}{n+1}$$

Theorem 2   *The dimension of $\Pi_m(\mathbb{R}^n)$ is $\binom{m+n}{n}$.*

Proof. The preceding theorem asserts that a basis for $\Pi_m(\mathbb{R}^n)$ is $\{x \mapsto x^\alpha : |\alpha| \leqslant m\}$. Here $x \in \mathbb{R}^n$. Using # to denote the number of elements in a set, we have only to prove

$$\#\{\alpha \in \mathbb{Z}_+^n : |\alpha| \leqslant m\} = \binom{m+n}{n}$$

We use induction on n. For $n = 1$, the formula is correct, since

$$\#\{\alpha \in \mathbb{Z}_+ : \alpha \leqslant m\} = m + 1 = \binom{m+1}{1}$$

Assume that the formula is correct for a particular n. For the next case we write

$$\#\{\alpha \in \mathbb{Z}_+^{n+1} : |\alpha| \leqslant m\} = \#\bigcup_{k=0}^{m}\left\{\alpha \in \mathbb{Z}_+^{n+1} : \alpha_{n+1} = k, \ \sum_{i=1}^{n}\alpha_i \leqslant m - k\right\}$$

$$= \sum_{k=0}^{m}\#\{\alpha \in \mathbb{Z}_+^n : |\alpha| \leqslant m - k\} = \sum_{k=0}^{m}\binom{m-k+n}{n}$$

$$= \sum_{k=0}^{m}\binom{k+n}{n} = \binom{m+n+1}{n+1}$$

In the last step, we applied Lemma 1.   ∎

Theorem 3. (The Leibniz Formula)   *If ϕ and ψ are test functions, then for any multi-index α we have*

(4)
$$D^\alpha(\phi\psi) = \sum_{\beta \leqslant \alpha}\binom{\alpha}{\beta}D^\beta\phi \cdot D^{\alpha-\beta}\psi$$

Proof.   (after Horváth)   We use induction on $|\alpha|$. If $|\alpha| = 0$, then $\alpha = (0,0,\ldots,0)$, and both sides of Equation (4) reduce to $\phi\psi$.

Now suppose that (4) has been established for all multi-indices α such that $|\alpha| \leqslant m$. Let γ be a multi-index of order $m + 1$. By renumbering the variables if necessary we can assume that $\gamma_1 \geqslant 1$. Let $\alpha = (\gamma_1 - 1, \gamma_2, \gamma_3, \ldots, \gamma_n)$. Then $D^\gamma = D_1 D^\alpha$, where D_1 denotes $\partial/\partial x_1$. Since $|\alpha| \leqslant m$, the induction hypothesis applies to D^α, and hence

(5)
$$D^\gamma(\phi\psi) = D_1 D^\alpha(\phi\psi) = D_1 \sum_{\beta \leqslant \alpha}\binom{\alpha}{\beta}D^\beta\phi \cdot D^{\alpha-\beta}\psi$$

$$= \sum_{\beta \leqslant \alpha}\binom{\alpha}{\beta}\left[D_1 D^\beta\phi \cdot D^{\alpha-\beta}\psi + D^\beta\phi \cdot D_1 D^{\alpha-\beta}\psi\right]$$

Now we set $\beta = (\beta_1, \beta_2, \ldots, \beta_n)$ and $\beta' = (\beta_1 + 1, \beta_2, \ldots, \beta_n)$. Observe that $\beta \leqslant \alpha$ if and only if $\beta' \leqslant \gamma$. Hence the first part of the sum in Equation (5) can be written as

$$\sum_{\beta' \leqslant \gamma}\binom{\alpha}{\beta}D^{\beta'}\phi \cdot D^{\gamma-\beta'}\psi$$

(6)
$$= \sum_{\beta' \leqslant \gamma}\binom{\gamma_1 - 1}{\beta_1' - 1}\binom{\gamma_2}{\beta_2'}\cdots\binom{\gamma_n}{\beta_n'}D^{\beta'}\phi \cdot D^{\gamma-\beta'}\psi$$

$$= \sum_{\beta \leqslant \gamma}\binom{\gamma_1 - 1}{\beta_1 - 1}\binom{\gamma_2}{\beta_2}\cdots\binom{\gamma_n}{\beta_n}D^\beta\phi \cdot D^{\gamma-\beta}\psi$$

$$= \sum_{\beta_1=0}^{\gamma_1}\binom{\gamma_1 - 1}{\beta_1 - 1}\sum_{\beta_2=0}^{\gamma_2}\cdots\sum_{\beta_n=0}^{\gamma_n}\binom{\gamma_2}{\beta_2}\cdots\binom{\gamma_n}{\beta_n}D^\beta\phi \cdot D^{\gamma-\beta}\psi$$

The second part of the sum in Equation (5) can be written as

$$(7) \qquad \sum_{\beta_1=0}^{\gamma_1} \binom{\gamma_1-1}{\beta_1} \sum_{\beta_2=0}^{\gamma_2} \cdots \sum_{\beta_n=0}^{\gamma_n} \binom{\gamma_2}{\beta_2} \cdots \binom{\gamma_n}{\beta_n} D^\beta \phi \cdot D^{\gamma-\beta}\psi$$

Now invoke the easily proved identity

$$\binom{m-1}{k-1} + \binom{m-1}{k} = \binom{m}{k}$$

Adding these two parts of the sum in Equation (5), we obtain

$$D^\gamma(\phi\psi) = \sum_{\beta_1=0}^{\gamma_1} \binom{\gamma_1}{\beta_1} \sum_{\beta_2=0}^{\gamma_2} \cdots \sum_{\beta_n=0}^{\gamma_n} \binom{\gamma_2}{\beta_2} \cdots \binom{\gamma_n}{\beta_n} D^\beta \phi \cdot D^{\gamma-\beta}\psi$$

$$= \sum_{\beta \leqslant \gamma} \binom{\gamma}{\beta} D^\beta \phi \cdot D^{\gamma-\beta}\psi \qquad\qquad \blacksquare$$

Since $\mathcal{D}'$ is a vector space, a distribution can be be multiplied by a constant to produce another distribution. Multiplication of a distribution T by a function $f \in C^\infty(\mathbb{R}^n)$ can also be defined:

$$(f \cdot T)(\phi) = T(f\phi).$$

Notice that if f is a constant function, say $f(x) = c$, then $f \cdot T$, as just defined, agrees with cT. In order to verify that $f \cdot T$ *is* a distribution, let $\phi_j \to 0$ in $\mathcal{D}$. Then there is a single compact set K containing the supports of all ϕ_j, and $D^\alpha\phi_j \to 0$ uniformly on K for all multi-indices. If $f \in C^\infty(\mathbb{R}^n)$ then by Leibniz's formula,

$$D^\alpha(f\phi_j) = \sum_{\beta \leqslant \alpha} \binom{\alpha}{\beta} D^\beta f \cdot D^{\alpha-\beta}\phi_j \to 0$$

This proves that $f\phi_j \to 0$ in $\mathcal{D}$. Since T is continuous,

$$(f \cdot T)(\phi_j) = T(f\phi_j) \to 0$$

Hence $f \cdot T$ is continuous. Its linearity is obvious.

Here is a theorem from elementary calculus, extended to products of functions and distributions.

Theorem 4.   *Let D be a simple partial derivative, say $D = \dfrac{\partial}{\partial x_i}$, and let ∂ be the corresponding distribution derivative. If $T \in \mathcal{D}'$ and $f \in C^\infty(\mathbb{R}^n)$, then*

$$\partial(fT) = Df \cdot T + f \cdot \partial T$$

Proof.

$$(Df \cdot T + f \cdot \partial T)(\phi) = T(Df \cdot \phi) + \partial T(f \cdot \phi) = T(Df \cdot \phi) - T(D(f\phi))$$
$$= T(Df \cdot \phi) - T(Df \cdot \phi + f \cdot D\phi)$$
$$= -T(f \cdot D\phi) = -(fT)(D\phi) = (\partial(fT))(\phi) \qquad \blacksquare$$

Theorem 5.   Let $n = 1$, let T be a distribution, and let u be an element of $C^\infty(\mathbb{R})$. If $\partial T + uT = \tilde{f}$, for some f in $C(\mathbb{R})$, then $T = \tilde{g}$ for some g in $C^1(\mathbb{R})$, and $g' + ug = f$.

Proof.   If $u = 0$, then $\partial T = \tilde{f}$. Write $f = h'$, where

$$h(x) = \int_a^x f(y)\, dy.$$

Then $h \in C^1(\mathbb{R})$. From the equation

$$\partial(T - \tilde{h}) = \partial T - \tilde{f} = 0$$

we conclude that $T - \tilde{h} = \tilde{c}$ for some constant c. (See Theorem 3 in Section 5.2, page 256.) Hence $T = \widetilde{h + c}$.

If u is not zero, let $v = \exp \int u\, dx$. Then $v' = vu$ and $v \in C^\infty(\mathbb{R})$. Then vT is well-defined, and by Theorem 4,

$$\partial(vT) = v'T + v\partial T = v(uT + \partial T) = v\tilde{f} = \widetilde{vf}$$

By the first part of the proof, we have $vT = \tilde{g}$ for some $g \in C^1(\mathbb{R})$. Hence $T = \widetilde{g/v}$. It is easily verified that $(g/v)' + u(g/v) = f$. $\qquad \blacksquare$

Theorem 6.   If $\phi \in \mathcal{D}$, then

$$\int_{\mathbb{R}^n} \phi(x)dx = \lim_{h \downarrow 0} h^n \sum_{\alpha \in \mathbb{Z}^n} \phi(h\alpha)$$

Proof.   The right side is just the limit of Riemann sums for the integral. In the case $n = 2$, we set up a lattice of points in $\mathbb{R}^2$. These points are of the form $(ih, jh) = h(i, j) = h\alpha$, where α runs over the set of all multi-integers, having positive or negative entries. Each square created by four adjacent lattice points has area h^2. $\qquad \blacksquare$

Problems 5.4

1. Prove that if $v \in C^\infty(\mathbb{R}^n)$ and if $f \in L^1_{\text{loc}}(\mathbb{R}^n)$, then $\widetilde{vf} = v\tilde{f}$.
2. For integers n and m, $\binom{n}{m} = \binom{n}{n-m}$. Is a similar result true for multi-indices?
3. Prove that if $T_j \to T$ in $\mathcal{D}'$ and if $f \in C^\infty(\mathbb{R}^n)$, then $fT_j \to fT$.
4. Let θ be a test function such that $\theta(0) \neq 0$. Prove that every test function is the sum of a multiple of θ and a test function that vanishes at 0.
5. (Continuation) Prove that if $f \in C^k(\mathbb{R})$ and $f(0) = 0$, then $f(x)/x$, when defined appropriately at 0, is in $C^{k-1}(\mathbb{R})$.

6. (Continuation) Let $n = 1$ and put $\psi(x) = x$. Prove that a distribution T that satisfies $\psi T = 0$ must be a scalar multiple of the Dirac distribution.
7. For fixed ψ in $C^\infty(\mathbb{R}^n)$ there is a multiplication operator M_ψ defined on $\mathcal{D}'$ by the equation $M_\psi T = \psi T$. Prove that M_ψ is linear and continuous.
8. Prove that the product of a C^∞-function and a regular distribution is a regular distribution.
9. Prove that if $h' = f$ and $h \in C^1(\mathbb{R})$, then $\partial \widetilde{h} = \widetilde{f}$.
10. Let $\phi \in C^\infty(\mathbb{R})$ and let H be the Heaviside function. Compute $\partial(\phi H)$.
11. Prove the Leibniz formula for the product of a C^∞-function and a distribution.
12. Define addition of multi-indices α and β by the formula $(\alpha + \beta)_i = \alpha_i + \beta_i$ for $1 \leqslant i \leqslant n$. If $\beta \leqslant \alpha$, we can define subtraction of β from α by $(\alpha - \beta)_i = \alpha_i - \beta_i$. Define also $\alpha! = \alpha_1! \alpha_2! \cdots \alpha_n!$. Prove that

$$\binom{\alpha}{\beta} = \frac{\alpha!}{\beta!(\alpha - \beta)!} \qquad \text{if } \beta \leqslant \alpha$$

13. (Continuation) Express $(1 + |x|^2)^m$ as a linear combination of "monomials" x^α. Here $x \in \mathbb{R}^n$, $m \in \mathbb{N}$, $\alpha \in \mathbb{N}^n$.
14. Prove that for any multi-index α,

$$D^\alpha(1 + |x|^2)^m = \sum_{|\beta| \leqslant |\alpha|} c_\beta x^\beta (1 + |x|^2)^{m - |\alpha|}$$

15. Verify the formal Taylor's expansion

$$f(x + h) = \sum_{\alpha \geqslant 0} \frac{1}{\alpha!} h^\alpha D^\alpha f(x)$$

16. The binomial coefficients are often displayed in "Pascal's triangle":

$$1$$
$$1 \quad 1$$
$$1 \quad 2 \quad 1$$
$$1 \quad 3 \quad 3 \quad 1$$
$$1 \quad 4 \quad 6 \quad 4 \quad 1$$

Each entry in Pascal's triangle (other than the 1's) is the sum of the two elements appearing above it to the right and left. Prove that this statement is correct, i.e., that

$$\binom{n + 1}{k} = \binom{n}{k} + \binom{n}{k - 1}$$

17. Prove that $\sum_{k=0}^{n} \binom{n}{k} = 2^n$. Is a similar result true for the sum $\sum_{|\beta| \leqslant |\alpha|} \binom{\alpha}{\beta}$?
18. Prove that the number of multi-indices β that satisfy the inequality $0 \leqslant \beta \leqslant \alpha$ is $(1 + \alpha_1) \cdots (1 + \alpha_n)$.
19. Find a single general proof that can establish the univariate and multivariate cases of the Binomial Theorem and the Leibniz Formula. This proof would exploit the similarity between the formulas

$$(x + y)x^i y^j = x^{i+1} y^j + x^i y^{j+1} \text{ and } D(D^i f D^j g) = D^{i+1} f D^j g + D^i f D^{j+1} g$$

20. Prove Lemma 1.

5.5 Convolutions

The **convolution** of two functions f and ϕ on $\mathbb{R}^n$ is a function $f * \phi$ whose defining equation is

$$(1) \qquad (f * \phi)(x) = \int_{\mathbb{R}^n} f(y)\phi(x - y)\, dy \qquad (x \in \mathbb{R}^n)$$

The integral will certainly exist if $\phi \in \mathcal{D}$ and if $f \in L^1_{\text{loc}}(\mathbb{R}^n)$, because for each x, the integration takes place over a compact subset of $\mathbb{R}^n$. With a change of variable in the integral, $y = x - z$, one proves that

$$(f * \phi)(x) = \int_{\mathbb{R}^n} f(x - z)\phi(z)\, dz = (\phi * f)(x)$$

In taking the convolution of two functions, one can expect that some favorable properties of one factor will be inherited by the convolution function. This vague concept will be illustrated now in several ways. Suppose that f is merely integrable, while ϕ is a test function. In Equation (1), suppose that $n = 1$, and that we wish to differentiate $f * \phi$ (with respect to x, of course). On the right side of the equation, x appears only in the function ϕ, and consequently

$$(f * \phi)'(x) = \int_{-\infty}^{\infty} f(y)\phi'(x - y)\, dy$$

The differentiability of the factor ϕ is inherited by the convolution product $f * \phi$. This phenomenon persists with higher derivatives and with many variables.

It follows from what has already been said that if ϕ is a polynomial of degree at most k, then so is $f * \phi$. This is because the $(k + 1)$st derivative of $f * \phi$ will be zero. Similarly, if ϕ is a periodic function, then so is $f * \phi$.

We shall see that convolutions are useful in approximating functions by smooth functions. Here the "mollifiers" of Section 5.1 play a role. Let ϕ be a mollifier; that is, $\phi \in \mathcal{D}$, $\phi \geq 0$, $\int \phi = 1$, and $\phi(x) = 0$ when $|x| \geq 1$. Define $\phi_j(x) = j^n \phi(jx)$. It is easy to verify that $\int \phi_j = 1$. (In this discussion j ranges over the positive integers.) Then

$$f(x) - (f * \phi_j)(x) = f(x) - \int f(x - z)\phi_j(z)\, dz$$

$$= \int f(x)\phi_j(z)\, dz - \int f(x - z)\phi_j(z)\, dz$$

$$= \int [f(x) - f(x - z)]\phi_j(z)\, dz$$

Since $\phi(x)$ vanishes outside the unit ball in $\mathbb{R}^n$, $\phi_j(x)$ vanishes outside the ball of radius $1/j$, as is easily verified. Hence in the equation above the only values of z that have any effect are those for which $|z| < 1/j$. If f is uniformly continuous, the calculation shows that $f * \phi_j(x)$ is close to $f(x)$, and we have therefore approximated f by the smooth function $f * \phi$. Variations on this idea will appear from time to time.

Using special linear operators B and E_x defined by

(2) $(E_x\phi)(y) = \phi(y - x)$

(3) $(B\phi)(y) = \phi(-y)$

we can write Equation (1) in the form

(4) $(f * \phi)(x) = \widetilde{f}(E_x B\phi)$

For $f \in L^1_{\text{loc}}(\mathbb{R}^n)$ and $\phi \in \mathcal{D}$ we have

(5)
$$\widetilde{E_x f}(\phi) = \int E_x f \cdot \phi = \int f(y - x)\phi(y)\, dy = \int f(z)\phi(z + x)\, dz$$
$$= \widetilde{f}(E_{-x}\phi)$$

Equations 4 and 5 suggest the following definition.

Definition. If T is a distribution, $E_x T$ is defined to be $T E_{-x}$. If $\phi \in \mathcal{D}$, then the convolution $T * \phi$ is defined by $(T * \phi)(x) = T(E_x B\phi)$.

Lemma 1.   For $T \in \mathcal{D}'$ and $\phi \in \mathcal{D}$,

(6) $E_x(T * \phi) = (E_x T) * \phi = T * E_x \phi$

Proof. Straightforward calculation, using some results in Problem 1, gives us

$$\big[E_x(T * \phi)\big](y) = (T * \phi)(y - x) = T(E_{y-x}B\phi)$$
$$\big[(E_x T) * \phi\big](y) = (E_x T)(E_y B\phi) = T(E_{-x}E_y B\phi) = T(E_{y-x}B\phi)$$
$$\big[T * E_x \phi\big](y) = T(E_y B E_x \phi) = T(E_y E_{-x}B\phi) = T(E_{y-x}B\phi) \qquad \blacksquare$$

Lemma 2.   If T is a distribution and if $\phi_j \twoheadrightarrow \phi$ in $\mathcal{D}$, then $T * \phi_i \to T * \phi$ pointwise.

Proof. By linearity (see Problem 3), it suffices to consider the case when $\phi = 0$. If $\phi_j \twoheadrightarrow 0$ in $\mathcal{D}$, then for all x,

$$(T * \phi_j)(x) = T(E_x B\phi_j) \to 0$$

by the continuity of B, E_x, and T (Problem 8). $\blacksquare$

Lemma 3.   Let $[x_j]$ be a sequence of points in $\mathbb{R}^n$ converging to x. For each $\phi \in \mathcal{D}$,

(7) $E_{x_j}\phi \twoheadrightarrow E_x \phi$ (convergence in $\mathcal{D}$)

Proof. If $K_1 = \{x, x_1, x_2, \dots \}$ and if K_2 is the support of ϕ, then (as is easily verified) the supports of $E_{x_j}\phi$ are contained in the compact set

$$K_1 + K_2 = \{u + v : u \in K_1, \ v \in K_2\}$$

Now we observe that

(8) $(E_{x_j}\phi)(y) \to (E_x\phi)(y)$ uniformly for $y \in K_1 + K_2$

Indeed, for a given $\varepsilon > 0$ there is a $\delta > 0$ such that

$$|u - v| < \delta \Longrightarrow |\phi(u) - \phi(v)| < \varepsilon \qquad (u, v \in K_1 + K_2 - K_1)$$

(This is uniform continuity of the continuous function ϕ on a compact set.) Hence if $|x_j - x| < \delta$, then $|\phi(y - x_j) - \phi(y - x)| < \varepsilon$. It now follows that

$$(D^\alpha E_{x_j}\phi)(y) \to (D^\alpha E_x\phi)(y) \quad \text{uniformly for} \quad y \in K_1 + K_2$$

because $D^\alpha E_{x_j}\phi = E_{x_j} D^\alpha \phi$, and (8) can be applied to $D^\alpha \phi$.   ∎

Lemma 4.   Let $e = (1, 0, \ldots, 0)$, $0 < |t| < 1$, and $F_t = t^{-1}(E_0 - E_{te})$. Then for each test function ϕ, $F_t\phi \to \dfrac{\partial\phi}{\partial x_1}$ as $t \to 0$. (This convergence is in the topology of $\mathcal{D}$.)

Proof.   Since $|t| < 1$, there is a single compact set K containing the supports of $F_t\phi$ and $\dfrac{\partial\phi}{\partial x_1}$. By the mean value theorem (used twice) we have (for $0 < \theta, \theta' < 1$)

$$\left|\frac{\partial\phi}{\partial x_1}(x) - (F_t\phi)(x)\right| = \left|\frac{\partial\phi}{\partial x_1}(x) - t^{-1}[\phi(x) - \phi(x - te)]\right|$$

$$= \left|\frac{\partial\phi}{\partial x_1}(x) - \frac{\partial\phi}{\partial x_1}(x - \theta te)\right|$$

$$= \left|\frac{\partial^2\phi}{\partial x_1^2}(x - \theta' te)\right| |\theta t|$$

$$\leqslant \left\|\frac{\partial^2\phi}{\partial x_1^2}\right\|_K |t|$$

The norm used here is the supremum norm on K. Our inequality shows that as $t \to 0$, $(F_t\phi)(x) \to \left(\dfrac{\partial\phi}{\partial x_1}\right)(x)$ uniformly in x on K. Since ϕ can be any test function, we can apply our conclusion to $D^\alpha\phi$, inferring that $F_t D^\alpha\phi$ converges uniformly to $\dfrac{\partial}{\partial x_1}D^\alpha\phi$ on K. Since D^α commutes with F_t (Problem 9) and with other derivatives, we conclude that $D^\alpha F_t\phi$ converges uniformly on K to $D^\alpha\dfrac{\partial\phi}{\partial x_1}$. This proves that the convergence of $F_t\phi$ is in accordance with the notion of convergence adopted in $\mathcal{D}$.   ∎

Theorem.   If T is a distribution and if ϕ is a test function, then for each multi-index α,

$$(9) \qquad\qquad D^\alpha(T * \phi) = (\partial^\alpha T) * \phi = T * D^\alpha\phi$$

Proof.   From Equations (3) and (2) we infer that

$$D^\alpha B = (-1)^{|\alpha|} B D^\alpha$$

$$D^\alpha E_x = E_x D^\alpha$$

Hence

$$(\partial^\alpha T * \phi)(x) = (\partial^\alpha T)(E_x B\phi) = (-1)^{|\alpha|} \, T \, D^\alpha E_x B\phi$$
$$= (-1)^{|\alpha|} T \, E_x \, D^\alpha B\phi = T \, E_x \, B \, D^\alpha \phi = T * D^\alpha \phi$$

This proves one part of Equation (9). For the other part, consider the special case $\alpha = (0, \ldots, 0, 1, 0, \ldots, 0)$. Thus $D = \dfrac{\partial}{\partial x_i}$. Let $e = (0, \ldots, 0, 1, 0, \ldots, 0) \in \mathbb{R}^n$, and $F_t = t^{-1}(E_0 - E_{te})$. Since

$$\frac{\partial}{\partial x_i}\phi(y) = \lim_{t \to 0} \frac{\phi(y) - \phi(y - te)}{t} = \lim_{t \to 0}\left[t^{-1}(E_0 - E_{te})\phi \right](y)$$

$$= \lim_{t \to 0}(F_t\phi)(y)$$

we have $D\phi = \lim\limits_{t \to 0} F_t\phi$, by Lemma 4. Using Lemma 1, we have

$$F_t(T * \phi) = T * F_t\phi$$

By Lemma 2, we can let $t \to 0$ to obtain

$$D(T * \phi) = T * D\phi$$

By iteration of this basic result we obtain, for any multi-index α,

$$D^\alpha(T * \phi) = T * D^\alpha \phi \qquad\qquad\blacksquare$$

Corollary.   If $T \in \mathcal{D}'$ and $\phi \in \mathcal{D}$, then $T * \phi \in C^\infty(\mathbb{R}^n)$.

Proof.   We have to prove that $D^\alpha(T * \phi) \in C(\mathbb{R}^n)$ for all multi-indices α. Put $\psi = D^\alpha\phi$. Then by the theorem,

$$D^\alpha(T * \phi) = T * D^\alpha\phi = T * \psi$$

To see that $T * \psi$ is continuous, let $[x_j]$ be a sequence in $\mathbb{R}^n$ tending to x. By Lemma 3,
$$(T * \psi)(x_j) = T \, E_{x_j} B\psi \to T \, E_x B\psi = (T * \psi)(x) \qquad\qquad\blacksquare$$

Problems 5.5

1. Prove that
    (a) $E_x E_y = E_{x+y}$
    (b) $B E_x = E_{-x} B$
    (c) $\phi * \psi = \psi * \phi$
    (d) $\delta_y E_x B = \delta_x E_y$
2. Prove that $E_x : \mathcal{D}' \to \mathcal{D}'$ is linear, continuous, injective, and surjective.
3. Prove that for fixed $T \in \mathcal{D}'$ the map $\phi \mapsto T * \phi$ is linear from $\mathcal{D}$ to $C^\infty(\mathbb{R}^n)$.
4. Prove that if $T \in \mathcal{D}'$ and $\phi \in \mathcal{D}$, then

$$T(\phi) = \delta(T * B\phi) \qquad (\delta = \text{Dirac distribution})$$

5. Fixing a distribution T, define the convolution operator C_T by $C_T\phi = T * \phi$. Show that $\delta_x C_T = T E_x B$.
6. Prove that the vector sum of two compact sets in $\mathbb{R}^n$ is compact. Show by example that the vector sum of two closed sets need not be closed. Show that the vector sum of a compact set and a closed set is closed.
7. For 2π-periodic functions, define $(f * g)(x) = \int_0^{2\pi} f(y)g(x - y)\, dy$. Compute the convolution of $f(x) = \sin x$ and $g(x) = \cos x$.
8. Prove that B and E_x are continuous linear maps of $\mathcal{D}$ into $\mathcal{D}$. Are they injective? Are they surjective?
9. Prove that $D^\alpha E_x = E_x D^\alpha$.
10. What is $\delta * \phi$?
11. Which of these equations is (or are) valid?
    (a) $B(E_x(\phi(y))) = \phi(x - y)$
    (b) $(B(E_x\phi))(y) = \phi(x - y)$
    (c) $E_x(B(\phi(y))) = \phi(-x - y)$
    (d) $E_x(B(\phi(y))) = \phi(x - y)$

5.6 Differential Operators

Definition.   A **linear differential operator** with constant coefficients is any finite sum of terms $c_\alpha D^\alpha$. Such an operator has the representation

$$A = \sum_{|\alpha| \leqslant m} c_\alpha D^\alpha$$

The constants c_α may be complex numbers. Clearly, A can be applied to any function in $C^m(\mathbb{R}^n)$.

Definition.   A distribution T is called a **fundamental solution** of the operator $\sum c_\alpha D^\alpha$ if $\sum c_\alpha \partial^\alpha T$ is the Dirac distribution.

Example 1.   What are the fundamental solutions of the operator D in the case $n = 1$? $(D = \dfrac{d}{dx})$. We seek all the distributions T that satisfy $\partial T = \delta$. We saw in Example 1 of Section 5.2 (page 254) that $\partial \widetilde{H} = \delta$, where H is the Heaviside function. Thus $\widetilde{H}$ is *one* of the fundamental solutions. Since the distributions sought are exactly those for which $\partial T = \partial \widetilde{H}$, we see by Theorem 3 in Section 5.2 (page 256) that $T = \widetilde{H} + \widetilde{c}$ for some constant c.

> **Theorem 1.   The Malgrange–Ehrenpreis Theorem.**   *Every operator $\sum c_\alpha D^\alpha$ has a fundamental solution.*

For the proof of this basic theorem, consult [Ho] page 189, or [Ru1] page 195. The next theorem reveals the importance of fundamental solutions in the study of partial differential equations.

Theorem 2.   Let A be a linear differential operator with constant coefficients, and let T be a distribution that is a fundamental solution of A. Then for each test function ϕ, $A(T * \phi) = \phi$.

Proof.   Let $A = \sum c_\alpha D^\alpha$. Then $\sum c_\alpha \partial^\alpha T = \delta$. The basic formula (the theorem of Section 5, page 271) states that

$$D^\alpha(T * \phi) = \partial^\alpha T * \phi$$

From this we conclude that

$$A(T * \phi) = \sum c_\alpha D^\alpha(T * \phi) = \left(\sum c_\alpha \partial^\alpha T \right) * \phi = \delta * \phi = \phi$$

In the last step we use the calculation

$$(\delta * \phi)(x) = \delta(E_x B \phi) = (E_x B \phi)(0) = (B \phi)(0 - x) = \phi(x) \qquad \blacksquare$$

Example 2.   We use the theory of distributions to find a solution of the differential equation $\dfrac{du}{dx} = \phi$, where ϕ is a test function. By Example 1, one fundamental solution is the distribution $\widetilde{H}$. By the preceding theorem, $\widetilde{H} * \phi$ will solve the differential equation. We have, with a simple change of variable,

$$u(x) = (\widetilde{H} * \phi)(x) = \int_{-\infty}^{\infty} H(y)\phi(x - y)\,dy = \int_{-\infty}^{x} \phi(z)\,dz \qquad \blacksquare$$

Example 3.   Let us search for a solution of the differential equation

$$u' + au = \phi$$

using distribution theory. First, we try to discover a fundamental solution, i.e., a distribution T such that $\partial T + aT = \delta$. If T is such a distribution and if $v(x) = e^{ax}$, then

$$\partial(v \cdot T) = Dv \cdot T + v \cdot \partial T = av \cdot T + v \cdot (\delta - aT) = v \cdot \delta = \delta$$

Consequently, by Example 1,

$$v \cdot T = \widetilde{H} + \widetilde{c} \quad \text{and} \quad T = \frac{1}{v}(\widetilde{H} + \widetilde{c})$$

Thus T is a regular distribution $\widetilde{f}$, and since c is arbitrary, we use $c = 0$, arriving at

$$f(x) = e^{-ax} H(x)$$

A solution to the differential equation is then given by

$$u(x) = (f * \phi)(x) = \int_{-\infty}^{\infty} e^{-ay} H(y)\phi(x - y)\,dy$$

$$= \int_{0}^{\infty} e^{-ay}\phi(x - y)\,dy$$

This formula produces a solution if ϕ is bounded and of class C^1. ∎

The Laplacian.   In the following paragraphs, a fundamental solution to the Laplacian operator will be derived. This operator, denoted by Δ or by ∇^2, is given by

$$\Delta = \frac{\partial^2}{\partial x_1^2} + \cdots + \frac{\partial^2}{\partial x_n^2}$$

At first, some elementary calculations need to be recorded. The notation is $x = (x_1, \ldots, x_n)$ and $|x| = (x_1^2 + \cdots + x_n^2)^{1/2}$.

Lemma 1.   For $x \neq 0$, $\dfrac{\partial}{\partial x_j}|x| = x_j|x|^{-1}$.

Proof.

$$\frac{\partial}{\partial x_j}|x| = \frac{\partial}{\partial x_j}(x_1^2 + \cdots + x_n^2)^{1/2} = \tfrac{1}{2}(x_1^2 + \cdots + x_n^2)^{-1/2}(2x_j) = x_j|x|^{-1} \quad ∎$$

Lemma 2.   For $x \neq 0$, $\dfrac{\partial^2}{\partial x_j^2}|x| = |x|^{-1} - x_j^2|x|^{-3}$.

Proof.

$$\frac{\partial^2}{\partial x_j^2}|x| = \frac{\partial}{\partial x_j}\big[x_j|x|^{-1}\big] = |x|^{-1} + x_j(-1)|x|^{-2}x_j|x|^{-1}$$

$$= |x|^{-1} - x_j^2|x|^{-3} \quad ∎$$

Lemma 3.   For $x \neq 0$, and $g \in C^2(0, \infty)$,

$$\Delta g(|x|) = g''(|x|) + (n-1)|x|^{-1}g'(x)$$

Proof.

$$\frac{\partial}{\partial x_j}g(|x|) = g'(|x|)\frac{\partial}{\partial x_j}|x| = g'(|x|)x_j|x|^{-1}$$

$$\frac{\partial^2}{\partial x_j^2}g(|x|) = g''(|x|)x_j^2|x|^{-2} + g'(|x|)\frac{\partial^2}{\partial x_j^2}|x|$$

$$= g''(|x|)x_j^2|x|^{-2} + g'(|x|)(|x|^{-1} - x_j^2|x|^{-3})$$

$$\Delta g(|x|) = \sum_{j=1}^{n}\frac{\partial^2}{\partial x_j^2}g(|x|)$$

$$= g''(|x|)|x|^2|x|^{-2} + g'(|x|)(n|x|^{-1} - |x|^2|x|^{-3})$$

$$= g''(|x|) + (n-1)|x|g'(x) \quad ∎$$

For reasons that become clear later, we require a function g (not a constant) such that $\Delta g(|x|) = 0$ throughout $\mathbb{R}^n$, with the exception of the singular point $x = 0$. By Lemma 3, we see that g must satisfy the following differential equation, in which the notation $r = |x|$ has been introduced:

$$g''(r) + \frac{n-1}{r} g'(r) = 0$$

This can be written in the form

$$g''(r)/g'(r) = (1-n)r^{-1}$$

and this can be interpreted as

$$\frac{d}{dr} \log g'(r) = (1-n)r^{-1}$$

From this we infer that

$$\log g'(r) = (1-n)\log r + \log c$$
$$g'(r) = cr^{1-n}$$

If $n \geqslant 3$, this last equation gives $g(r) = r^{2-n}$ as the desired solution. Thus we have proved the following result:

Theorem 3.   *If $n \geqslant 3$ then $\Delta|x|^{2-n} = 0$ at all points of $\mathbb{R}^n$ except $x = 0$.*

Of course, this theorem can be proved by a direct verification that $|x|^{2-n}$ satisfies Laplace's equation, except at 0. The fact that Laplace's equation is *not* satisfied at 0 is of special importance in what follows. Let $f(x) = |x|^{2-n}$. As usual, $\tilde{f}$ will denote the corresponding distribution.

In accordance with the definition of derivative of a distribution, we have

$$\Delta\tilde{f} = \sum_{i=1}^n \frac{\partial^2}{\partial x_i^2}\tilde{f} = \sum_{i=1}^n (-1)^2 \tilde{f} \circ \frac{\partial^2}{\partial x_i^2} = \tilde{f} \circ \Delta$$

For any test function ϕ,

(1) $$(\Delta\tilde{f})(\phi) = \tilde{f}(\Delta\phi) = \int_{\mathbb{R}^n} |x|^{2-n}(\Delta\phi)(x)\,dx$$

The integral on the right is improper because of the singular point at 0. It is therefore defined to be

(2) $$\lim_{\varepsilon\downarrow 0} \int_{|x|\geqslant\varepsilon} |x|^{2-n}(\Delta\phi)(x)\,dx$$

For sufficiently small ε, the support of ϕ will be contained in $\{x : |x| \leqslant \varepsilon^{-1}\}$. The integral in (2) can be over the set

$$A_\varepsilon = \{x : \varepsilon \leqslant |x| \leqslant \varepsilon^{-1}\}$$

An appeal will be made to Green's Second Identity, which states that for regions Ω satisfying certain mild hypotheses,

$$\int_{\Omega} (u\Delta v - v\Delta u) = \int_{\partial\Omega} (u\nabla v - v\nabla u) \cdot N$$

The three–dimensional version of this can be found in [Hur] page 489, [MT] page 449, [Tay1] page 459, or [Las] page 118. The n-dimensional form can be found in [Fla] page 83. In the formula, N denotes the unit normal vector to the surface $\partial\Omega$. Applying Green's formula to the integral in Equation (2), we notice that $\Delta|x|^{2-n} = 0$ in A_ε. Hence the integral is

(3)
$$\int_{A_\varepsilon} |x|^{2-n}\Delta\phi = \int_{\partial A_\varepsilon} (|x|^{2-n}\nabla\phi - \phi\nabla|x|^{2-n}) \cdot N$$

The boundary of A_ε is the union of two spheres whose radii are ε and ε^{-1}. On the outer boundary, $\phi = \nabla\phi = 0$ because the support of ϕ is interior to A_ε. The following computation will also be needed:

$$\nabla|x|^{2-n} \cdot N = \sum_{j=1}^{n}\left(\frac{\partial}{\partial x_j}|x|^{2-n}\right)\frac{x_j}{|x|} = \sum_{j=1}^{n}(2-n)|x|^{1-n}\left(\frac{x_j}{|x|}\right)^2 = (2-n)|x|^{1-n}$$

The first term on the right in Equation (3) is estimated as follows

$$\left|\int_{|x|=\varepsilon} |x|^{2-n}\nabla\phi \cdot N\right| dS \leqslant \varepsilon^{2-n} \max_{|x|=\varepsilon}|\nabla\phi(x)| \int_{|x|=\varepsilon} dS$$

$$= c\varepsilon^{2-n}\sigma_n\varepsilon^{n-1} = O(\varepsilon)$$

Hence when $\varepsilon \to 0$, this term approaches 0. The symbol σ_n represents the "area" of the unit sphere in $\mathbb{R}^n$. As for the other term,

$$\left|\int_{|x|=\varepsilon} [\phi(x) - \phi(0)]\nabla|x|^{2-n} \cdot N\right| dS \leqslant (n-2)\int_{|x|=\varepsilon} |x|^{1-n}|\phi(x) - \phi(0)|\, dS$$

$$\leqslant (n-2)\varepsilon^{1-n}\max_{|x|=\varepsilon}|\phi(x) - \phi(0)| \int_{|x|=\varepsilon} dS$$

$$= (n-2)\varepsilon^{1-n}\omega(\varepsilon)\sigma_n\varepsilon^{n-1} \to 0$$

In this calculation, $\omega(\varepsilon)$ is the maximum of $|\phi(x) - \phi(0)|$ on the sphere defined by $|x| = \varepsilon$. Obviously, $\omega(\varepsilon) \to 0$, because ϕ is continuous. Thus the integral in Equation (3) is

$$(2 - n)\sigma_n\phi(0) \equiv (2 - n)\sigma_n\delta(\phi)$$

Hence this is the value of the integral in Equation (1). We have established, therefore, that $\Delta\widehat{f} = (2 - n)\sigma_n\delta$. Summarizing, we have the following result.

Theorem 4.   *A fundamental solution of the Laplacian operator in dimension $n \geqslant 3$ is the distribution corresponding to $\dfrac{|x|^{2-n}}{(2-n)\sigma_n}$, where σ_n denotes the area of the unit sphere in $\mathbb{R}^n$.*

Example 4.   Find a fundamental solution of the operator A defined (for $n = 1$) by the equation
$$A\phi = \phi'' + 2a\phi' + b\phi \qquad (\phi \in \mathcal{D})$$

We seek a distribution T such that $AT = \delta$. Let us look for a regular distribution, $T = \tilde{f}$. Using the definition of derivatives of distributions, we have

$$(A\tilde{f})(\phi) = \tilde{f}(\phi'' - 2a\phi' + b\phi)$$
$$= \int_{-\infty}^{\infty} f(x)[\phi''(x) - 2a\phi'(x) + b\phi(x)]\, dx$$

Guided by previous examples, we guess that f should have as its support the interval $[0, \infty)$. The integral above then is restricted to the same interval. Using integration by parts, we obtain

$$f\phi'\Big|_0^\infty - \int_0^\infty f'\phi' - 2af\phi\Big|_0^\infty + 2a\int_0^\infty f'\phi + b\int_0^\infty f\phi$$

$$= -f(0)\phi'(0) - f'\phi\Big|_0^\infty + \int_0^\infty f''\phi + 2af(0)\phi(0) + \int_0^\infty (2af' + bf)\phi$$

$$= -f(0)\phi'(0) + f'(0)\phi(0) + 2af(0)\phi(0) + \int_0^\infty (f'' + 2af' + bf)\phi$$

The easiest way to make this last expression simplify to $\phi(0)$ is to define f on $[0, \infty)$ in such a way that

$$\text{(i)} \qquad f'' + 2af' + bf = 0$$
$$\text{(ii)} \qquad f(0) = 0$$
$$\text{(iii)} \qquad f'(0) = 1$$

This is an initial-value problem, which can be solved by writing down the general solution of the equation in (i) and adjusting the coefficients in it to achieve (ii) and (iii). The characteristic equation of the differential equation in (i) is

$$\lambda^2 + 2a\lambda + b = 0$$

Its roots are $-a \pm \sqrt{a^2 - b}$. Let $d = \sqrt{a^2 - b}$. If $d \neq 0$, then the general solution of (i) is
$$c_1 e^{-ax} e^{dx} + c_2 e^{-ax} e^{-dx}$$

Upon imposing the conditions (ii) and (iii) we find that

$$f(x) = \begin{cases} d^{-1} e^{-ax} \sinh\, dx & x \geqslant 0 \\ 0 & x < 0 \end{cases}$$

The case when $d = 0$ is left as Problem 14.   ∎

A linear differential operator with nonconstant coefficients is typically of the form

(4) $$A = \sum_\alpha c_\alpha D^\alpha$$

In order for this to interact properly with distributions, it is necessary to assume that $c_\alpha \in C^\infty(\mathbb{R}^n)$. Then AT is defined, when T is a distribution, by

(5) $$AT = \sum c_\alpha(D^\alpha T) = \sum(-1)^{|\alpha|} c_\alpha(T \circ D^\alpha)$$

Remember that $T \circ D^\alpha$ is a distribution; multiplication of this distribution by the C^∞-function c_α is well-defined (as in Section 5.5). The result of applying this to a test function ϕ is therefore

(6) $$(AT)(\phi) = \sum(-1)^{|\alpha|}(T \circ D^\alpha)(c_\alpha\phi)$$

Notice that the parentheses in Equation (5) are necessary because $c_\alpha T \circ D$ is ambiguous; it could mean $(c_\alpha T) \circ D$.

It is useful to define the **formal adjoint** of the operator A in Equation (4). It is

(7) $$A^*\phi = \sum_\alpha (-1)^{|\alpha|} D^\alpha(c_\alpha\phi) \qquad \phi \in \mathcal{D}$$

Notice that this definition is in harmony with the definition of adjoint for operators on Hilbert space, for we have

$$(AT)(\phi) = T(A^*\phi) \qquad (T \in \mathcal{D}', \ \phi \in \mathcal{D})$$

and this can be written in the notation of linear functionals as

$$\langle AT, \phi \rangle = \langle T, A^*\phi \rangle \qquad (T \in \mathcal{D}', \ \phi \in \mathcal{D})$$

Using Example 4 as a model, we can now prove a theorem about fundamental solutions of *ordinary* differential operators (i.e., $n = 1$).

Theorem 5.   *Consider the operator*

$$A = \sum_{j=0}^{m} c_j(x)\, \frac{d^j}{dx^j}$$

in which $c_j \in C^\infty(\mathbb{R})$ and $c_m(x) \neq 0$ for all x. This operator has a fundamental solution that is a regular distribution.

Proof. We find a function f defined on $[0, \infty)$ such that

(i) $$\sum_{j=0}^{m} c_j(x)f^{(j)}(x) = 0$$

(ii) $$c_{m-1}(0)f^{(m-1)}(0) = 1$$

(iii) $$c_j(0)f^{(j)}(0) = 0 \qquad (0 \leqslant j \leqslant m - 2)$$

Such a function exists by the theory of ordinary differential equations. In particular, an initial-value problem has a unique solution that is defined on any interval $[0, b]$, provided that the coefficient functions are continuous there and the leading coefficient does not have a zero in $[0, b]$. We also extend f to all of $\mathbb{R}$ by setting $f(x) = 0$ on the interval $(-\infty, 0)$. With the function f in hand, we must verify that $A\tilde{f} = \delta$. This is done as in Example 4.   ∎

Problems 5.6

1. If the coefficients c_α are constants, then the **formal adjoint** of the operator $\sum c_\alpha D^\alpha$ is $\sum (-1)^{|\alpha|} c_\alpha D^\alpha$. If the former is denoted by A, then the latter is denoted by A^*. Prove that for any distribution T, $AT = T \circ A^*$.

2. (Continuation)   Prove that the Laplacian

$$\Delta = \sum_{i=1}^{n} \left(\frac{\partial}{\partial x_i} \right)^2$$

   is self-adjoint; i.e., $\Delta^* = \Delta$.

3. Solve the equation $Y'' + 2Y' + Y = \delta + \delta'$ in the distribution sense, using a function of the form $Y(x) = H(x) f(x)$.

4. If P is a polynomial in n variables, say $P = \sum c_\alpha x^\alpha$, and if D is the n-tuple $\left(\dfrac{\partial}{\partial x_1}, \dfrac{\partial}{\partial x_2}, \dots, \dfrac{\partial}{\partial x_n} \right)$, then what should we mean by $P(D)$?

5. Fix $y \in \mathbb{R}^n$ and let $f(x) = e^{\langle x, y \rangle}$ for $x \in \mathbb{R}^n$. Prove that $P(D)f = P(y)f$. Express this result in the language of eigenvalues and eigenvectors.

6. If the functions v_α belong to $C^\infty(\mathbb{R}^n)$, then a differential operator $\sum v_\alpha D^\alpha$ has a counterpart $\sum v_\alpha \partial^\alpha$ that acts on *distributions*. Consider the operator

$$(1 + x^2)^{-1} \frac{\partial}{\partial x} + e^y \frac{\partial}{\partial y}$$

   and compute its effect on the Dirac distribution on $\mathbb{R}^2$.

7. Let $n = 2$ and find a fundamental solution of the operator $\dfrac{\partial^2}{\partial x_1 \partial x_2}$. (Try an analogue of the Heaviside distribution.)

8. What is the null space of the operator in Problem 7, interpreting it as a map on $C^\infty(\mathbb{R}^2)$?

9. What is the null space of the operator in Problem 7 if we interpret it as a map on $\mathcal{D}'$?

10. Let $n = 1$, and find a fundamental solution of the operator $\dfrac{d^2}{dx^2}$. Use it to give a solution to $u'' = \phi$ in the form of an integral.

11. Let $n = 1$ and $f(x) = e^{ik|x|}$. Show that a multiple of $\tilde{f}$ is a fundamental solution of the operator $\dfrac{d^2}{dx^2} + k^2 I$. Give an integral that solves the equation $u'' + k^2 u = \phi$.

12. Let p be a function such that p and $\frac{1}{p}$ belong to $C^1(\mathbb{R})$. Define $f(x)$ to be $\int_0^x dt/p(t)$ if $x > 0$ and to be 0 if $x \leqslant 0$. Show that, in the distributional sense, $(pf')' = \delta$.

13. In Examples 2 and 3 find more general solutions by retaining the constant in $H + c$.

14. Complete Example 4 by obtaining the fundamental solution when $d = 0$.

5.7 Distributions with Compact Support

In this section we prove a theorem on partitions of unity in the space $\mathcal{D}$ of test functions. Then we define the support of a distribution, and study the distributions whose support is compact. In particular, the convolution $S * T$ of two distributions can be defined if one of S and T has compact support. Recall the more fundamental notion of the **support** of a function f. It means the closure of the set $\{x \: : \: f(x) \neq 0\}$.

Lemma 1.   There is a function $f \in C^{\infty}(\mathbb{R})$ such that $0 \leqslant f \leqslant 1$, $f(x) = 0$ on $(-\infty, 0]$, and $f(x) = 1$ on $[1, \infty)$.

Proof. Define

$$g(x) = \begin{cases} \exp[x^2/(x^2 - 1)] & |x| < 1 \\ 0 & |x| \geqslant 1 \end{cases}$$

and

$$f(x) = \begin{cases} g(x - 1) & x \leqslant 1 \\ 1 & \text{otherwise} \end{cases} \qquad \blacksquare$$

The graphs of f and g are shown in Figure 5.3.

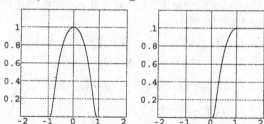

Figure 5.3

Lemma 2.   If $x_0 \in \mathbb{R}^n$ and $\rho > r > 0$, then there is a test function ϕ such that

(i) $0 \leqslant \phi \leqslant 1$
(ii) $\phi(x) = 1$ if $|x - x_0| \leqslant r$
(iii) $\phi(x) = 0$ if $|x - x_0| \geqslant \rho$.

Proof. Use the function f from the preceding lemma, and define

$$\phi(x) = 1 - f\big(a|x - x_0|^2 - b\big)$$

with $a = (\rho^2 - r^2)^{-1}$ and $b = r^2 a$. If $|x - x_0| \leqslant r$, then $a|x - x_0|^2 - b \leqslant ar^2 - b = 0$, so $\phi(x) = 1$. If $|x - x_0| \geqslant \rho$, then $a|x - x_0|^2 - b \geqslant a\rho^2 - b = a(\rho^2 - r^2) = 1$, so $\phi(x) = 0$. $\qquad \blacksquare$

Theorem 1. Partitions of Unity.   *Let $\mathcal{A}$ be a collection of open sets in $\mathbb{R}^n$, whose union is denoted by Ω. Then there is a sequence $[\phi_i]$ in $\mathcal{D}$ (called a "partition of unity subordinate to $\mathcal{A}$") such that:*

    a. $0 \leqslant \phi_i \leqslant 1$ *for* $i = 1, 2, \ldots$
    b. *For each i there is an $\mathcal{O}_i \in \mathcal{A}$ such that* $\operatorname{supp}(\phi_i) \subset \mathcal{O}_i$.
    c. *For each compact subset K of Ω there is an index m such that*

$$\phi_1(x) + \cdots + \phi_m(x) = 1$$

on a neighborhood of K.

Proof.   (Rudin)   Let $\left[\overline{B}(x_i, r_i)\right]$ denote the sequence of all closed balls in $\mathbb{R}^n$ having rational center x_i, rational radius r_i, and contained in a member of $\mathcal{A}$. By the preceding lemma, there exists for each i a test function ψ_i such that $0 \leqslant \psi_i \leqslant 1$, $\psi_i(x) = 1$ on $\overline{B}(x_i, r_i/2)$, and $\psi_i(x) = 0$ outside of $\overline{B}(x_i, r_i)$. Put $\phi_1 = \psi_1$ and

(1) $$\phi_i = (1 - \psi_1)(1 - \psi_2) \cdots (1 - \psi_{i-1})\psi_i \qquad (i \geqslant 2)$$

It is clear that on the complement of $B(x_i, r_i)$, we have $\psi_i(x) = 0$ and $\phi_i(x) = 0$. By induction we now prove that

(2) $$\phi_1 + \cdots + \phi_i = 1 - \left[(1 - \psi_1)(1 - \psi_2) \cdots (1 - \psi_i)\right]$$

Equation (2) is obviously correct for $i = 1$. If it is correct for the index $i - 1$, then it is correct for i because

$$\phi_1 + \cdots + \phi_i = 1 - \left[(1 - \psi_1) \cdots (1 - \psi_{i-1})\right] + \left[(1 - \psi_1) \cdots (1 - \psi_{i-1})\right]\psi_i$$
$$= 1 - \left[(1 - \psi_1) \cdots (1 - \psi_{i-1})(1 - \psi_i)\right]$$

Since $0 \leqslant \psi_i \leqslant 1$ for all i, we see from Equation (2) that

$$\sum_{i=1}^{\infty} \phi_i(x) \leqslant 1 \qquad\qquad (x \in \mathbb{R}^n)$$

On the other hand, if $x \in \bigcup_{i=1}^{m} B(x_i, r_i/2)$, then $\psi_i(x) = 1$ for some i in $\{1, \ldots, m\}$. Then $\phi_1(x) + \cdots + \phi_m(x) = 1$ from Equation (2). Since the open balls $B(x_i, r_i/2)$ cover Ω, each compact set K in Ω is contained in a finite union $\bigcup_{i=1}^{m} B(x_i, r_i/2)$. This establishes (c).   ∎

Fixing a distribution T, we consider a closed set F in $\mathbb{R}^n$ having this property:

(3) $\qquad T(\phi) = 0$ for all test functions ϕ satisfying $\operatorname{supp}(\phi) \subset \mathbb{R}^n \smallsetminus F$

Theorem 2.   *Let $\operatorname{supp}(T)$ denote the intersection of all closed sets having property (3). Then $\operatorname{supp}(T)$ is the smallest closed set having property (3).*

Proof.   Let $\mathcal{F}$ be the family of all closed sets F having property (3). Then

$$\operatorname{supp}(T) = \bigcap\{F : F \in \mathcal{F}\}$$

Being an intersection of closed sets, supp(T) is itself closed. The only question is whether it has property (3). To verify this, let ϕ be a test function such that supp(ϕ) $\subset \mathbb{R}^n \smallsetminus$ supp(T). It is to be shown that $T(\phi) = 0$. By De Morgan's Law,

$$\text{supp}(\phi) \subset \mathbb{R}^n \smallsetminus \bigcap \{F \ : \ F \in \mathcal{F}\} = \bigcup \{\mathbb{R}^n \smallsetminus F \ : \ F \in \mathcal{F}\}$$

By the preceding theorem, there is a partition of unity $[\psi_j]$ subordinate to the family of open sets $\{\mathbb{R}^n \smallsetminus F : F \in \mathcal{F}\}$. Since supp($\phi$) is compact, there exists (by Theorem 1) an index m such that

$$\sum_{i=1}^{m} \psi_i(x) = 1 \quad \text{on a neighborhood of supp}(\phi)$$

Notice that $\phi = \phi \sum_{i=1}^{m} \psi_i$, because if $\phi(x) = 0$, the equation is obviously true, while if $\phi(x) \neq 0$, then $x \in$ supp(ϕ) and $\sum_{i=1}^{m} \psi_i(x) = 1$. Hence, by the linearity of T,

$$T(\phi) = T \left(\sum_{i=1}^{m} \phi\psi_i \right) = \sum_{i=1}^{m} T(\phi\psi_i)$$

Again by Theorem 1, there exists for each i an $F_i \in \mathcal{F}$ such that

$$\text{supp}(\phi\psi_i) \subset \text{supp}(\psi_i) \subset \mathbb{R}^n \smallsetminus F_i$$

Since $F_i \in \mathcal{F}$, F_i has property (3), and we conclude that $T(\phi\psi_i) = 0$ for $1 \leqslant i \leqslant m$. By Equation (4), $T(\phi) = 0$. ∎

Notice that supp() has two different meanings: one for functions on $\mathbb{R}^n$ and another for distributions. This is the conventional practice. By Problem 10, the two definitions are compatible.

Example 1. The support of the Dirac distribution δ is the set $\{0\}$. If ϕ is a test function for which supp(ϕ) $\subset \mathbb{R}^n \smallsetminus \{0\}$ then clearly $\delta(\phi) = 0$. ∎

Example 2. The support of the Heaviside distribution $\widetilde{H}$ is the set $[0, \infty)$. ∎

Definition. The space $\mathcal{E}$ is defined to be the space $C^\infty(\mathbb{R}^n)$ with convergence defined as follows: $\phi_j \to 0$ if for each multi-index α, $D^\alpha\phi_j(x)$ converges uniformly to 0 on every compact set.

Theorem 3.   *Each distribution having a compact support has an extension to $\mathcal{E}$ that is continuous.*

Proof. Let T be a distribution for which supp(T) is compact. By the theorem on partitions of unity, there is a test function ψ such that $\psi(x) = 1$ on a neighborhood of supp(T). Define $\overline{T}$ on $\mathcal{E}$ by the equation $\overline{T}(\phi) = T(\phi\psi)$. This is meaningful because $\phi\psi \in \mathcal{D}$. Now we wish to establish that $\overline{T}$ is continuous on $\mathcal{E}$. To this end, let $\phi_j \in \mathcal{E}$ and suppose that $\phi_j \to 0$, the convergence being as prescribed in $\mathcal{E}$. All the functions $\phi_j\psi$ vanish outside of supp(ψ), and for each multi-index α, $D^\alpha(\phi_j\psi)$ converges uniformly to 0 by the Leibniz formula. Hence $\phi_j\psi \twoheadrightarrow 0$ in $\mathcal{D}$. By the continuity of T and the definition of $\overline{T}$,

$$\overline{T}(\phi_j) = T(\phi_j\psi) \to 0$$

Finally, we must prove that $\overline{T}$ is an extension of T. Let ϕ be any test function; we want to show that $\overline{T}(\phi) = T(\phi)$. Equivalent equations are $T(\phi\psi) = T(\phi)$ and $T(\phi\psi - \phi) = 0$. To establish the latter, it suffices to show that

$$\text{supp}(\phi\psi - \phi) \subset \mathbb{R}^n \smallsetminus \text{supp}(T)$$

(Here we have used Theorem 2.) Since $\phi\psi - \phi = \phi \cdot (\psi - 1)$, it is enough to prove that

$$\text{supp}(\psi - 1) \subset \mathbb{R}^n \smallsetminus \text{supp}(T)$$

To this end, let $x \in \text{supp}(\psi - 1)$. By definition of a support, we can write $x = \lim x_j$, where $(\psi - 1)(x_j) \neq 0$. Since $\psi(x_j) \neq 1$, we have $x_j \notin \mathcal{N}$, where $\mathcal{N}$ is an open neighborhood of $\text{supp}(T)$ on which ψ is identically 1. Since $x_j \in \mathbb{R}^n \smallsetminus \mathcal{N}$, we have $x \in \mathbb{R}^n \smallsetminus \mathcal{N}$ because the latter is closed. Hence $x \in \mathbb{R}^n \smallsetminus \text{supp}(T)$.   ∎

Theorem 4.   *Each continuous linear functional on $\mathcal{E}$ is an extension of some distribution having compact support.*

Proof. Let L be a continuous linear functional on $\mathcal{E}$. Let $T = L \mid \mathcal{D}$, which denotes the restriction of L to $\mathcal{D}$. It is easily seen that T is a distribution. In order to prove that the support of T is compact, suppose otherwise. Then for each k there is a test function ϕ_k whose support is contained in $\{x : |x| > k\}$ such that $T(\phi_k) = 1$. It follows that $\phi_k \to 0$ in $\mathcal{E}$, whereas $L(\phi_k) = 1$. In order to prove that $L = \overline{T}$, as in the preceding proof select $\gamma_j \in \mathcal{D}$ so that $\gamma_j(x) = 1$ if $|x| \leqslant j$ and $\gamma_j(x) = 0$ if $|x| \geqslant 2j$. If $\phi \in \mathcal{E}$, then $\gamma_j\phi \to \phi$ in $\mathcal{E}$. Hence

$$L(\phi) = \lim L(\gamma_j\phi) = \lim T(\gamma_j\phi) = \overline{T}(\phi)$$

because $\text{supp}(T) \subset \text{supp}(\gamma_j)$ for all sufficiently large j.   ∎

The preceding two theorems say in effect that the space $\mathcal{E}'$ consisting of all continuous linear functionals on $\mathcal{E}$ can be identified with a subset of $\mathcal{D}'$, viz. the set of all distributions having compact support.

Recall that the convolution of a test function ϕ with a function $f \in L^1_{\text{loc}}(\mathbb{R}^n)$ has been defined by

$$(4) \qquad (f * \phi)(x) = \int_{\mathbb{R}^n} f(y)\phi(x - y)\, dy$$

The convolution of a distribution T with a test function ϕ has been defined by

$$(5) \qquad\qquad (T * \phi)(x) = T(E_x B\phi)$$

where $(B\phi)(x) = \phi(-x)$ and $(E_x\psi)(y) = \psi(y - x)$.

Now observe that if T has compact support, then T (or more properly, its extension $\overline{T}$) can operate (as a linear functional) on any element of $\mathcal{E}$. Consequently, in this case, (5) is meaningful not only for $\phi \in \mathcal{D}$ but also for $\phi \in \mathcal{E}$. Equation (5) is adopted as the definition of the convolution of a distribution having compact support with a function in $C^\infty(\mathbb{R}^n)$.

Theorem 5.   *If T is a distribution with compact support and if $\phi \in \mathcal{E}$, then $T * \phi \in \mathcal{E}$.*

Proof.   See [Ru1], Theorem 6.35, page 159.   ∎

Now let S and T be two distributions, at least one of which has compact support. We define $S * T$ to be a distribution whose action is given by the following formula:

$$(6) \qquad\qquad (S * T)(\phi) = \delta\big(S * (T * B\phi)\big) \qquad (\phi \in \mathcal{D})$$

Here $\delta(\phi) = \phi(0)$ and $(B\phi)(x) = \phi(-x)$. We first verify that (6) is meaningful, i.e., that each argument is in the domain of the operator that is applied to it. Obviously, $B\phi \in \mathcal{D}$ and $T * B\phi \in \mathcal{E}$ by the corollary in Section 5.5, page 272. If S has compact support, then by the preceding lemma, $S * (T * B\phi) \in \mathcal{E}$. Hence δ can be applied. On the other hand, if T has compact support, then $T * B\phi$ is an element of $\mathcal{E}$ having compact support; in other words, an element of $\mathcal{D}$. Then $S * (T * B\phi)$ belongs to $\mathcal{E}$, and again δ can be applied.

It is a fact that we do not stop to prove that $S * T$ is a *continuous* linear functional on $\mathcal{D}$; thus it is a distribution. (See [Ru1], page 160].)

Finally, we indicate the source of the definition in Equation (6). If S and T are regular distributions, then they correspond to functions f and g in $L^1_{\text{loc}}(\mathbb{R}^n)$. In that case,

$$(S * T)(\phi) = (\widetilde{f * g})(\phi) = \int (f * g)(x)\phi(x)\,dx$$

$$= \iint f(y)g(x - y)\phi(x)\,dy\,dx$$

On the other hand,

$$\delta\Big(f * \big(g * (B\phi)\big)\Big) = \Big(f * \big(g * (B\phi)\big)\Big)(0) = \int f(y)(g * B\phi)(-y)\,dy$$

$$= \iint f(y)g(z)(B\phi)(-y - z)\,dz\,dy$$

$$= \iint f(y)g(z)\phi(y + z)\,dz\,dy$$

$$= \iint f(y)g(x - y)\phi(x)\,dx\,dy$$

Problems 5.7

1. Refer to the theorem concerning partitions of unity, page 282, and prove that for each x there is an index j such that $\phi_i(x) = 0$ for all $i > j$.
2. Let $[x_i]$ be a list of all the rational points in $\mathbb{R}^n$. Define T by $T(\phi) = \sum_{i=1}^{\infty} 2^{-i}\phi(x_i)$, where ϕ is any test function. Prove that T is a distribution and $\text{supp}(T) = \mathbb{R}^n$.
3. For a distribution T, let $\mathcal{F}_1$ be the family of all closed sets F such that $T(\phi) = 0$ when $\phi \mid F = 0$. Let $\mathcal{F}_2$ be the family of all closed sets F such that $T(\phi) = 0$ when $\text{supp}(\phi) \subset \mathbb{R}^n \smallsetminus F$. Show that $\mathcal{F}_1$ is generally a proper subset of $\mathcal{F}_2$.

4. Refer to the theorem on partitions of unity and prove that $\psi \sum_{i=1}^{j} \phi_i \to \psi$ as $j \to \infty$, provided that supp$(\psi) \subset \Omega$.
5. Prove that the extension of T as defined in the proof of Theorem 3 is independent of the particular ψ chosen in the proof.
6. If $\phi \in \mathcal{D}$, $T \in \mathcal{D}'$, and supp$(\phi) \cap$ supp$(T) = \varnothing$ (the empty set), then $T(\phi) = 0$.
7. If $T \in \mathcal{D}'$ and supp$(T) = \varnothing$, what conclusion can be drawn?
8. Show that if $\phi \in C^\infty(\mathbb{R}^n)$, if $T \in \mathcal{D}'$, and if $\phi(x) = 1$ on a neighborhood of supp(T), then $\phi T = T$.
9. Why, in proving Lemma 1, can we not take g to be a multiple of the function ρ introduced in Section 7.1?
10. Let $f \in C(\mathbb{R}^n)$. Show that supp$(\tilde{f}) =$ supp(f).
11. Let T be an arbitrary distribution, and let K be a compact set. Show that there exists a distribution S having compact support such that $S(\phi) = T(\phi)$ for all test functions ϕ that satisfy supp$(\phi) \subset K$.
12. Prove that a distribution can have at most one continuous extension on $\mathcal{E}$.
13. Prove that if a distribution does not have compact support, then it cannot have a continuous extension on $\mathcal{E}$.
14. Let T be a distribution and $\mathcal{N}$ a neighborhood of supp(T). Prove that for any test function ϕ, $T(\phi)$ depends only on $\phi \mid \mathcal{N}$.
15. Prove or disprove: If two test functions ϕ and ψ take the same values on the support of a distribution T, then $T(\phi) = T(\psi)$.
16. Refer to the theorem on partitions of unity and prove that the balls $B(x_i, r_i/2)$ cover the complement of the support of T.
17. Prove that if f and g are in $L^1_{loc}\mathbb{R}^n$, then for all test functions ϕ,

$$\widetilde{(f * g)}(\phi) = \tilde{f}(B(g * B\phi))$$

Chapter 6

The Fourier Transform

6.1   Definitions and Basic Properties   287
6.2   The Schwartz Space $\mathcal{S}$ 294
6.3   The Inversion Theorems   301
6.4   The Plancherel Theorem   305
6.5   Applications of the Fourier Transform   310
6.6   Applications to Partial Differential Equations   318
6.7   Tempered Distributions   321
6.8   Sobolev Spaces   325

6.1 Definitions and Basic Properties

The concept of an integral transform is undoubtedly familiar to the reader in its manifestation as the Laplace transform. This is a useful mechanism for handling certain differential equations. In general, integral transforms are helpful in problems where there is a function f to be determined from an equation that it satisfies. A judiciously chosen transform is then applied to that equation, the result being a simpler equation in the transformed function F. After this simpler equation has been solved for F, the inverse transform is applied to obtain f. We illustrate with the Laplace transform.

Example. Consider the initial value problem

$$(1) \qquad f'' - f' - 2f = 0 \qquad f(0) = \alpha \qquad f'(0) = \beta$$

The Laplace transform of f is the function F defined by

$$F(s) = \int_0^\infty f(t)e^{-st}\,dt$$

The theory of this transform enables us to write down the equation satisfied by F:

$$(2) \qquad (s^2 - s - 2)F(s) + (1 - s)\alpha - \beta = 0$$

Thus the Laplace transform has turned a *differential* equation (1) into an *algebraic* equation (2). The solution of (2) is

$$F(s) = (\beta + \alpha s - \alpha)/(s^2 - s - 2)$$

By taking the inverse Laplace transform, we obtain f :

$$f(t) = \tfrac{1}{3}(\alpha + \beta)e^{2t} + \tfrac{1}{3}(2\alpha - \beta)e^{-t} \qquad\blacksquare$$

The Fourier transform, now to be taken up, has applications of the type just outlined as well as a myriad of other uses in mathematics, especially in partial differential equations. The Fourier transform can be defined on any locally compact Abelian group, but we confine our attention to $\mathbb{R}^n$ (which *is* such a group). The material presented here is accessible in many authoritative sources, such as [Ru1], [Ru2], [Ru3], [SW], and [Fol].

The reader should be aware that in the literature there is very little uniformity in the definition of the Fourier transform. We have chosen to use here the definition of Stein and Weiss [SW]. It has a number of advantages, not the least of which is harmony with their monograph. It is the same as the definition used by Horváth [Horv], Dym and McKean [DM], and Lieb and Loss [LL]. Other favorable features of this definition are the simplicity of the inversion formula, elegance in the Plancherel Theorem, and its suitability in approximation theory.

We define a set of functions called **characters** e_y by the formula

$$e_y(x) = e^{2\pi i x y} \qquad y \in \mathbb{R}^n, \quad x \in \mathbb{R}^n$$

Here we have written

$$xy = \langle x, y \rangle = x \cdot y = x_1 y_1 + x_2 y_2 + \cdots + x_n y_n$$

where the x_i and y_i are components of the vectors x and y. Notice that each character maps $\mathbb{R}^n$ to the unit circle in the complex plane. Some convenient properties of the characters are summarized in the next result, the proof of which has been relegated to the problems.

Theorem 1. *The characters satisfy these equations:*

(a) $e_y(u + v) = e_y(u)e_y(v)$

(b) $E_u e_y = e_y(-u)e_y$ where $(E_u f)(x) = f(x - u)$

(c) $e_y(x) = e_x(y)$

(d) $e_y(\lambda x) = e_{\lambda y}(x)$ $(\lambda \in \mathbb{C})$

The **Fourier transform** of a function f in $L^1(\mathbb{R}^n)$ is the function $\widehat{f}$ defined by the equation

$$\widehat{f}(y) = \int_{\mathbb{R}^n} e^{-2\pi i x y} f(x)\, dx \qquad (y \in \mathbb{R}^n)$$

In this equation, f can be complex-valued. The kernel $e^{-2\pi i x y}$ is obviously complex-valued, but x and y run over $\mathbb{R}^n$. Notice that

$$\widehat{f}(y) = \langle f, e_y \rangle$$

since in dealing with complex-valued functions, the conjugate of the second function appears under the integral defining the inner product.

Example.   Let $n = 1$ and let f be given by

$$f(x) = \begin{cases} 1 & \text{on } [-1, 1] \\ 0 & \text{elsewhere} \end{cases}$$

Then

$$\hat{f}(y) = \int_{-1}^{1} e^{-2\pi i x y}\, dx = \frac{e^{-2\pi i x y}}{-2\pi i y}\bigg|_{x=-1}^{x=1}$$

$$= \frac{e^{-2\pi i y} - e^{2\pi i y}}{-2\pi i y} = \frac{1}{\pi y}\frac{e^{2\pi i y} - e^{-2\pi i y}}{2i} = \frac{\sin(2\pi y)}{\pi y} \qquad \blacksquare$$

The function $x \longmapsto \sin(2\pi x)/(\pi x)$ is called the **sinc** function. It plays an important rôle in signal processing and approximation theory. See [CL], Chapter 29, and the further references given there.

If $f \in L^1(\mathbb{R}^n)$, what can be said of $\hat{f}$? Later, we shall prove that it is continuous and vanishes at ∞. For the present we simply note that it is bounded. Indeed, $\|\hat{f}\|_\infty \leqslant \|f\|_1$, because

$$(3) \qquad |\hat{f}(y)| \leqslant \int |e^{-2\pi i x y}|\, |f(x)|\, dx = \int |f(x)|\, dx = \|f\|_1$$

In order to use the Fourier transform effectively, it is essential to know how it interacts with other linear operators on functions, such as translation and differentiation. The next theorem begins to establish results of that type.

Theorem 2.   *Let E denote the translation operator, defined by $(E_y f)(x) = f(x - y)$. Then we have $\widehat{E_y f} = e_{-y}\hat{f}$ and $\widehat{e_y f} = E_y \hat{f}$.*

Proof.   We verify the first equation and leave the second to the problems. We have

$$\widehat{E_y f}(x) = \int f(u - y)e^{-2\pi i x u}\, du = \int f(v)e^{-2\pi i x(v+y)}\, dv$$

$$= e^{-2\pi i x y} \int f(v)e^{-2\pi i x v}\, dv = e_{-y}(x)\hat{f}(x) \qquad \blacksquare$$

Recall the definition of the convolution of two functions, as given in Section 5.5, page 269:

$$(f * g)(x) = \int_{\mathbb{R}^n} f(y)g(x - y)\, dy$$

Theorem 3.   If f and g belong to $L^1(\mathbb{R}^n)$, then the same is true of $f * g$, and

$$\|f * g\|_1 \leqslant \|f\|_1 \cdot \|g\|_1$$

Proof.   ([Smi])   Define a function h on $\mathbb{R}^n \times \mathbb{R}^n$ by the equation

$$h(x, y) = g(x - y)$$

Let us prove that h is measurable. It is not enough to observe that the map $(x, y) \longmapsto x - y$ is continuous and that g is measurable, because the composition of a measurable function with a continuous function need not be measurable. For any open set $\mathcal{O}$ we must show that $h^{-1}(\mathcal{O})$ is measurable. Define a linear transformation A by $A(x, y) = (x - y, x + y)$. The following equivalences are obvious:

$$
\begin{aligned}
(x, y) \in h^{-1}(\mathcal{O}) \quad &\Longleftrightarrow \quad h(x, y) \in \mathcal{O} \\
&\Longleftrightarrow \quad g(x - y) \in \mathcal{O} \\
&\Longleftrightarrow \quad x - y \in g^{-1}(\mathcal{O}) \\
&\Longleftrightarrow \quad (x - y, x + y) \in g^{-1}(\mathcal{O}) \times \mathbb{R}^n \\
&\Longleftrightarrow \quad A(x, y) \in g^{-1}(\mathcal{O}) \times \mathbb{R}^n \\
&\Longleftrightarrow \quad (x, y) \in A^{-1}\big[g^{-1}(\mathcal{O}) \times \mathbb{R}^n\big]
\end{aligned}
$$

This shows that

$$h^{-1}(\mathcal{O}) = A^{-1}\big[g^{-1}(\mathcal{O}) \times \mathbb{R}^n\big]$$

Since g is measurable, $g^{-1}(\mathcal{O})$ and $g^{-1}(\mathcal{O}) \times \mathbb{R}^n$ are measurable sets. Since A is invertible, A^{-1} is a linear transformation; it carries each measurable set to another measurable set. Hence $h^{-1}(\mathcal{O})$ is measurable. Here we use the theorem that a function of class C^1 from $\mathbb{R}^n$ to $\mathbb{R}^n$ maps measurable sets into measurable sets, and apply that theorem to A^{-1}.

The function $F(x, y) = f(y)g(x - y)$ is measurable, and

$$\iint |F(x, y)|\, dx\, dy = \int |f(y)| \int |g(x - y)|\, dx\, dy$$

$$= \int |f(y)| \, \|g\|_1 \, dy = \|f\|_1 \, \|g\|_1$$

By Fubini's Theorem (See Chapter 8, page 426), F is integrable (i.e., $F \in L^1(\mathbb{R}^n \times \mathbb{R}^n)$). By the Fubini Theorem again,

$$\|f * g\|_1 = \int |(f * g)(x)|\, dx \leqslant \iint |F(x, y)|\, dy\, dx = \|f\|_1 \, \|g\|_1 \qquad \blacksquare$$

Theorem 3 can be found in many references, such as [Smi] page 334, [Ru3] page 156, [Gol] page 19.

Theorem 4.   If f and g belong to $L^1(\mathbb{R}^n)$, then

$$\widehat{f * g} = \hat{f}\hat{g}$$

Proof.   We use the Fubini Theorem again:

$$\widehat{f * g}(x) = \int e_{-x}(y)(f * g)(y)\, dy = \int e_{-x}(y) \int f(u)g(y - u)\, du\, dy$$

$$= \iint e_{-x}(u + y - u)f(u)g(y - u)\, du\, dy$$

$$= \int e_{-x}(u)f(u) \int e_{-x}(y - u)g(y - u)\, du\, dy$$

$$= \int e_{-x}(u)f(u) \int e_{-x}(y - u)g(y - u)\, dy\, du$$

$$= \int e_{-x}(u)f(u)\, du \int e_{-x}(z)g(z)\, dz = \hat{f}(x)\hat{g}(x) \qquad \blacksquare$$

Theorem 5.   If $f \in L^1(\mathbb{R}^n)$, then $\hat{f} \in C_0(\mathbb{R}^n)$. Thus, f is continuous and "vanishes at ∞."

Proof.   From the definition of $\hat{f}$,

$$|\hat{f}(x) - \hat{f}(y)| \leqslant \int |e^{-2\pi i x z} - e^{-2\pi i y z}|\, |f(z)|\, dz$$

If y converges to x through a sequence of values y_j, then the integrand is bounded above by $2|f(z)|$, and converges to 0 pointwise (i.e., for each z). By the Lebesgue dominated convergence theorem (Chapter 8, page 406), the integral tends to 0. Hence $\hat{f}(y_j) \to \hat{f}(x)$.

In order to see that $\hat{f}$ vanishes at infinity, we note that $-1 = e^{-\pi i}$ and compute $\hat{f}$ as follows, using $r = 2|x|^2$:

$$\hat{f}(x) = \int f(y)e^{-2\pi i x y}\, dy = \int -f(u)e^{-2\pi i x(u + x/r)}\, du$$

With the change of variable $y = u + x/r$ this becomes

$$\hat{f}(x) = \int -f\left(y - \frac{x}{r}\right)e^{-2\pi i x y}\, dy$$

It follows that

$$2\hat{f}(x) = \int e^{-2\pi i x y}\left[f(y) - f\left(y - \frac{x}{r}\right)\right] dy$$

and that

$$2|\hat{f}(x)| \leqslant \int \left|f(y) - f\left(y - \frac{x}{r}\right)\right| dy$$

At this point we want to say that as x tends to infinity, $x/r \to 0$ (because $r = 2|x|^2$), and the right-hand side of the previous inequality tends to zero. That assertion is justified by Lemma 3 of Section 6.4, page 306. ∎

Suggested references for this chapter are [Ad], [BN], [Bac], [Br], [CL], [DM], [Fol], [GV], [Gol], [Gre], [Gri], [Hel], [Ho], [Horv], [Kat], [Ko], [Lan1], [LL], [Loo], [RS], [Ru1], [Ru2], [Ru3], [Schl], [SW], [Ti1], [Ti2], [Wal], [Wie], [Yo], [Ze], and [Zem].

The table of Fourier transforms presented next uses some definitions and proofs that emerge in later sections.

Table of Fourier Transforms

Function	Its Fourier Transform	Definitions
f	$\widehat{f}(x) = \displaystyle\int_{\mathbb{R}^n} f(y) e^{-2\pi i x y}\, dy$	$xy \equiv \displaystyle\sum_{j=1}^{n} x_j y_j$
$E_v f$	$e_{-v}\widehat{f}$	$(E_v f)(x) \equiv f(x - v)$
$e_v f$	$E_{-v}\widehat{f}$	$e_v(x) \equiv e^{2\pi i x v}$
$P(D)f$	$P^+ \widehat{f}$	$P^+(x) \equiv P(2\pi i x)$
$f * g$	$\widehat{f}\widehat{g}$	$(f * g)(x) \equiv \displaystyle\int f(y) g(x - y)\, dy$
Pf	$P^-(D)\widehat{f}$	$P^-(x) \equiv P\left(\dfrac{-x}{2\pi i}\right)$
$S_\lambda f$	$\lambda^{-n} S_{1/\lambda}\widehat{f}$	$(S_\lambda f)(x) \equiv f(\lambda x)$
$\mathcal{X}_{[-1,1]}$	$x \longmapsto \dfrac{\sin(2\pi x)}{\pi x}$	$\mathcal{X}_A \equiv \begin{cases} 1 & x \in A \\ 0 & x \notin A \end{cases}$
$x \longmapsto e^{-\pi x^2}$	$x \longmapsto e^{-\pi x^2}$	
$x \longmapsto e^{-|x|}$	$x \longmapsto 2(1 + 4\pi^2 x^2)^{-1}$	
$x \longmapsto e^{-a x^2}$	$x \longmapsto \sqrt{\pi/a}\, e^{-\pi^2 x^2 / a}$	
$x \longmapsto (x^2 + a^2)^{-1}$	$x \longmapsto \dfrac{\pi}{a} e^{-2\pi a |x|}$	
$x \longmapsto \dfrac{\sin(2\pi x)}{\pi x}$	$\mathcal{X}_{[-1,1]}$	

Problems 6.1

1. Prove Theorem 1.
2. Does the group $\mathbb{R}^n$ have any continuous characters other than those described in the text?
3. Express $\delta(f * e_t)$ in terms of a Fourier transform.
4. What are the characters of the additive group $\mathbb{Z}$?
5. Find the Fourier transform of the function

$$f(x) = \begin{cases} \cos x & 0 \leqslant x \leqslant 1 \\ 0 & \text{elsewhere.} \end{cases}$$

6. Prove the second assertion of Theorem 2.
7. Prove that the mapping $\mathcal{F}$ that takes a function f into $\widehat{f}$ is linear and continuous from $L^1(\mathbb{R}^n)$ to $C(\mathbb{R}^n)$. (**Note:** $\mathcal{F}$ is also called the Fourier transform.)
8. Prove that if $\lambda > 0$ and $h(x) = f(x/\lambda)$, then $\widehat{h}(x) = \lambda^n \widehat{f}(\lambda x)$.
9. Prove that if $h(x) = \overline{f(-x)}$, then $\widehat{h}(x) = \overline{\widehat{f}(x)}$.
10. Prove that $L^1(\mathbb{R}^n)$ with convolution as product is a commutative Banach algebra. A Banach algebra is a Banach space in which a multiplication has been introduced such that $x(yz) = (xy)z$, $x(y + z) = xy + xz$, $(x + y)z = xz + yz$, $\lambda(xy) = (\lambda x)y = x(\lambda y)$, and $\|xy\| \leqslant \|x\| \, \|y\|$.
11. Does the Banach algebra described in Problem 10 have a unit element? That is, does there exist an element u such that $u * f = f * u = f$ for all f?
12. Show that the function $f(x) = (1 + ix)^{-2}$, $x \in \mathbb{R}$, has the property that $\widehat{f}(s) > 0$ for $s < 0$ and $\widehat{f}(s) = 0$ otherwise.
13. Prove that the function $\phi(x) = x^2$ has the property that for each $f \in L^1(\mathbb{R}^n)$, $\phi \circ \widehat{f}$ is the Fourier transform of some function in $L^1(\mathbb{R}^n)$. What other functions ϕ have this property?
14. Prove that the function $f(x) = \exp(x - e^x)$ has as its Fourier transform the function $\widehat{f}(s) = \Gamma(1 - is)$, and that $\widehat{f}(s)$ is never 0. The Gamma function is expounded in [Wid1].
15. (This problem relates to the proof of Theorem 4.) Let $f : X \to Y$ and $g : Y \to Z$ be two functions. Show that for any subset A in Z, $(g \circ f)^{-1}(A) = f^{-1}(g^{-1}(A))$. Now let $X = Y = \mathbb{R}$. Show that if g is continuous and f is measurable, then $g \circ f$ is measurable. Explain why $f \circ g$ need not be measurable.
16. How are the Fourier transforms of f and $\overline{f}$ related?
17. Prove that if $f_j \to f$ in $L^1(\mathbb{R}^n)$, then $\widehat{f_j}(x) \to \widehat{f}(x)$ uniformly in $\mathbb{R}^n$.
18. What logic can there be for the following approximate formula?

$$\widehat{f}(x) \approx \sum_{k=-N}^{k=N} f(k)e^{-2\pi ikx}$$

Under what conditions does the approximate equation become an exact equation?
19. (The Autocorrelation Theorem) Prove that if

$$g(x) = \int_{-\infty}^{\infty} \overline{f(u)} f(u + x) \, du$$

then $\widehat{g} = |\widehat{f}|^2$.
20. Prove that $\int f * g = \int f \int g$.
21. Assume that f is real-valued and prove that the maximum value of $f * Bf$ occurs at the origin. The definition of B is $(Bf)(x) = f(-x)$.
22. Recall the Heaviside function H from Section 5.1, page 250. Define $f(x) = e^{-ax}H(x)$ and $g(x) = e^{-bx}H(x)$. Compute $f * g$, assuming that $0 < a < b$.
23. Prove that if f is real-valued, then $|\widehat{f}|^2$ is an even function.
24. We have adopted the following definition of the Fourier transform:

$$(\mathcal{F}_1 f)(y) = \int_{\mathbb{R}^n} e^{-2\pi ixy} f(x) \, dx$$

Other books and papers sometimes use an alternative definition:

$$(\mathcal{F}_2 f)(y) = \frac{1}{(2\pi)^{n/2}} \int_{\mathbb{R}^n} e^{-ixy} f(x)\, dx$$

Find the relationship between these two transforms.

25. (Continuation) Prove that the inverse Fourier transforms obey this formula:

$$\mathcal{F}_1^{-1} = (2n)^{-n/2} \mathcal{F}_2^{-1} \circ S_{1/2\pi} \qquad \text{where } (S_\lambda f)(x) = f(\lambda x)$$

26. (Generalization of Theorem 3) Prove that if $f \in L^p(\mathbb{R}^n)$ and $g \in L^q(\mathbb{R}^n)$, then the convolution $f * g$ is well-defined, and $\|f * g\|_\infty \leqslant \|f\|_p \|g\|_q$.

27. Let f be the characteristic function of the interval $[-1/2, 1/2]$. Thus, $f(x) = 1$ if $|x| \leqslant 1/2$, and $f(x) = 0$ otherwise. Define $g(x) = (1 - |x|)f(x/2)$ and show that $f * f = g$.

28. Prove the "Modulation Theorem": If $g(x) = f(x)\cos(ax)$, then

$$\widehat{g}(x) = \frac{1}{2}\widehat{f}\Big(\frac{a}{2\pi} + x\Big) + \frac{1}{2}\widehat{f}\Big(\frac{a}{2\pi} - x\Big)$$

29. Define the operator B by the equation $(Bf)(x) = f(-x)$, and prove that $f * Bf$ is always even if f is real-valued.

6.2 The Schwartz Space

The space $\mathcal{S}$, also denoted by $\mathcal{S}(\mathbb{R}^n)$, is the set of all ϕ in $C^\infty(\mathbb{R}^n)$ such that $P \cdot D^\alpha \phi$ is a bounded function, for each polynomial P and each multi-index α. Functions with this property are said to be "rapidly decreasing," and the space itself is called the **Schwartz space**. In the case $n = 1$, membership in the Schwartz space simply requires $\sup_x |x^m \phi^{(k)}(x)|$ to be finite for all m and k.

Example 1. The Gaussian function ϕ defined by

$$\phi(x) = e^{-|x|^2}$$

belongs to $\mathcal{S}$.   ∎

It is easily seen, with the aid of Leibniz's formula, that if $\phi \in \mathcal{S}$, then $P \cdot \phi \in \mathcal{S}$ for any polynomial P, and $D^\alpha \phi \in \mathcal{S}$ for any multi-index α.

We note that $\mathcal{S}(\mathbb{R}^n)$ is a subspace of $L^1(\mathbb{R}^n)$. This is because functions in $\mathcal{S}$ decrease with sufficient rapidity to be integrable. Specifically, if $\phi \in \mathcal{S}$, then the function $x \longmapsto (1 + |x|^2)^n \phi(x)$ is bounded, say by M. Then

$$\int_{\mathbb{R}^n} |\phi(x)|\, dx \leqslant M \int_{\mathbb{R}^n} (1 + |x|^2)^{-n}\, dx$$

$$= M \int_0^\infty r^{n-1} \omega_n (1 + r^2)^{-n}\, dr < \infty$$

In this calculation we used "polar" coordinates and the "method of shells." The thickness of the shell is dr, the radius of the shell is r, and the area of the shell is $r^{n-1}\omega_n$, where ω_n denotes the area of the unit sphere in $\mathbb{R}^n$.

Definition. In $\mathcal{S}$, convergence is defined by saying that $\phi_j \twoheadrightarrow 0$ if and only if $P(x) \cdot D^\alpha \phi_j(x) \to 0$ uniformly in $\mathbb{R}^n$ for each multi-index α and for each polynomial P. In other terms, $\|P \cdot D^\alpha \phi_j\|_\infty \to 0$ for every multi-index α and every polynomial P, the sup-norm being computed over $\mathbb{R}^n$.

Lemma 1.   *If P is a polynomial, then the mapping $\phi \longmapsto P \cdot \phi$ is linear and continuous from $\mathbf{S}$ into $\mathbf{S}$.*

Proof.   Let $\phi_j \twoheadrightarrow 0$. We ask whether $Q \cdot D^\beta (P \cdot \phi_j) \to 0$ uniformly for each polynomial Q and multi-index β. By using the Leibniz formula, this expression can be exhibited as a sum of terms $Q_\gamma \cdot D^\alpha \phi_j$, where the Q_γ are polynomials and α is a multi-index such that $\alpha \leqslant \beta$. Each of these terms individually converges uniformly to zero, because that is a consequence of $\phi_j \twoheadrightarrow 0$ in $\mathbf{S}$. Therefore, their sum also converges to 0.   ∎

Lemma 2.   *If $g \in \mathbf{S}$, then the mapping $\phi \longmapsto g\phi$ is linear and continuous from $\mathbf{S}$ into $\mathbf{S}$*

Proof.   This is left to the problems.   ∎

Lemma 3.   *For any multi-index α, the mapping $\phi \longmapsto D^\alpha \phi$ is linear and continuous from $\mathbf{S}$ into $\mathbf{S}$.*

Proof.   This is left to the problems.   ∎

In studying how the Fourier transform interacts with differential operators, it is convenient to adopt the following definitions. Let P be a polynomial in n variables. Then P has a representation as a finite sum $P(x) = \sum c_\alpha x^\alpha$, in which each α is a multi-index, $x = (x_1, \ldots, x_n)$, c_α is a complex number, and $x^\alpha = x_1^{\alpha_1} x_2^{\alpha_2} \cdots x_n^{\alpha_n}$. Each function $x \longmapsto x^\alpha$ is called a **monomial**. We define also

$$D = \left(\frac{\partial}{\partial x_1}, \frac{\partial}{\partial x_2}, \ldots, \frac{\partial}{\partial x_n} \right)$$

$$D^\alpha = \left(\frac{\partial}{\partial x_1} \right)^{\alpha_1} \left(\frac{\partial}{\partial x_2} \right)^{\alpha_2} \cdots \left(\frac{\partial}{\partial x_n} \right)^{\alpha_n}$$

$$P(D) = \sum c_\alpha D^\alpha$$

Lemma 4.   *The function e_y defined by $e_y(x) = e^{2\pi i x y}$ obeys the equation $P(D)e_y = P(2\pi i y)e_y$ for any polynomial P.*

Proof.   It suffices to deal with the case of one monomial and establish that $D^\alpha e_y = (2\pi i y)^\alpha e_y$. We have

$$\frac{\partial}{\partial x_j} e_y(x) = \frac{\partial}{\partial x_j} e^{2\pi i (y_1 x_1 + \cdots + y_n x_n)}$$

$$= e^{2\pi i (y_1 x_1 + \cdots + y_n x_n)} (2\pi i y_j) = (2\pi i y_j) e_y(x)$$

Thus, by induction, we have

$$\left(\frac{\partial}{\partial x_j}\right)^{\alpha_j} e_y = (2\pi i y_j)^{\alpha_j} e_y$$

Consequently, $D^{\alpha} e_y = (2\pi i y_1)^{\alpha_1}(2\pi i y_2)^{\alpha_2}\cdots(2\pi i y_n)^{\alpha_n} e_y = (2\pi i y)^{\alpha} e_y$ ∎

The next result illustrates how the Fourier transform can simplify certain processes, such as differentiation. In this respect, it mimics the performance of the Laplace transform.

Theorem 1.   If $\phi \in \mathbf{S}$, and if P is a polynomial, then $[P(D/(2\pi i))\phi]^{\wedge} = P \cdot \hat{\phi}$. Equivalently, $[P(D)\phi]^{\wedge} = P^{+}\hat{\phi}$, where $P^{+}(x) = P(2\pi i x)$.

Proof.   We have to show that

$$\left[\sum c_{\alpha}\left(\frac{D}{2\pi i}\right)^{\alpha}\phi\right]^{\wedge}(y) = \sum c_{\alpha} y^{\alpha}\hat{\phi}(y)$$

Since the Fourier map $f \longmapsto \hat{f}$ is linear, it suffices to prove that

$$\left[\left(\frac{D}{2\pi i}\right)^{\alpha}\phi\right]^{\wedge}(y) = y^{\alpha}\hat{\phi}(y)$$

Equivalently, we must prove that

$$\left(\frac{1}{2\pi i}\right)^{|\alpha|}\widehat{[D^{\alpha}\phi]}(y) = y^{\alpha}\hat{\phi}(y)$$

Thus we must prove that

$$\left(\frac{1}{2\pi i}\right)^{|\alpha|}\int_{\mathbb{R}^n}(D^{\alpha}\phi)(x)e_{-y}(x)\,dx = y^{\alpha}\hat{\phi}(y)$$

In this integral we can use integration by parts repeatedly to transfer all derivatives from ϕ to the kernel function e_{-y}. Each use of integration by parts will introduce a factor of -1. Observe that no boundary values enter during the integration by parts, since $\phi \in \mathbf{S}$. Using also the preceding lemma, we find that the integral becomes successively

$$\left(\frac{1}{2\pi i}\right)^{|\alpha|}(-1)^{|\alpha|}\int \phi(x)\left[D^{\alpha}e_{-y}(x)\right](x)\,dx$$

$$= (-1)^{|\alpha|}\left(\frac{1}{2\pi i}\right)^{|\alpha|}\int \phi(x)(-2\pi i y)^{\alpha}e_{-y}(x)\,dx$$

$$= y^{\alpha}\int \phi(x)e_{-y}(x)\,dx = y^{\alpha}\hat{\phi}(y)$$ ∎

Example 2.   Let Δ denote the Laplacian operator; i.e.,

$$\Delta = \sum_{j=1}^{n} \left(\frac{\partial}{\partial x_j}\right)^2$$

Then $\Delta = P(D/(2\pi i))$ if P is defined to be

$$P(x) = \left(-4\pi^2\right)\left(x_1^2 + x_2^2 + \cdots + x_n^2\right) = -4\pi^2|x|^2$$

Hence, for $\phi \in \mathcal{S}$,

$$\widehat{\Delta\phi} = \left[P\left(\frac{D}{2\pi i}\right)\phi\right]^{\wedge} = P\widehat{\phi}$$

Equivalently,

$$\widehat{\Delta\phi}(x) = -4\pi^2|x|^2\widehat{\phi}(x)$$ ∎

Theorem 2.   *If $\phi \in \mathcal{S}$ and P is a polynomial, then $P(-D/(2\pi i))\widehat{\phi} = \widehat{P\phi}$. Equivalently, $P(D)\widehat{\phi} = \widehat{P^*\phi}$, where $P^*(y) = P(-2\pi i y)$.*

Proof.   We insert the variables, and interpret $P(D)$ as differentiating with respect to the variable x. Thus, with the help of Lemma 4, we have

$$\left[P(D)\widehat{\phi}\right](x) = P(D)\int e_{-x}(y)\phi(y)\,dy = P(D)\int e_{-y}(x)\phi(y)\,dy$$

$$= \int [P(D)e_{-y}](x)\phi(y)\,dy = \int P(-2\pi i y)e_{-y}(x)\phi(y)\,dy$$

$$= \int P^*(y)e_{-x}(y)\phi(y)\,dy = \widehat{P^*\phi}(x)$$ ∎

In the preceding proof, one requires the following theorem from calculus. See, for example, [Wid1] page 352, or [Bart] page 271.

Theorem 3.   *If f and $\partial f/\partial x$ are continuous functions on $\mathbb{R}^2$, then, provided that the integral on the right converges, we have*

$$\frac{d}{dx}\int_0^\infty f(x,t)\,dt = \int_0^\infty \frac{\partial}{\partial x}f(x,t)\,dt$$

Theorem 4.   *The mapping $\phi \longmapsto \widehat{\phi}$ is continuous and linear from $\mathcal{S}$ into $\mathcal{S}$.*

Proof.   First we must prove that $\widehat{\phi} \in \mathcal{S}$ when $\phi \in \mathcal{S}$. It is to be shown that $\widehat{\phi}$ is a C^∞-function and that $P \cdot D^\alpha\widehat{\phi}$ is bounded for each polynomial P and for each multi-index α. In Theorem 5 of Section 6.1 (page 291), we noted that $\widehat{\phi}$ is continuous. By Theorem 2 above, $D^\alpha\widehat{\phi} = \widehat{Q \cdot \phi}$ for an appropriate polynomial Q. Since $Q \cdot \phi \in \mathcal{S}$ (by Lemma 1), we know that $\widehat{Q \cdot \phi}$ is continuous, and can therefore conclude that $D^\alpha\widehat{\phi}$ is continuous. Hence $\widehat{\phi} \in C^\infty$.

Now we ask whether $P \cdot D^\alpha \widehat{\phi}$ is bounded. By the preceding remarks and Theorem 1,

(1) $$P \cdot D^\alpha \widehat{\phi} = P \cdot \widehat{Q \cdot \phi} = \left[P\left(\frac{D}{2\pi i} \right)(Q \cdot \phi) \right]^\wedge$$

Since $P\big(D/(2\pi i)\big)(Q \cdot \phi) \in \mathcal{S}$, its Fourier transform is bounded, as indicated in Equation (3) of Section 6.1, page 289.

For the continuity of the map, let $\phi_j \twoheadrightarrow 0$ in $\mathcal{S}$. We want to prove that $\widehat{\phi}_j \twoheadrightarrow 0$ in $\mathcal{S}$. That means that $P \cdot D^\alpha \widehat{\phi}_j \to 0$ uniformly for any polynomial P and any multi-index α. By Equation (1) above, the question to be addressed is whether $[P(D/(2\pi i))(Q \cdot \phi_j)]^\wedge \to 0$ uniformly. If we put $\psi_j = P(D/(2\pi i))(Q \cdot \phi_j)$, we ask whether $\widehat{\psi}_j(t) \to 0$ uniformly. Now, $\psi_j \in \mathcal{S}$, and $\psi_j \twoheadrightarrow 0$ in $\mathcal{S}$ by Lemmas 1 and 2. Hence $(1 + |x|^2)^n \psi_j(x) \to 0$ uniformly. It follows that for a given $\varepsilon > 0$ there is an integer m such that $(1 + |x|^2)^n |\psi_j(x)| < \varepsilon$ whenever $j > m$. For such j,

$$\int |\psi_j(x)| \, dx < \varepsilon \int (1 + |x|^2)^{-n} \, dx = c\varepsilon$$

and this shows that $\int |\psi_j| \to 0$. From the inequality

$$|\widehat{\psi}_j(x)| = \left| \int \psi_j(y) e_x(y) \, dy \right| \leqslant \int |\psi_j(y)| \, dy$$

we infer that $\widehat{\psi}_j(x) \to 0$ uniformly.   ∎

This section concludes with a proof of the Poisson summation formula. This important result states that under suitable hypotheses on the function f, the following equation is valid:

(1) $$\sum_{\nu \in \mathbb{Z}^n} f(\nu) = \sum_{\nu \in \mathbb{Z}^n} \widehat{f}(\nu)$$

A variety of hypotheses can be adopted for this result. See, for example, [SW] page 252, [Lan1] page 373, [Yo] page 149, [Gri] page 32, [Wal] page 60, [Kat] page 129, [Fri] page 104, [Ho] page 177, [DM] page 111, [Fol] page 337, [Ti1] page 60.

Theorem 5. Poisson Summation Formula.   *If $f \in C(\mathbb{R}^n)$ and if*

$$\sup_x \left(|f(x)| + |\widehat{f}(x)| \right)(1 + |x|)^{n+\varepsilon} < \infty$$

for some $\varepsilon > 0$, then $\sum_{\nu \in \mathbb{Z}^n} f(\nu) = \sum_{\nu \in \mathbb{Z}^n} \widehat{f}(\nu)$.

Proof.   Let c equal the supremum in the hypotheses. Then for $\|x\|_\infty \leqslant 1$ and $\nu \neq 0$ we have

$$|f(x + \nu)| \leqslant c\left(1 + |x + \nu| \right)^{-n-\varepsilon} \leqslant c\left(1 + \|x + \nu\|_\infty \right)^{-n-\varepsilon}$$

$$\leqslant c\left(1 + \|\nu\|_\infty - \|x\|_\infty \right)^{-n-\varepsilon} \leqslant c\|\nu\|_\infty^{-n-\varepsilon}$$

(In verifying these calculations, notice that the exponents are negative.) Then
we have

$$\sum_{\nu \neq 0} |f(x+\nu)| \leqslant c \sum_{\nu \neq 0} \|\nu\|_\infty^{-n-\varepsilon} = c \sum_{j=1}^{\infty} \sum_{\|\nu\|_\infty = j} \|\nu\|_\infty^{-n-\varepsilon}$$

$$= c \sum_{j=1}^{\infty} j^{-n-\varepsilon} \#\{\nu : \|\nu\| = j\}$$

$$= c \sum_{j=1}^{\infty} j^{-n-\varepsilon}(c_1 j^{n-1}) = c_2 \sum_{j=1}^{\infty} j^{-1-\varepsilon} < \infty$$

By a theorem of Weierstrass (the "M-Test", page 373)), this proves that the
function $F(x) = \sum_\nu f(x+\nu)$ is continuous, for its series is absolutely and
uniformly convergent. The function F is integer-periodic: For $\mu \in \mathbb{Z}^n$,

$$F(x+\mu) = \sum_\nu f(x+\mu+\nu) = \sum_\nu f(x+\nu) = F(x)$$

Let $Q = [0,1)^n$, the unit cube in $\mathbb{R}^n$. The Fourier coefficients of the periodic
function F are

$$A_\nu = \int_Q F(x)e^{-2\pi i\nu x}\,dx = \int_Q \sum_\mu f(x+\mu)e^{-2\pi i\nu x}\,dx$$

$$= \sum_\mu \int_{Q+\mu} f(y)e^{-2\pi i\nu(y-\mu)}\,dy = \sum_\mu \int_{Q+\mu} f(y)e^{-2\pi i\nu y}\,dy$$

$$= \int_{\mathbb{R}^n} f(y)e^{-2\pi i\nu y}\,dy = \widehat{f}(\nu)$$

Our hypotheses on f are strong enough to imply that $f \in L^1(\mathbb{R}^n)$. Hence
$\widehat{f} \in C_0(\mathbb{R}^n)$. The hypothesis on $\widehat{f}$ shows that the series $\sum \widehat{f}(x+\nu)$ is absolutely
and uniformly convergent. Hence

$$\sum_\nu |A_\nu| = \sum_\nu |\widehat{f}(\nu)| < \infty$$

From this we see that the Fourier series of F, $\sum_\nu A_\nu e^{2\pi i\nu x}$, is uniformly and
absolutely convergent. By the classical theory of Fourier series (such as in [Zy],
vol. II, page 300) we have

$$\sum_\nu A_\nu e^{2\pi i\nu x} = F(x)$$

It follows that

$$\sum_\nu f(\nu) = F(0) = \sum_\nu A_\nu = \sum_\nu \widehat{f}(\nu)$$

■

Problems 6.2

1. Prove Lemma 2.
2. Prove Lemma 3.
3. Let f be an even function in $L^1(\mathbb{R})$. Show that

$$\tfrac{1}{2}\widehat{f}(t) = \int_0^\infty f(x)\cos(2\pi tx)\,dx$$

The right-hand side of this equation is known as the Fourier Cosine Transform of f.

4. Let B be the operator such that $(Bf)(x) = f(-x)$. Find formulas for $\widehat{Bf}$ and $B\widehat{f}$. How are these related?
5. What is the Leibniz formula appropriate for the operator $(D/(2\pi i))^\alpha$?
6. Let $f \in L^1(\mathbb{R}^n)$ and $g(x) = f(Ax)$, where A is a nonsingular $n \times n$ matrix. Find the relation between $\widehat{f}$ and $\widehat{g}$.
7. Prove that, after Fourier transforms have been taken, the differential equation $f'(x) + xf(x) = 0$ becomes $(\widehat{f})'(t) + 4\pi^2 t\widehat{f}(t) = 0$.
8. Give a complete formal proof that $P \cdot f \in \mathcal{S}$ whenever P is a polynomial and $f \in \mathcal{S}$.
9. Prove that for a function of n variables having the special form

$$f(x_1,\dots,x_n) = \prod_{j=1}^n f_j(x_j)$$

we have

$$\widehat{f}(t_1,\dots,t_n) = \prod_{j=1}^n \widehat{f_j}(t_j)$$

10. Prove that if $f \in L^1(\mathbb{R})$ and $f > 0$, then

$$|\widehat{f}(t)| < \widehat{f}(0) \qquad (t \neq 0)$$

11. Let $f_m(x) = 1$ if $|x| \leqslant m$, and $f_m(x) = 0$ otherwise. (Here $x \in \mathbb{R}$, and $m = 1, 2, \dots$.) Compute $f_m * f_1$ and show that it is the Fourier transform of a function in $L^1(\mathbb{R})$.
12. Interpret Lemma 4 as a statement about eigenvalues and eigenvectors of a differential operator.
13. Prove, by using Fubini's Theorem, that for functions f and g in $L^1(\mathbb{R}^n)$, $\int \widehat{f}g = \int f\widehat{g}$.
14. Explain why $e^{-|x|}$ is not in $\mathcal{S}$.
15. Prove that if $\phi \in \mathcal{S}$, then $\widehat{D^\alpha \phi}$ exists for any multi-index α.
16. Let P be a polynomial on $\mathbb{R}^n$ and let g be an element of $C^\infty(\mathbb{R}^n)$ such that $|g(x)| \leqslant |P(x)|$ for all $x \in \mathbb{R}^n$. Is the mapping $f \longmapsto gf$ continuous from $\mathcal{S}$ into $\mathcal{S}$?
17. Prove that $\mathcal{S}$ is the subspace of $C^\infty(\mathbb{R}^n)$ consisting of all functions ϕ such that for each α, the map $x \longmapsto x^\alpha(D^\alpha \phi)(x)$ is bounded.
18. Show that $\phi_j \to 0$ in $\mathcal{S}$ if and only if $P(D)(Q\phi_j) \to 0$ uniformly in $\mathbb{R}^n$ for all polynomials P and Q.
19. Prove that if P is a polynomial and c is a scalar, then $P(cD)e_y = P(2\pi icy)e_y$.
20. Prove that if P is a polynomial and c is a scalar, then $P(cD)\widehat{\phi} = \widehat{P_c \phi}$, where $P_c(x) = P(-2\pi icx)$.
21. Prove that $P(-D)\widehat{\phi} = \widehat{P^+ \phi}$, where $P^+(x) = P(2\pi ix)$.
22. Prove that for $x \in \mathbb{R}^n$, $|x^\alpha| \leqslant |x|^{|\alpha|}$.
23. Using the operator B in Problem 4, prove that $\widehat{\widehat{f}} = B\widehat{f}$.
24. Prove that

$$\frac{d^k}{dx^k}e^{x^2} = e^{x^2}p_k(x)$$

where the polynomials p_k are defined recursively by the equations $p_0(x) = 1$ and $p_{k+1}(x) = 2xp_k(x) + p_k'(x)$.

25. Prove that if f obeys the differential equation

$$f''(x) - 4\pi^2 x^2 f(x) = \lambda f(x)$$

then the same is true of $\widehat{f}$.

26. Prove that for each n there is a constant c_n such that

$$\#\{\nu \in \mathbb{Z}^n : \|\nu\|_\infty = j\} = c_n j^{n-1}$$

Suggestion: Compute $\#\{\nu \in \mathbb{Z}^n : \|\nu\|_\infty \leqslant j\}$ first.

27. Prove that for $\phi \in \mathbf{S}(\mathbb{R}^n)$ and $\lambda \neq 0$,

$$\sum_{\nu \in \mathbb{Z}^n} \phi(x + \lambda \nu) = \sum_{\nu \in \mathbb{Z}^n} \lambda^{-n} \widehat{\phi}\left(\frac{\nu}{\lambda}\right) e^{2\pi i \nu x / \lambda}$$

28. Evaluate $\sum_{k=-\infty}^{\infty}(1 + k^2)^{-1}$ by using Theorem 5.
29. Evaluate $\sum_{k=1}^{\infty}(k^4 + a^4)^{-1}$ by using Theorem 5.
30. The first moment of a function f is defined to be

$$\int_{-\infty}^{\infty} x f(x)\, dx$$

Prove that under suitable hypotheses, the first moment is $(\widehat{f})'(0)/(-2\pi i)$.

6.3 The Inversion Theorems

In the previous section it was shown that the operator $\mathcal{F}$ defined by $\mathcal{F}(\phi) = \widehat{\phi}$ is linear and continuous from $\mathbf{S}$ into $\mathbf{S}$. In this section our goal is to prove that $\mathcal{F}$ is surjective and invertible, and to give an elegant formula for $\mathcal{F}^{-1}$.

Theorem 1.   *The function θ defined on $\mathbb{R}^n$ by $\theta(x) = e^{-\pi x^2}$ is a fixed point of the Fourier transform. Thus, $\widehat{\theta} = \theta$.*

Proof. First observe that the notation is

$$x^2 = xx = x \cdot x = \langle x, x \rangle = \sum_{j=1}^{n} x_j^2 = |x|^2$$

We prove our result first when $n = 1$ and then derive the general case. Define, for $x \in \mathbb{R}$, the analogous function $\psi(x) = e^{-\pi x^2}$. Since $\psi'(x) = e^{-\pi x^2}(-2\pi x) = -2\pi x \psi(x)$, we see that ψ is the unique solution of the initial-value problem

$$(1) \qquad\qquad \psi'(x) + 2\pi x \psi(x) = 0 \qquad\qquad \psi(0) = 1$$

By Problem 7 in Section 6.2, or the direct use of Theorems 1 and 2 (pages 296–297), we obtain, by taking Fourier transforms in Equation (1),

$$(\widehat{\psi})'(x) + 2\pi x \widehat{\psi}(x) = 0$$

The initial value of $\widehat{\psi}$ is

$$\widehat{\psi}(0) = \int_{-\infty}^{\infty} \psi(x)\, dx = \int_{-\infty}^{\infty} e^{-\pi x^2}\, dx = 1$$

(See Problem 10 for this.) We have seen that ψ and $\widehat{\psi}$ are two solutions of the initial-value problem (1). By the theory of ordinary differential equations, $\psi = \widehat{\psi}$. This proves the theorem for $n = 1$. Now we notice that

$$\theta(x) = \exp\left[-\pi(x_1^2 + \cdots + x_n^2)\right]$$

$$= e^{-\pi x_1^2} e^{-\pi x_2^2} \cdots e^{-\pi x_n^2}$$

$$= \psi(x_1)\psi(x_2)\cdots\psi(x_n)$$

By Problem 9 of Section 6.2, page 300,

$$\widehat{\theta}(x) = \prod_{j=1}^{n} \widehat{\psi}(x_j) = \prod_{j=1}^{n} \psi(x_j) = \theta(x) \qquad ∎$$

Theorem 2. First Inversion Theorem.   *If $\phi \in \mathcal{S}(\mathbb{R}^n)$, then*

$$\phi(x) = \int_{\mathbb{R}^n} \widehat{\phi} \cdot e_x = \int_{\mathbb{R}^n} \widehat{\phi}(y) e^{2\pi i y x}\, dy$$

Proof. We use the conjurer's tricks of smoke and mirrors. Let θ be the function in the preceding theorem, and put $g(x) = \theta(x/\lambda)$. Then $\widehat{g}(y) = \lambda^n \widehat{\theta}(\lambda y)$. (Problem 8 in Section 6.1, page 293.) By Problem 13 in Section 6.2, page 300,

$$\int \widehat{\phi}(y)\theta\left(\frac{y}{\lambda}\right) dy = \int \widehat{\phi}(y) g(y)\, dy = \int \phi(y)\widehat{g}(y)\, dy = \lambda^n \int \phi(y)\widehat{\theta}(\lambda y)\, dy$$

$$= \int \phi\left(\frac{u}{\lambda}\right)\widehat{\theta}(u)\, du$$

In the preceding calculation, let $\lambda = k$, where $k \in \mathbb{N}$, and contemplate letting $k \to \infty$. In order to use the Dominated Convergence Theorem (Section 8.6, page 406), we must establish L^1-bounds on the integrands. Here they are:

$$\left|\widehat{\phi}(y)\theta\left(\frac{y}{k}\right)\right| \leqslant |\widehat{\phi}(y)|\,\|\theta\|_\infty \qquad \widehat{\phi} \in L^1(\mathbb{R}^n)$$

$$\left|\phi\left(\frac{u}{k}\right)\right| \leqslant \|\phi\|_\infty |\widehat{\theta}(u)| \qquad \widehat{\theta} \in L^1(\mathbb{R}^n)$$

Then by the Dominated Convergence Theorem,

(2) $$\theta(0) \int \widehat{\phi}(y)\, dy = \phi(0) \int \widehat{\theta}(u)\, du$$

But we have, by the special properties of θ,

$$1 = \theta(0) = \hat{\theta}(0) = \int \theta(x)\,dx = \int \hat{\theta}(x)\,dx$$

Thus Equation (2) becomes

(3)
$$\int \hat{\phi}(y)\,dy = \phi(0)$$

This result is now applied to the shifted function $E_{-x}\phi$:

$$\int \widehat{E_{-x}\phi}(y)\,dy = (E_{-x}\phi)(0)$$

By Theorem 2 in Section 6.1, page 289, this is equivalent to

$$\int \hat{\phi}(y) \cdot e_x(y)\,dy = \phi(x) \qquad \blacksquare$$

Theorem 3.   *The Fourier transform operator $\mathcal{F}$ from $\mathbf{S}(\mathbb{R}^n)$ to $\mathbf{S}(\mathbb{R}^n)$ is a continuous linear bijection, and $\mathcal{F}^{-1} = \mathcal{F}^3$.*

Proof. The continuity and linearity of $\mathcal{F}$ were established by Theorem 4 in Section 6.2, page 297. The fact that $\mathcal{F}$ is surjective is established by writing the basic inversion formula from the preceding theorem as

$$\phi(x) = \int \hat{\phi}(y)e_x(y)\,dy = \int \hat{\phi}(y)e_{-x}(-y)\,dy = \int \hat{\phi}(-u)e_{-x}(u)\,du = [\mathcal{F}(B\hat{\phi})](x)$$

Here B is the operator such that $(B\phi)(x) = \phi(-x)$. The inversion formula also shows that $\mathcal{F}$ is injective, for if $\hat{\phi} = 0$, then obviously $\phi = 0$. Again by the inversion formula,

$$(\widehat{\hat{\phi}})(y) = \int \hat{\phi}(x) \cdot e_{-y}(x)\,dx = \phi(-y) = (B\phi)(y)$$

Thus $\mathcal{F}^2 = B$. It follows that $\mathcal{F}^4 = I$ and $\mathcal{F}^3\mathcal{F} = I$. $\blacksquare$

Theorem 4. Second Inversion Theorem.   *If f and $\hat{f}$ belong to $L^1(\mathbb{R}^n)$, then for almost all x,*

$$f(x) = \int_{\mathbb{R}^n} \hat{f}(y)e^{2\pi ixy}\,dy$$

Proof. Assume that f and $\hat{f}$ are in $L^1(\mathbb{R}^n)$. Let $\phi \in \mathbf{S}$. Then by Theorem 2, $\phi(x) = \int e_x\hat{\phi}$. By Problem 13 in Section 6.2, page 300, $\int \hat{f}\phi = \int f\hat{\phi}$. Hence if we put $F(y) = \int \hat{f}e_y$, then we have (with the help of the Fubini theorem)

$$\int \hat{\phi}(x)f(x)\,dx = \int \phi(x)\hat{f}(x)\,dx = \iint e_x(y)\hat{\phi}(y)\,dy\,\hat{f}(x)\,dx$$

$$= \int \hat{\phi}(y)\left[\int \hat{f}(x)e_y(x)\,dx\right]dy = \int \hat{\phi}(y)F(y)\,dy$$

Thus $\int \psi(x)(f - F)(x)\,dx = 0$ for all $\psi \in \mathcal{S}$, because $\widehat{\phi}$ can be any element of $\mathcal{S}$. The same equation is true for all $\psi \in \mathcal{D}$, since $\mathcal{D}$ is a subset of $\mathcal{S}$. Now apply Theorem 2 of Section 5.1, page 251, according to which $g = 0$ when $\tilde{g} = 0$. The conclusion is that that $f(x) = F(x)$ almost everywhere. ∎

Problems 6.3

1. Does $\mathcal{F}$ commute with the operators B and E_x?
2. Find the inverse Fourier transform of the function
$$f(t) = \begin{cases} \sin t & |t| \leqslant \pi \\ 0 & |t| > \pi \end{cases}$$

3. Explain why, for $\phi \in \mathcal{S}$,
$$\int \int e^{-2\pi i x(y+z)} \phi(y)\,dy\,dx = \phi(-z)$$

4. Let $f \in L^1(\mathbb{R})$ and define $h(x) = \int_a^x f(t)\,dt$. Prove that if $h \in L^1(\mathbb{R})$, then $\widehat{h}(t) = (2\pi i t)^{-1}\widehat{f}(t)$.
5. For the function $f(x) = e^{-|x|}$, show that $\widehat{f}(x) = 2/(1 + 4\pi^2 x^2)$. Show that $\widehat{f}$ is analytic in a horizontal strip in the complex plane described by the inequality $|\mathcal{I}m(z)| < 1/(4\pi^2)$. (Here $n = 1$.)
6. Let $f(x) = e^{-x}$ for $x \geqslant 0$ and let $f(x) = 0$ for $x < 0$. Find $\widehat{f}$ and verify by direct integration that $f(x) = \int \widehat{f} \cdot e_x$.
7. In Section 6.1 we saw that the following is a Fourier transform pair:
$$f(x) = \begin{cases} 1 & -1 \leqslant x \leqslant 1 \\ 0 & \text{otherwise} \end{cases} \qquad \widehat{f}(t) = \frac{\sin 2\pi t}{\pi t}$$

Prove that f belongs to $L^1(\mathbb{R})$ but $\widehat{f}$ does not. Explain why this does not violate the inversion theorem.
8. Using Theorem 1 of this section and Problem 8 in Section 6.1, page 293, prove that the Fourier transform of the function $\phi(x) = e^{-ax^2}$ is
$$\widehat{\phi}(t) = \left(\frac{\pi}{a}\right)^{1/2} e^{-\pi^2 x^2/a}$$

Prove also that $\mathcal{F}^{-1}\phi = \mathcal{F}\phi$. Prove that this last equation follows from the sole fact that ϕ is an even function.
9. Prove that if f is odd and belongs to $L^1(\mathbb{R})$, then
$$\frac{i}{2}\widehat{f}(t) = \int_0^\infty f(x)\sin(2\pi t x)\,dx$$

The right-hand side of this equation defines the Fourier Sine Transform of f.
10. Prove that
$$\int_{-\infty}^\infty e^{-x^2}\,dx = \frac{1}{\sqrt{\pi}}$$

This can be accomplished by considering the square of this integral, which can be written as the (double) integral of $e^{-(x^2+y^2)}$ over $\mathbb{R}^2$. This double integral can be computed by polar coordinates.

6.4 The Plancherel Theorem

This section is devoted to extending the Fourier operator from the Schwartz space $\mathcal{S}(\mathbb{R}^n)$ to $L^2(\mathbb{R}^n)$. It turns out that the extended operator has a number of endearing properties, leading one to conclude that $L^2(\mathbb{R}^n)$ is the "natural" setting for this important operator.

Lemma 1.   *If f and g belong to the Schwartz space $\mathcal{S}(\mathbb{R}^n)$, then $f * g$ also belongs to $\mathcal{S}(\mathbb{R}^n)$, and furthermore, $\widehat{fg} = \widehat{f} * \widehat{g}$.*

Proof.   Since f and g belong to $\mathcal{S}$, so does fg by Lemma 2 in Section 6.2, page 295. By Theorem 4 of Section 6.2, page 297, $\widehat{f}$, $\widehat{g}$, and $\widehat{fg}$ belong to $\mathcal{S}$. Consequently, $\widehat{fg}$ belongs to $\mathcal{S}$. By Theorem 4 of Section 6.1, page 290, $\widehat{f * g} = \widehat{f}\widehat{g}$. Hence $\widehat{f * g} \in \mathcal{S}$, and by the inversion theorem (Theorem 2 in the preceding section), $f * g \in \mathcal{S}$. Using the operator $\mathcal{F}$ such that $\mathcal{F}(f) = \widehat{f}$ and the operator B such that $(Bf)(x) = f(-x)$, we have

$$\widehat{f} * \widehat{g} = \mathcal{F}^{-1}\mathcal{F}(\widehat{f} * \widehat{g}) = \mathcal{F}^{-1}(\mathcal{F}^2 f \cdot \mathcal{F}^2 g)$$
$$= \mathcal{F}^{-1}(Bf \cdot Bg) = \mathcal{F}^{-1}B(fg) = \mathcal{F}^{-1}\mathcal{F}^2(fg) = \widehat{fg} \qquad \blacksquare$$

Lemma 2.   *If $f \in L^1(\mathbb{R}^n)$ and $\phi \in \mathcal{D}(\mathbb{R}^n)$, then $f * \phi \in C^\infty(\mathbb{R}^n)$.*

Proof.   By the theorem in Section 5.5, page 271,

$$D^\alpha(T * \phi) = T * D^\alpha\phi \qquad (T \in \mathcal{D}', \phi \in \mathcal{D})$$

In particular, for $f \in L^1(\mathbb{R}^n)$,

$$(1) \qquad D^\alpha(f * \phi) = f * D^\alpha\phi$$

(Recall that the definition of convolution involving distributions was made to conform to the ordinary convolution if the distribution arises from a function.) Now $f * g$ is continuous for any continuous g with compact support, as is easily seen from writing

$$(f * g)(x) - (f * g)(y) = \int f(u)\left[g(x - u) - g(y - u)\right] du$$

Applying this to the right side of Equation (1), we see that $D^\alpha(f * g)$ is continuous for every multi-index α. $\qquad \blacksquare$

The Lebesgue space $L^p(\mathbb{R}^n)$, where $1 \leqslant p < \infty$, has as elements all measurable functions f such that $|f|^p \in L^1(\mathbb{R}^n)$. The norm is $\|f\|_p = \||f|^p\|_1^{1/p}$. Further information about these spaces is found in Section 8.7, pages 409ff.

Lemma 3.   The translation operator E_x has the following continuity property: If $1 \leqslant p < \infty$ and $f \in L^p(\mathbb{R}^n)$, then the mapping $x \longmapsto E_x f$ is continuous from $\mathbb{R}^n$ to $L^p(\mathbb{R}^n)$.

Proof.   The continuous functions with compact support form a dense set in L^p, if $1 \leqslant p < \infty$. Hence, if $\varepsilon > 0$, then there exists such a continuous function h for which $\left\| f - h \right\|_p \leqslant \varepsilon$. Let the support of h be contained in the ball B_r of radius r centered at 0. By the uniform continuity of h there is a $\delta > 0$ such that

$$|x - y| < \delta \qquad \Longrightarrow \qquad |h(x) - h(y)| < \varepsilon$$

There is no loss of generality in supposing that $\delta < r$. If $|x - y| < \delta$, then

$$\left\| E_x h - E_y h \right\|_p^p = \int |h(z - x) - h(z - y)|^p \, dz \leqslant \varepsilon^p \operatorname{vol}(B_{2r}) \leqslant \varepsilon^p (4r)^n$$

From the triangle inequality it follows that

$$
\begin{aligned}
\left\| E_x f - E_y f \right\|_p &\leqslant \left\| E_x f - E_x h \right\|_p + \left\| E_x h - E_y h \right\|_p + \left\| E_y h - E_y f \right\|_p \\
&= \left\| E_x (f - h) \right\|_p + \left\| E_x h - E_y h \right\|_p + \left\| E_y (h - f) \right\|_p \\
&\leqslant \left\| f - h \right\|_p + \varepsilon (4r)^{n/p} + \left\| h - f \right\|_p \\
&\leqslant 2\varepsilon + \varepsilon (4r)^{n/p}
\end{aligned}
$$ ∎

In the next result we use a function $\rho \in \mathcal{D}$ such that $\rho \geqslant 0$ and $\int \rho = 1$. This is a "mollifier." Then ρ_k is defined by $\rho_k(x) = k^n \rho(kx)$, for $k = 1, 2, \ldots$

Theorem 1.   If $f \in L^1(\mathbb{R}^n)$, then $f * \rho_k \to f$ in the metric of $L^1(\mathbb{R}^n)$. Furthermore, $f * \rho_k \in C^\infty(\mathbb{R}^n)$.

Proof.   Since $\int \rho_k = 1$,

$$(f * \rho_k)(x) - f(x) = \int \left[f(x - z) - f(x) \right] \rho_k(z) \, dz$$

Hence by Fubini's Theorem (Chapter 8, page 426)

$$
\begin{aligned}
\int |f * \rho_k - f| &\leqslant \iint |f(x - z) - f(x)| \rho_k(z) \, dz \, dx \\
&= \iint |f(x - z) - f(x)| \, dx \, \rho_k(z) \, dz \\
&= \int \left\| E_z f - f \right\|_1 \rho_k(z) \, dz
\end{aligned}
$$

Here we need Lemma 3: If $f \in L^1(\mathbb{R}^n)$ and $\varepsilon > 0$, then there is a $\delta > 0$ such that $\left\| E_z f - f \right\|_1 \leqslant \varepsilon$ whenever $|z| \leqslant \delta$. If $\rho(x) = 0$ when $|x| > r$, then $\rho_k(x) = 0$ when $|x| > r/k$. Hence when $r/k \leqslant \delta$ we will have $\left\| f * \rho_k - f \right\|_1 \leqslant \varepsilon$. Lemma 2 shows that $f * \rho_k \in C^\infty(\mathbb{R}^n)$. ∎

Theorem 2.   *The space of test functions $\mathcal{D}(\mathbb{R}^n)$ is a dense subspace of $L^1(\mathbb{R}^n)$.*

Proof.   Let $f \in L^1(\mathbb{R}^n)$, and let $\varepsilon > 0$. We wish to find an element of $\mathcal{D}(R^n)$ within distance ε of f. The function $f * \rho_k$ from the preceding theorem would be a candidate, but it need not have compact support. So, we do the natural thing, which is to define

$$f_m(x) = \begin{cases} f(x) & \text{if } |x| \leqslant m \\ 0 & \text{elsewhere} \end{cases}$$

Then $f_m(x) \to f(x)$ pointwise, and the Dominated Convergence Theorem (page 406) gives us $\int |f_m| \to \int |f|$. Consequently, we can select an integer m such that $\|f\|_1 - \|f_m\|_1 < \varepsilon/2$. Then

$$\int_{|x|>m} |f(x)|\, dx < \varepsilon/2$$

Now select a "mollifier" ρ; i.e., ρ is a nonnegative test function such that $\int \rho = 1$. As usual, let $\rho_k(x) = k^n \rho(kx)$. By Theorem 1, there is an index k such that $\|f_m * \rho_k - f_m\|_1 < \varepsilon/2$. Hence

$$\|f_m * \rho_k - f\|_1 \leqslant \|f_m * \rho_k - f_m\|_1 + \|f_m - f\|_1 < \varepsilon$$

Observe that $f_m * \rho_k$ has compact support and belongs to $C^\infty(\mathbb{R}^n)$, by Lemma 2. Hence $f_m * \rho_k$ is in $\mathcal{D}$.   ∎

Plancherel's Theorem.   *The Fourier operator $\mathcal{F}$ defined originally on $\mathcal{S}(\mathbb{R}^n)$ has a unique continuous extension defined on $L^2(\mathbb{R}^n)$, and this extended operator is an isometry of $L^2(\mathbb{R}^n)$ onto $L^2(\mathbb{R}^n)$.*

Proof.   For two functions in $\mathcal{S}$ we have the Parseval formula:

$$\langle f, g \rangle = \langle \hat{f}, \hat{g} \rangle \quad \text{or} \quad \int f\, \bar{g} = \int \hat{f}\, \bar{\hat{g}}$$

This is proved with the following calculation, in which the inversion theorem is used:

$$\langle \hat{f}, \hat{g} \rangle = \int \hat{f}(y)\, \overline{\hat{g}(y)}\, dy = \int \int f(x) e^{-2\pi i x y}\, \overline{\hat{g}(y)}\, dx\, dy$$

$$= \int f(x) \int \overline{e^{2\pi i x y}\, \hat{g}(y)}\, dy\, dx = \int f(x) \overline{g(x)}\, dx$$

This leads to the isometry property for functions in $\mathcal{S}$:

$$\|f\|_2^2 = \int f\, \bar{f} = \int \hat{f}\, \bar{\hat{f}} = \|\hat{f}\|_2^2$$

Since $\mathcal{S}$ is dense in $L^2(\mathbb{R}^n)$ (Problem 4), $\mathcal{F}$ has a unique continuous extension with the same bound, $\|\mathcal{F}\| \leqslant 1$ (Problem 6). It is then easily seen that the extension is also an isometry (Problem 7).   ∎

The extension of $\mathcal{F}$ referred to in this theorem is sometimes called the Fourier–Plancherel transform. One must not assume that the usual formula for $\widehat{f}$ can be used for $f \in L^2(\mathbb{R}^n)$, because the integrand in the usual formula

$$\widehat{f}(x) = \int f(y)e^{-2\pi ixy}\,dy$$

need not be integrable (i.e., in L^1). However, f is an L^2 limit of L^1 functions, because $L^1(\mathbb{R}^n)$ is a dense subset of $L^2(\mathbb{R}^n)$. For example, we can use this sequence:

$$f_m(x) = \begin{cases} f(x) & \text{if } |x| \leqslant m \\ 0 & \text{if } |x| > m \end{cases}$$

Since f belongs to L^2, f_m belongs to L^1. Indeed, letting $\mathcal{X}_m$ be the characteristic function of the ball $\{x : |x| \leqslant m\}$, we have by the Cauchy–Schwarz inequality

$$\|f_m\|_1 = \int |f(x)|\mathcal{X}_m(x)\,dx \leqslant \|f\|_2\|\mathcal{X}_m\|_2$$

The sequence $[f_m]$ converges to f in the metric of L^2 because

$$\|f - f_m\|_2^2 = \int |f(x) - f_m(x)|^2\,dx = \int_{|x|>m} |f(x)|^2\,dx \to 0$$

It now follows that $\widehat{f} = \lim \widehat{f}_m$, the limit being taken in the L^2 sense. This state of affairs is often expressed by writing

$$\widehat{f}(y) = \underset{m\to\infty}{\text{L.I.M.}} \int_{|x|\leqslant m} f(x)e^{-2\pi ixy}\,dy$$

In this equation, L.I.M. stands for "limit in the mean," and this refers to a limit in the space L^2.

Another procedure for generating a sequence that converges to $\widehat{f}$ in L^2 is to select an orthonormal basis $[u_n]$ for L^2, to express f in terms of the basis, and to take Fourier transforms:

$$f = \sum_{m=0}^{\infty} \langle f, u_m \rangle u_m$$

$$\widehat{f} = \sum_{m=0}^{\infty} \langle f, u_m \rangle \widehat{u_m}$$

This manipulation is justified by the linearity and continuity of the Fourier transform operator acting on L^2. In order to be practical, this formula must employ an orthonormal basis of functions whose Fourier transforms are known.

An example in $\mathbb{R}$ is given by the Hermite functions h_m of Problem 12. They form an orthogonal basis for $L^2(\mathbb{R})$. The functions $H_m = h_m / \|h_m\|$ provide an orthonormal basis, and by Problem 12,

$$\widehat{f} = \sum_{m=0}^{\infty} \langle f, H_m \rangle \widehat{H_m} = \sum_{m=0}^{\infty} \langle f, H_m \rangle (-i)^m H_m$$

Having seen that the Fourier operator $\mathcal{F}$ is continuous from $L^2(\mathbb{R}^n)$ to $L^2(\mathbb{R}^2)$ and from $L^1(\mathbb{R}^n)$ to $L^\infty(\mathbb{R}^n)$, one might ask whether it is continuous from $L^p(\mathbb{R}^n)$ to $L^q(\mathbb{R}^n)$ in general, when $\dfrac{1}{p} + \dfrac{1}{q} = 1$. The answer is "Yes" if $p \leqslant q$. We quote without proof the Hausdorff–Young Theorem: If $1 \leqslant p \leqslant 2$, and if $\dfrac{1}{p} + \dfrac{1}{q} = 1$, then the Fourier operator is continuous from $L^p(\mathbb{R}^n)$ to $L^q(\mathbb{R}^n)$, and its norm is not greater than 1. For further information, consult [RS] and [SW]. The exact value of the norm has been established by Beckner [Bec].

Problems 6.4

1. In Lemma 2, can we conclude that $f * g$ belongs to $C_0(\mathbb{R}^n)$?
2. Explain why Lemma 3 is not true for the space $L^\infty(\mathbb{R}^n)$.
3. Prove that $\mathcal{S} \subset L^2(\mathbb{R}^n)$.
4. Prove that $\mathcal{S}$ is dense in $L^2(\mathbb{R}^n)$.
5. Prove that if $f, g \in \mathcal{S}$, then $\int f\,\overline{g} = \int \widehat{f}\,\overline{\widehat{g}}$.
6. Prove this theorem: Let Y be a dense subspace of a normed linear space X. Let $A \in \mathcal{L}(Y, Z)$, where Z is a Banach space. Then there is a unique $\overline{A} \in \mathcal{L}(X, Z)$ such that $\overline{A}\,|\,Y = A$ and $\|\overline{A}\| = \|A\|$. Suggestions: If $x \in X$, then there is a sequence $y_k \in Y$ such that $y_k \to x$. Put $\overline{A}x = \lim A y_k$. Show that the limit exists and is independent of the sequence y_k.
7. In the situation of Problem 6, show that if A is an isometry, then so is $\overline{A}$.
8. Show that *neither* of the inclusions $L^1(\mathbb{R}^n) \subset L^2(\mathbb{R}^n)$, $L^2(\mathbb{R}^n) \subset L^1(\mathbb{R}^n)$ is valid.
9. Find an element of $L^2(\mathbb{R}) \smallsetminus L^1(\mathbb{R})$ and compute its Fourier–Plancherel transform.
10. Prove that the Fourier transform of the function

$$f(x) = \begin{cases} e^{-ax} & x \geqslant 0 \\ 0 & x < 0 \end{cases}$$

is $(2\pi i x + a)^{-1}$. Here, $a > 0$. Show that $\widehat{f} \in L^2(\mathbb{R}) \smallsetminus L^1(\mathbb{R})$.
11. Prove that the equation $\widehat{fg} = \widehat{f} * \widehat{g}$ holds for functions in $L^2(\mathbb{R}^n)$.
12. The eigenvalues of $\mathcal{F} : L^2(\mathbb{R}) \to L^2(\mathbb{R})$ are $\pm 1, \pm i$, and no others. Show that $\widehat{h_m} = (-i)^m h_m$, where h_m is the Hermite function

$$h_m(x) = \exp(\pi x^2) D^m \exp(-2\pi x^2)$$

Suggestions: Prove that $h_{m+1}(x) = h'_m(x) - 2\pi x h_m(x)$. Then prove that $\widehat{h_{m+1}}(x) = -i(\widehat{h_m})'(x) + 2\pi i x \widehat{h_m}(x)$. Show that the functions $(-1)^m h_m$ obey the same recurrence relation as $\widehat{h_m}$.
13. For f and g in $\mathcal{S}$, prove that

$$\mathcal{F}^{-1}(fg) = (\mathcal{F}^{-1}f) * (\mathcal{F}^{-1}g)$$

Then prove this for functions in $L^2(\mathbb{R}^n)$.
14. Generalize Lemma 2 so that it applies to $f \in L^1_{\text{loc}}(\mathbb{R}^n)$. Can the hypotheses on f be relaxed?

15. Define $f(x) = 0$ for $x < 1$ and $f(x) = x^{-1}$ for $x \geqslant 1$. Find $\widehat{f}$. Useful reference: Chapter 5 in [AS].

16. Is this reasoning correct? If $f \in L^1$, then $\widehat{f}$ is continuous. Since L^1 is dense in L^2, the same conclusion must hold for $f \in L^2$.

17. The **variance** of a function f is defined to be $\|uf\|/\|f\|$, where $u(x) = x$. Prove this version of the Uncertainty Principle: The product of the variances of f and $\widehat{f}$ cannot be less than $1/(4\pi)$.

6.5 Applications of the Fourier Transform

We will give some representative examples to show how the Fourier transform can be used to solve differential equations and integral equations. Then, an application to multi–variate interpolation will be presented. These are what might be called *direct* applications, as contrasted with applications to other branches of abstract mathematics.

Example 1.   Let $n = 1$ and $D = \dfrac{d}{dx}$. If P is a polynomial, say $P(\lambda) = \sum_{j=0}^{m} c_j \lambda^j$, then $P(D)$ is a linear differential operator with constant coefficients:

$$(1) \qquad P(D) = \sum_{j=0}^{m} c_j D^j = \sum_{j=0}^{m} c_j (2\pi i)^j \left(\frac{D}{2\pi i} \right)^j$$

Consider the ordinary differential equation

$$(2) \qquad P(D)u = g \qquad -\infty < x < \infty$$

in which g is given and is assumed to be an element of $L^1(\mathbb{R})$. Apply the Fourier transform $\mathcal{F}$ to both sides of Equation (2). Then use Theorem 1 of Section 6.2 (page 296), which asserts that if $u \in \mathbf{S}$, then

$$(3) \qquad \mathcal{F}\big[P(D)u\big] = P^{+}\mathcal{F}(u)$$

where $P^{+}(x) = P(2\pi i x)$. The transformed version of Equation (2) is therefore

$$(4) \qquad P^{+}\mathcal{F}(u) = \mathcal{F}(g)$$

The solution of Equation (4) is

$$(5) \qquad \mathcal{F}(u) = \mathcal{F}(g)/P^{+}$$

The function u is recovered by taking the inverse transformation, if it exists:

$$(6) \qquad u = \mathcal{F}^{-1}\big[\mathcal{F}(g)/P^{+}\big]$$

Theorem 4 in Section 6.1, page 291, states that

$$(7) \qquad \mathcal{F}(\phi * \psi) = \mathcal{F}(\phi) \cdot \mathcal{F}(\psi)$$

An equivalent formulation, in terms of $\mathcal{F}^{-1}$, is

(8) $$\phi * \psi = \mathcal{F}^{-1}\left[\mathcal{F}(\phi) \cdot \mathcal{F}(\psi)\right]$$

If h is a function such that $\widehat{h} = 1/P^+$, then Equations (6) and (8) yield

(9) $$u = \mathcal{F}^{-1}\left[\frac{\widehat{g}}{P^+}\right] = \mathcal{F}^{-1}[\widehat{g}\widehat{h}] = g * h$$

In detail,

(10) $$u(x) = \int_{-\infty}^{\infty} g(y)h(x-y)\,dy$$

The function h must be obtained by the equation $h = \mathcal{F}^{-1}(1/P^+)$. ∎

Example 2.   This is a concrete case of Example 1, namely

(11) $$u'(x) + bu(x) = e^{-|x|} \qquad (b > 0, \quad b \neq 1)$$

The Fourier transform of the function $g(x) = e^{-|x|}$ is $\widehat{g}(t) = 2/(1 + 4\pi^2 t^2)$ (Problem 5 of Section 6.3, page 304). Hence the Fourier transform of Equation (11) is

$$2\pi it\,\widehat{u}(t) + b\,\widehat{u}(t) = 2/(1 + 4\pi^2 t^2)$$

Solving for $\widehat{u}$, we have

$$\widehat{u}(t) = \frac{2}{(1 + 4\pi^2 t^2)(b + 2\pi it)}$$

By the Inversion Theorem,

$$u(t) = \int_{-\infty}^{\infty} \frac{2e^{2\pi ixt}\,dx}{(1 + 4\pi^2 x^2)(b + 2\pi ix)}$$

To simplify this, substitute $z = 2\pi x$, to obtain

$$u(t) = \frac{1}{\pi}\int_{-\infty}^{\infty} \frac{e^{itz}\,dz}{(1 + z^2)(b + iz)}$$

The integrand, call it $f(z)$, has poles at $z = +i, -i$, and ib. In order to evaluate this integral, we use the residue calculus, as outlined at the end of this section. Let the complex variable be expressed as $z = x + iy$. Then

$$|e^{itz}| = |e^{it(x+iy)}| = |e^{-ty+itx}| = e^{-ty}$$

For $t > 0$ we see that

$$\lim_{r\to\infty}\sup\{|zf(z)| \; : \; |z| = r,\; Im(z) \geqslant 0\} = 0$$

Hence by Theorem 4 at the end of this section,

$$\int_{-\infty}^{\infty} f(z)\,dz = 2\pi i \times (\text{residue at } i \; + \;\; \text{residue at } ib)$$

By partial fraction decomposition we obtain

$$f(z) = e^{itz}\left[\frac{(2ib-2i)^{-1}}{z-i} - \frac{(2ib+2i)^{-1}}{z+i} + \frac{(i-ib^2)^{-1}}{z-ib}\right]$$

Hence the residues at $i, -i,$ and ib are respectively

$$\frac{e^{-t}}{2i(b-1)} \qquad \frac{-e^{t}}{2i(b+1)} \qquad \frac{e^{-bt}}{i(1-b^2)}$$

Thus for $t > 0$,

$$u(t) = \pi^{-1} 2\pi i \left[\frac{e^{-t}}{2i(b-1)} + \frac{e^{-bt}}{i(1-b^2)}\right]$$

$$= \frac{e^{-t}}{b-1} + \frac{2e^{-bt}}{1-b^2}$$

Similarly, for $t < 0$,

$$u(t) = \frac{-e^{t}}{1+b} \qquad\qquad\qquad ■$$

Example 3.   Consider the integral equation

$$\int_{-\infty}^{\infty} k(x-s)u(s)\,ds = g(x)$$

in which k and g are given, and u is an unknown function. We can write

$$u * k = g$$

After taking Fourier transforms and using Theorem 4 in Section 6.1 (page 290) we have

$$\widehat{u}\widehat{k} = \widehat{g}$$

whence $\widehat{u} = \widehat{g}/\widehat{k}$ and $u = \mathcal{F}^{-1}(\widehat{g}/\widehat{k})$.

For a concrete case, contemplate this integral equation:

$$\int_{-\infty}^{\infty} e^{-|x-s|}u(s)\,ds = e^{-x^2/2}$$

Here, the functions k and g in the above discussion are

$$k(x) = e^{-|x|} \qquad\qquad g(x) = e^{-x^2/2}$$

From Problems 6.3.5 and 6.3.8 (page 304), we have these Fourier transforms:

$$\widehat{k}(x) = \frac{2}{(1 + 4\pi^2 x^2)} \qquad \widehat{g}(x) = (2\pi)^{1/2} e^{-2\pi^2 x^2}$$

(It turns out that we do not require $\widehat{g}$.) Hence

$$\widehat{u}(x) = \widehat{g}(x)/\widehat{k}(x) = \widehat{g}(x)(1 + 4\pi^2 x^2)/2$$

To take the inverse transform, use the principle in Theorem 1 of Section 6.2 (page 296) that $\widehat{P(D)g} = P^+ \cdot \widehat{g}$. We let $P(x) = (1 - x^2)/2$, so that $P^+(x) = P(2\pi i x) = (1 + 4\pi^2 x^2)/2$. Then

$$\widehat{u} = P^+ \widehat{g} = \widehat{P(D)g}$$

The inverse transform then gives us

$$u(x) = [P(D)g](x) = \frac{1}{2}(g - g'')(x) = \frac{1}{2}e^{-x^2/2}(2 - x^2) \qquad \blacksquare$$

As another example of applications of the Fourier transform, we consider a problem of multi–variate interpolation. First, what is meant by "multi–variate interpolation"? Let us work, as usual, in $\mathbb{R}^n$. Suppose that at a finite set of points called "nodes" we have data, interpreted as the values of some unknown function. We will assume that the nodes are all different from one another. Since we will not need the components of the nodes, we can use the notation $x_1, x_2, \ldots, x_m$ for the set of nodes. Let the corresponding data values be real numbers $\lambda_1, \lambda_2, \ldots, \lambda_m$. We seek now an "interpolating" function for this information. That will be some nice, smooth function that is defined everywhere and takes the values λ_i at the nodes x_i. (Polynomials are *not* recommended for this task.) One way of obtaining a simple interpolating function is to start with a suitable function f, and use linear combinations of its translates to do the job. Thus, we will try to accomplish the interpolation with a function of the form

$$x \longmapsto \sum_{j=1}^{m} c_j f(x - x_j)$$

When the interpolation conditions are imposed, we arrive at the equations

$$\sum_{j=1}^{m} c_j f(x_i - x_j) = \lambda_i \qquad (1 \leqslant i \leqslant m)$$

This is a system of m linear equations in the m unknowns c_j. How can we be sure that the system has a solution? Since we want to be able to solve this problem for any λ_i, we must have a nonsingular coefficient matrix. This can be called the "interpolation matrix"; it is the matrix $A_{ij} = f(x_i - x_j)$. A striking theorem gives us an immense class of useful functions f to play the role described above.

Theorem 1.　　If f is the Fourier transform of a positive function in $L^1(\mathbb{R}^n)$, then for any finite set of points $x_1, x_2, \ldots, x_m$ in $\mathbb{R}^n$ the matrix having elements $f(x_i - x_j)$ will be positive definite (and hence nonsingular).

Proof.　Let $f = \widehat{g}$, where $g \in L^1(\mathbb{R}^n)$ and $g(x) > 0$ everywhere. The interpolation matrix in question must be shown to be positive definite. This means that $u^* A u > 0$ for all nonzero vectors u in $\mathbb{C}^m$. We undertake a calculation of this quadratic form:

$$u^* A u = \sum_{k=1}^m \sum_{j=1}^m \overline{u}_k A_{kj} u_j = \sum \sum \overline{u}_k u_j f(x_k - x_j)$$

$$= \sum \sum \overline{u}_k u_j \int_{\mathbb{R}^n} g(y) e^{-2\pi i y(x_k - x_j)} \, dy$$

$$= \int_{\mathbb{R}^n} g(y) \sum \overline{u}_k e^{-2\pi i y x_k} \sum u_j e^{2\pi i y x_j} \, dy$$

$$= \int_{\mathbb{R}^n} g(y) \big| h(y) \big|^2 \, dy \geqslant 0$$

Here we have written

$$h(y) = \sum_{j=1}^m u_j e^{2\pi i y x_j} \qquad (y \in \mathbb{R}^n)$$

So far, we have proved only that the interpolation matrix A is nonnegative definite. How can we conclude that the final integral above is positive? It will suffice to establish that the functions $y \longmapsto e^{2\pi i y x_j}$ form a linearly independent set, for in our computation, the vector u was not zero. Once we have the linear independence, it will follow that $|h(y)|^2$ is positive somewhere in $\mathbb{R}^n$, and by continuity will be positive on an open set. Since g is positive everywhere, the final integral above would have to be positive. The linear independence is proved separately in two lemmas.　　　　　　　　　　　　　　　　　　　　　■

Lemma 1.　Let $\lambda_1, \ldots, \lambda_m$ be m distinct complex numbers, and let $c_1, \ldots, c_m$ be complex numbers. If $\sum_{j=1}^m c_j e^{\lambda_j z} = 0$ for all z in a subset of $\mathbb{C}$ that has an accumulation point, then $\sum_{j=1}^m |c_j| = 0$.

Proof.　Use induction on m. If $m = 1$, the result is obvious, because $e^{\lambda_1 z}$ is not zero for any $z \in \mathbb{C}$. If the lemma has been established for a certain integer $m - 1$, then we can prove it for m as follows. Let $f(z) = \sum_1^m c_j e^{\lambda_j z}$, and suppose that $f(z_k) = 0$ for some convergent sequence $[z_k]$. Since f is an entire function, we infer that $f(z) = 0$ for all z in $\mathbb{C}$. (See, for example, [Ti2] page 88, or [Ru3] page 226.) Consider now the function

$$F(z) = \frac{d}{dz}\Big[e^{-\lambda_m z} f(z) \Big] = \frac{d}{dz} \sum_{j=1}^m c_j e^{(\lambda_j - \lambda_m)z} = \sum_{j=1}^{m-1} c_j \big(\lambda_j - \lambda_m\big) e^{(\lambda_j - \lambda_m)z}$$

Since $f = 0$, we have $F = 0$. By the induction hypothesis, $c_j(\lambda_j - \lambda_m) = 0$ for $1 \leqslant j \leqslant m-1$. Since the λ_j are distinct, we infer that $c_1 = \cdots = c_{m-1} = 0$. The function f then reduces to $f(z) = c_m e^{\lambda_m z}$. Since $f = 0$, $c_m = 0$.   ∎

Lemma 2.   *Let $w_1, \ldots, w_m$ be m distinct points in $\mathbb{C}^n$. Let $c_1, \ldots, c_m$ be complex numbers. If $\sum_{j=1}^{m} c_j e^{w_j x} = 0$ for all x in a nonempty open subset of $\mathbb{R}^n$, then $\sum_{j=1}^{m} |c_j| = 0$.*

Proof.   Let $\mathcal{O}$ be an open set in $\mathbb{R}^n$ having the stated property. Select $\xi \in \mathcal{O}$ such that the complex inner products $w_j \xi$ are all different. This is possible by the following reasoning. The condition on ξ can be expressed in the form $w_j \xi \neq w_k \xi$ for $1 \leqslant j < k \leqslant m$. This, in turn, means that ξ does not lie in any of the sets

$$H_{jk} = \{x \in \mathbb{R}^n \ : \ (w_j - w_k)x = 0\} \qquad (1 \leqslant j < k \leqslant m)$$

Each set H_{jk} is the intersection of two hyperplanes in $\mathbb{R}^n$. (See Problem 4.) Hence each H_{jk} is a set of Lebesgue measure 0 in $\mathbb{R}^n$, and the same is true of any countable union of such sets. The finite family of sets H_{jk} therefore cannot cover the open set $\mathcal{O}$, which must have positive measure. Now define, for $t \in \mathbb{C}$, the function $f(t) = \sum_{1}^{m} c_j e^{(w_j \xi)t}$. Since $\xi \in \mathcal{O}$, our hypothesis gives us $f(1) = 0$. Let U be a neighborhood of 1 in $\mathbb{C}$ such that $t\xi \in \mathcal{O}$ when $t \in U$. Since $f(t) = 0$ on U, Lemma 1 shows that $\sum_{j=1}^{m} |c_j| = 0$.   ∎

More information on the topic of interpolation can be found in the textbook [CL]. Functions of the type f, as in Theorem 1, are said to be "strictly positive definite on $\mathbb{R}^n$." They are often used in neural networks, in the "hidden layers," where most of the heavy computing is done.

The remainder of this section is devoted to a review of the residue calculus. This group of techniques is often needed in evaluating the integrals that arise in inverting the Fourier transform.

Theorem 2. Laurent's Theorem.   *Let f be a function that is analytic inside and on a circle C in the complex plane, except for having an isolated singularity at the center ζ. Then at each point inside C with the exception of ζ we have*

$$f(z) = \sum_{n=-\infty}^{\infty} c_n(z - \zeta)^n \qquad c_n = \frac{1}{2\pi i} \int_C \frac{f(z)\, dz}{(z - \zeta)^{n+1}}$$

The coefficient c_{-1} is called the **residue of f at** ζ. By Laurent's theorem, the residue is also given by

(12) $$c_{-1} = \frac{1}{2\pi i} \int_C f(z)\, dz$$

Example 4.   The integral $\int_C e^z/z^4\,dz$, where C is the unit circle, can be computed with the principle in Equation (12). Indeed, the given integral is $2\pi i$ times the residue of e^z/z^4 at 0. Since

$$e^z/z^4 = \left(1 + z + \frac{z^2}{2!} + \frac{z^3}{3!} + \cdots\right)\Big/z^4$$

$$= z^{-4} + z^{-3} + \tfrac{1}{2}z^{-2} + \tfrac{1}{6}z^{-1} + \cdots$$

we see that the residue is $\tfrac{1}{6}$ and the integral is $\tfrac{1}{3}\pi i$.   ∎

> **Theorem 3 The Residue Theorem.**   Let C be a simple closed curve inside of which f is analytic with the exception of isolated singularities at the points $\zeta_1, \ldots, \zeta_m$. Then $\dfrac{1}{2\pi i}\displaystyle\int_C f(z)dz$ is the sum of the residues of f at $\zeta_1, \ldots, \zeta_m$.

Proof.   Draw mutually disjoint circles $C_1, \ldots, C_m$ around the singularities and contained within C. The integral around the path shown in the figure is zero, by Cauchy's integral theorem. (Figure 6.1a depicts the case $m = 2$.) Therefore,

$$0 = \int_C f(z)\,dz - \int_{C_1} f(z)\,dz - \cdots - \int_{C_m} f(z)\,dz$$

In this equation, divide by $2\pi i$ and note that the negative terms are the residues of f at $\zeta_1, \ldots, \zeta_m$.   ∎

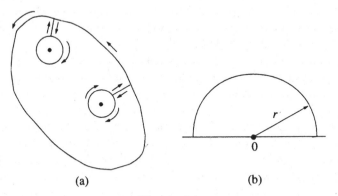

(a)   (b)

Figure 6.1

Example 5.   Let us compute $\displaystyle\int_C \frac{dz}{z^2+1}$, where C is the circle described by $|z - i| = 1$. By the preceding theorem, the integral is $2\pi i$ times the sum of the residues inside C. We have

$$f(z) = \frac{1}{z^2 + 1} = \frac{1}{(z+i)(z-i)} = \frac{i/2}{z+i} - \frac{i/2}{z-i}$$

The residue at i is therefore $-i/2$, and the value of the integral is π.   ∎

Example 6.   Let us compute the integral in Example 5 when C is the circle $|z - i| = 3$. This circle is large enough to enclose both singularities. The residue at $-i$ is $i/2$, and the sum of the residues is 0. The integral is therefore 0. (This illustrates the next theorem.)   ∎

>   **Theorem 4.**   *If f is a proper rational function and if the curve C encloses all the poles of f, then $\int_C f(z)dz = 0$.*

Proof.   Write $f = p/q$, where p and q are polynomials. Since f is proper, the degree of p is less than that of q. Hence the point at ∞ is not a singularity of f. Now, C is the boundary of one region containing the poles, and it is also the boundary of the complementary region in which f is analytic. Hence $\int_C f(z)dz = 0$.   ∎

>   **Theorem 5.**   *Let f be analytic in the closed upper half–plane with the exception of a finite number of poles, none of which are on the real axis. Define*
>
> $$M_r \equiv \sup \{|zf(z)| : |z| = r,\ \mathcal{I}(z) \geqslant 0\}$$
>
>   *If M_r converges to 0 as $r \to \infty$, then $\dfrac{1}{2\pi i} \displaystyle\int_{-\infty}^{\infty} f(z)\, dz$ is the sum of the residues at the poles in the upper half–plane.*

Proof.   Consider the region shown in Figure 6.1b, where C is the semicircular arc and r is chosen so large that all the poles of f lying in the upper half-plane are contained in the semicircular region. On C we have $z = re^{i\theta}$ and $dz = ire^{i\theta} d\theta$. Hence

$$\int_C |f(z)|\, dz = \int_0^\pi |f(re^{i\theta}) \cdot r| d\theta \leqslant \pi M_r \longrightarrow 0$$

By Theorem 3,

$$\int_{-r}^{r} f(z)\, dz + \int_C f(z)\, dz = 2\pi i \times \text{(sum of residues)}$$

By taking the limit as $r \to 0$, we obtain the desired result.   ∎

Problems 6.5

1. Solve this integral equation

$$\int_{-\infty}^{\infty} e^{-|x-s|} u(s)\, ds = f(x)$$

2. Solve this integral equation

$$\int_{-\infty}^{x} u(s)\, ds - \int_{x}^{\infty} u(s)\, ds + u(x) = f(x)$$

3. What happens in Example 2 if $b = 1$? Is the solution $u(t) = \sinh t$?

4. Prove that if $w \in \mathbb{C}^n$, then $w = u + iv$ for suitable points u and v in $\mathbb{R}^n$. We will write $u = \mathcal{R}(w)$ and $v = \mathcal{I}(v)$. Prove, then, that for $x \in \mathbb{R}^n$, the equation $xw = 0$ holds if and only if $x\mathcal{R}(w) = 0$ and $x\mathcal{I}(w) = 0$.

5. Let g be analytic in a circle C centered at z_0, and let f be analytic in C except for having a simple pole at z_0. What is the residue of gf at z_0?

6. Prove that arbitrary data at arbitrary nodes $x_1, x_2, \ldots, x_m$ in $\mathbb{R}^n$ can be interpolated by a function of the form $x \longmapsto \sum_{k=1}^{m} c_k \exp(-|x - x_k|^2)$. (These are Gaussian functions.)

6.6 Applications to Partial Differential Equations

Example 1. The simplest case of the heat equation is

$$(1) \qquad\qquad u_{xx} = u_t$$

in which the subscripts denote partial derivatives. The distribution of heat in an infinite bar would obey this equation for $\infty < x < \infty$ and $t \geqslant 0$. A fully defined practical problem would consist of the differential equation (1) and some auxiliary conditions. To illustrate, we consider (1) with *initial* condition

$$(2) \qquad\qquad u(x,0) = f(x) \qquad -\infty < x < \infty$$

The function f gives the initial temperature distribution in the rod. We define $\widehat{u}(y,t)$ to be the Fourier transform of u *in the space variable*. Thus

$$\widehat{u}(y,t) = \int_{-\infty}^{\infty} u(x,t) e^{-2\pi i x y}\, dx$$

Taking the Fourier transform in Equations (1) and (2) with respect to the space variable, we obtain

$$(3) \qquad \begin{cases} -4\pi^2 y^2 \widehat{u}(y,t) = \widehat{u}_t(y,t) \\[2em] \widehat{u}(y,0) = \widehat{f}(y) \end{cases}$$

Here, again, we use the principle of Theorem 1 in Section 6.2, page 296: $\widehat{P(D)u} = P^+\widehat{u}$, where $P^+(x) = P(2\pi i x)$.

Equation (3) defines an initial-value problem involving a first-order linear ordinary differential equation for the function $\widehat{u}(y, \cdot)$. (The variable y can be ignored, or interpreted as a parameter.) We note that $(\widehat{u})_t = \widehat{(u_t)}$. The phenomenon just observed is typical: Often, a Fourier transform will lead us from a partial differential equation to an ordinary differential equation. The solution of (3) is

$$(4) \qquad\qquad \widehat{u}(y,t) = \widehat{f}(y) e^{-4\pi^2 y^2 t}$$

Now let us think of t as a parameter, and ignore it. Write Equation (4) as $\widehat{u}(y,t) = \widehat{f}(y)\widehat{G}(y,t)$, where $\widehat{G}(y,t) = e^{-4\pi^2 y^2 t}$. Using the principle that $\widehat{\phi\psi} = \widehat{\phi} * \widehat{\psi}$ (Theorem 4 in Section 6.1, page 291), we have

$$(5) \qquad\qquad u(\cdot,t) = f(\cdot) * G(\cdot,t)$$

where $G(\cdot,t)$ is the inverse transform of $y \longmapsto e^{-4\pi^2 y^2 t}$. This inverse is $G(x,t) = (4\pi t)^{-1/2} e^{-x^2/(4t)}$, by Problem 8 of Section 6.3, page 304. Consequently,

$$(6) \qquad\qquad u(x,t) = (4\pi t)^{-1/2} \int_{-\infty}^{\infty} f(x-z) e^{-z^2/(4t)}\, dz \qquad\qquad ∎$$

Example 2.   We consider the problem

$$(7) \qquad \begin{cases} u_{xx} = u_t & x \geqslant 0, \ t \geqslant 0 \\ u(x,0) = f(x), \ u(0,t) = 0 & x \geqslant 0, \ t \geqslant 0 \end{cases}$$

This is a minor modification of Example 1. The bar is "semi-infinite," and one end remains constantly at temperature zero. It is clear that f should have the property $f(0) = u(0,0) = 0$. Suppose that we extend f somehow into the interval $(-\infty, 0)$, and then use the solution of the previous example. Then at $x = 0$ we have

$$(8) \qquad\qquad u(0,t) = (4\pi t)^{-\frac{1}{2}} \int_{-\infty}^{\infty} f(-z) e^{-z^2/4t}\, dz$$

The easiest way to ensure that this will be zero (and thus satisfy the boundary condition in our problem) is to extend f to be an odd function. Then the integrand in Equation (8) is odd, and $u(0,t) = 0$ automatically. So we define $f(-x) = -f(x)$ for $x > 0$, and then Equation (6) gives the solution for Equation (7). ∎

Example 3.   Again, we consider the heat equation with boundary conditions:

$$(9) \qquad \begin{cases} u_{xx} = u_t & x \geqslant 0, \ t \geqslant 0 \\ u(x,0) = f(x) & u(0,t) = g(t) \end{cases}$$

Because the differential equation is linear and homogeneous, the method of superposition can be applied. We solve two related problems, viz.,

$$(10) \qquad v_{xx} = v_t \qquad v(x,0) = f(x) \qquad v(0,t) = 0$$
$$(11) \qquad w_{xx} = w_t \qquad w(x,0) = 0 \qquad w(0,t) = g(t)$$

The solution of (9) will then be $u = v + w$. The problem in (10) is solved in Example 2. In (11), we take the sine transform in the space variable, using w^S to denote the transformed function. With the aid of Problem 1, we have

$$2\pi y g(t) - 4\pi^2 y^2 w^S(y,t) = w_t^S(y,t) \qquad w^S(y,0) = 0$$

Again this is an ordinary differential equation, linear and of the first order. Its solution is easily found to be

$$w^S(y,t) = 2\pi y e^{-4\pi^2 y^2 t} \int_0^t e^{4\pi^2 y^2 \sigma} g(\sigma)\, d\sigma$$

If w is made into an odd function by setting $w(x,t) = -w(-x,t)$ when $x < 0$, then we know from Problem 9 in Section 6.3 (page 304) that

$$\widehat{w}(y,t) = -2i w^S(y,t)$$

Therefore by the Inversion Theorem (Section 6.3, page 303)

$$w(x,t) = \int_{-\infty}^{\infty} \widehat{w}(y,t) e^{2\pi i x y}\, dy$$

or

$$w(x,t) = -4\pi i \int_{-\infty}^{\infty} e^{2\pi i x y} y e^{-4\pi^2 y^2 t} \int_0^t e^{4\pi^2 y^2 \sigma} g(\sigma)\, d\sigma\, dy$$

To simplify this, let $z = 2\pi y$. Then

$$w(x,t) = \frac{-i}{\pi} \int_{-\infty}^{\infty} z e^{ixz} \int_0^t e^{-z^2(t-\sigma)} g(\sigma)\, d\sigma\, dz \qquad \blacksquare$$

Example 4.   The **Helmholtz Equation** is

$$\Delta u - gu = f$$

in which Δ is the Laplacian, $\sum_{k=1}^{n} \partial^2/\partial x_k^2$. The functions f and g are prescribed on $\mathbb{R}^n$, and u is the unknown function of n variables. We shall look at the special case when g is the constant 1. To illustrate some variety in approaching such problems, let us simply try the hypothesis that the problem can be solved with an appropriate convolution: $u = f * h$. Substitution of this form for u in the differential equation leads to

$$\Delta(f * h) - f * h = f$$

Carrying out the differentiation under the integral that defines the convolution, we obtain

$$f * \Delta h - f * h = f$$

Is there a way to cancel the three occurrences of f in this equation? After all, L^1 is a Banach algebra, with multiplication defined by convolution. But there are pitfalls here, since there is no unit element, and therefore there are no inverses. However, the Fourier transform converts the convolutions into ordinary products, according to Theorem 4 in Section 6.1 (page 291):

$$(\widehat{f})(\Delta h)^\wedge - \widehat{f}\,\widehat{h} = \widehat{f}$$

From this equation cancel the factor $\widehat{f}$, and then express $(\Delta h)^{\wedge}$ as in Example 2 in Section 6.2 (page 297):

$$-4\pi^2|x|^2\,\widehat{h}(x) - \widehat{h}(x) = 1$$

$$\widehat{h}(x) = \frac{-1}{1 + 4\pi^2|x|^2}$$

The formula for h itself is obtained by use of the inverse Fourier transform, which leads to

$$h(x) = \pi^{n/2}\int_0^\infty t^{-n/2}\exp\left(-t - |\pi x|^2/t\right)dt$$

The calculation leading to this is given in [Ev], page 187. In that reference, a different definition of the Fourier transform is used, and Problem 6.1.24, page 293, can be helpful in transferring results among different systems.   ∎

Problems 6.6

1. Define the Sine Transform by the equation

$$f^S(t) = \int_0^\infty f(x)\sin 2\pi xt\,dx$$

Show that $(f'')^S(t) = 2\pi tf(0) - 4\pi^2 t^2 f^S(t)$. (Two integrations by parts are needed, in addition to the assumption that $f \in L^1$.)

2. Define the Cosine Transform by the equation

$$f^C(t) = \int_0^\infty f(x)\cos 2\pi xt\,dx$$

Show that $(f'')^C(t) = -f'(0) - 4\pi^2 t^2 f^C(t)$.

3. A function can be decomposed into odd and even parts by writing $f = f_o + f_e$, in which

$$f_o(x) = \tfrac{1}{2}f(x) - \tfrac{1}{2}f(-x) \qquad f_e(x) = \tfrac{1}{2}f(x) + \tfrac{1}{2}f(-x)$$

Show that $\widehat{f} = 2f_e^C - 2if_o^S$.

4. If w^S is the sine transform in the first variable in the function $(x,t) \longmapsto w(x,t)$, what is the difference between $(w^S)_t$ and $(w_t)^S$?

5. Define a scaling operator by the equation $(S_\lambda f)(x) = f(\lambda x)$. Prove that $D^\alpha \circ S_\lambda = \lambda^{|\alpha|}S_\lambda \circ D^\alpha$.

6. If the problem $\Delta u - u = f$ is solved by the formula $u = f * h$, for a certain function h, how can we solve the problem $\Delta u - c^2 u = f$, assuming $c > 0$?

6.7 Tempered Distributions

Let us recall the definitions of two important spaces. The space $\mathcal{D}$, the space of "test functions," consists of all functions in $C^\infty(\mathbb{R}^n)$ that have compact support. (Of course, $\mathcal{D}$ depends on n, but the notation does not show this.) In $\mathcal{D}$ we define convergence by saying that $\phi_j \to 0$ in $\mathcal{D}$ if there is one compact set containing the supports of all ϕ_j and if $(D^\alpha \phi_j)(x)$ converges uniformly to 0 for each multi-index α.

The space $\mathcal{S}$ consists of all functions ϕ in $C^\infty(\mathbb{R}^n)$ such that the function $P \cdot D^\alpha \phi$ is bounded, for all polynomials P and for all multi-indices α. In $\mathcal{S}$ we defined $\phi_j \twoheadrightarrow 0$ to mean that $P \cdot D^\alpha \phi_j$ converges uniformly to 0, for each P and for each α.

It is clear that $\mathcal{D} \subset \mathcal{S}$. A distribution T, being a continuous linear functional on $\mathcal{D}$, may or may not possess a continuous linear extension to $\mathcal{S}$. If it does possess such an extension, the distribution T is said to be **tempered**.

Theorem 1.   *Every distribution having compact support is tempered.*

Proof. Let T be a distribution with compact support K. Select $\psi \in \mathcal{D}$ so that $\psi(x) = 1$ for all x in an open neighborhood of K. We extend T by defining $\overline{T}(\phi) = T(\phi\psi)$ when $\phi \in \mathcal{S}$. Is $\overline{T}$ an extension of T? In other words, do we have $\overline{T}(\phi) = T(\phi)$ for $\phi \in \mathcal{D}$? An equivalent question is whether $T(\psi\phi - \phi) = 0$ for $\phi \in \mathcal{D}$. We use the definition of the support of T to answer this. We must verify only that the support of $(1 - \psi)\phi$ is contained in $\mathbb{R}^n \smallsetminus K$. This is true because $1 - \psi$ is zero on a neighborhood of K. The linearity of $\overline{T}$ is trivial. For the continuity, suppose that $\phi_j \twoheadrightarrow 0$ in $\mathcal{S}$. Then for any α, $D^\alpha \phi_j$ tends uniformly to 0, and $D^\alpha(\phi_j \psi)$ tends uniformly to 0 by Leibniz's Rule. Since there is one compact set containing the supports of all $\psi\phi_j$, we can conclude that $\psi\phi_j \twoheadrightarrow 0$ in $\mathcal{D}$. By the continuity of T, $T(\psi\phi_j) \to 0$ and $\overline{T}(\phi_j) \to 0$. ∎

If T is a tempered distribution, it is customary to make no notational distinction between T and its extension to $\mathcal{S}$. If V is a continuous linear functional on $\mathcal{S}$, then its restriction $V|\mathcal{D}$ is a distribution. Indeed, the linearity of $V|\mathcal{D}$ is obvious, and the continuity is verified as follows. Let $\phi_j \twoheadrightarrow 0$ in $\mathcal{D}$. Since there is one compact set K containing the supports of all ϕ_j and since any polynomial is bounded on K, we see that $P(x)(D^\alpha \phi_j)(x) \to 0$ uniformly for any multi-index α and for any polynomial P. Hence $\phi_j \twoheadrightarrow 0$ in $\mathcal{S}$. Since V is continuous, $(V|\mathcal{D})(\phi_j) \to 0$. This proves that the space $\mathcal{S}'$ of all continuous linear functionals on $\mathcal{S}$ can be identified with the space of all tempered distributions.

Lemma.   *The set $\mathcal{D}(\mathbb{R}^n)$ is dense in the space $\mathcal{S}(\mathbb{R}^n)$.*

Proof. Given an element ϕ in $\mathcal{S}$, we must construct a sequence in $\mathcal{D}$ converging to ϕ (in the topology of $\mathcal{S}$). For this purpose, select $\psi \in \mathcal{D}$ such that $\psi(x) = 1$ whenever $|x| \leqslant 1$. For $j = 1, 2, \ldots$, let $\psi_j(x) = \psi(x/j)$. It is obvious that $\psi_j\phi$ belongs to $\mathcal{D}$. In order to show that these elements converge to ϕ in $\mathcal{S}$, we must prove that for any polynomial P and any multi-index α,

(1) $P \cdot D^\alpha(\phi - \phi\psi_j) \to 0$ uniformly in $\mathbb{R}^n$

In the following, P and α are fixed. By the Leibniz Formula, the expression in (1) equals

(2) $P \cdot \displaystyle\sum_{\beta \leqslant \alpha} \binom{\alpha}{\beta} D^{\alpha - \beta}\phi \cdot D^\beta(1 - \psi_j)$

Since $\phi \in \mathbf{S}$, we must have $P \cdot D^{\alpha-\beta}\phi \in \mathbf{S}$ also. Hence $|x|^2 |P(x) \cdot D^{\alpha-\beta}\phi(x)|$ is bounded, say by M. This bound can be chosen to serve for all β in the range $0 \leqslant \beta \leqslant \alpha$. Increase M if necessary so that for all β in that same range,

$$|D^\beta(1 - \psi_j)| = \left|\frac{1}{j}\right|^{|\beta|} |D^\beta(1 - \psi)| \leqslant |D^\beta(1 - \psi)| \leqslant |D^\beta 1| + |D^\beta \psi| \leqslant M$$

Fix an index j. If $|x| \leqslant j$, then $1 - \psi_j(x) = 0$, and the expression in (2) is 0 at x. If $|x| \geqslant j$, then

$$|P(x) \cdot D^{\alpha-\beta}\phi(x)| \leqslant M/|x|^2 \leqslant M/j^2$$

Also, for $|x| \geqslant j$, $|D^\beta(1 - \psi_j)| \leqslant M$. Hence the expression in (2) has modulus no greater than

$$\sum_{0 \leqslant \beta \leqslant \alpha} \binom{\alpha}{\beta} M j^{-2} M = c j^{-2}$$

This establishes (1).   ∎

Theorem 2.   *Let f be a measurable function such that $f/P \in L^1(\mathbb{R}^n)$ for some polynomial P. Then $\tilde{f}$ is a tempered distribution.*

Proof.   For $\phi \in \mathbf{S}$, we have

$$\tilde{f}(\phi) = \int_{\mathbb{R}^n} f(x)\phi(x)\,dx$$

Suppose that P is a polynomial such that $f/P \in L^1$. Write

$$\tilde{f}(\phi) = \int (f/P)(P \cdot \phi)$$

Since $\phi \in \mathbf{S}$, $P\phi$ is bounded, and the integral exists. If $\phi_j \to 0$ in $\mathbf{S}$, then $P(x)\phi_j(x) \to 0$ uniformly on $\mathbb{R}^n$, and consequently,

$$|\tilde{f}(\phi_j)| \leqslant \sup_x \left| P(x)\phi_j(x) \right| \int \left| f/P \right| \to 0 \qquad ∎$$

Definition.   The Fourier transform of a tempered distribution T is defined by the equation $\widehat{T}(\phi) = T(\widehat{\phi})$ for all $\phi \in \mathbf{S}$. An equivalent equation is $\widehat{T} = T \circ \mathcal{F}$, where $\mathcal{F}$ is the Fourier operator mapping ϕ to $\widehat{\phi}$.

Theorem 3.   *If T is a tempered distribution, then so is $\widehat{T}$. Moreover, the map $T \longmapsto \widehat{T}$ is linear, injective, surjective, and continuous from $\mathbf{S}'$ to $\mathbf{S}'$.*

Proof.   The Fourier operator $\mathcal{F}$ is a continuous linear bijection from $\mathbf{S}$ onto $\mathbf{S}$ by Theorem 3 in Section 6.3, page 303. Also, $\mathcal{F}^{-1} = \mathcal{F}^3$. Since $\widehat{T} = T \circ \mathcal{F}$, we see that $\widehat{T}$ is the composition of two continuous linear maps, and is therefore itself continuous and linear. Hence $\widehat{T}$ is a member of $\mathbf{S}'$.

For the linearity of the map in question we write

$$(aT + bU)^\wedge = (aT + bU) \circ \mathcal{F} = aT \circ \mathcal{F} + bU \circ \mathcal{F} = a\widehat{T} + b\widehat{U}$$

For the injectivity, suppose $\widehat{T} = 0$. Then $T \circ \mathcal{F} = 0$ and $T(\phi) = 0$ for all ϕ in the range of $\mathcal{F}$. Since $\mathcal{F}$ is surjective from $\mathcal{S}$ to $\mathcal{S}$, the range of $\mathcal{F}$ is $\mathcal{S}$. Hence $T(\phi) = 0$ for all ϕ in $\mathcal{S}$; i.e., $T = 0$.

For the surjectivity, let T be any element of $\mathcal{S}'$. Then $T = T \circ \mathcal{F}^4 = (T \circ \mathcal{F}^3) \circ \mathcal{F}$. Note that $T \circ \mathcal{F}^3$ is in $\mathcal{S}'$ by the first part of this proof.

For the continuity, let $T_j \in \mathcal{S}'$ and $T_j \to 0$. This means that $T_j(\phi) \to 0$ for all ϕ in $\mathcal{S}$. Consequently, $\widehat{T_j}(\phi) = T_j(\widehat{\phi}) \to 0$ and $\widehat{T_j} \to 0$.   ∎

Example. Since the Dirac distribution δ has compact support, it is a tempered distribution. What is its Fourier transform? We have, for any $\phi \in \mathcal{S}$,

$$\widehat{\delta}(\phi) = \delta(\widehat{\phi}) = \widehat{\phi}(0) = \int \phi(x)\, dx = \widetilde{1}(\phi)$$

(Remember that the tilde denotes the distribution corresponding to a function.) Thus, $\widehat{\delta} = \widetilde{1}$.   ∎

Theorem 4.   *If T is a tempered distribution and P is a polynomial, then*

$$\widehat{PT} = P\left(\frac{-\partial}{2\pi i}\right)\widehat{T} \qquad \text{and} \qquad P \cdot \widehat{T} = \widehat{P\left(\frac{\partial}{2\pi i}\right)T}$$

Proof. For ϕ in $\mathcal{S}$ we have

$$\widehat{PT}(\phi) = (PT)(\widehat{\phi}) = T(P\widehat{\phi}) = T\left[(P(\frac{D}{2\pi i})\phi)^\wedge\right] = \widehat{T}\left(P(\frac{D}{2\pi i})\phi\right) = \left[P(\frac{-\partial}{2\pi i})\widehat{T}\right](\phi)$$

We used Theorem 1 in Section 6.2, page 296, in this calculation. The other equation is left as Problem 6.   ∎

Problems 6.7

1. Prove that every f in $L^1(\mathbb{R}^n)$ is a tempered distribution.
2. Prove that every polynomial is a tempered distribution.
3. Prove that if f is measurable and satisfies $|f| \leqslant |P|$ for some polynomial P, then f is a tempered distribution.
4. Prove that the function f defined by $f(x) = e^x$, $(n = 1)$ is not a tempered distribution. Note that f is a distribution, however,
5. Let $f(x) = (|x| + 1)^{-1}$. Explain how it is possible for $\widehat{f}$ to belong to $L^2(\mathbb{R})$ in spite of the fact that the integral $\int_{-\infty}^{\infty} f(x)e^{-2\pi i x t}\, dx$ is meaningless.
6. Prove the remaining part of Theorem 4.
7. Under what circumstances is the reciprocal of a polynomial a tempered distribution?
8. Define δ_a by $\delta_a(\phi) = \phi(a)$. Compute $\widehat{\delta_a}$.
9. Prove that our definition of the Fourier transform of a tempered distribution is consistent with the classical Fourier transform of a function.
10. Is the function $f(x) = e^{-|x|}$ a member of $\mathcal{S}$? Is f a tempered distribution?
11. What flaw is there in defining $\widehat{T}\phi = T\widehat{\phi}$ for $T \in \mathcal{D}'$ and $\phi \in \mathcal{D}$? ·
12. Find the Fourier transforms of these functions, interpreted as tempered distributions:
    (a) $g(x) = x$ $(x \in \mathbb{R})$

(b) $g(x) = 1$ $(x \in \mathbb{R}^n)$
(c) $g(x) = x^2$ $(x \in \mathbb{R}^n)$.

13. Let $T \in \mathcal{D}'(\mathbb{R}^n)$. Why can we not prove that T is tempered by using the density of $\mathcal{D}$ in $\mathcal{S}$ and Problem 6.4.6 on page 309?

6.8 Sobolev Spaces

This section provides an introduction to Sobolev spaces. These are Banach spaces that have become essential in the study of differential equations and the numerical processes for solving them, such as the finite element method. It is therefore not surprising that the elements in these Sobolev spaces are functions possessing derivatives of certain orders. The theory relies on distribution theory and capitalizes on the fact that distributions have derivatives of all orders.

Our first step is to generalize the theory of distributions slightly by considering the domain of our test functions to be an arbitrary open set Ω in $\mathbb{R}^n$. Usually, Ω will remain fixed in any application of the theory. Our test functions are C^∞ functions defined on Ω and having compact support. The support of a test function is, then, a compact subset of Ω. The space of all these test functions is denoted by $\mathcal{D}(\Omega)$. Convergence in this space has the expected meaning: The assertion $\phi_j \twoheadrightarrow 0$ means that there is a single compact subset K in Ω containing the supports of all ϕ_j, and that on K we have $\partial^\alpha \phi_j \to 0$ uniformly for each multi-index α. The use of the distinctive symbol $\twoheadrightarrow$ is to remind us of the very special concept of convergence employed in this context.

The dual space of $\mathcal{D}(\Omega)$ is the space of distributions, denoted by $\mathcal{D}'(\Omega)$. Its elements are the continuous linear functionals defined on $\mathcal{D}(\Omega)$. Continuity of a distribution T means that T preserves limits: From the hypothesis "ϕ_j converges to ϕ" we may conclude that "$T(\phi_j)$ converges to $T(\phi)$."

As before, we can create distributions from locally integrable functions. Local integrability of a function f defined on Ω means that for every compact set K in Ω, the integral $\int_\Omega |f|$ is finite. We then write $f \in L^1_{\text{loc}}(\Omega)$. All these definitions are in harmony with the definitions first seen on $\mathbb{R}^n$, and indeed, we include $\Omega = \mathbb{R}^n$ as a special case. The distribution corresponding to a locally integrable function f has been denoted by $\widetilde{f}$; its definition is now

$$(1) \qquad \widetilde{f}(\phi) = \int_\Omega f(x)\phi(x)\,dx \qquad (\phi \in \mathcal{D}(\Omega))$$

Sometimes we do not belabor the distinction between f and $\widetilde{f}$, and think of each f in $L^1_{\text{loc}}(\Omega)$ as a distribution. The clear advantage of this is that such functions will possess derivatives of all orders (in the distribution sense). Derivatives of this type are called "weak derivatives" or "distribution derivatives" to distinguish them from the classical derivatives, which are then called "strong" derivatives. Thus if $f \in L^1_{\text{loc}}(\Omega)$ and if α is a multi-index, $D^\alpha f$ need not exist in the classical sense, but $\partial^\alpha f$ will always exist. Recall that in this book the symbol ∂^α was reserved for distributions, and

$$(2) \qquad (\partial^\alpha T)(\phi) = (-1)^{|\alpha|} T(D^\alpha \phi) \qquad (\phi \in \mathcal{D}(\Omega))$$

Equation (2) can be written more succinctly as $\partial^\alpha T = T \circ (-D)^\alpha$. Then, for any polynomial P, we have $P(\partial)T = T \circ P(-D)$.

The classical spaces $L^p(\Omega)$, for $1 \leqslant p < \infty$, are defined as follows. The elements of $L^p(\Omega)$ are the Lebesgue measurable functions f defined on Ω for which

$$(3) \qquad \|f\|_p \equiv \left\{ \int_\Omega |f(x)|^p \, dx \right\}^{1/p} < \infty$$

The norm here can also be denoted by $\|f\|_{L^p(\Omega)}$. The resulting normed linear space is *complete*. A fine point that complicates matters is that, in fact, the elements of L^p are *equivalence classes* of functions, two functions being regarded as equivalent (i.e., belonging to the same equivalence class) if they differ only on a set of measure zero. Section 8.7 (pages 409ff) provides more information about the L^p spaces.

Definition.   To say that a distribution T belongs to $L^p(\Omega)$ means that $T = \widetilde{g}$ for some $g \in L^p(\Omega)$. When this circumstance occurs, we write $\|T\|_p = \|g\|_p$. With suitable caution, one can also write $T \in L^p(\Omega)$.

Let $k \in \mathbb{Z}_+$ and $1 \leqslant p \leqslant \infty$. The **Sobolev space** $W^{k,p}(\Omega)$ consists of all functions f in $L^p(\Omega)$ such that $\partial^\alpha \widetilde{f} \in L^p(\Omega)$ for all multi-indices α satisfying $|\alpha| \leqslant k$. More precisely, this space is

$$\left\{ f \in L^p(\Omega) \; : \;\; \text{there exist } g_\alpha \in L^p(\Omega) \;\; \text{such that} \;\; \partial^\alpha \widetilde{f} = \widetilde{g}_\alpha, \;\; \text{for } |\alpha| \leqslant k \right\}$$

Observe that we need the fact (indicated in Problem 2) that each member of $L^p(\Omega)$ is in $L^1_{\text{loc}}(\Omega)$. In the space $W^{k,p}(\Omega)$, a norm is defined by putting

$$(4) \qquad \|f\|_{k,p} = \left(\sum_{|\alpha| \leqslant k} \|\partial^\alpha \widetilde{f}\|_p^p \right)^{1/p} = \left(\sum_{|\alpha| \leqslant k} \|g_\alpha\|_p^p \right)^{1/p}$$

The verification of the norm axioms is relegated to the problems. Notice that in Equation (4) the conventions of the above definition are being used.

Theorem 1.   *The Sobolev spaces $W^{k,p}(\Omega)$ are complete.*

Proof.   We begin by observing a useful implication:

$$(5) \qquad \Big[h_j \to 0 \;\; \text{in} \;\; L^p(\Omega) \Big] \;\; \Longrightarrow \;\; \Big[\widetilde{h}_j \to 0 \;\; \text{in} \;\; \mathcal{D}'(\Omega) \Big]$$

To prove this, let ϕ be a test function. By the Hölder Inequality (Section 8.7, page 409),

$$(6) \qquad |\widetilde{h}_j(\phi)| = \left| \int_\Omega h_j(x)\phi(x) \right| \leqslant \left\{ \int_\Omega |h_j(x)|^p \right\}^{1/p} \left\{ \int_\Omega |\phi(x)|^q \right\}^{1/q}$$

Here $1 \leqslant p < \infty$ and $1/p + 1/q = 1$. Inequality (6) establishes (5).

Now let $[f_j]$ be a Cauchy sequence in $W^{k,p}(\Omega)$. By the definition of this space, there exist functions $g_j^\alpha \in L^p(\Omega)$ such that $\partial^\alpha \widetilde{f_j} = \widetilde{g_j^\alpha}$ for $|\alpha| \leqslant k$. By Problem 2, each g_j^α belongs to $L_{\text{loc}}^1(\Omega)$, and therefore $\widetilde{g_j^\alpha}$ is a distribution. From the definition of the norm (4), we see that for each α the sequence $[g_j^\alpha]$ has the Cauchy property in $L^p(\Omega)$. Since $L^p(\Omega)$ is complete (by the Riesz–Fischer Theorem, page 411), there exist functions $f^\alpha \in L^p(\Omega)$ such that $g_j^\alpha \to f^\alpha$ in $L^p(\Omega)$. By (5), we conclude that $\widetilde{g_j^\alpha} \to \widetilde{f^\alpha}$. In particular, for $\alpha = 0$, $g_j^0 \to f^0$. Consider the equation

$$(7) \qquad \widetilde{g_j^\alpha}(\phi) = (\partial^\alpha \widetilde{f_j})(\phi) = (-1)^{|\alpha|} \widetilde{f_j}(D^\alpha \phi) = (-1)^{|\alpha|} \widetilde{g_j^0}(D^\alpha \phi)$$

By letting $j \to \infty$ in (7) and using the convergence of $\widetilde{g_j^\alpha}$, we obtain

$$\widetilde{f^\alpha}(\phi) = (-1)^{|\alpha|} \widetilde{f^0}(D^\alpha \phi) = \partial^\alpha \widetilde{f^0}(\phi)$$

This proves that $\partial^\alpha \widetilde{f^0} = \widetilde{f^\alpha}$, and shows that for $|\alpha| \leqslant k$, $\partial^\alpha \widetilde{f^0} \in L^p(\Omega)$. By the definition of the Sobolev space, $f^0 \in W^{k,p}(\Omega)$. Finally, we write

$$\|f_j - f^0\|_{k,p}^p = \sum_{|\alpha| \leqslant k} \|\partial^\alpha \widetilde{f_j} - \partial^\alpha \widetilde{f^0}\|_p^p = \sum_{|\alpha| \leqslant k} \|g_j^\alpha - f^\alpha\|_p^p \to 0 \qquad \blacksquare$$

The test function space $\mathcal{D}(\Omega)$ is, in general, not a dense subspace of the Sobolev space $W^{k,p}(\Omega)$. This is easy to understand: Each ϕ in $\mathcal{D}(\Omega)$ has compact support in Ω. Consequently, $\phi(x) = 0$ on the boundary of Ω. The closure of $\mathcal{D}(\Omega)$ in $W^{k,p}(\Omega)$ can therefore contain only functions that vanish on the boundary of Ω. However, the special case $\Omega = \mathbb{R}^n$ is satisfactory from this standpoint:

Theorem 2.   *The test function subspace $\mathcal{D}(\mathbb{R}^n)$ is dense in $W^{k,p}(\mathbb{R}^n)$.*

For the proof of Theorem 2, consult [Hu]. Some closely related theorems are given in this section.

In many proofs we require a mollifier, which is a test function ψ having these additional properties: $\psi \geqslant 0$, $\psi(x) = 0$ when $\|x\| \geqslant 1$, and $\int \psi = 1$. Then one puts $\psi_j(x) = j^n \psi(jx)$. A "mollification of f with radius ϵ" is then $\psi_j * f$ with $1/j < \epsilon$. These matters are discussed in Section 5.1 (pages 246ff) and Section 5.5 (pages 269ff).

Lemma 1.   *If $f \in L^p(\mathbb{R}^n)$, and if ψ_j is as described above, then $f * \psi_j \to f$ in $L^p(\mathbb{R}^n)$, as $j \to \infty$.*

Proof. The case $p = 1$ is contained in the proof of Theorem 1 in Section 6.4, page 306. Let B_j be the support of ψ_j (i.e., the ball at 0 of radius $1/j$). By familiar calculations and Hölder's inequality (Section 8.7, page 409) we have

$$\left|(f * \psi_j)(x) - f(x)\right| = \left|\int_{B_j} [f(x - y) - f(x)] \psi_j(y) \, dy\right|$$

$$\leq \left\{\int_{B_j} |f(x - y) - f(x)|^p \, dy\right\}^{1/p} \|\psi_j\|_q$$

(Here q is the index conjugate to p: $pq = p + q$.) Hence,

$$\left|(f * \psi_j)(x) - f(x)\right|^p \leq \|\psi_j\|_q^p \int_{B_j} |f(x - y) - f(x)|^p \, dy .$$

Thus, using the Fubini theorem (page 426), we have

$$\int_{\mathbb{R}^n} |(f * \psi_j)(x) - f(x)|^p \, dx \leq \|\psi_j\|_q^p \int_{B_j} \int_{\mathbb{R}^n} |f(x - y) - f(x)|^p \, dx \, dy$$

We can write this in the form

$$\|f * \psi_j - f\|_p^p \leq \|\psi_j\|_q^p \int_{B_j} \|E_y f - f\|_p^p \, dy$$

where E_y denotes the translation operator defined by $(E_y \phi)(x) = \phi(x - y)$. Recall, from Lemma 3 in Section 6.4 (page 306), that for a fixed element f in $L^p(\mathbb{R}^n)$, the map $y \longmapsto E_y f$ is continuous from $\mathbb{R}^n$ to $L^p(\mathbb{R}^n)$. Hence there corresponds to any positive ε a positive δ such that

$$|y| \leq \delta \quad \Longrightarrow \quad \|E_y f - f\|_p < \varepsilon$$

Thus if $1/j \leq \delta$, we shall have, from the above inequalities,

$$\|f * \psi_j - f\|_p^p \leq \varepsilon^p \mu(B_j) \|\psi_j\|_q^p$$

where $\mu(B_j)$ is the Lebesgue measure of the ball of radius $1/j$. By enclosing that ball in a "cube" of side $2/j$, we see that $\mu(B_j) \leq (2/j)^n$. Thus,

$$\|f * \psi_j - f\|_p \leq \varepsilon (2/j)^{n/p} \|\psi_j\|_q$$

In order to estimate the right-hand side in the above inequality, use Problem 6 to get

$$\varepsilon \left(\frac{2}{j}\right)^{n/p} \|\psi_j\|_q = \varepsilon \left(\frac{2}{j}\right)^{n/p} j^{n(q-1)/q} \|\psi\|_q = \varepsilon 2^n j^{n(1-1/q-1/p)} \|\psi\|_q = \varepsilon 2^n \|\psi\|_q \quad \blacksquare$$

Theorem 3.   *The set of functions in $W^{k,p}(\Omega)$ that are of class C^∞ is dense in $W^{k,p}(\Omega)$.*

Proof.   Let $B_1, B_2, \ldots$ be a sequence of open balls such that $\overline{B_i} \subset \Omega$ for all i and $\bigcup B_i = \Omega$. The center and radius of B_i are indicated by writing $B_i = B(x_i, r_i)$. Appealing to Theorem 1 in Section 5.7 (page 282), we obtain a partition of unity subordinate to the collection of open balls. Thus, we have test functions ϕ_i satisfying $0 \leqslant \phi_i \leqslant 1$. Further, $\mathrm{supp}(\phi_i) \subset B_i$, and for any compact set K in Ω, there exists an integer m such that $\sum_1^m \phi_i = 1$ on a neighborhood of K. Now suppose that $f \in W^{k,p}(\Omega)$. Let $0 < \epsilon < 1/2$. Eventually, we shall find a C^∞-function g in $W^{k,p}(\Omega)$ such that $\|f - g\| < 2\epsilon$.

Select a sequence $\delta_i \downarrow 0$ such that $\overline{B(x_i, (1 + \delta_i)r_i)} \subset \Omega$ for each i. Define $f_i = \phi_i f$. Let g_i be a mollification of f with radius $\delta_i r_i$. At the same time, we decrease δ_i if necessary to obtain the inequality $\|g_i - f_i\|_{W^{k,p}(\Omega)} < \epsilon/2^i$. (This step requires the preceding lemma.) Define $g = \sum g_i$. If $\mathcal{O}$ is a bounded open set in Ω, then $\overline{\mathcal{O}}$ is compact, and for some integer m, $\sum_{i=1}^m \phi_i = 1$ on a neighborhood of $\overline{\mathcal{O}}$. On $\mathcal{O}$, we have

$$\sum_{i=1}^m f_i = \sum_{i=1}^m \phi_i f = f \sum_{i=1}^m \phi_i = f$$

Then we can perform the following calculation, in which the norm in the space $W^{k,p}(\mathcal{O})$ is employed (until the last step, where the domain Ω enters):

$$\|f - g\| = \|\sum_{i=1}^m f_i - \sum_{i=1}^\infty g_i\| = \|\sum_{i=1}^\infty (f_i - g_i)\|$$

$$\leqslant \sum_{i=1}^\infty \|f_i - g_i\| \leqslant \sum_{i=1}^\infty \|f_i - g_i\|_{W^{k,p}(\Omega)}$$

$$\leqslant \epsilon/2 + \epsilon/4 + \cdots = \epsilon \qquad \blacksquare$$

Another way of interpreting the space $W^{k,p}(\Omega)$ will be described here. It allows one to understand this space without an appeal to distributions. Notice, to start with, that the set

$$V = \{f \in C^k(\Omega) \ : \ \|f\|_{k,p} < \infty\}$$

is a linear subspace of $W^{k,p}(\Omega)$. In drawing this conclusion it is necessary to identify any classical derivative $D^\alpha f$ with its distributional counterpart $\partial^\alpha(\tilde{f})$. Indeed, the elements of $W^{k,p}(\Omega)$ have been defined to be distributions.

Since $V \subset W^{k,p}(\Omega)$, the closure of V is also a subspace, and it is denoted here by $V^{k,p}(\Omega)$. We can also characterize $V^{k,p}(\Omega)$ as the completion of V in the norm $\| \cdot \|_{k,p}$. The true state of affairs is quite simple, as stated in the next theorem, the proof of which we refer the reader to [Ad] or [Maz].

Theorem 4. The Meyers–Serrin Theorem.

$$V^{k,p}(\Omega) = W^{k,p}(\Omega) \qquad\qquad 1 \leqslant p < \infty$$

Embedding Theorems. Here we explore the relations that may exist between two Sobolev spaces, in particular, the relation of one such space being continuously embedded in another.

For general normed linear spaces $(E, \|\cdot\|_E)$ and $(F, \|\cdot\|_F)$, we say that F is **embedded** in E (and write $F \hookrightarrow E$) if

(a) $F \subset E$;

(b) There is a constant c such that $\|f\|_E \leqslant c\|f\|_F$ for all $f \in F$.

Part (a) of this definition is algebraic: It asserts that F is a linear subspace of the linear space E. Part (b) is topological: It asserts that the identity map $I : F \to E$ is continuous (i.e., bounded). Indeed, if

$$\|I\| = \sup\{\|If\|_E \ : \ \|f\|_F = 1\} = c$$

then the inequality in Part (b) follows.

Example 1. Every continuous function on the interval $[a, b]$ is integrable. Hence, this simple containment relation is valid: $C[a, b] \subset L^1[a, b]$. Is this an embedding? We seek a constant c such that

$$\|f\|_1 \leqslant c\|f\|_\infty \qquad (f \in C[a, b])$$

The constant $c = b - a$ obviously serves:

$$\|f\|_1 = \int_a^b |f(x)|\, dx \leqslant \int_a^b \|f\|_\infty = (b - a)\|f\|_\infty \qquad\qquad \blacksquare$$

Example 2. If $1 \leqslant s < r < \infty$ and if the domain Ω has finite Lebesgue measure, then $L^r(\Omega) \hookrightarrow L^s(\Omega)$. To prove this, start with an f in $L^r(\Omega)$ and write $r = ps$. We may assume that $f \geqslant 0$. Then f^s is in $L^p(\Omega)$ because $\int f^{sp} = \int f^r$. Use the Hölder Inequality (page 409) with conjugate indices p and $q = p/(p-1)$:

$$\int f^s \cdot 1 \leqslant \|f^s\|_p \cdot \|1\|_q$$

Taking the $1/s$ power in this inequality gives us

$$\|f\|_s \leqslant \|f^s\|_p^{1/s}\|1\|_q^{1/s} = \|f\|_r \mu(\Omega)^{(1/s)-(1/r)} \qquad\qquad \blacksquare$$

Theorem 5. $W^{1,2}(\mathbb{R}) \hookrightarrow W^{0,\infty}(\mathbb{R})$.

Proof. (In outline. For details, see [LL], Chapter 8.) Let f be an element of $W^{1,2}(\mathbb{R})$. Since $\mathcal{D}(\mathbb{R})$ is dense in $W^{1,2}(\mathbb{R})$, there exists a sequence $[f_i]$ in $\mathcal{D}(\mathbb{R})$ converging to f in the norm of $W^{1,2}$. Each f_i has compact support and therefore satisfies $f_i(\pm\infty) = 0$. Since $f_i f_i' = (f_i^2)'/2$, we have

$$f_i^2(x) = \frac{1}{2}[f_i^2(x) - f_i^2(-\infty)] - \frac{1}{2}[f_i^2(\infty) - f_i^2(x)] = \int_{-\infty}^x f_i f_i' - \int_x^\infty f_i f_i'$$

By taking the limit of a suitable subsequence, we obtain the same equation for f, at almost all points x. Then, with the aid of the Cauchy–Schwarz inequality and the inequality between the geometric and arithmetic means, we have

$$f^2(x) \leqslant \int_{-\infty}^{x} |ff'| + \int_{x}^{\infty} |ff'| = \int_{-\infty}^{\infty} |ff'| \leqslant \|f\|_2 \|f'\|_2 \leqslant \frac{1}{2}\|f\|_2^2 + \frac{1}{2}\|f'\|_2^2$$

Consequently,

$$|f(x)| \leqslant \frac{1}{\sqrt{2}}\sqrt{\|f\|_2^2 + \|f'\|_2^2}$$

This establishes the embedding inequality:

$$\|f\|_{0,\infty} \leqslant \frac{1}{\sqrt{2}}\|f\|_{1,2} \qquad\blacksquare$$

The next theorem is one of many embedding theorems, and is given here as just a sample from this vast landscape. It involves one of the spaces $W_0^{k,p}(\Omega)$. This space is defined to be the closed subspace of $W^{k,p}(\Omega)$ generated by the set of test functions $\mathcal{D}(\Omega)$. For this theorem and many others in the same area, consult [Zie] pages 53ff, or [Ad] pages 97ff.

Theorem 6. Let Ω be an open set in $\mathbb{R}^n$. Let k and j be nonnegative integers. If $1 \leqslant p < \infty$, $kp \leqslant n$, and $p \leqslant r \leqslant np/(n - kp)$, then

$$W_0^{j+k,p}(\Omega) \hookrightarrow W^{j,r}(\Omega)$$

There are in the literature many theorems concerning *compact* embeddings of Sobolev spaces. This means, naturally, that the identity map that arises in the definition is a compact operator, i.e., it maps bounded sets to sets having compact closure. An example of such a theorem is the next one, known by the names Rellich and Kondrachov. It is obviously a counterpart of Theorem 6.

Theorem 7. Let Ω_0 be an open and bounded subset of an open domain Ω in $\mathbb{R}^n$. If $j \geqslant 0$, $k \geqslant 1$, $1 \leqslant p < \infty$, $0 < n - kp \leqslant n$, $kp \leqslant n$, and $1 \leqslant r < np/(n - kp)$, then there is a compact embedding

$$W_0^{j+k,p}(\Omega) \hookrightarrow W^{j,r}(\Omega_0)$$

For describing embeddings into spaces of continuous functions, define $C_b^m(\Omega)$ to be the set of all functions defined on Ω such that the derivatives $D^\alpha f$ exist, are continuous, and are bounded on Ω, for all multi-indices α satisfying $|\alpha| \leqslant m$. The norm adopted for this space is

$$\|f\|_{C_b^m(\Omega)} = \max_{|\alpha|\leqslant m} \sup_{x\in\Omega} |D^\alpha f(x)|$$

Theorem 8. If $k > nm/2$, then $W_0^{k,p}(\Omega) \hookrightarrow C_b^m(\Omega)$.

Theorem 8 (and others like it) can be used to establish that a distributional solution of a partial differential equation is in fact a classical solution.

The Sobolev–Hilbert Spaces. The spaces $W^{k,2}(\Omega)$ are Hilbert spaces and are conventionally denoted by $H^k(\Omega)$. For the special case $\Omega = \mathbb{R}^n$, we can follow Friedlander [Fri], and define them for arbitrary real indices s as follows. The space $H^s(\mathbb{R}^n)$ consists of all tempered distributions T for which

$$(1 + |x|^2)^{s/2}\widehat{T} \quad \in \quad L^2(\mathbb{R}^n)$$

Matters not touched upon here: (1) The importance of conditions on the boundary of Ω for more powerful embeddings. (2) The Sobolev spaces for non-integer values of k. (3) The duality theory of Sobolev spaces; i.e., identifying their conjugate spaces as function spaces.

Problems 6.8

1. Prove that the norm defined in Equation (4) satisfies all the postulates for a norm.
2. Prove that $L^p(\Omega) \subset L^1_{\mathrm{loc}}(\Omega)$.
3. Prove that for $1 \leqslant p < \infty$, $\mathcal{D}(\Omega) \subset L^p(\Omega) \subset \mathcal{D}'(\Omega)$. Show that the embedding of $L^p(\Omega)$ in $\mathcal{D}'(\Omega)$ is continuous and injective.
4. Show that the function
$$f(x) = \begin{cases} 1 & |x| < 1 \\ 0 & |x| \geqslant 1 \end{cases}$$
   belongs to $W^{0,p}(\mathbb{R})$ but not to $W^{1,p}(\mathbb{R})$.
5. Prove this theorem of W.H. Young. If $f \in L^p(\mathbb{R})$ and $g \in L^1(\mathbb{R})$, then $f * g \in L^p(\mathbb{R})$, and
$$\|f * g\|_p \leqslant \|f\|_p \, \|g\|_1$$
   (See [HewS], page 414, for a stronger result.)
6. Prove that if $\phi \in \mathcal{D}(\mathbb{R}^n)$ and $\phi_j(x) = j^n\phi(jx)$, then $\|\phi_j\|_q = j^{n(q-1)/q}\|\phi\|_q$.
7. Why can we not define a more general Sobolev space, say $W^p_\alpha(\Omega)$, where α is a multi-index, and admit all functions such that $\partial^\beta \widetilde{f} \in L^p(\Omega)$ for all multi-indices $\beta \leqslant \alpha$? What would the norm be? Would the space be complete?
8. Find the norm of the identity operator for these embeddings: (a) $C[a,b] \hookrightarrow L^2[a,b]$; (b) $\ell^1 \hookrightarrow \ell^2$; (c) $(\mathbb{R}^n, \|\cdot\|_\infty) \hookrightarrow (\mathbb{R}^n, \|\cdot\|_2)$.
9. Prove that if $m \geqslant k$, then $W^{m,p}(\Omega) \subset W^{k,p}(\Omega)$, and this set inclusion is actually a continuous embedding. That is, the identity map of $W^{m,p}(\Omega)$ into $W^{k,p}(\Omega)$ is continuous. What is the relationship between the norms in these two spaces?
10. If $\widetilde{f} \in L^p(\Omega)$, does it follow that $f \in L^p(\Omega)$?
11. Let $\mathcal{O}$ be an open set in $\mathbb{R}^n$ that contains the closed ball $\overline{B(x,r)}$. Prove that for some $\rho > r$, $B(x,\rho)$ is contained in $\mathcal{O}$. (This was used in the proof of Theorem 2.)
12. Prove that the following formula defines an inner product in the space $W^{k,2}(\Omega)$:
$$\langle f, g \rangle = \sum_{\alpha \leqslant k} \int_\Omega (D^\alpha f)(D^\alpha g)\, dx$$
13. Let g and h be locally integrable functions on the open set Ω. If
$$\int_\Omega g(x)\phi(x)\, dx = \int_\Omega h(x)D^\alpha \phi(x)\, dx$$
   for all $\phi \in \mathcal{D}(\Omega)$, what conclusion can be drawn?
14. Prove that if $\phi \in \mathcal{D}(\Omega)$, then extending this function to $\mathbb{R}^n$ by setting $\phi(x) = 0$ on $\mathbb{R}^n \smallsetminus \Omega$ produces a function in $\mathcal{D}(\mathbb{R}^n)$.
15. Prove that if $|\alpha| \leqslant k$, then D^α is a continuous linear transformation from $W^{k,p}(\Omega)$ into $L^p(\Omega)$.

Chapter 7

Additional Topics

7.1 Fixed-Point Theorems   333
7.2 Selection Theorems   339
7.3 Separation Theorems   342
7.4 The Arzelà–Ascoli Theorem   347
7.5 Compact Operators and the Fredholm Theory   351
7.6 Topological Spaces   361
7.7 Linear Topological Spaces   367
7.8 Analytic Pitfalls   373

7.1 Fixed-Point Theorems

The Contraction Mapping Theorem was proved in Section 4.2, and was accompanied by a number of applications that illustrate its power. In the literature, past and present, there are many other fixed-point theorems, based upon a variety of hypotheses. We shall sample some of these theorems here.

In reading this chapter, refer, if necessary, to Section 7.6 for topological spaces, and to Section 7.7 for linear topological spaces.

Let us say that a topological space X has the *fixed-point property* if every continuous map $f : X \to X$ has a fixed point (that is, a point p such that $f(p) = p$). An important problem, then, is to identify all the topological spaces that have the fixed-point property. A celebrated theorem of Brouwer (1910) begins this program.

Theorem 1. Brouwer's Fixed-Point Theorem. *Every compact convex set in $\mathbb{R}^n$ has the fixed-point property.*

We shall not prove this theorem here, but refer the reader to proofs in [DS] page 468, [Vic] page 28, [Dug] page 340, [Schj] page 74, [Lax], [KA] page 636, [Sma] page 11, [Gr] page 149, and [Smi] page 406.

Theorem 2.   *If a topological space has the fixed-point property,*
then the same is true of every space homeomorphic to it.

Proof. Let spaces X and Y be homeomorphic. This means that there is a
homeomorphism $h : X \twoheadrightarrow Y$ (a continuous map having a continuous inverse).
Suppose that X has the fixed-point property. To prove that Y has the fixed-
point property, let f be a continuous map of Y into Y. Then the map $h^{-1} \circ$
$f \circ h$ is continuous from X to X, and thus has a fixed point x. The equation
$h^{-1}(f(h(x))) = x$ leads immediately to $f(h(x)) = h(x)$, and $h(x)$ is a fixed point
of f.   ∎

Lemma.   *If K is a compact set in a locally convex linear topological*
space, and if U is a symmetric, convex, open neighborhood of 0, then
there is a finite set F in K and a continuous map P from K to the
convex hull of F such that $x - Px \in U$ for all $x \in K$.

Proof. The family $\{x + U : x \in K\}$ is an open cover of K, and by compactness
there must exist points $x_1, \ldots, x_n$ in K such that $K \subset \bigcup_{i=1}^{n}(x_i + U)$. Let h be
the **Minkowski functional** of U, defined by the equation

$$h(x) = \inf\left\{\lambda : \frac{x}{\lambda} \in U , \ \lambda > 0\right\}$$

(See [KN], page 15.) Define

$$g_i(x) = \max\{0, 1 - h(x - x_i)\} \qquad (1 \leqslant i \leqslant n)$$

It is elementary to verify that the inequality $g_i(x) > 0$ is equivalent to the
assertion that $x - x_i \in U$. Since each x in K belongs to at least one of the sets
$x_i + U$, we have $\sum_{i=1}^{n} g_i(x) > 0$ for all $x \in K$. Define

$$Px = \frac{\sum\limits_{i=1}^{n} g_i(x)x_i}{\sum\limits_{j=1}^{n} g_j(x)} \equiv \sum_{i=1}^{n} \theta_i(x)x_i$$

Since $\theta_i(x) \geqslant 0$ and $\sum \theta_i(x) = 1$, we see that Px is in the convex hull of
$\{x_1, x_2, \ldots, x_n\}$ whenever $x \in K$. Since the condition $\theta_i(x) > 0$ occurs if and
only if $x_i \in x + U$, we see that Px is a convex combination of points in $x + U$.
By the convexity of this set, $Px \in x + U$.   ∎

Theorem 3.   The Schauder–Tychonoff Fixed-Point Theorem.
Every compact convex set in a locally convex linear topological Haus-
dorff space has the fixed-point property.

Proof. ([Day], [Sma]) Let K be such a set, and let f be a continuous map of
K into K. We denote the family of all convex, symmetric, open neighborhoods
of 0 by $\{U_\alpha : \alpha \in A\}$. The set A is simply an index set, which we partially
order by writing $\alpha \geqslant \beta$ when $U_\alpha \subset U_\beta$. Thus ordered, A becomes a *directed set,*

suitable as the domain of a net. Since K is compact, the map f is uniformly continuous, and there corresponds to any $\alpha \in A$ an $\alpha' \in A$ such that $U_{\alpha'} \subset U_\alpha$ and $f(x) - f(y) \in U_\alpha$ whenever $x - y \in U_{\alpha'}$.

For any $\alpha \in A$, the preceding lemma provides a continuous map P_α such that $P_\alpha(K)$ is a compact, convex, finite-dimensional subset of K. This map has the further property that $x - P_\alpha x \in U_\alpha$ for each x in K. The composition $P_\alpha \circ f$ maps $P_\alpha(K)$ into itself. Hence, by the Brouwer Fixed-Point Theorem (Theorem 1 above), $P_\alpha \circ f$ has a fixed point z_α in $P_\alpha(K)$. By the compactness of K, the net $[z_\alpha : \alpha \in A]$ has a cluster point z in K. In order to see that z is a fixed point of f, write

$$(1) \qquad f(z) - z = \big[f(z) - f(z_\alpha)\big] + \big[f(z_\alpha) - P_\alpha f(z_\alpha)\big] + \big[z_\alpha - z\big]$$

For any $\beta \in A$, we can select $\alpha \in A$ such that $\alpha \geqslant \beta$ and $z - z_\alpha \in U_{\beta'}$. Then $f(z) - f(z_\alpha) \in U_\beta$. Also, $f(z_\alpha) - P_\alpha f(z_\alpha) \in U_\alpha \subset U_\beta$. Finally, $z - z_\alpha \in U_{\beta'} \subset U_\beta$. Equation (1) now shows that $f(z) - z \in 3U_\beta$. Since β is any element of A, $f(z) = z$. Theorem 1 in Section 7.7 (page 368) justifies this last conclusion. ∎

Corollary.   *If a continuous map is defined on a domain D in a locally convex linear topological Hausdorff space and takes values in a compact, convex subset of D, then it has a fixed point.*

Proof.   Let $F : D \to K$, where K is a compact, convex set in D. Then the restriction of F to K is a continuous map of K to K. By the Schauder–Tychonoff Theorem, $F|K$ has a fixed point. ∎

In Section 4.2 it was shown how fixed-point theorems can lead to existence proofs for solutions of differential equations. This topic is taken up again here. We consider an initial-value problem for a system of first-order differential equations:

$$\begin{cases} x_i'(t) = f_i(t, x_1(t), \ldots, x_n(t)) & (1 \leqslant i \leqslant n) \\ x_i(0) = 0 & (1 \leqslant i \leqslant n) \end{cases}$$

This is written more compactly in the form

$$(2) \qquad \begin{cases} \mathbf{x}'(t) = \mathbf{f}(t, \mathbf{x}(t)) \\ \mathbf{x}(0) = 0 \end{cases}$$

where $\mathbf{x} = (x_1, x_2, \ldots, x_n)$ and $\mathbf{f} = (f_1, f_2, \ldots, f_n)$.

Although the choice of initial values $x_i(0) = 0$ may seem to sacrifice generality, these initial values can always be obtained by making simple changes of variable. Changing t to $t - a$ shifts the initial point, and changing x_i to $x_i - c_i$ shifts the initial values.

The space $C_n[a, b]$ consists of n-tuples of functions in $C[a, b]$. If $\mathbf{x} = (x_1, \ldots, x_n) \in C_n[a, b]$, we write

$$\|\mathbf{x}\|_\infty = \sup_{a \leqslant t \leqslant b} \big\|(x_1(t), \ldots, x_n(t))\big\|_1 = \sup_{a \leqslant t \leqslant b} \|\mathbf{x}(t)\|_1$$

where $\|\ \|_1$ denotes the ℓ^1-norm on $\mathbb{R}^n$. That is,

$$\|u\|_1 = \sum_{i=1}^n |u_i| \quad \text{if} \quad u = (u_1, u_2, \ldots, u_n) \in \mathbb{R}^n$$

Theorem 4.   Let $\mathbf{f}(t, \mathbf{u})$ be defined for $0 \leqslant t \leqslant a$ and for $\mathbf{u} \in \mathbb{R}^n$ such that $\|\mathbf{u}\|_1 \leqslant r$. Assume that on this domain $\mathbf{f}$ is continuous and satisfies $\|\mathbf{f}(t, \mathbf{u})\|_1 \leqslant r/a$. Then the initial-value problem (2) has a solution $\mathbf{x}$ in $C_n[0, a]$, and $\|\mathbf{x}\|_\infty \leqslant r$.

Proof.   Refer to Section 4.2, page 179, where an initial-value problem is shown to be equivalent to an integral equation. In the present circumstances, the integral equation arising from Equation (2) is

$$(3) \qquad\qquad \mathbf{x}(t) = \int_0^t \mathbf{f}(s, \mathbf{x}(s))\, ds$$

Equation (3) presents us with a fixed-point problem for the nonlinear operator A defined by

$$(A\mathbf{x})(t) = \int_0^t \mathbf{f}(s, \mathbf{x}(s))\, ds$$

The domain of A is taken to be

$$D = \{\mathbf{x} \in C_n[a, b] \ : \ \|\mathbf{x}\|_\infty \leqslant r\}$$

First, we shall prove that A maps D into D. Let $\mathbf{x} \in D$ and $\mathbf{y} = A\mathbf{x}$. Since $\|\mathbf{x}\|_\infty \leqslant r$, the inequality $\|\mathbf{x}(s)\|_1 \leqslant r$ follows for all s in the interval $[0, a]$. Hence

$$\|\mathbf{y}(t)\|_1 = \sum_{i=1}^n |y_i(t)| = \sum_{i=1}^n \left| \int_0^t f_i(s, \mathbf{x}(s))\, ds \right| \leqslant \sum_{i=1}^n \int_0^a |f_i(s, \mathbf{x}(s))|\, ds$$

$$= \int_0^a \sum_{i=1}^n |f_i(s, \mathbf{x}(s))|\, ds = \int_0^a \|\mathbf{f}(s, \mathbf{x}(s))\|_1\, ds \leqslant a\left(\frac{r}{a}\right) = r$$

This shows that $\|\mathbf{y}\|_\infty \leqslant r$.

The next step is to prove that $A(D)$ is equicontinuous. If $\mathbf{x}$ and $\mathbf{y}$ are as in the preceding paragraph, and if $0 \leqslant t_1 \leqslant t_2 \leqslant a$, then

$$\|\mathbf{y}(t_2) - \mathbf{y}(t_1)\|_1 = \sum_{i=1}^n |y_i(t_2) - y_i(t_1)|$$

$$= \sum_{i=1}^n \left| \int_0^{t_2} f_i(s, \mathbf{x}(s))\, ds - \int_0^{t_1} f_i(s, \mathbf{x}(s))\, ds \right|$$

$$\leqslant \sum_{i=1}^n \int_{t_1}^{t_2} |f_i(s, \mathbf{x}(s))|\, ds = \int_{t_1}^{t_2} \sum_{i=1}^n |f_i(s, \mathbf{x}(s))|\, ds$$

$$= \int_{t_1}^{t_2} \|\mathbf{f}(s, \mathbf{x}(s))\|_1\, ds \leqslant \frac{r}{a}(t_2 - t_1)$$

The set $A(D)$ is an equicontinuous subset of the bounded set D in $C_n[0, a]$. By the Ascoli Theorem (Section 7.4, page 349), the closure of $A(D)$ is compact. By Mazur's Theorem (Theorem 10, below), the closed convex hull H of $A(D)$ is compact. Since D is closed and convex, $H \subset D$. The preceding corollary is therefore applicable, and A has a fixed point $\mathbf{x}$ in H. Then $\|\mathbf{x}\|_\infty \leqslant r$, and $\mathbf{x}$ solves the initial-value problem. ∎

Theorem 5.   *There is no continuous mapping of the closed unit ball in $\mathbb{R}^n$ to its boundary that leaves all boundary points fixed. (In other words, there is no "retraction" of the unit ball in $\mathbb{R}^n$ onto its boundary.)*

Proof. Let B^n be the ball and S^{n-1} the sphere that is its boundary. Suppose that $f : B^n \to S^{n-1}$, that f is continuous, and that $f(x) = x$ for all $x \in S^{n-1}$. Let g be the antipodal map on S^{n-1}, given by $g(x) = -x$. Then $g \circ f$ has no fixed point (in violation of the Brouwer Fixed-Point Theorem). To see this, suppose $g(f(z)) = z$. Then $f(z) = -z$ and $1 = \|f(z)\| = \| - z\| = \|z\|$. Thus $z \in S^{n-1}$. The point z contradicts our assumption that $f(x) = x$ on S^{n-1}. ∎

The next theorem is a companion to the corollary of Theorem 3. Notice that the hypothesis of convexity has been transferred from the range to the domain of f.

Theorem 6.   *Let D be a convex set in a locally convex linear topological Hausdorff space. If f maps D continuously into a compact subset of D, then f has a fixed point.*

Proof. As in the proof of Theorem 3, we use the family of neighborhoods U_α. Let K be a compact subset of D that contains $f(D)$. Proceed as in the proof of Theorem 3, using the same set of neighborhoods U_α. By the lemma, for each α there is a finite set F_α in K and a continuous map $P_\alpha K \to \text{co}(F_\alpha)$ such that $x - P_\alpha x \in U_\alpha$ for each $x \in K$. If $x \in \text{co}(F_\alpha)$, then $x \in D$, $f(x) \in K$, and $P_\alpha(f(x)) \in \text{co}(F_\alpha)$. Thus $P_\alpha \circ f$ maps the compact, convex, finite-dimensional set $\text{co}(F_\alpha)$ into itself. By the Brouwer Theorem, $P_\alpha \circ f$ has a fixed point z_α in $\text{co}(F_\alpha)$. Then $f(z_\alpha)$ lies in the compact set K, and the net $[f(z_\alpha) : \alpha \in A]$ has a cluster point y in K. We will show that $f(y) = y$ by establishing that $f(y) - y \in U_\alpha$ for all α. Theorem 1 in Section 7.7, page 368, applies here.

Let α be given. Select $\beta \geqslant \alpha$ so that $U_\beta + U_\beta \subset U_\alpha$. By the continuity of f at y, select $\gamma \geqslant \beta$ so that $f(y) - f(x) \in U_\beta$ whenever $x \in K$ and $y - x \in U_\gamma$. Select $\delta \geqslant \gamma$ so that $U_\delta + U_\delta \subset U_\gamma$. Select $\varepsilon \geqslant \delta$ so that $f(z_\varepsilon) \in y + U_\delta$. Then we have

$$y - z_\varepsilon = [y - f(z_\varepsilon)] + [f(z_\varepsilon) + P_\varepsilon f(z_\varepsilon)] \in U_\delta + U_\varepsilon \subset U_\delta + U_\delta \subset U_\gamma$$

Hence $f(y) - f(z_\varepsilon) \in U_\beta$. Furthermore,

$$f(y) - y = [f(y) - f(z_\varepsilon)] + [f(z_\varepsilon) - y] \in U_\beta + U_\delta \subset U_\beta + U_\beta \subset U_\alpha \qquad ∎$$

Theorem 7. Rothe's Theorem.   *Let B denote the closed unit ball of a normed linear space X. If f maps B continuously into a compact subset of X and if $f(\partial B) \subset B$, then f has a fixed point.*

Proof.   Let r denote the radial projection into B defined by $r(x) = x$ if $\|x\| \leqslant 1$ and $r(x) = x/\|x\|$ if $\|x\| > 1$. This map is continuous (Problem 1). Hence $r \circ f$ maps B into a compact subset of B. By Theorem 6, $r \circ f$ has a fixed point x in B. If $\|x\| = 1$, then $\|f(x)\| = 1$ by hypothesis, and we have $x = r(f(x)) = f(x)$ by the definition of r. If $\|x\| < 1$, then $\|r(f(x))\| < 1$ and $x = r(f(x)) = f(x)$, again by the definition of r.   ∎

Theorem 8.   *Let B denote the closed unit ball in a normed space X. Let $\{f_t : 0 \leqslant t \leqslant 1\}$ be a family of continuous maps from B into one compact subset of X. Assume that*

   (i) $f_0(\partial B) \subset B$.
   (ii) *The map $(t, x) \mapsto f_t(x)$ is continuous on $[0, 1] \times B$.*
   (iii) *No f_t has a fixed point in ∂B.*
Then f_1 has a fixed point in B.

Proof.   (From [Sma]) If $0 < \varepsilon < 1$, define

$$
g_\varepsilon(x) = \begin{cases} f_1\left(\dfrac{x}{1-\varepsilon}\right) & \|x\| \leqslant 1 - \varepsilon \\[2em] f_{(1-\|x\|)/\varepsilon}\left(\dfrac{x}{\|x\|}\right) & 1 - \varepsilon \leqslant \|x\| \leqslant 1 \end{cases}
$$

Notice that g_ε is continuous, since the two formulas agree when $\|x\| = 1 - \varepsilon$. If $x \in \partial B$, then $\|x\| = 1$ and $g_\varepsilon(x) = f_0(x) \in B$. Thus f maps ∂B into B. If K is a compact set containing all the images $f_t(B)$, then $g_\varepsilon(B) \subset K$, by the definition of g_ε. The map g_ε satisfies the hypotheses of Theorem 7, and g_ε has a fixed point x_ε in B.

We now shall prove that for all sufficiently small ε, $\|x_\varepsilon\| \leqslant 1 - \varepsilon$. If this is not true, then we can let ε converge to zero through a suitable sequence of values and have, for each ε in the sequence, $\|x_\varepsilon\| > 1 - \varepsilon$. Since $g_\varepsilon(x_\varepsilon) = x_\varepsilon$, we see that x_ε is in K. By compactness, we can assume that the sequence of ε's has the further properties $x_\varepsilon \to x_o$ and $(1 - \|x_\varepsilon\|)/\varepsilon \to t$, where $x_o \in K$ and $t \in [0, 1]$. By the definition of g_ε,

$$
f_{(1-\|x_\varepsilon\|)/\varepsilon}\left(\frac{x_\varepsilon}{\|x_\varepsilon\|}\right) = x_\varepsilon
$$

In the limit, we have $f_t(x_o) = x_o$ and $\|x_o\| = 1$, in contradiction of hypothesis (iii).

We now know that $\|x_\varepsilon\| \leqslant 1 - \varepsilon$ for all sufficiently small ε. Thus, for such values of ε,

$$
x_\varepsilon = g_\varepsilon(x_\varepsilon) = f_1\left(\frac{x_\varepsilon}{1-\varepsilon}\right)
$$

The points x_ε belong to K, and for any cluster point we will have $x = f_1(x)$. ∎

Problems 7.1

1. Prove that the radial projection defined in the proof of Theorem 7 is continuous.

2. Prove Theorem 7 for an arbitrary closed convex set that contains 0 as an interior point. Hint: Replace the norm by a Minkowski functional as in the proof of the lemma.

3. In Theorem 6 assume that D is closed. Show that the theorem is now an easy corollary of Theorem 3, by using the closed convex hull of K and Mazur's Theorem.

4. Prove that the unit ball in $\ell^2(\mathbb{Z})$ does not have the fixed-point property by following this outline. Points in $\ell^2(\mathbb{Z})$ are functions on $\mathbb{Z}$ such that $\sum |x(n)|^2 < \infty$. Let δ be the element in $\ell^2(\mathbb{Z})$ such that $\delta(0) = 1$, and $\delta(n) = 0$ otherwise. Let A be the linear operator defined by $(Ax)(n) = x(n+1)$. Define $f(x) = (1 - \|x\|)\delta + Ax$. This function maps the unit ball into itself continuously but has no fixed point. This example is due to Kakutani.

5. In $\mathbb{R}^n$, define $B = \{x \,:\, 0 < \|x\| \leqslant 1\}$ and $S = \{x \,:\, \|x\| = 1\}$. Is there a continuous map $f : B \to S$ such that $f(x) = x$ when $x \in S$? (Cf. Theorem 5.)

6. In an alternative exposition of fixed-point theory, Theorem 5 is established first, and then the Brouwer theorem is proved from it. Fill in this outline of such a proof. Suppose $f : B^n \to B^n$ is continuous and has no fixed point. Define a retraction g of B^n onto S^{n-1} as follows. Let $g(x)$ be the point where the ray from $f(x)$ through x pierces S^{n-1}.

7. In 1904, Bohl proved that the "cube" $K = \{x \in \mathbb{R}^n : \|x\|_\infty \leqslant 1\}$ has this property: If f maps K continuously into K and maps no point to 0, then for some x on the boundary of K, $f(x)$ is a negative multiple of x. Using Bohl's Theorem, prove that the boundary of K is not a retract of K (and thus substantiate the claim that Bohl deserves much credit for the Brouwer Theorem).

7.2 Selection Theorems

Let X and Y be two topological spaces. The notation 2^Y denotes the family of all subsets of Y. Let $\Phi : X \to 2^Y$. Thus, for each $x \in X$, $\Phi(x)$ is a *subset* of Y. Such a map is said to be **set-valued**. A **selection** for Φ is a map $f : X \to Y$ such that $f(x) \in \Phi(x)$ for each $x \in X$. Thus f "selects" an element of $\Phi(x)$, namely $f(x)$. If $\Phi(x)$ is a *nonempty* subset of Y for each $x \in X$, then a selection f must exist. This is one way of expressing the axiom of choice. In the setting adopted above, one can ask whether Φ has a **continuous** selection. The Michael Selection Theorem addresses this question.

Here is a concrete situation in which a good selection theorem can be used. Let X be a Banach space, and Y a finite-dimensional subspace in X. For each $x \in X$, we define the distance from x to Y by the formula

$$\text{dist}(x, Y) = \inf_{y \in Y} \|x - y\|$$

Since Y is finite-dimensional, an easy compactness argument shows that for each x, the set

$$\Phi(x) = \{y \in Y : \|x - y\| = \text{dist}(x, Y)\}$$

is nonempty. That is, each x in X has at least one nearest point (or "best approximation") in Y. In general, the nearest point will not be unique. See the sketch in Figure 7.1 for the reason.

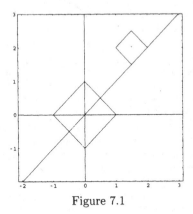

Figure 7.1

In the sketch, the box with center at 0 is the unit ball. The line of slope 1 represents a subspace Y. The small box is centered at a point x outside Y. That box is the ball of least radius centered at x that intersects Y. The intersection is $\Phi(x)$. The set $\Phi(x)$ is the set of all best approximations to x in Y. In this case $\Phi(x)$ is convex, since it is the intersection of a subspace with a ball. It is also closed, by the definition of Φ and the continuity of the norm. Now we ask, is there a *continuous* map $f : X \to Y$ such that for each x, $f(x)$ is a nearest point to x in Y? One way to answer such questions is to invoke Michael's theorem, to which we now turn.

First some definitions are required. An **open covering** of a topological space X is a family of open sets whose union is X. One covering $\mathcal{B}$ is a **refinement** of another $\mathcal{A}$ if each member of $\mathcal{B}$ is contained in some member of $\mathcal{A}$. A covering $\mathcal{B}$ is said to be **locally finite** if each point of X has a neighborhood that intersects only finitely many members of $\mathcal{B}$. A Hausdorff space X is **paracompact** if each open covering of X has a refinement that is an open and locally finite covering of X ([Kel], page 156). Clearly, a compact Hausdorff space is paracompact.

It is a nontrivial and useful fact that all metric spaces are paracompact ([Kel], page 156). In many applications this obviates the proving of paracompactness by means of special arguments.

Given the set-valued mapping $\Phi : X \to 2^Y$ and a subset $\mathcal{U}$ in Y, we put

$$\Phi^-(\mathcal{U}) = \{x \in X : \Phi(x) \cap \mathcal{U} \text{ is nonempty}\}$$

Finally, we declare Φ to be **lower semicontinuous** if $\Phi^-(\mathcal{U})$ is open in X whenever $\mathcal{U}$ is open in Y.

Theorem 1. The Michael Selection Theorem.   *Let* Φ *be a lower semicontinuous set-valued map defined on a paracompact topological space and taking as values nonempty closed convex sets in a Banach space. Then* Φ *has a continuous selection.*

For the proof of this theorem, we refer the reader to [Mich1] and [Mich2]. As an application of Michael's theorem, we give a result about approximating possibly discontinuous maps by continuous ones.

Theorem 2   *Let* X *be a paracompact space,* Y *a Banach space, and* H *a closed subspace in* Y. *Suppose that* $f : X \to Y$ *is continuous and* $g : X \to H$ *is bounded. Then for each* $\varepsilon > 0$ *there is a continuous map* $\overline{g} : X \to H$ *that satisfies*

(1) $$\sup_{x \in X} \left\| f(x) - \overline{g}(x) \right\| \leqslant \sup_{x \in X} \left\| f(x) - g(x) \right\| + \varepsilon$$

Thus when approximating the continuous map f *by the bounded map* g, *we can find a continuous map* $\overline{g}$ *that is almost as good as* g.

Proof.   Let λ denote the number on the right in Inequality (1). For each $x \in X$, define

$$\Phi(x) = \left\{ h \in H : \left\| f(x) - h \right\| \leqslant \lambda \right\}$$

This set is nonempty because $g(x) \in \Phi(x)$. (Notice that g is a selection for Φ but not necessarily a continuous selection.) The set $\Phi(x)$ is closed and convex in the Banach space H.

We shall prove that Φ is lower semicontinuous. Let $\mathcal{U}$ be open in H. It is to be shown that $\Phi^-(\mathcal{U})$ is open in X. Let $x \in \Phi^-(\mathcal{U})$. Then $\Phi(x) \cap \mathcal{U}$ is nonempty. Select h in this set. Then $h \in \mathcal{U}$ and $\left\| f(x) - h \right\| \leqslant \lambda$. Also $\left\| f(x) - g(x) \right\| < \lambda$. So, by considering the line segment from h to $g(x)$, we conclude that there is an $h' \in \mathcal{U}$ such that $\left\| f(x) - h' \right\| < \lambda$. Since f is continuous at x, there is a neighborhood $\mathcal{N}$ of x such that

$$\left\| f(u) - f(x) \right\| < \lambda - \left\| f(x) - h' \right\| \qquad (u \in \mathcal{N})$$

By the triangle inequality, $\left\| f(s) - h' \right\| < \lambda$ when $s \in \mathcal{N}$. This proves that $h' \in \Phi(s)$, that $\Phi(s) \cap \mathcal{U}$ is nonempty, that $s \in \Phi^-(\mathcal{U})$, that $\mathcal{N} \subset \Phi^-(\mathcal{U})$, that $\Phi^-(\mathcal{U})$ is open, and that Φ is lower semicontinuous.

Now apply Michael's theorem to obtain a continuous selection $\overline{g}$ for Φ. Then $\overline{g}$ is a continuous map of X into H and satisfies $\overline{g}(x) \in \Phi(x)$ for all x. Hence $\overline{g}$ satisfies (1). ∎

Another important theorem that follows readily from Michael's is the theorem of Bartle and Graves:

Theorem 3. The Bartle–Graves Theorem.   *A continuous linear map of one Banach space onto another must have a continuous (but not necessarily linear) right inverse.*

Proof. Let $A : X \to Y$, as in the hypotheses. Since A is surjective, the equation $Ax = y$ has solutions x for each $y \in Y$. At issue, then, is whether a continuous choice of x can be made. It is clear that we should set

$$\Phi(y) = \big\{ x \in X : Ax = y \big\}$$

Obviously, each set $\Phi(y)$ is closed, convex, and nonempty. Is Φ lower semicontinuous? Let $\mathcal{O}$ be open in X. We must show that the set $\Phi^-(\mathcal{O})$ is open in Y. But $\Phi^-(\mathcal{O}) = A(\mathcal{O})$ by a short calculation. By the Interior Mapping Theorem (Section 1.8, page 48), $A(\mathcal{O})$ is open. Thus Φ is lower semicontinuous, and by Michael's theorem, a continuous selection f exists. Thus $f(y) \in \Phi(y)$, or $A(f(y)) = y$.   ∎

In the literature there are many selection theorems that involve *measurable* functions instead of *continuous* ones. If X is a measurable space and Y a topological space, a map $\Phi : X \to 2^Y$ is said to be **weakly measurable** if the set

$$\big\{ x \in X : \Phi(x) \cap \mathcal{O} \text{ is not empty} \big\}$$

is measurable in X for each open set $\mathcal{O}$ in Y. (For a discussion of measurable spaces, see Section 8.1, pages 381ff.) The measurable selection theorem of Kuratowski and Ryll-Nardzewski follows. Its proof can be found in [KRN], [Part], and [Wag].

Theorem 4. Kuratowski and Ryll-Nardzewski Theorem.   *Let Φ be a weakly measurable map of X to 2^Y, where X is a measurable space and Y is a complete, separable metric space. Assume that for each x, $\Phi(x)$ is closed and nonempty. Then Φ has a measurable selection. Thus, there exists a function $f : X \to Y$ such that $f(x) \in \Phi(x)$ for all x, and $f^{-1}(\mathcal{O})$ is measurable for each open set $\mathcal{O}$ in Y.*

7.3 Separation Theorems

The next three theorems are called "separation theorems." They pertain to disjoint pairs of convex sets, and to the positioning of a hyperplane so that the convex sets are on opposite sides of the hyperplane. In $\mathbb{R}^2$, the hyperplanes are lines, and simple figures show the necessity of convexity in carrying out this separation. In Theorem 3, one can see the necessity of compactness by considering one set to be the lower half plane and the other to be the set of points (x, y) for which $y \geqslant x^{-1}$ and $x > 0$.

Theorem 1.   Let X be a normed linear space and let K be a convex subset of X that contains 0 as an interior point. If $z \in X \smallsetminus K$, then there is a continuous linear functional ϕ defined on X such that for all $x \in K$, $\phi(x) \leqslant 1 \leqslant \phi(z)$.

Proof. Again, we need the **Minkowski functional** of K. It is

$$p(x) = \inf\{\lambda : \lambda > 0 \text{ and } x/\lambda \in K\}$$

We prove now that $p(x + y) \leqslant p(x) + p(y)$ for all x and y. Select $\lambda, \mu > 0$ so that x/λ and y/μ are in K. By the convexity of K,

$$\frac{x+y}{\lambda+\mu} \equiv \frac{\lambda}{\lambda+\mu}\frac{x}{\lambda} + \frac{\mu}{\lambda+\mu}\frac{y}{\mu} \quad \in \quad K$$

Hence $p(x + y) \leqslant \lambda + \mu$. Taking the infima of λ and μ, we obtain $p(x + y) \leqslant p(x) + p(y)$. Next we prove that for $\lambda \geqslant 0$ the equation $p(\lambda x) = \lambda p(x)$ is true. Select $\mu > 0$ so that $x/\mu \in K$. Then $\lambda x/\lambda\mu \in K$ and $p(\lambda x) \leqslant \lambda\mu$. Taking the infimum of μ, we conclude that $p(\lambda x) \leqslant \lambda p(x)$. From this we obtain the reverse inequality by writing $\lambda p(x) = \lambda p(\lambda^{-1}\lambda x) \leqslant \lambda\lambda^{-1}p(\lambda x) = p(\lambda x)$.

Now define a linear functional ϕ on the one-dimensional subspace generated by z by writing

$$\phi(\lambda z) = \lambda p(z) \qquad (\lambda \in \mathbb{R})$$

If $\lambda \geqslant 0$, then $\phi(\lambda z) = p(\lambda z)$. If $\lambda < 0$, then $\phi(\lambda z) = \lambda p(z) \leqslant 0 \leqslant p(\lambda z)$. Hence $\phi \leqslant p$. By the Hahn–Banach Theorem (Section 1.6, page 32), ϕ has a linear extension (denoted also by ϕ) that is dominated by p. For each $x \in K$ we have $\phi(x) \leqslant p(x) \leqslant 1$. As for z, we have $\phi(z) = p(z) \geqslant 1$, because if $p(z) < 1$, then $z/\lambda \in K$ for some $\lambda \in (0, 1)$, and by convexity the point

$$z \equiv \lambda(z/\lambda) + (1 - \lambda)0$$

would belong to K.

Lastly, we prove that ϕ is continuous. Select a positive r such that the ball $B(0, r)$ is contained in K. For $\|x\| < 1$ we have $rx \in B(0, r)$ and $rx \in K$. Hence $p(rx) \leqslant 1$, $\phi(rx) \leqslant 1$, and $\phi(x) \leqslant 1/r$. Thus $\|\phi\| \leqslant 1/r$. ∎

Theorem 2.   Let K_1, K_2 be a disjoint pair of convex sets in a normed linear space X. If one of them has an interior point, then there is a nonzero functional $\phi \in X^*$ such that

$$\sup_{x \in K_1} \phi(x) \leqslant \inf_{x \in K_2} \phi(x)$$

Proof. By performing a translation and by relabeling the two sets, we can assume that 0 is an interior point of K_1. Fix a point z in K_2 and consider the set $K_1 - K_2 + z$. This set is convex and contains 0 as an interior point. Also, $z \notin K_1 - K_2 + z$ because K_1 is disjoint from K_2. By the preceding theorem, there is a $\phi \in X^*$ such that for $u \in K_1$ and $v \in K_2$ we have $\phi(u - v + z) \leqslant 1 \leqslant \phi(z)$. Hence $\phi(u) \leqslant \phi(v)$. ∎

Theorem 3.   Let K_1, K_2 be a disjoint pair of closed convex sets in a normed linear space X. Assume that at least one of the sets is compact. Then there is a $\phi \in X^*$ such that

$$\sup_{x \in K_2} \phi(x) < \inf_{x \in K_1} \phi(x)$$

Proof.   The set $K_1 - K_2$ is closed and convex. (See Problems 1.2.19 on page 12 and 1.4.17 on page 23.) Also, $0 \notin K_1 - K_2$, and consequently there is a ball $B(0, r)$ that is disjoint from $K_1 - K_2$. By the preceding theorem, there is a nonzero continuous functional ϕ such that

$$\sup_{\|x\| \leqslant r} \phi(x) \leqslant \inf_{x \in K_1 - K_2} \phi(x)$$

Since ϕ is not zero, there is an $\varepsilon > 0$ such that for $u \in K_1$ and $v \in K_2$, $\varepsilon \leqslant \phi(u) - \phi(v)$.   ∎

Separation theorems have applications in optimization theory, game theory, approximation theory, and in the study of linear inequalities. The next theorem gives an example of the latter.

Theorem 4.   Let U be a compact set in a real Hilbert space. In order that the system of linear inequalities

(1) $\langle u, x \rangle > 0$ $(u \in U)$

be consistent (i.e., have a solution, x) it is necessary and sufficient that 0 not be in the closed convex hull of U.

Proof.   For the sufficiency of the condition, assume the condition to be true. Thus, $0 \notin \overline{\text{co}}(U)$. By Theorem 3, there is a vector x and a real number λ such that $\overline{\text{co}}(U)$ and 0 are on opposite sides of the hyperplane

$$\{y : \langle y, x \rangle = \lambda\}$$

We can suppose that $\langle y, x \rangle > \lambda$ for $y \in \overline{\text{co}}(U)$ and that $\langle 0, x \rangle < \lambda$. Obviously, $\lambda > 0$ and x solves the system (1).

Now assume that system (1) is consistent and that x is a solution of it. By continuity and compactness, there exists a positive ε such that $\langle u, x \rangle \geqslant \varepsilon$ for all $u \in U$. For any $v \in \text{co}(U)$ we can write a convex combination $v = \sum \theta_i u_i$ and then compute

$$\langle v, x \rangle = \left\langle \sum \theta_i u_i, x \right\rangle = \sum \theta \langle u_i, x \rangle \geqslant \sum \theta_i \varepsilon = \varepsilon$$

Then, by continuity, $\langle w, x \rangle \geqslant \varepsilon$ for all $w \in \overline{\text{co}}(U)$. Obviously, $0 \notin \overline{\text{co}}(U)$.   ∎

In order to prove a representative result in game theory, some notation will be useful. The standard n-**dimensional simplex** is the set

$$S_n = \left\{ x \in \mathbb{R}^n \ : \ x \geqslant 0 \text{ and } \sum_{i=1}^{n} x_i = 1 \right\}$$

Theorem 5.   *For an $m \times n$ matrix A, either $Ax \geqslant 0$ for some $x \in S_n$, or $y^T A < 0$ for some $y \in S_m$.*

Proof.   Suppose that there is no x in the simplex S_n for which $Ax \geqslant 0$. Then $A(S_n)$ contains no point in the nonnegative orthant,

$$P_m = \{ y \in \mathbb{R}^m \ : \ y \geqslant 0 \}$$

Consequently, the convex sets $A(S_n)$ and P_m can be separated by a hyperplane. Suppose, then, that

$$P_m \subset \{ y \in \mathbb{R}^m \ : \ \langle u, y \rangle > \lambda \}$$
$$A(S_n) \subset \{ y \in \mathbb{R}^m \ : \ \langle u, y \rangle < \lambda \}$$

Since $0 \in P_m$, $\lambda < 0$. Let e_i denote the i-th standard unit vector in $\mathbb{R}^m$. For positive t, $te_i \in P_m$. Hence $\langle u, te_i \rangle > \lambda$, $tu_i > \lambda$, $u_i > \lambda/t$, and $u_i \geqslant 0$. Thus $u \in P_m$ and $\langle u, Ax \rangle < 0$ for all $x \in S_n$. Obviously, $u \neq 0$, so we can assume $u \in S_m$. Since $u^T Ax < 0$ for all $x \in S_n$, we have $u^T Ae_i < 0$ for $1 \leqslant i \leqslant n$, or, in other terms, $u^T A < 0$. (In this last argument, e_i was a standard unit vector in $\mathbb{R}^n$.)   ∎

This section concludes with a brief discussion of a fundamental topic in game theory. A **rectangular two-person game** depends on an $m \times n$ matrix of real numbers. Player 1 selects in secret an integer i in the range $1 \leqslant i \leqslant m$. Likewise, Player 2 selects in secret an integer j in the range $1 \leqslant j \leqslant n$. The two chosen integers i and j are now revealed, and the payoff to Player 1 is the quantity a_{ij} in the matrix. If this payoff is positive, Player 2 pays Player 1. If the payoff is negative, Player 1 pays Player 2.

Both players have full knowledge of the matrix A. If Player 1 chooses i, then he can assure himself of winning at least the quantity $\min_j a_{ij}$. His *best* choice for i ensures that he will win $\max_i \min_j a_{ij}$. Player 2 reasons similarly: By choosing j, he limits his loss to $\max_i a_{ij}$. His best choice for j will minimize the worst loss and that number is then $\min_j \max_i a_{ij}$. If

$$\min_j \max_i a_{ij} = \max_i \min_j a_{ij}$$

then this common number is the amount that one player can be sure of winning and is the limit on what the other can lose.

In the more interesting case (which will include the case just discussed) the players will make random choices of the two integers (following carefully assigned probability distributions) and play the game over and over. Player 1 will assign specific probabilities to each possible choice from the set $\{1, 2, \ldots, m\}$. The probabilities can be denoted by x_i. We will then want $x_i \geqslant 0$ for each i as well as $\sum_{i=1}^{m} x_i = 1$. In brief, $x \in S_m$. Similarly, Player 2 assigns probabilities y_j to the choices in $\{1, 2, \ldots, n\}$. Thus $y \in S_n$. When the game is played just once, the expected payoff to Player 1 can be computed to be $\sum_{i=1}^{m} \sum_{j=1}^{n} a_{ij} x_i y_j$. Player 1 seeks to maximize this with an appropriate choice of x in S_m, while Player 2 seeks to minimize this by a suitable choice of y in S_n. The principal theorem in this subject is as follows.

Theorem 6. The Min-Max Theorem of Von Neumann *Let A be any $m \times n$ matrix. Then*

$$\max_{x \in S_m} \min_{y \in S_n} x^T A y = \min_{y \in S_n} \max_{x \in S_m} x^T A y$$

Proof. It is easy to prove an inequality $\leqslant$ between the terms in the above equation. To do so, let $u \in S_n$ and $v \in S_m$. Then

$$\min_{y \in S_n} v^T A y \leqslant v^T A u \leqslant \max_{x \in S_m} x^T A u$$

Since u and v were arbitrary in the sets S_n and S_m, respectively, we can choose them so that we get

$$(2) \qquad \max_{x \in S_m} \min_{y \in S_n} x^T A y \leqslant \min_{y \in S_n} \max_{x \in S_m} x^T A y$$

Now suppose that a strict inequality holds in Inequality (2). Select a real number r such that

$$(3) \qquad \max_{x \in S_m} \min_{y \in S_n} x^T A y < r < \min_{y \in S_n} \max_{x \in S_m} x^T A y$$

Consider the matrix A' whose generic element is $a_{ij} - r$. By Theorem 5 (applied actually to $-A'$), either $A'u \leqslant 0$ for some $u \in S_n$ or $v^T A' \geqslant 0$ for some $v \in S_m$.

If the first of these alternatives is true, then for all $x \in S_m$, we have $x^T A' u \leqslant 0$. In quick succession, one concludes that $\max_x x^T A' u \leqslant 0$, $\min_y \max_x x^T A' y \leqslant 0$, and $\min_y \max_x x^T A y \leqslant r$. In the last inequality we simply compute the bilinear form $x^T A' y$, remembering that $x \in S_m$ and $y \in S_n$. The concluding inequality here is a direct contradiction of Inequality (3).

Similarly, if there exists $v \in S_m$ for which $v^T A' \geqslant 0$, then we have for all $y \in S_n$, $v^T A' y \geqslant 0$, $\min_y v^T A' y \geqslant 0$, $\max_x \min_y x^T A' y \geqslant 0$, and $\max_x \min_y x^T A y \geqslant r$, contradicting Inequality (3) again. ∎

In the language of game theory, Theorem 6 asserts that each player of a rectangular game has an optimal strategy. These are the "probability vectors" $\bar{x} \in S_m$ and $\bar{y} \in S_n$ such that

$$\bar{x}^T A \bar{y} = \max_{x \in S_m} \min_{y \in S_n} x^T A y = \min_{y \in S_n} \max_{x \in S_m} x^T A y$$

The common value of these three quantities is called the **value of the game**. Convenient references for these matters and for the theory of games in general are [McK] and [Mor].

Problems 7.3

1. Let $f : X \times Y \to \mathbb{R}$, where X and Y are arbitrary sets. Prove that

$$\sup_{x} \inf_{y} f(x, y) \leqslant \inf_{y} \sup_{x} f(x, y)$$

(In order for this to be universally valid, one must admit $+\infty$ as a permissible value for the supremum and $-\infty$ for the infimum.)

2. Prove for any $u \in \mathbb{R}^n$ that $\max_{x \in S_n} \langle u, x \rangle = \max_{1 \leqslant i \leqslant n} u_i$.

3. Let P_n denote the set of x in $\mathbb{R}^n$ that satisfy $x \geqslant 0$. Prove that for $u \in \mathbb{R}^n$ and $\lambda \in \mathbb{R}$ these properties are equivalent:

   a. $P_n \subset \{x : \langle u, x \rangle > \lambda\}$

   b. $u \in P_n$ and $\lambda < 0$.

4. **Saddle points.** If there is a pair of integers (r, s) such that $a_{is} \leqslant a_{rs} \leqslant a_{rj}$ for all i and j, then a_{rs} is called a "saddle point" for the rectangular game. Prove that if such a point exists, each player has an optimal strategy of the form $(0, \ldots, 0, 1, 0, \ldots, 0)$.

5. (A variation on Theorem 4) Let X be any linear space, Φ a set of linear functionals on X. Prove that the system of linear inequalities

$$\phi(x) < 0 \qquad (\phi \in \Phi)$$

has a finite inconsistent subsystem if and only if $0 \in \text{co}(\Phi)$.

6. (A result in approximation theory) Let K be a closed convex set in a normed linear space X. Let u be a point not in K, and set $r = \text{dist}(u, K)$. Prove that there exists a functional $\phi \in X^*$ such that

$$\sup_{\|x - u\| \leqslant r} \phi(x) \leqslant \inf_{x \in K} \phi(x)$$

7. (Separation theorem in Hilbert space) Let K be a closed convex set in Hilbert space, and u a point outside K. Then there is a unique point v in K such that for all x in K,

$$\langle x, u - v \rangle \leqslant \langle v, u - v \rangle < \langle u, u - v \rangle$$

8. Let $\phi_1, \phi_2, \ldots, \phi_n$ be continuous linear functionals on a normed space. Let $a_1, a_2, \ldots, a_n$ be scalars, and define affine functionals $\psi_i(x) = \phi_i(x) + a_i$. Define $F(x) = \max_i \psi_i(x)$. Prove that F is bounded below if and only if the inequality $\max_i \phi_i(x) \geqslant 0$ is true for all x of norm 1.

7.4 The Arzelà–Ascoli Theorems

The hypothesis of compactness is often present in important theorems of analysis. For this reason, much attention has been directed to the problem of characterizing the compact sets in various Banach spaces. The Arzelà–Ascoli Theorem does this for spaces of continuous functions. The Dunford–Pettis Theorem does this for L^1-spaces, and the Fréchet–Kolmogorov Theorem does it for the L^p-spaces. The most extensive source for results on this topic is [DS], Chapter 4. See also [Yo] for the Fréchet–Kolmogorov Theorem.

We begin with spaces of continuous functions. Let (X, d) and (Y, ρ) be compact metric spaces. For example, X and Y could be compact intervals on the real line. We denote by $C(X, Y)$ the space of all continuous maps from X into Y. It is known that continuity and uniform continuity are the same for maps of X into Y. Thus, a map f of X into Y belongs to $C(X, Y)$ if and only if there corresponds to each positive ε a positive δ such that $\rho(f(u), f(v)) < \varepsilon$ whenever $d(u, v) < \delta$.

The space $C(X, Y)$ is made into a metric space by defining its distance function Δ by the equation

(1)
$$\Delta(f, g) = \sup_{x \in X} \rho(f(x), g(x))$$

A first goal is to characterize the compact sets in $C(X, Y)$.

Let K be a subset of $C(X, Y)$. We say that K is **equicontinuous** if to each positive ε there corresponds a positive δ such that this implication is valid:

(2)
$$\Big[f \in K \text{ and } d(u, v) < \delta \Big] \implies \rho\Big(f(u), f(v)\Big) < \varepsilon$$

Theorem 1. First Arzelà–Ascoli Theorem.   *Let X and Y be compact metric spaces. A subset of $C(X, Y)$ is compact if and only if it is closed and equicontinuous.*

Proof. Let K be the subset in question. First, suppose that K is compact. Then it is closed, by the theorem in general topology that asserts that compact sets in Hausdorff spaces are closed ([Kel], page 141). In order to prove that K is equicontinuous, let ε be a prescribed positive number. Since K is compact, it is totally bounded ([Kel], page 198). Consequently, there exist elements $f_1, f_2, \ldots, f_n$ in K such that

$$K \subset \bigcup_{i=1}^{n} B(f_i, \varepsilon)$$

where $B(f, \varepsilon)$ is the ball $\{g : \Delta(f, g) < \varepsilon\}$. A finite set of continuous functions is obviously equicontinuous, and therefore there exists a δ for which this implication is valid:

$$\Big[1 \leqslant i \leqslant n \text{ and } d(u, v) < \delta \Big] \implies \rho(f_i(u), f_i(v)) < \varepsilon$$

If $g \in K$ and $d(u, v) < \delta$, then for a suitable value of j we have

$$\rho(g(u), g(v)) \leqslant \rho(g(u), f_j(u)) + \rho(f_j(u), f_j(v)) + \rho(f_j(v), g(v)) \leqslant 3\varepsilon$$

The index j is chosen so that $\Delta(g, f_j) < \varepsilon$. The above inequality establishes the equicontinuity of K.

Now suppose that K is closed and equicontinuous. The space $M(X,Y)$ of all maps from X into Y with the metric Δ as defined in Equation (1) is a metric space that contains $C(X,Y)$ as a closed subset. It suffices then to prove that K is compact in $M(X,Y)$.

Let ε be a prescribed positive number. Select a positive δ such that the implication (2) above is valid. Since X and Y are totally bounded, we can arrange that

$$X \subset \bigcup_{i=1}^{n} B(x_i, \delta) \qquad Y \subset \bigcup_{i=1}^{m} B(y_i, \varepsilon)$$

In order to have a disjoint cover of X, let

$$A_i = B(x_i, \delta) \smallsetminus [B(x_1, \delta) \cup \cdots \cup B(x_{i-1}, \delta)] \qquad (1 \leqslant i \leqslant n)$$

Notice that if $x \in A_i$, then it follows that $x \in B(x_i, \delta)$, that $d(x_i, x) < \delta$, and that $\rho(f(x_i), f(x)) < \varepsilon$ for all $f \in K$.

Now consider the functions g from X to Y that are constant on each A_i and are allowed to assume only the values $y_1, y_2, \ldots, y_m$. (There are exactly m^n such functions.) The balls $B(g, 2\varepsilon)$ cover K. To verify this, let $f \in K$. For each i, select y_{j_i} so that $\rho(f(x_i), y_{j_i}) < \varepsilon$. Then let g be a function of the type described above whose value on A_i is y_{j_i}. For each $x \in X$ there is an index i such that $x \in A_i$. Then

$$\rho\big(f(x), g(x)\big) \leqslant \rho\big(f(x), f(x_i)\big) + \rho\big(f(x_i), g(x)\big) < 2\varepsilon$$

Hence $\Delta(f, g) < 2\varepsilon$. This proves that K is totally bounded. Since a closed and totally bounded set is compact, K is compact.   ∎

As usual, if X is a compact metric space, $C(X)$ will denote the Banach space of all continuous real-valued functions on X, normed by writing

$$\|f\| = \sup_{x \in X} |f(x)|$$

Theorem 2. Arzelà–Ascoli Theorem II.   *Let X be a compact metric space. A subset of $C(X)$ is compact if and only if it is closed, bounded, and equicontinuous.*

Proof.   Suppose that K is a compact set in $C(X)$. Then it is closed. It is also totally bounded, and can be covered by a finite number of balls of radius 1:

$$K \subset \bigcup_{i=1}^{n} B(f_i, 1)$$

For any $g \in K$ there is an index i for which $g \in B(f_i, 1)$. Then

$$\|g\| \leqslant \|g - f_i\| + \|f_i\| \leqslant 1 + \max_i \|f_i\| \equiv M$$

Thus K is bounded. Let $Y = [-M, M]$. Then

$$K \subset C(X, Y)$$

The preceding theorem now is applicable, and K is equicontinuous.

For the other half of the proof let K be a closed, bounded, and equicontinuous set. Since K is bounded, we have again $K \subset C(X, Y)$, where Y is a suitable compact interval. The preceding theorem now shows that K is compact.   ∎

Theorem 3. Dini's Theorem.   *Let $f_1, f_2, \ldots$ be continuous real-valued functions on a compact topological space. For each x assume that $|f_n(x)| \downarrow 0$. Then this convergence is uniform.*

Proof.   Given $\varepsilon > 0$, put $S_k = \{x : |f_k(x)| \geqslant \varepsilon\}$. Then each S_k is closed, and $S_{k+1} \subset S_k$. For each x there is an index k such that $x \notin S_k$. Hence $\bigcap_{k=1}^{\infty} S_k$ is empty. By compactness and the finite intersection property, we conclude that $\bigcap_{k=1}^{n} S_k$ is empty for some n. This means that S_n is empty, and that $|f_n(x)| < \varepsilon$ for all x. Thus $|f_k(x)| < \varepsilon$ for all $k \geqslant n$. This is uniform convergence.   ∎

We conclude this section by quoting some further compactness theorems. In the spaces $L^p(\mathbb{R})$, the following characterization of compact sets holds. Here, $1 \leqslant p < \infty$. A precursor of this theorem was given by Riesz, and a generalization to locally compact groups with their Haar measure has been proved by Weil. See [Edw] page 269, [DS] page 297, and [Yo] page 275.

Theorem 4. The Fréchet–Kolmogorov Theorem.   *A closed and bounded set K in the space $L^p(\mathbb{R})$ is compact if and only if the following two limits hold true, uniformly for f in K:*

$$\lim_{h \to 0} \int_{\mathbb{R}} |f(x+h) - f(x)|^p \, dx = 0$$

$$\lim_{M \to \infty} \int_{|x| > M} |f(x)|^p \, dx = 0$$

Theorem 5.   *A closed and bounded set K in the space c_0 (defined in Section 1.1) is compact if and only if for each positive ε there corresponds an integer n such that $\sup_{x \in K} \sup_{i > n} |x(i)| < \varepsilon$.*

Theorem 6.   *A closed and bounded set K in ℓ^2 is compact if and only if*

$$\lim_{n \to \infty} \sup_{x \in K} \sum_{i \geqslant n} x(i)^2 = 0$$

Problems 7.4

1. Define $f_n \in C[0,1]$ by $f_n(x) = nx/(nx+1)$. Is the set $\{f_n : n \in \mathbb{N}\}$ equicontinuous? Is it bounded? Is it closed?

2. Let $f_\lambda(x) = e^{\lambda x}$. Show that $\{f_\lambda : \lambda \leqslant b\}$ is equicontinuous on $[0, a]$.

3. In the space $C[a, b]$ let K be the set of all polynomials of degree at most n that satisfy $\|p\| \leqslant 1$. (Here n is fixed.) Is K equicontinuous? Is it compact?

4. Let α and λ be fixed positive numbers. Let K be the set of all functions f on $[a, b]$ that satisfy the Lipschitz condition

$$|f(x) - f(y)| \leqslant \lambda |x - y|^\alpha$$

Is K closed? compact? equicontinuous? bounded?

5. Let K be a set of continuously differentiable functions on $[a, b]$. Put $K' = \{f' : f \in K\}$. Prove that if K' is bounded, then K is equicontinuous.

6. Let K be an equicontinuous set in $C[a, b]$. Prove that if there exists a point x_0 in $[a, b]$ such that $\{f(x_0) : f \in K\}$ is bounded, then K is bounded.

7. Define an operator L on the space $C[a, b]$ by

$$(Lf)(x) = \int_a^b k(x, y) f(y) \, dy$$

where k is continuous on $[a, b] \times [a, b]$. Prove that L is a compact operator; that is, it maps the unit ball into a compact set.

8. Prove or disprove: Let $[f_n]$ be a sequence of continuous functions on a compact space. Let f be a function such that $|f(x) - f_n(x)| \downarrow 0$ for all x. Then f is continuous and the convergence $f_n \to f$ is uniform.

9. Select an element $a \in \ell^2$, and define

$$K = \{x \in \ell^2 \ : \ |x_i| \leqslant |a_i| \text{for all } i\}$$

Prove that K is compact. The special case when $a_i = 1/i$ gives the so-called **Hilbert cube**.

10. Prove that not every compact set in ℓ^2 is of the form described in the preceding problem.

11. Reconcile the compactness theorems for L^2 and ℓ^2, in the light of the isometry between these spaces.

7.5 Compact Operators and the Fredholm Theory

This section is devoted to operators that we think of as "perturbations of the identity," meaning operators $I + A$, where I is the identity and A is a **compact** operator. The definition and elementary properties of compact operators were given in Section 2.3, page 85. In particular, we found that operators with finite-dimensional range are compact, and that the set of compact members of $\mathcal{L}(X, Y)$ is closed if X and Y are Banach spaces. Thus, a limit of operators, each having finite-dimensional range, is necessarily a compact operator. In many (but not all) Banach spaces, every compact operator is such a limit. This fact can be exploited in practical problems involving compact operators; one begins by approximating the operator by a simpler one having finite-dimensional range. Typically, this leads to a system of linear equations that must be solved numerically. (Examples of problems involving operators with finite-dimensional range occur in Problems 20, 21, 22, and 29 in Section 2.1, pages 68–69.)

Here, however, we consider a related class of operators, namely those of the form $I + A$ (where A is compact), and find that such operators have favorable properties too. Intuitively, we expect such operators to be well behaved, because they are close to the identity operator. For example, we shall prove the famous Fredholm Theorem, which asserts that for such operators the property of injectivity (one-to-oneness) is equivalent to surjectivity (being "onto"). This is a theorem familiar to us in the context of linear operators from $\mathbb{R}^n$ to $\mathbb{R}^n$:

For an $n \times n$ matrix, the properties of having a 0-dimensional kernel and an n-dimensional range are equivalent. The proof of the Fredholm Theorem is given in several pieces, and then further properties of such operators are explored. We have relied heavily on the exposition in [Jam], and recommend this reference to the reader.

Lemma 1.   *Let A be a compact operator on a normed linear space. If $I + A$ is surjective, then it is injective.*

Proof.   Let $B = I + A$ and $X_n = \ker(B^n)$. Suppose that B is surjective but not injective. We shall be looking for a contradiction. Note that $0 \subset X_1 \subset X_2 \subset \cdots$ It is now to be proved that these inclusions are *proper*. Select a nonzero element y_1 in X_1. Since B is surjective, there exist points $y_2, y_3, \ldots$ such that $B y_{n+1} = y_n$ for $n = 2, 3, \ldots$ We have

$$B^n y_n = B^{n-1} B y_n = B^{n-1} y_{n-1} = \cdots = B^2 y_2 = B y_1 = 0$$

Furthermore,

$$B^{n-1} y_n = B^{n-2} B y_n = B^{n-2} y_{n-1} = \cdots = B^2 y_3 = B y_2 = y_1 \neq 0$$

These two equations prove that $y_n \in X_n \smallsetminus X_{n-1}$ and that those inclusions mentioned above are proper.

By the Riesz Lemma, (Section 1.4, page 22), there exist points x_n such that $x_n \in X_n$, $\|x_n\| = 1$, and $\mathrm{dist}(x_n, X_{n-i}) \geqslant 1/2$. If $m > n$, then we have $B^m x_m = 0$ because $x_m \in X_m = \ker(B^m)$. Also, $B^{m-1} x_n = 0$ because $x_n \in X_n \subset X_{m-1}$. Finally, $B^m x_n = 0$ because $x_n \in X_n \subset X_m$. These observations show that

$$B^{m-1}(B x_m - x_n - B x_n) = B^m x_m - B^{m-1} x_n - B^m x_n = 0$$

Now we can write

$$\begin{aligned}
\|A x_n - A x_m\| &= \|(B-I)x_n - (B-I)x_m\| = \|B x_n - x_n - B x_m + x_m\| \\
&= \|x_m - (B x_m + x_n - B x_n)\| \geqslant \mathrm{dist}(x_m, X_{m-1}) \geqslant 1/2
\end{aligned}$$

The sequence $[A x_n]$ therefore can have no Cauchy subsequence, contradicting the compactness property of A.   ∎

Lemma 2.   *If A is a compact operator on a Banach space, then the range of $I + A$ is closed.*

Proof.   Let $B = I + A$. Take a convergent sequence $[y_n]$ in the range of B, and write $y = \lim y_n$. We want to prove that y is in the range of B. Since this is obvious if $y = 0$, we assume that $y \neq 0$. Denote the kernel (null space) of B by K. Let $y_n = B x_n$ for suitable points x_n.

If $[x_n]$ contains a bounded subsequence, then (because A is compact) $[A x_n]$ contains a convergent subsequence, say $A x_{n_i} \to u$. Since $A x_{n_i} + x_{n_i} = B x_{n_i} =$

$y_{n_i} \to y$, we infer that $x_{n_i} = y_{n_i} - Ax_{n_i} \to y - u$. Then $y = \lim Bx_{n_i} = B(y-u)$, and y is in the range of B. This completes the proof in this case.

If $[x_n]$ contains no bounded subsequence, then $\|x_n\| \to \infty$. Since $y \neq 0$, we can discard a finite number of terms from the sequence $[x_n]$ and assume that $x_n \notin K$ for all n. Using Riesz's Lemma, construct vectors $v_n = k_n + \alpha_n x_n$ so that $\|v_n\| = 1$, $k_n \in K$, and $\text{dist}(v_n, K) \geq 1/2$. Note that

$$(1) \qquad\qquad Bv_n = \alpha_n Bx_n = \alpha_n y_n$$

Since $\|\alpha_n y_n\| = \|Bv_n\| \leq \|B\|$ and $y_n \to y \neq 0$, we see that $[\alpha_n]$ is bounded. Since $[v_n]$ is bounded, $[Av_n]$ contains a convergent subsequence. Using the boundedness of $[\alpha_n]$, we can arrange that

$$Av_{n_i} \to z \qquad \text{and} \qquad \alpha_{n_i} \to \alpha$$

From Equation (1), we conclude that $(I + A)v_{n_i} = \alpha_{n_i} y_{n_i}$ and

$$v_{n_i} = \alpha_{n_i} y_{n_i} - Av_{n_i} \to \alpha y - z$$

If α were 0, we would have $v_{n_i} \to -z$ and $-Bz = \lim Bv_{n_i} = \lim(v_{n_i} + Av_{n_i}) = -z + z = 0$. This would show that $z \in K$. This cannot be true because it would imply

$$1/2 \leq \text{dist}(v_{n_i}, K) \leq \|v_{n_i} + z\| = 0$$

Hence $\alpha \neq 0$. Since $Bv_{n_i} \to \alpha y$, we have

$$B(\alpha^{-1} v_{n_i}) \to y$$

Consequently, $B(y - \alpha^{-1}z) = y$, and y is in the range of B. $\blacksquare$

Lemma 3. Let A be a compact operator on a Banach space. If $I + A$ is injective, then it is surjective.

Proof. Let $B = I + A$ and let X_n denote the range of B^n. We have

$$B^n = (I + A)^n = \sum_{k=0}^{n} \binom{n}{k} A^k = I + \sum_{k=1}^{n} \binom{n}{k} A^k$$

Since each A^k is compact (for $k \geq 1$), B^n is the identity plus a compact operator. Thus X_n is closed by Lemma 2.

If $x \in X_n$ for some n, then for an appropriate u we have

$$x = B^n u = B^{n-1} Bu \in X_{n-1}$$

Thus

$$(2) \qquad\qquad X = X_0 \supset X_1 \supset X_2 \supset \cdots$$

Now our objective is to establish that $X_1 = X_0$.

If all the inclusions in the list (2) are proper, we can use Riesz's Lemma to select $x_n \in X_n$ such that $\|x_n\| = 1$ and $\text{dist}(x_n, X_{n+1}) \geqslant 1/2$. Then, for $n < m$, we have

$$\|Ax_m - Ax_n\| = \|(B - I)x_m - (B - I)x_n\| = \|x_n - (x_m + Bx_n - Bx_m)\|$$
$$\geqslant \text{dist}(x_n, X_{n+1}) \geqslant 1/2$$

because $x_m \in X_m \subset X_{n+1}$, $Bx_m \in X_{m+1} \subset X_{n+1}$, and $Bx_n \in X_{n+1}$. This argument shows that $[Ax_n]$ can contain no Cauchy subsequence, contradicting the compactness of A.

Thus, not all the inclusions in the list (2) are proper, and for some n, $X_n = X_{n+1}$. We define n to be the *first* integer having this property. All we have to do now is prove that $n = 0$.

If $n > 0$, let x be any point in X_{n-1}. Then $x = B^{n-1}y$ for some y, and

$$Bx = B^n y \in X_n = X_{n+1}$$

It follows that $Bx = B^{n+1}z$ for some z. Since B is injective by hypothesis, $x = B^n z \in X_n$. Since x was an arbitrary point in X_{n-1}, this shows that $X_{n-1} \subset X_n$. But the inclusion $X_{n-1} \supset X_n$ also holds. Hence $X_{n-1} = X_n$, contrary to our choice of n. Hence $n = 0$. ∎

Theorem 1. The Fredholm Alternative.   *Let A be a compact linear operator on a Banach space. The operator $I + A$ is surjective if and only if it is injective.*

Proof.   This is the result of putting Lemmas 1 and 3 together. ∎

The name attached to this theorem is derived from its traditional formulation, which states that one and only one of the these alternatives holds: (1) $I + A$ is surjective; (2) $I + A$ is not injective. A stronger result is known, and we refer the reader to [BN] or [KA] for its proof:

Theorem 2.   *If A is a bounded linear operator on a Banach space and if A^n is compact for some natural number n, then the properties of surjectivity and injectivity of $I + A$ imply each other.*

An easy extension of Theorem 1 is important:

Theorem 3.   *Let B be a bounded linear invertible operator, and let A be a compact operator, both defined on one Banach space and taking values in another. Then $B + A$ is surjective if and only if it is injective.*

Proof.   Suppose that $B + A$ is injective. Then so are $B^{-1}(B + A)$ and $I + B^{-1}A$. Now, the product of a compact operator with a bounded operator is compact. (See Problem 7.) Thus, Theorem 1 is applicable, and $I + B^{-1}A$ is surjective. Hence so are $B(I + B^{-1}A)$ and $B + A$. The proof of the reverse implication is similar. ∎

Theorem 4.   *A compact linear transformation operating from one normed linear space to another maps weakly convergent sequences into strongly convergent sequences.*

Proof. Let A be such an operator, $A : X \to Y$. Let $x_n \rightharpoonup x$ (weak convergence) in X. It suffices to consider only the case when $x = 0$. Thus we want to prove that $Ax_n \to 0$. By the weak convergence, $\phi(x_n) \to 0$ for all $\phi \in X^*$. Interpret $\phi(x_n)$ as a sequence of linear maps x_n acting on an element $\phi \in X^*$. Since X^* is complete even if X is not, the Uniform Boundedness Theorem (Section 1.7, page 42) is applicable in X^*. One concludes that $\|x_n\|$ is bounded. For any $\psi \in Y^*$,

$$\psi(Ax_n) = (\psi \circ A)x_n \to 0$$

because $\psi \circ A \in X^*$. Thus $Ax_n \rightharpoonup 0$. If Ax_n does not converge strongly to 0, there will exist a subsequence such that $\|Ax_{n_i}\| \geqslant \varepsilon > 0$. By the compactness of A, and by taking a further subsequence, we may assume that $Ax_{n_i} \to y$ for some y. Obviously, $\|y\| \geqslant \varepsilon$. Now we have the contradiction $Ax_{n_i} \rightharpoonup y$ and $Ax_{n_i} \rightharpoonup 0$. ∎

Lemma 4.   *Let $[A_n]$ be a bounded sequence of continuous linear transformations from one normed linear space to another. If $A_n x \to 0$ for each x in a compact set K, then this convergence is uniform on K.*

Proof. Suppose that the convergence in question is not uniform. Then there exist a positive ε, a sequence of integers n_i, and points $x_{n_i} \in K$ such that $\|A_{n_i} x_{n_i}\| \geqslant \varepsilon$. Since K is compact, we can assume at the same time that x_{n_i} converges to a point x in K. Then we have a contradiction of pointwise convergence from this inequality:

$$\|A_{n_i} x\| = \|A_{n_i} x_{n_i} + (A_{n_i} x - A_{n_i} x_{n_i})\|$$
$$\geqslant \|A_{n_i} x_{n_i}\| - \|A_{n_i} x - A_{n_i} x_{n_i}\|$$
$$\geqslant \varepsilon - \|A_{n_i}\| \|x - x_{n_i}\| \qquad ∎$$

For some Banach spaces X, each compact operator $A : X \to X$ is a limit of operators of finite rank. This is true of $X = C(T)$ and $X = L^2(T)$, but not of all Banach spaces. One positive general result in this direction is as follows.

Theorem 5.   *Let X and Y be Banach spaces. If Y has a (Schauder) basis, then every compact operator from X to Y is a limit of finite-rank operators.*

Proof. If $[v_n]$ is a basis for Y, then each y in Y has a unique representation of the form

$$y = \sum_{k=1}^{\infty} \lambda_k(y) v_k$$

(See Problems 24–26 in Section 1.6, pages 38–39.) The functionals λ_k are continuous, linear, and satisfy $\sup_k \|\lambda_k\| < \infty$. By taking the partial sum of the first n terms, we define a projection P_n of Y onto the linear span of the first n vectors v_k. Now let A be a compact linear transformation from X to Y, and let S denote the unit ball in X. The closure of $A(S)$ is compact in Y, and $P_n - I$ converges pointwise to 0 in Y. By the preceding lemma, this convergence is uniform on $A(S)$. This implies that $(P_n A - A)(x)$ converges uniformly to 0 on S. Since each $P_n A$ has finite-dimensional range, this completes the proof. ∎

Theorem 6. *Let A be a compact operator acting between two Banach spaces. If the range of A is closed, then it is finite dimensional.*

Proof. Since A is compact, it is continuous and has a closed graph. Assume that $A : X \to Y$ and that $A(X)$ is closed in Y. Then $A(X)$ is a Banach space. Let S denote the unit ball in X. By the Interior Mapping Theorem (Section 1.8, page 48), $A(S)$ is a neighborhood of 0 in $A(X)$. On the other hand, by its compactness, A maps S into a compact subset of $A(X)$. Since $A(X)$ has a compact neighborhood of 0, $A(X)$ is finite dimensional, by Theorem 2 in Section 1.4, page 22. ∎

Let us consider the very practical problem of solving an equation $Ax - \lambda x = b$ when A is of finite rank. In other words, A has a finite-dimensional range. Let $\{v_1, \ldots, v_n\}$ be a basis for the range of A. Let $b \in X$, $Ab = \sum \beta_i v_i$, and $Av_j = \sum_i a_{ij} v_i$.

We must assume that the numerical values of a_{ij} and β_i are available to us. Determining the unknown x will now reduce to a standard problem in numerical linear algebra. The case when $\lambda = 0$ is somewhat different, and we dispose of that first.

If $\lambda = 0$, we find u_i such that $Au_i = v_i$. Since the equation $Ax = b$ can be solved only if b is in the range of A, we write $b = \sum \gamma_i v_i$. Then the solution is $x = \sum_{i=1}^{n} \gamma_i u_i$, because with that definition of x we have $Ax = \sum_{i=1}^{n} \gamma_i Au_i = \sum_{i=1}^{n} \gamma_i v_i = b$.

Now assume that λ is not zero. The following two assertions are equivalent:

(a) There is an $x \in X$ such that $Ax - \lambda x = b$.
(b) There exist $c_1, \ldots, c_n \in \mathbb{R}$ such that $\sum_j a_{ij} c_j - \lambda c_i = \beta_i$ (for $i = 1, \ldots, n$).

To prove this equivalence, first assume that **(a)** is true. Define c_i by $Ax = \sum c_i v_i$. Then one has successively

$$\sum_{j=1}^{n} c_j v_j - \lambda x = b$$

$$\sum_{j=1}^{n} c_j Av_j - \lambda Ax = Ab$$

$$\sum_{j=1}^{n} c_j \sum_{i=1}^{n} a_{ij} v_i - \lambda \sum_{i=1}^{n} c_i v_i = \sum_{i=1}^{n} \beta_i v_i$$

Since $\{v_1, \ldots, v_n\}$ is linearly independent,

$$(3) \qquad \sum_{j=1}^{n} a_{ij} c_j - \lambda c_i = \beta_i \quad \text{for } i = 1, \ldots, n$$

This proves **(b)**.

For the converse, suppose that **(b)** is true. Define $x = (\sum c_j v_j - b)/\lambda$. Then we verify **(a)** by calculating

$$Ax - \lambda x = \frac{1}{\lambda} \left[\sum_{j=1}^{n} c_j A v_j - Ab - \lambda \sum_{i=1}^{n} c_i v_i \right] + b$$

$$= \frac{1}{\lambda} \left[\sum_{i=1}^{n} \sum_{j=1}^{n} a_{ij} c_j v_i - \sum_{i=1}^{n} \beta_i v_i - \lambda \sum_{i=1}^{n} c_i v_i \right] + b = b$$

This analysis has established that the original problem is equivalent to a matrix problem of order n, where n is the dimension of the range of the operator A. The actual numerical calculations to obtain x involve solving the Equation (3) for the unknown coefficients c_j. We have not yet made any assumptions to guarantee the solvability of Equation (3).

Now one can prove the Fredholm Alternative for this case by elementary linear algebra. Indeed we have these equivalences (in which $\lambda \neq 0$):

(i) $A - \lambda I$ is surjective.
(ii) For each b there is an x such that $Ax - \lambda x = b$.
(iii) For all $(\beta_1, \ldots, \beta_n)$ the system $\sum_{j=1}^{n} a_{ij} c_j - \lambda c_i = \beta_i$ $(i = 1, \ldots, n)$ is soluble.
(iv) The system $\sum_{j=1}^{n} a_{ij} c_j - \lambda c_i = 0$ has only the trivial solution.
(v)   The equation $Ax - \lambda x = 0$ has only the trivial solution.
(vi) λ is not an eigenvalue of the operator A.
(vii) $A - \lambda I$ is injective.

Integral Equations.   The theory of linear operators is well illustrated in the study of integral equations. These have arisen in earlier parts of this book, such as in Sections 2.3, 2.5, 4.1, 4.2, and 4.3. A special type of integral equation has what is known as a degenerate kernel. A kernel is called **degenerate** or **separable** if it is of the form $k(s, t) = \sum_{1}^{n} u_i(s) v_i(t)$. The corresponding integral operator is

$$(Kx)(t) = \int k(s, t) x(s) \, ds = \int \sum_{i=1}^{n} u_i(s) v_i(t) x(s) \, ds$$

$$= \sum_{i=1}^{n} v_i(t) \int u_i(s) x(s) \, ds$$

If we use the inner-product notation for the integrals in the above equation, we have the simpler form

$$Kx = \sum_{i=1}^{n} \langle x, u_i \rangle v_i$$

It is clear that K has a finite-dimensional range and is therefore a compact operator. (Various spaces are suitable for the discussion.)

Lemma 5. *In the definition of the degenerate kernel $k = \sum_{i=1}^{n} u_i v_i$, there is no loss of generality in supposing that $\{u_1, \ldots, u_n\}$ and $\{v_1, \ldots, v_n\}$ are linearly independent sets.*

Proof. Suppose that $\{v_1, \ldots, v_n\}$ is linearly dependent. Then one vector is a linear combination of the others, say $v_n = \sum_{i=1}^{n-1} a_i v_i$. Then we can write the kernel with a sum of fewer terms as follows:

$$
\begin{aligned}
Kx &= \sum_{i=1}^{n} \langle x, u_i \rangle v_i = \sum_{i=1}^{n-1} \langle x, u_i \rangle v_i + \langle x, u_n \rangle v_n \\
&= \sum_{i=1}^{n-1} \langle x, u_i \rangle v_i + \langle x, u_n \rangle \sum_{i=1}^{n-1} a_i v_i \\
&= \sum_{i=1}^{n-1} \left[\langle x, u_i \rangle + a_i \langle x, u_n \rangle \right] v_i = \sum_{i=1}^{n-1} \langle x, u_i + a_i u_n \rangle v_i
\end{aligned}
$$

A similar argument applies if $\{u_1, \ldots, u_n\}$ is dependent. ∎

To solve the integral equation $Kx - \lambda x = b$ when K has a separable kernel (as above), we assume $\lambda \neq 0$ and that $\{v_1, \ldots, v_n\}$ is independent. Then $\{v_1, \ldots, v_n\}$ is a basis for the range of K, and the theory of the preceding pages applies. If the integral equation has a solution, then the system of linear equations

$$
\sum_{j=1}^{n} a_{ij} c_j - \lambda c_i = \beta_i \qquad (1 \leqslant i \leqslant n)
$$

has a solution, where $Kv_j = \sum_i a_{ij} v_i$ and $Kb = \sum \beta_i v_i$. It follows that the solution is $x = \lambda^{-1}(\sum c_i v_i - b)$.

The **spectrum** of a linear operator A on a normed linear space is the set of all complex numbers λ for which $A - \lambda I$ is *not* invertible.

Theorem 7. *Let A be a compact operator on a Banach space. Each nonzero element of the spectrum of A is an eigenvalue of A.*

Proof. Let $\lambda \neq 0$ and suppose that λ is not an eigenvalue of A. We want to show that λ is not in the spectrum, or equivalently, that $A - \lambda I$ is invertible. Since λ is not an eigenvalue, the equation $(A - \lambda I)x = 0$ has only the solution $x = 0$. Hence $A - \lambda I$ is injective. By the Fredholm Alternative, $A - \lambda I$ is surjective. Hence $(A - \lambda I)^{-1}$ exists as a linear map. The only question is whether it is a *bounded* linear map. The affirmative answer comes immediately from the Interior Mapping Theorem and its corollaries in Section 1.8, page 48ff. That would complete the proof. There is an alternative that avoids use of the Interior Mapping Theorem but uses again the compactness of A. To follow this path, assume that $(A - \lambda I)^{-1}$ is not bounded. We can find x_n such that $\|x_n\| = 1$ and $\|(A - \lambda I)^{-1} x_n\| \to \infty$. Put $y_n = (A - \lambda I)^{-1} x_n$. Then $\|y_n\| / \|(A - \lambda I) y_n\| \to \infty$.

Put $z_n = y_n / \|y_n\|$, so that $\|z_n\| = 1$ and $\|(A - \lambda I) z_n\| \to 0$. Since A is compact, there is a convergent subsequence $A z_{n_k} \to w$. Then

$$z_{n_k} = \lambda^{-1} \Big[A z_{n_k} - (A - \lambda I) z_{n_k} \Big] \to \lambda^{-1} w$$

Hence $A(\lambda^{-1} w) = w$ or $(A - \lambda I) w = 0$. Since $\|w\| = |\lambda| \neq 0$, we have contradicted the injective property of $A - \lambda I$. ∎

For an integral equation having a more general kernel, there is a possibility of approximating the kernel by a separable (degenerate) one, and solving the resulting simpler integral equation. The approximation is possible, as we shall see. We begin by recalling the important Stone–Weierstrass Theorem:

Theorem 8. Stone–Weierstrass Theorem. *Let T be a compact topological space. Every subalgebra of $C(T)$ that contains the constant functions and separates the points of T is dense in $C(T)$.*

Definition. A family of functions on a set is said to **separate the points** of that set if for every pair of distinct points in the set there exists a function in the family that takes different values at the two points.

Example 1. Let $T = [a, b] \subset \mathbb{R}$. Let $\mathcal{A}$ be the algebra of all polynomials in $C(T)$. Then $\mathcal{A}$ is dense in $C(T)$, by the Stone–Weierstrass Theorem. This implies that for any continuous function f defined on $[a, b]$ and for any $\epsilon > 0$ there is a polynomial p such that

$$\|f - p\| \equiv \max\{|f(t) - p(t)| : a \leqslant t \leqslant b\} < \epsilon$$

Example 2. Let S and T be compact spaces. Then the set

$$\left\{ f : f(s, t) = \sum_{i=1}^{n} a_i(s) b_i(t) \quad \text{for some } n, \ a_i \in C(S), \ b_i \in C(T) \right\}$$

is dense in $C(S \times T)$.

The next theorem shows that theoretically we obtain a solution to the original problem as a limit of solutions to the simpler ones involving operators of finite rank.

Theorem 9.   *Let $A_0, A_1, \ldots$ be compact operators on a Banach space, and suppose $\lim_n A_n = A_0$. If λ is not an eigenvalue of A_0 and if for each n there is a point x_n such that $A_n x_n - \lambda x_n = b$, then for all sufficiently large n,*

$$\|x_0 - x_n\| \leqslant \|(A_n - \lambda I)^{-1}\| \, \|A_0 - A_n\| \, \|x_0\|$$

Proof. Since λ is not an eigenvalue of A_0, it is not in the spectrum of A_0, by Theorem 7. Hence $A_0 - \lambda I$ is invertible. Select m such that for $n \geqslant m$

$$\|(A_n - \lambda I) - (A_0 - \lambda I)\| = \|A_n - A_0\| < \|(A_0 - \lambda I)^{-1}\|^{-1}$$

By Problem 2 in Section 4.3 (page 189), $(A_n - \lambda I)^{-1}$ exists (when $n \geqslant m$). Now write

$$
\begin{aligned}
x_n - x_0 &= \left[(A_n - \lambda I)^{-1} - (A_0 - \lambda I)^{-1}\right]b \\
&= (A_n - \lambda I)^{-1}\left[I - (A_n - \lambda I)(A_0 - \lambda I)^{-1}\right]b \\
&= (A_n - \lambda I)^{-1}\left[I - \{A_0 - \lambda I - (A_0 - A_n)\}(A_0 - \lambda I)^{-1}\right]b \\
&= (A_n - \lambda I)^{-1}(A_0 - A_n)(A_0 - \lambda I)^{-1}b \\
&= (A_n - \lambda I)^{-1}(A_0 - A_n)x_0
\end{aligned}
$$

∎

Problems 7.5

1. Supply the "similar" argument omitted from the proof of Lemma 1.

2. Let $k(s,t) = \sum_1^n u_i(s)v_i(t)$, where $u_i \in L_2(S)$ and $v_i \in L_2(T)$. Show that $k(s,t)$ can also be represented as $\sum_1^n \widetilde{u}_i(s)\widetilde{v}_i(t)$, where $\{\widetilde{v}_i\}$ an orthonormal set.

3. On page 358 we saw how to solve $Kx - \lambda x = b$ if K is an integral operator having a separable kernel and $\lambda \neq 0$. Give a complete analysis of the case when $\lambda = 0$.

4. Is the set of polynomials of the form $c_0 + c_1 t^{17} + c_2 t^{34} + c_3 t^{51} + \cdots$ dense in $C[a,b]$? Generalize.

5. Solve the integral equation
$$
\int_0^1 (t-s)x(s)\,ds - \lambda x(t) = b(t)
$$

6. Solve the integral equation
$$
\int_0^1 [a(s) + b(t)]x(s)\,ds - x(t) = c(t)
$$

7. Prove that the set of compact operators in $\mathcal{L}(X,X)$ is an **ideal**. This means that if A is compact and B is any bounded linear operator, then AB and BA are compact. (This property is in addition to the subspace axioms.)

8. Let A be a compact operator on a Banach space. Prove that if $I - A$ is injective, then it is invertible.

9. Let $A, B \in \mathcal{L}(X,X)$. Assume that $AB = BA$ and that AB is invertible. Prove that (a) $A(AB)^{-1} = (AB)^{-1}A$; (b) $B^{-1} = A(AB)^{-1}$; (c) $A^{-1} = B(AB)^{-1}$; (d) $A(AB)^{-1} = (AB)^{-1}A$; (e) $(BA)^{-1} = A^{-1}B^{-1}$.

10. Let A be a linear operator from X to Y. Suppose that we are in possession of elements $u_1, u_2, \ldots, u_n$ whose images under A span the range of A. Describe how to solve the equation $Ax = b$.

11. Prove that if A is a compact operator on a normed linear space, then for some natural number n, the ranges of $(I+A)^n, (I+A)^{n+1}, \ldots$ are all identical.

12. Let A be a linear transformation defined on and taking values in a linear space. Prove that if A is surjective but not injective, then $\ker(A^n)$ is a proper subset of $\ker(A^{n+1})$, for $n = 1, 2, 3, \ldots$

13. Let A be a bounded linear operator defined on and taking values in a normed linear space. Suppose that for $n = 1, 2, 3, \ldots$, the range of A^n properly contains the range of A^{n+1}. Prove that the sum $A + K$ is never invertible when K is a compact operator.

14. Prove that if A_n are continuous linear transformations acting between two Banach spaces, and if $A_n x \to 0$ for all x, then this convergence is uniform on all compact sets. (Cf. Lemma 4.)

15. Give examples of operators on the space c_0 that have one but not both of the properties injectivity and surjectivity.

16. (More general form of Lemma 5) Let A be an operator defined by the equation $Ax = \sum_{i=1}^{n} \phi_i(x) v_i$, in which $x \in X$, $v_i \in Y$, and $\phi_i \in X^*$. Prove that there is no loss of generality in supposing that $\{v_1, \dots, v_n\}$ and $\{\phi_1, \dots, \phi_n\}$ are linearly independent sets.

17. Show how to solve the equation $Ax - x = b$ if the range of A is spanned by $\{Au_1, \dots, Au_n\}$, for some $u_i \in X$. Prove that the equation is solvable if 1 is not an eigenvalue of A.

18. Let A and B be members of $\mathcal{L}(X, Y)$, where X and Y are Banach spaces. Suppose that B is invertible and that $(B^{-1}A)^m$ is compact for some natural number m. Prove that $B + A$ is surjective if and only if it is injective.

19. Provide the details for the assertions in Example 2.

20. Use the Stone–Weierstrass Theorem to prove this result of Diaconis and Shahshahani [DiS]: If X is a normed linear space and f is a continuous function from X to $\mathbb{R}$, then for any compact set K in X and for any positive ε there exist $\phi_i \in X^*$ and coefficients c_i such that

$$\sup_{x \in K} \left| f(x) - \sum_{i=1}^{n} c_i e^{\phi_i(x)} \right| < \varepsilon$$

21. Prove this for an arbitrary compact operator A: The transformation $I + A$ is surjective if and only if -1 is not an eigenvalue of A.

22. Prove the finite-dimensional version of the Fredholm alternative, which we formulate as follows, for an arbitrary matrix A and vector b: The system $Ax = b$ is consistent if and only if the system $y^T A = 0$ $y^T b \neq 0$ is inconsistent.

23. Discuss the existence and uniqueness of solutions to the integral equation

$$\int_0^\pi \cos(s - t) u(s) \, ds = f(t)$$

where f is a prescribed function, and u is the unknown function.

7.6 Topological Spaces

In this section we provide an abbreviated introduction to topological notions—hardly more than enough to bring us to the Tychonoff Theorem.

So far in this book we have been dealing routinely with topological spaces, but of a special kind, usually metric spaces or normed linear spaces. Now we require a more general discussion so that there will be a suitable framework for the weak topologies on linear spaces and for other examples.

A good starting point is the question, What is a **topology**? It is a family of sets such that:

    a. The empty set, $\varnothing$, is a member of the family.

    b. The intersection of any two members of the family is also in the family.

    c. The union of any subfamily is also in the family.

If $\mathcal{T}$ is a topology, we define X to be the union of all members of $\mathcal{T}$. We say that X is the **space** of $\mathcal{T}$ and that $\mathcal{T}$ is a topology on X. We call the pair

$(X, \mathcal{T})$ a **topological space**. Each member of $\mathcal{T}$ is called an **open set** in X. By Axiom **c**, X is open. So is the empty set, $\varnothing$. The use of the word "open" will be unambiguous if there is only one topology being discussed. If there are several, a more exact terminology will be needed. For example, one could refer to the $\mathcal{T}$-open sets and the $\mathcal{S}$-open sets, if $\mathcal{T}$ and $\mathcal{S}$ are topologies. In any case, X and $\varnothing$ will always be open, no matter what topology has been assigned to X.

Example 1. Let X be any set and let $\mathcal{T} = 2^X$. That notation signifies that $\mathcal{T}$ is the family of all subsets of X, including the empty set and X itself. Obviously, the axioms are fulfilled in this example. This topology is the largest one that can be defined on X, and is called the **discrete** topology. Every singleton $\{x\}$ is an open set in this topological space. Every topology on X is contained in the discrete topology on X. ∎

Example 2. Let X be any set, and define $\mathcal{T}$ to consist of only the empty set and the given set X. This is the smallest topology on X, and is called whimsically the **indiscrete** topology on X. Every topology on X contains the indiscrete topology. ∎

Example 3. In $\mathbb{R}$ define a set V to be open if every point x in V is the center of an interval $(x-\varepsilon, x+\varepsilon)$ that lies wholly in V. This defines the **usual** topology on $\mathbb{R}$. We do not stop to prove that this definition of "open" leads to a family satisfying the axioms (but this is a good exercise for the reader). ∎

Example 4. In any set X let a notion of **distance** between points be introduced. The distance from x to y can be denoted by $d(x, y)$, and this function should satisfy three axioms:
  a. If $x \neq y$, then $d(x, y) = d(y, x) > 0$.
  b. For each x, $d(x, x) = 0$.
  c. For all x, y, z, $d(x, z) \leqslant d(x, y) + d(y, z)$.

Then a topology can be defined as in Example 3. Namely, a set V is open if each point x in V is the center of a "ball"

$$B(x, \varepsilon) = \{y \; : \; d(x, y) < \epsilon\}$$

that lies wholly in V. The pair (X, d) is a **metric space**, and is a topological space, it being understood that its topology is the one just described. All normed linear spaces are metric spaces, because the equation $d(x, y) = \|x - y\|$ defines a metric. ∎

A topological space is said to be a **Hausdorff space** if for any pair of distinct points x and y there is a disjoint pair of open sets U and V such that $x \in U$ and $y \in V$. Every metric space is a Hausdorff space, since $B(x, \varepsilon)$ and $B(y, \varepsilon)$ will be disjoint from each other if ε is sufficiently small. The Hausdorff property is one of a number of **separation** axioms that topological spaces may satisfy. It is useful in questions of convergence, for it ensures that a sequence (or net) can converge to at most one limit.

A **base** for a topology $\mathcal{T}$ is any subfamily $\mathcal{B}$ of $\mathcal{T}$ such that every open set is a union of sets in $\mathcal{B}$. For example, the open intervals with rational endpoints form a base for the usual topology on $\mathbb{R}$. In a discrete space, the singletons $\{x\}$ form a base.

A **subbase** for a topology $\mathcal{T}$ is a subfamily $\mathcal{S}$ of $\mathcal{T}$ such that the finite intersections of sets in $\mathcal{S}$ form a base for $\mathcal{T}$. For example, the intervals of the form (a, ∞) and $(-\infty, b)$ provide a subbase for the (usual) topology of $\mathbb{R}$. This is evident, since by intersecting two such intervals we can obtain an interval (a, b).

A topology on a space X can be defined by specifying any family $\mathcal{S}$ of subsets of X as a subbase for the topology. A base $\mathcal{B}$ is then the family of all finite intersections of sets in $\mathcal{S}$, and the topology itself consists of all unions of sets in $\mathcal{B}$. An easy proof then establishes that the resulting family satisfies the axioms for a topology.

If A is any set in a topological space, we can form the largest open set contained in A by simply taking the union of all open sets that are subsets of A. The resulting set is called the **interior** of A and is often denoted by A°. The reader can verify that a set is open if and only if it equals its interior.

In a topological space, a **neighborhood** of a point is any set whose interior contains the given point. It is easily verified that a set is open if and only if it is a neighborhood of each of its points.

In a topological space, the **closed** sets are the complements of the open sets. Further properties are easily proved:

a. The empty set and the space itself are closed.

b. The intersection of any family of closed sets is closed.

c. The union of any finite collection of closed sets is closed.

As a consequence of the preceding definitions, each set can be enclosed in a smallest closed set, called the **closure** of that set. Namely, we take as the closure of A the intersection of all closed sets containing A. Then a set is closed if and only if it equals its closure. One quickly proves that a point x belongs to the closure of a set A if and only if each neighborhood of x intersects A.

Another basic notion in general topology is that of the **relative topology** in a subset of a topological space. If Y is a subset of a topological space X, and if $\mathcal{T}$ is the topology on X, then we take as the "relative" topology on Y the family

$$\mathcal{U} = \{Y \cap \mathcal{O} : \mathcal{O} \in \mathcal{T}\}$$

A set can be open in Y (meaning that it belongs to $\mathcal{U}$) without necessarily being open in X. For example, if Y is a non-open subset of X, then $Y \in \mathcal{U}$, while $Y \notin \mathcal{T}$.

Another ingredient of general topology is an extended concept of convergence. It turns out that sequential convergence is inadequate for describing topological notions, in general. Sequential convergence suffices in some spaces, such as metric spaces, but not in all spaces. The generalization of sequences consists principally in allowing an ordered set other than the natural numbers to serve as the domain of the indices. The definitions pertaining to this topic are as follows.

A **partially ordered set** is a pair $(D, \prec)$ in which D is a set and $\prec$ is a relation obeying these axioms:

    a. $\alpha \prec \alpha$
    b. If $\alpha \prec \beta$ and $\beta \prec \gamma$, then $\alpha \prec \gamma$.

A **directed set** is a partially ordered set in which an additional axiom is required:

    c. Given α and β in D, there is $\gamma \in D$ such that $\alpha \prec \gamma$ and $\beta \prec \gamma$.

The reader will recognize $\mathbb{N}$ as a familiar example of a directed set, it being understood that $\prec$ is the ordinary relation $\leqslant$. Another important example is the set of all neighborhoods of a point in a topological space, where $\prec$ is interpreted as $\supset$.

A **net** or **generalized sequence** is a function on a directed set. This is obviously more general than a sequence, which is a function on $\mathbb{N}$. We can use the notation $[x, D, \prec]$ for a net, specifying the function x, the directed set D, and the relation $\prec$. When we need not concern ourselves with niceties, the notation $[x_\alpha]$ can be used, just as we abuse notation for sequences and write $[x_n]$.

Useful conventions are as follows. A net $[x_\alpha]$ is **eventually** in a set V if there is a β such that $x_\alpha \in V$ whenever $\beta \prec \alpha$. If a net is eventually in every neighborhood of a point y, then we say that the net **converges** to y. Let us illustrate with one example of a theorem employing nets.

Theorem 1.   *A point y is in the closure of a set S in a topological space if and only if some net in S converges to y.*

Proof. If the net $[x_\alpha]$ is in S and converges to y, then to each neighborhood U of y there corresponds an index β such that $x_\alpha \in U$ whenever $\beta \prec \alpha$. In particular, $x_\beta \in U$. Thus each neighborhood of y contains a point of S, and y is in the closure of S. Conversely, suppose that y is in the closure of S. Let D be the family of all neighborhoods of y, ordered by inclusion: $\alpha \prec \beta$ means $\beta \subset \alpha$. Since y is in the closure of S, there exists for each $\alpha \in D$ a point $x_\alpha \in \alpha \cap S$. The net $[x_\alpha]$ thus defined (with the aid of the Axiom of Choice) is in S and converges to y. ∎

In the preceding proof, had we known in advance that the point y possessed a countable neighborhood base, we could have used a sequential argument. However, there exist spaces in which some points do not have a countable neighborhood base. (A base for the neighborhoods of a point x is a family of neighborhoods of x such that every neighborhood of x contains one of the sets in the family.)

A family $\mathcal{A}$ of sets in a topological space X is said to be an **open cover** of X if all the sets in $\mathcal{A}$ are open and if X is contained in the union of the sets in $\mathcal{A}$. If every open cover of X has a finite subfamily that is also an open cover of X, then X is said to be **compact**. The compact sets on the real line are precisely the closed and bounded sets. The same assertion is true for any finite-dimensional normed linear space. These matters were investigated in Section 1.4, pages 19–22. But in that part of the book we adopted a sequential definition of compactness that is inadequate for general topology.

Alexander's Theorem.   *Let a subbase be specified for a topology on a space. If every cover of the space by subbase elements has a finite subcover, then the space is compact.*

Proof. Let X be the space, $\mathcal{T}$ the topology, and $\mathcal{S}$ the subbase in question. Assume that every cover of X by elements of $\mathcal{S}$ has a finite subcover. Suppose that X is not compact. We seek a contradiction.

The family of open covers that do not have finite subcovers is a nonempty family. Partially order this family by inclusion, and invoke Zorn's Lemma (Section 1.6, page 32). This maneuver produces an open cover $\mathcal{A}$ of X that is maximal with respect to the property of possessing no finite subcover. Define $\mathcal{A}' = \mathcal{S} \cap \mathcal{A}$. Certainly, no finite subfamily of $\mathcal{A}'$ covers X. Since all sets in $\mathcal{A}'$ are members of the subbase, our hypotheses imply that $\mathcal{A}'$ itself does not cover X.

This last assertion implies that there exists a point x that is contained in no member of $\mathcal{A}'$. Since $\mathcal{A}$ is an open cover of X, we can select $U \in \mathcal{A}$ such that $x \in U$. By the properties of a subbase, there exist sets $S_1, \ldots, S_n$ in $\mathcal{S}$ such that $x \in \bigcap_{i=1}^{n} S_i \subset U$. Since x is contained in no member of $\mathcal{A}'$, one concludes that $S_i \notin \mathcal{A}'$. Hence $S_i \notin \mathcal{A}$. By the maximal property of $\mathcal{A}$, each enlarged family $\mathcal{A} \cup \{S_i\}$ contains a finite subcover of X, for $i = 1, 2, \ldots, n$. Hence, for each i in $\{1, 2, \ldots, n\}$, there is an open set $\mathcal{O}_i$ that is a union of finitely many sets in $\mathcal{A}$ and has the property $\mathcal{O}_i \cup S_i = X$. Define $B = \mathcal{O}_1 \cup \cdots \cup \mathcal{O}_n$. Then $B \cup S_i = X$ for each i, and $\bigcap_{i=1}^{n}(B \cup S_i) = X$. It follows that

$$X \subset B \cup (S_1 \cap S_2 \cap \cdots \cap S_n) \subset B \cup U$$

Since B is the union of finitely many sets in $\mathcal{A}$, and since $U \in \mathcal{A}$, we see that a finite subfamily of $\mathcal{A}$ covers X, contradicting a property of $\mathcal{A}$ established previously.   ∎

If we have two topological spaces, say $(X_1, \mathcal{T}_1)$ and $(X_2, \mathcal{T}_2)$, then we can topologize the Cartesian product $X_1 \times X_2$ in a standard way: We take as a base for the topology of $X_1 \times X_2$ the family of all sets $A \times B$, where A is open in X_1 and B is open in X_2. (The topology itself consists of all unions of sets in the base.)

The notion of a product extends to any family (finite, countable, or uncountable) of topological spaces $(X_i, \mathcal{T}_i)$, where $i \in I$. The index set I can be of arbitrary cardinality. The product space is denoted by ΠX_i or $\Pi \{X_i : i \in I\}$ and is defined to be the set of all functions x on I such that $x(i) \in X_i$ for all $i \in I$. In this context, we usually write $x_i = x(i)$. This is exactly the process by which we construct $\mathbb{R}^n$ from $\mathbb{R}$. We take n factors, all equal to $\mathbb{R}$. The generic element of the product space is a "vector" that we write as $x = [x_1, x_2, \ldots, x_n]$. Thus, x is a function on the index set $\{1, 2, \ldots, n\}$.

For each $i \in I$ there is a **projection** P_i from the product space $X = \Pi X_i$ to X_i. It is defined by $P_i(x) = x_i$. The topology on X is taken to be the weakest one that makes each of these projections continuous. One then must require that each set $P_i^{-1}[\mathcal{O}]$ be open when $\mathcal{O}$ is an open set in X_i. The family of all these sets is taken as a subbase for the product topology on X.

Tychonoff Theorem.   *A topological product of compact spaces is compact.*

Proof. For each i in an index set I, let $(X_i, \mathcal{T}_i)$ be a compact topological space. We form the product space $X = \Pi X_i$, and give it the product topology as described above. Use the projections P_i described above also. A subbase for the product topology is the family $\mathcal{S}$ of all sets of the form $P_i^{-1}(\mathcal{O})$, where i ranges over I and $\mathcal{O}$ ranges over $\mathcal{T}_i$. In order to take advantage of Alexander's Theorem, let $\mathcal{W}$ be a cover of X by subbase sets. Thus

$$\mathcal{W} \subset \mathcal{S} \quad \text{and} \quad X = \bigcup \{\mathcal{O} : \mathcal{O} \in \mathcal{W}\}$$

For each i let $\mathcal{V}_i$ be the family of all open sets in X_i whose inverse images by P_i are in $\mathcal{W}$:

$$\mathcal{V}_i = \{\mathcal{O} \in \mathcal{T}_i : P_i^{-1}(\mathcal{O}) \in \mathcal{W}\}$$

Assertion: For some i, $\mathcal{V}_i$ covers X_i. To prove this, assume that it is false. Then for each i, $\mathcal{V}_i$ fails to cover X_i, and consequently, there exists a point $x_i \in X_i$ such that $x_i \notin \bigcup \{\mathcal{O} : \mathcal{O} \in \mathcal{V}_i\}$. By the Axiom of Choice we can select these points x_i simultaneously and thereby construct an x in X such that

$$P_i x = x(i) = x_i \in X_i \smallsetminus \bigcup_{i \in I} \{\mathcal{O} : \mathcal{O} \in \mathcal{V}_i\}$$

Consequently, we have for each i and for each open set $\mathcal{O}$ in X_i the following implications:

$$P_i^{-1}(\mathcal{O}) \in \mathcal{W} \Longrightarrow \mathcal{O} \in \mathcal{V}_i \Longrightarrow x_i \notin \mathcal{O} \Longrightarrow P_i x \notin \mathcal{O} \Longrightarrow x \notin P_i^{-1}(\mathcal{O})$$

However, $\mathcal{W}$ consists exclusively of sets having the form $P_i^{-1}(\mathcal{O})$, and so the above implication reads as follows:

$$U \in \mathcal{W} \Longrightarrow x \notin U$$

This contradicts the fact that $\mathcal{W}$ is a cover of X, and proves the assertion.

Now select an index $j \in I$ such that $\mathcal{V}_j$ covers X_j. By the compactness of X_j, a finite subfamily of $\mathcal{V}_j$ covers X_j, say $\mathcal{O}_1, \ldots, \mathcal{O}_n \in \mathcal{V}_j$ and $X_j = \mathcal{O}_1 \cup \cdots \cup \mathcal{O}_n$. It follows (by using P_j^{-1}) that $X = P_j^{-1}(\mathcal{O}_1) \cup \cdots \cup P_j^{-1}(\mathcal{O}_n)$. Since these n sets are in $\mathcal{W}$ (by the definition of $\mathcal{V}_j$), we have found a finite subcover in $\mathcal{W}$, as desired. ∎

A particular case of the topological product is especially useful. It occurs when all the factors X_i are equal, say to X. In the general theory, we can then take each X_i to be a copy of X. The notation X^I now is more natural than $\Pi\{X : i \in I\}$. This space consists of all functions from I to X. We still have the projections P_i from X^I to X, and $P_i(x) = x(i)$ for all $i \in I$.

7.7 Linear Topological Spaces

Although normed linear spaces have served us well in this book, there are some matters of importance in applied mathematics that require a more general topologized linear space. (The theory of distributions is a pertinent example.) What is needed is a linear space in which the topological notions of continuity, compactness, completeness, etc., do not necessarily arise from a norm and its induced metric. The appropriate definition follows.

Definition.   A linear topological space is a pair $(X, \mathcal{T})$ in which X is a linear space and $\mathcal{T}$ is a topology on X such that the algebraic operations in X are continuous.

Being more specific about the continuity, we say that the two maps

$$(x, y) \mapsto x + y \qquad\qquad (\lambda, x) \mapsto \lambda x$$

are continuous, the first being defined on $X \times X$ and the second being defined on $\mathbb{R} \times X$. There is a corresponding definition if the scalar field is taken to be $\mathbb{C}$ rather than $\mathbb{R}$.

We remind the reader that the sets belonging to the family $\mathcal{T}$ are called the **open** sets, and a neighborhood of a point x is any set U such that for some open set $\mathcal{O}$ we have $x \in \mathcal{O} \subset U$. The continuity axioms above can be stated in terms of neighborhoods like this:

a. If U is a neighborhood of $x + y$, then there exist neighborhoods V of x and W of y such that $v + w$ is in U whenever $v \in V$ and $w \in W$.

b. If U is a neighborhood of λx, then there are neighborhoods V of λ and W of x such that $\alpha w \in U$ whenever $\alpha \in V$ and $w \in W$.

A very useful fact is that the topology is completely determined by the neighborhoods of 0. This is formally stated in the next lemma.

Lemma 1   *In a linear topological space, a set V is a neighborhood of a point z if and only if $-z + V$ is a neighborhood of 0.*

Proof.   Hold z fixed, and define $f(x) = x + z$. This mapping sends 0 to z. Let V be a neighborhood of z. Since f is continuous, $f^{-1}(V)$ is a neighborhood of 0. Observe, now, that $f^{-1}(V) = \{x : f(x) \in V\} = \{x : x + z \in V\} = -x + V$. Conversely, assume that $-z + V$ is a neighborhood of 0. We have $f^{-1}(x) = x - z$, and f^{-1} is also continuous. It maps z to 0. Hence $(f^{-1})^{-1}$ carries $-z + V$ to a neighborhood of z. But

$$(f^{-1})^{-1}(-z + V) = \{x : f^{-1}(x) \in -z + V\} = \{x : x - z \in -z + V\} = V \quad \blacksquare$$

In any topological space, a family $\mathcal{U}$ of neighborhoods of a point x is called a **base for the neighborhoods of** x if each neighborhood of x contains a member of $\mathcal{U}$. For example, one base for the neighborhoods of a point x on the real line is the set of intervals $[x - \frac{1}{n}, x + \frac{1}{n}]$, where n ranges over $\mathbb{N}$.

Since we often want the Hausdorff axiom to hold in our linear topological spaces, we record here the appropriate condition.

Theorem 1. *A linear topological space is a Hausdorff space if and only if 0 is the only element common to all neighborhoods of 0.*

Proof. The Hausdorff property is that for any pair of points $x \neq y$ there must exist neighborhoods U and V of x and y respectively such that the pair U, V is disjoint. Select a neighborhood W of 0 such that $x - y \notin W$. Then (using the continuity of subtraction) select another neighborhood W' of zero such that $W' - W' \subset W$. Then $x + W'$ is disjoint from $y + W'$, for if z is a point in their intersection, we could write $z = x + w_1 = y + w_2$, with $w_i \in W'$. Then $x - y \doteq w_2 - w_1 \in W' - W' \subset W$. The other half of the proof is even easier: just separate any nonzero point from 0 by selecting a neighborhood of zero that excludes the nonzero point. ∎

At this juncture, we should alert the reader to the fact that some authors assume the Hausdorff property as part of the definition of a linear topological space.

Let X be a linear space (preferably without a topology, so that confusion between two topologies can be avoided in what we are about to discuss). The notation X' signifies the algebraic dual of X, i.e., the space of all linear maps from X into the scalar field ($\mathbb{R}$ or $\mathbb{C}$). We can use X to define a "weak" topology on X', and we can use X to define a weak topology on X'. There is an abstract description that includes both of these constructions, but let us proceed in a more pedestrian manner. What we have in mind is rather simple: We want the topologies to lead to pointwise convergence in both cases. Although we did not discuss it here, a topology can be defined by specifying the meaning of convergence of nets. The topic is addressed in [Kel], pages 73–76.

The topology on X induced by X' can be called the **weak** topology. A base for the neighborhood system at 0 is given by all sets of the form

$$V(\varepsilon; \phi_1, \phi_2, \ldots, \phi_n) = \{x \in X \ : \ |\phi_i(x)| < \varepsilon, \ 1 \leqslant i \leqslant n\}$$

In this equation ε is any positive number, and $\{\phi_1, \ldots, \phi_n\}$ is any finite subset of X'. Convergence in this topology means the following: A net $[x_\alpha]$ in X converges to a point x if $[\phi(x_\alpha)]$ converges to $\phi(x)$ for every $\phi \in X'$.

The topology on X' induced by X is often called the **weak*** topology. A base for the neighborhood system of 0 is given by

$$V(\varepsilon; x_1, x_2, \ldots, x_n) = \{\phi \in X' : |\phi(x_i)| < \varepsilon, \ 1 \leqslant i \leqslant n\}$$

Here, ε is any positive number and $\{x_1, \ldots, x_n\}$ is any finite set in X. With this topology, a net $[\phi_\alpha]$ in X^* converges to ϕ if and only if $[\phi_\alpha(x)]$ converges to $\phi(x)$ for all $x \in X$. This is pointwise convergence.

The topologies just described are both Hausdorff topologies, as is easily deduced from Theorem 1. Also, one sees immediately that the space X' is a subspace of $\mathbb{R}^X$, since the latter is the space of *all* functions from X to $\mathbb{R}$, while the former contains only *linear* functions. This observation leads one to surmise that the Tychonoff Theorem can help in understanding compactness in X'. The result that carries this out requires one further notion: In a linear topological space, a set A is **bounded** if for any neighborhood of zero, say U, we have $A \subset \lambda U$ for some real λ.

Compactness Theorem. *Let X be a linear space, and let X' be its algebraic dual. Give X' the weak* topology. Then the compact sets in X' are precisely the closed and bounded sets.*

Proof. Let K be a compact set in X'. The set K is closed because in any Hausdorff space compact sets are closed. (See [Kel], page 141.) To prove that K is bounded, let U be any neighborhood of 0 in X'. It is to be shown that $K \subset \lambda U$ for some λ. First, select a "basic" neighborhood $V = V(\varepsilon; x_1, \ldots, x_m)$ contained in U. Then

$$K \subset \bigcup \{\phi + V : \phi \in K\}$$

Since K is compact, this covering has a finite subcover:

$$K \subset \bigcup_{i=1}^{n} (\phi_i + V)$$

(Here $\phi_i \in K$.) Select λ so that all ϕ_i are in λV. Then

$$\phi_i + V \subset \lambda V + V = (\lambda + 1)V \subset (\lambda + 1)U$$

(An easy calculation justifies the equation in this string of inclusions.) Consequently, $K \subset (\lambda + 1)U$ and K is bounded.

For the converse, let K be a closed and bounded set in X'. For each x in X, define

$$U_x = \{\phi \in X' : |\phi(x)| \leqslant 1\}$$

Thus U_x is a neighborhood of 0 in X'. Since K is bounded, there exists (for each x in X) a positive scalar r_x such that $K \subset r_x U_x$. Put $D_x = \{c : |c| \leqslant r_x\}$. The set D_x is either a disk in $\mathbb{C}$ or an interval in $\mathbb{R}$. In either case, D_x is a compact set in the scalar field. If $\phi \in K$, then $\phi \in r_x U_x$ for all x. Hence $\phi(x) \in D_x$ for all x, and ϕ is in the product space $\Pi\{D_x : x \in X\}$. Consequently, K is a subset of this product. That K is a closed subset therein is easily proved. By the Tychonoff Theorem, this product of compact sets is compact in the product topology of $\mathbb{R}^X$ (or $\mathbb{C}^X$). This is the weak* topology in X'. Since K is a closed subset of a compact space, K is compact. ∎

Naturally, we are more interested in the spaces that are already linear topological spaces. In this case, there will be two topologies on X and three on X'. (Notice that X' will have a weak* topology coming from X and a weak topology coming from X''.) The originally given topologies can be called the "strong" topologies, in contrast to the "weak" ones discussed above. (Rudin argues for the term "original topology" instead of "strong topology," because elsewhere in functional analysis, "strong topology" means something else.)

Theorem 2.   *Let X be a linear topological space, and U a neighborhood of 0. Then the polar set*

$$U^\circ = \{\phi \in X^* : |\phi(x)| \leqslant 1 \text{ for all } x \in U\}$$

is compact in the weak* topology of X^*.

Proof.   The linear space X^* (whose elements are continuous linear functionals) is a subspace of X' (whose elements are linear functionals). The weak* topology in X^* is the relative topology in X^* derived from the weak* topology on X'. By the preceding theorem, we need only prove that U° is closed and bounded in the weak* sense in X'. If we have a net $[\phi_\alpha]$ in U° and $\phi_\alpha \rightharpoonup \phi$, then $\phi_\alpha(x) \to \phi(x)$ for all $x \in U$. Consequently, $|\phi(x)| \leqslant 1$ for all $x \in U$, and $\phi \in U^\circ$. Thus U° is closed in the weak* topology of X'. If W is any neighborhood of 0 in X', then W contains a set of the form

$$V(\varepsilon; x_1, \ldots, x_n) = \{\phi \in X' : |\phi(x_i)| < \varepsilon, \quad 1 \leqslant i \leqslant n\}$$

Select r so that $rx_i \in U$. Then for $\phi \in U^\circ$ we have $|\phi(x)| \leqslant 1$ for all $x \in U$. Furthermore, $|\phi(rx_i)| \leqslant 1$ and $\phi \in (1/r\varepsilon)W$. Thus U° is bounded.   ∎

Theorem 3. (The Banach–Alaoglu Theorem)   *The unit ball in the conjugate space of a normed linear space is compact in the weak* topology.*

Proof.   In the preceding theorem, take U to be the unit ball of X. The polar of U will then be the unit ball in X^*.   ∎

     If the neighborhoods of 0 in a linear topological space have a base consisting of *convex* sets, the space is said to be **locally convex**. It is these spaces that we shall emphasize in the following discussion. Among such spaces we find the normed spaces and the pseudo-normed spaces. A **pseudo-norm** or **seminorm** is a real-valued function p defined on a linear space X such that:

     1. $p(\lambda x) = |\lambda| p(x)$ for all $\lambda \in \mathbb{R}, x \in X$
     2. $p(x + y) \leqslant p(x) + p(y)$ for all $x, y \in X$

It follows that $p(0) = 0$ and that $p(x) \geqslant 0$ for all $x \in X$. If p is a seminorm on a linear space X, then in a standard way X receives a locally convex topology. Namely, a base for the neighborhoods of 0 is taken to be the family of sets

$$V(\epsilon) = \{x \in X : p(x) < \epsilon\} \qquad (\epsilon > 0)$$

It is easy to see that this set is convex.

     In many spaces, the topology is not quite so simple; their topologies are defined not by a single seminorm but by a family of seminorms. Let P be a family of seminorms on a linear space X. We define the topology by giving a neighborhood base for 0. The base consists of all sets

$$V(\varepsilon; p_1, p_2, \ldots, p_n) = \{x \ : \ p_i(x) < \varepsilon \text{ for } 1 \leqslant i \leqslant n\}$$

in which $\varepsilon > 0$ and $\{p_1, \ldots, p_n\}$ is any finite subset of P.

In a linear space topologized with a family P of seminorms, it is easy to prove that a sequence (or generalized sequence) of points x_k converges to zero if and only if $p(x_k)$ converges to zero for each $p \in P$.

Notice that these basic neighborhoods of 0 are convex, and X, thus topologized, is a locally convex linear topological space. A remarkable theorem now can be stated.

Theorem 4. *For any locally convex linear topological space there is a family of continuous seminorms that induces the topology.*

Proof. Let P be the family of all continuous seminorms defined on the given space. Let U be a neighborhood of 0 in the original topology. First we must prove that U contains one of the sets $V(\varepsilon; p_1, \ldots, p_n)$. Since the space is locally convex, U contains a convex neighborhood U_1 of 0. By the continuity of scalar multiplication, there exists a convex neighborhood U_2 of 0 and a number $\delta > 0$ such that $cx \in U_1$ whenever $x \in U_2$ and $|c| < \delta$. The set $U_3 = \bigcup\{\lambda U_2 : |\lambda| < 1\}$ is a convex neighborhood of 0 contained in U. Its Minkowski functional p is continuous because for any $r > 0$ and any $x \in rU_3$, we have $p(x) \leqslant r$. Thus, $V(\frac{1}{2}; p) \subset W_3 \subset U$. (Minkowski functionals were defined in the the proof of Theorem 1 in Section 7.3, page 343.)

Now let V be any "basic" neighborhood of 0 in the new topology. Say, $V = V(\varepsilon; p_1, \ldots, p_n)$. Since each p_i is continuous in the original topology, V is open in the original topology. It therefore contains a convex neighborhood of 0 from the original topology. ∎

One of the main justifications for emphasizing locally convex linear topological spaces is that such spaces have useful conjugate spaces. For any linear topological space X, one can define X^* to be the linear space of all continuous linear functionals on X. Without further assumptions, X^* may have only one element, namely 0! A good example of this phenomenon is the space ℓ^p in which $0 < p < 1$. The topology is given by a norm–like functional that is actually not a norm (since it fails the triangle inequality):

$$\|x\| = \Big(\sum_{n=1}^{\infty} |x_n|^p\Big)^{1/p}$$

The only continuous linear functional is 0.

The principal corollary of the Hahn–Banach Theorem is valid for locally convex spaces, and takes the following form.

Theorem 5. *Every continuous linear functional defined on a subspace of a locally convex space has a continuous linear extension defined on the entire space.*

Example. Consider the space $\mathcal{D}$ of test functions on $\mathbb{R}^n$. This space was defined in Chapter 5, page 247. Its elements are C^∞ functions having compact support. The convergence to zero of a sequence $[\phi_k]$ in $\mathcal{D}$ was defined to mean that there was one compact set containing the supports of all ϕ_k, and on that compact set, $D^\alpha \phi_k$ converged uniformly to zero, for every α. This notion of convergence can be defined with a sequence of seminorms. For $j = 1, 2, 3 \ldots$, define

$$p_j(\phi) = \sup\{|(D^\alpha \phi)(x)| \; : \; x \in \mathbb{R}^n, \; \|x\| \leqslant j, \; |\alpha| \leqslant j\}$$

Thus topologized, the space of test functions becomes a locally convex linear topological space. Its conjugate space is the space of distributions. ∎

In a linear topological space, a set A is **totally bounded** if, for any neighborhood U of 0, A can be covered by a finite number of translates of U:

$$A \subset \bigcup_{i=1}^{m}(x_i + U)$$

More succinctly, $A \subset F + U$, for some finite set F. From the definition of compactness in terms of coverings, it is obvious that a compact set in a linear topological space is totally bounded. We shall use these ideas to prove Mazur's Theorem, to the effect that $\overline{\text{co}}(K)$ is compact when K is compact. We shall require the following result, for which we refer the reader to [KN].

Theorem 6.   *A set in a linear topological space is compact if and only if it is both totally bounded and complete.*

Lemma 2.   *In a locally convex linear topological space, the convex hull of a totally bounded set is totally bounded.*

Proof. Let Y be such a set and let U be any neighborhood of 0. Select a convex neighborhood V of 0 such that $V + V \subset U$. Since Y is totally bounded, there is a finite set F such that $Y \subset F + V$. Let $Z = \text{co}(F)$. The set Z is compact, being the image of a compact set under a continuous map of the form $(\theta_1, \ldots, \theta_n) \mapsto \sum_{i=1}^{n} \theta_i z_i$, where $\{z_1, \ldots, z_n\} = F$. It follows that Z is totally bounded, and that $Z \subset F' + V$ for another finite set F'. By the convexity of V we have

$$\text{co}(Y) \subset \text{co}(F + V) = \text{co}(F) + V = Z + V \subset F' + V + V \subset F' + U \quad ∎$$

Theorem 7.   Mazur's Theorem.   *The closed convex hull of a totally bounded set in a complete locally convex linear topological space is compact.*

Proof. Let K be such a set in such a space. By the preceding lemma, $\text{co}(K)$ is totally bounded. Hence $\overline{\text{co}}(K)$ is closed and totally bounded. Since the ambient space is complete, $\overline{\text{co}}(K)$ is complete and totally bounded. Hence, by Theorem 6, it is compact. ∎

7.8 Analytic Pitfalls

The purpose of this section is to frighten (or amuse) the reader by exhibiting some examples where erroneous conclusions are reached through an analysis that seems at first glance to be sound. In every case, however, some theorem pertinent to the situation has been overlooked. The relevant theorems are all quoted somewhere in this section or elsewhere in the book. Proofs or references are given for each of them. A connecting thread for many of these examples is the question of whether interchanging the order of two limit processes is justified. We begin with some matters from the subject of Calculus.

Here is an elementary example to show what can go wrong:

$$\lim_{x \to 0} \lim_{y \to 0} \frac{x - y}{x + y} = \lim_{x \to 0} \frac{x}{x} = 1$$

$$\lim_{y \to 0} \lim_{x \to 0} \frac{x - y}{x + y} = \lim_{y \to 0} \frac{-y}{y} = -1$$

A theorem governing this situation (and many others) is E.H. Moore's theorem, proved later.

It is natural to think that if a function is defined by a series of analytic functions, then the resulting function should be continuous, continuously differentiable, and so on. (This was a commonly held view until the mid-1850s.) For example, the series

(1) $$f(x) = \sum_{n=1}^{\infty} \frac{1}{2^n} \cos(3^n x)$$

consists of analytic terms, and the function f should be a "nice" one. We think that the function defined by the series should inherit the good properties of the terms in the series. Indeed, in this example, f *is* continuous, by the Weierstrass M-Test. This test, or theorem, goes as follows.

Theorem 1.   (Weierstrass M-Test.)   *If the functions g_n are continuous on a compact Hausdorff space X and if*

(2) $$|g_n(x)| \leqslant M_n \quad \text{(for all } x \in X) \quad \text{and} \quad \sum_{n=1}^{\infty} M_n < \infty$$

then the series $\sum_{n=1}^{\infty} g_n(x)$ converges uniformly on X and defines a function that is continuous on X.

The hypotheses in display (2) constitute the "M-Test." In modern notation, we could write instead $\sum_{n=1}^{\infty} \|g_n\|_\infty < \infty$. In the example of Equation (1), one can set $g_n(x) = 2^{-n} \cos(3^n x)$ and see immediately that the constants $M_n = 2^{-n}$ serve in Weierstrass's Theorem.

The Weierstrass M-Test gives us some hypotheses under which we can interchange two limits:

$$\lim_{h\to 0}\lim_{m\to\infty}\sum_{n=1}^{m}g_n(x+h) = \lim_{m\to\infty}\lim_{h\to 0}\sum_{n=1}^{m}g_n(x+h)$$

Returning to the function f in Equation (1), we propose to compute f' by differentiating term by term in the series, getting

$$f'(x) = -\sum_{n=1}^{\infty} 3^n 2^{-n}\sin(3^n x)$$

But here there is an alarming difference, as the factors $3^n 2^{-n}$ are *growing*, not shrinking. The very convergence of the series is questionable.

This example, f, is the famous Non-Differentiable Function of Weierstrass. It is not differentiable at any point whatsoever! A detailed proof can be found in [Ti2] or [Ch]. A sketch showing a partial sum of the series is in Figure 7.2.

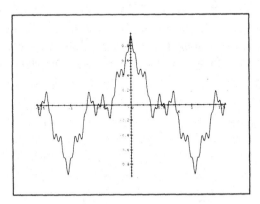

Figure 7.2 A partial sum in the non-differentiable function

When we take more terms and blow up the picture, we see more or less the same behavior, which reminds us of fractals. See Figure 7.3, where a magnification factor of about 15 has been used.

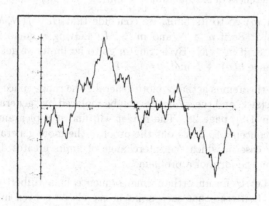

Figure 7.3 Another partial sum, magnified

Now for the positive side of this question concerning differentiating a series term by term: A classical theorem that can be found, for example, in [Wi] is as follows.

Theorem 2.   *If the functions g_n are continuously differentiable on a closed and bounded interval, if the series $\sum_n g_n(x)$ converges on that interval, and if the series $\sum_n g_n'(x)$ converges uniformly on that interval, then $(\sum_n g_n)' = \sum_n g_n'$.*

Since differentiation involves a limiting process, the theorem just quoted is again providing hypotheses to justify the interchange of two limits.

What can be said, in general, to legitimate interchanging limits? A famous theorem of Eliakim Hastings Moore gives one possible answer to this question.

Theorem 3.   *Let $f : \mathbb{N} \times \mathbb{N} \to \mathbb{R}$. Assume that $\lim_{n \to \infty} f(n, m)$ exists for each m and that $\lim_{m \to \infty} f(n, m)$ exists for each n, uniformly in n. Then the two limits $\lim_n \lim_m f(n, m)$ and $\lim_m \lim_n f(n, m)$ exist and are equal.*

Proof.   Define $g(m) = \lim_n f(n, m)$ and $h(n) = \lim_m f(n, m)$. Let $\varepsilon > 0$. Find a positive integer M such that

$$m \geqslant M \implies \big|f(n, m) - h(n)\big| < \varepsilon \quad \text{for all } n$$

Notice that the uniformity hypothesis is being used at this step. A consequence is that $\big|f(n, M) - h(n)\big| < \varepsilon$, and by the triangle inequality $|f(n, m) - f(n, M)| < 2\varepsilon$ when $m \geqslant M$. Find N such that

$$n \geqslant N \implies |f(n, M) - g(M)| < \varepsilon$$

No uniformity of the limit in m is needed here, as M has been fixed. Now we have $|f(N, M) - g(M)| < \varepsilon$ and $|f(N, M) - f(n, M)| < 2\varepsilon$ when $n \geqslant N$. We next conclude that $|f(n, m) - f(N, M)| < 4\varepsilon$ when $n \geqslant N$ and $m \geqslant M$. This establishes that the doubly indexed sequence $f(n, m)$ has the Cauchy property. By the completeness of $\mathbb{R}$, the limit $\lim\limits_{(n,m)\to(\infty,\infty)} f(n, m)$ exists. Call it L. Then, by letting (n, m) go to its limit, we conclude that $|L - f(N, M)| \leqslant 4\varepsilon$. Also, $|L - f(n, m)| < 8\varepsilon$ if $n \geqslant N$ and $m \geqslant M$. Letting n go to its limit, we get $|L - g(m)| < 8\varepsilon$ if $m \geqslant M$. By letting m go to its limit, we get $|L - h(n)| \leqslant 8\varepsilon$ if $n \geqslant N$. Hence $h(n) \to L$ and $g(m) \to L$. ∎

Moore's theorem is actually more general: The range space can be any complete metric space, and the sequences can be replaced by "generalized" sequences ("nets"). See [DS], page 28. The reader will find it a pleasant exercise in the use of these concepts to carry out the proof in the more general case.

Another case in which the interchange of limits creates difficulties is presented next in the form of a problem.

Problem. Let U be an orthonormal sequence in a Hilbert space, say $U = \{u_1, u_2, \ldots\}$. Is it true that each point in the closed convex hull of U is representable as an infinite series $\sum_{n=1}^{\infty} a_n u_n$, in which $a_n \geqslant 0$ and $\sum a_n = 1$?

At first, this seems to be almost obvious: We are simply allowing an "infinite" convex combination of elements from U in order to represent points in the closure of the convex hull of U. A proof might proceed as follows. (Here we use "co" for the convex hull and $\overline{\text{co}}$ for the closed convex hull.) Suppose that $x \in \overline{\text{co}}(U)$. Then there exists a sequence $x_n \in \text{co}(U)$ such that $x_n \to x$. With no loss of generality, we may suppose that

$$x_n = \sum_{i=1}^{n} a_{ni} u_i \quad \text{where} \quad a_{ni} \geqslant 0 \text{ and } \sum_{i=1}^{n} a_{ni} = 1 \text{ for all } n$$

Letting n tend to ∞, we arrive at $x = \sum_{i=1}^{\infty} a_i u_i$, where $a_i = \lim_n a_{ni}$. This limit is justified by the Hilbert space structure. Indeed, by the properties of an orthonormal sequence, we must have $a_{ni} = \langle x_n, u_i \rangle$, and therefore $\lim_n a_{ni} = \lim_n \langle x_n, u_i \rangle = \langle x, u_i \rangle$.

After examining the proof and discovering the flaw in it, the reader should contemplate the following special case. Define $x_n = (u_1 + \cdots + u_n)/n$. Certainly, x_n is in the convex hull of U. Since x_n is given by an orthonormal expansion, we have $\|x_n\|^2 = n(1/n^2) = 1/n$. This calculation shows that $x_n \to 0$. Hence 0 is in the closure of the convex hull of U. But it is not possible to represent 0 as an "infinite" convex combination of the vectors u_n. The only representation of 0 is the trivial one, and those coefficients do not add up to 1.

The foregoing example shows that in general, for a series of constants,

$$\lim_{n\to\infty} \sum_{i=1}^{n} c_{ni} \neq \sum_{i=1}^{\infty} \lim_{n\to\infty} c_{ni}$$

The same phenomenon can be illustrated by a more familiar example. Consider the contrast between approximating a function with a polynomial and expanding that function in a Taylor series. The expansion in powers of the variable is not obtained simply by allowing more and more terms in a polynomial and appealing to the Weierstrass approximation theorem. Indeed, only a select few of the continuous functions will have Taylor series. If f is continuous, say on $[-1, 1]$, then there is a sequence of polynomials p_n such that $\|f - p_n\|_\infty \to 0$. If it is desired to represent f as a series, one can write

$$f = p_1 + (p_2 - p_1) + (p_3 - p_2) + (p_4 - p_3) + \cdots = \sum_{n=1}^{\infty} q_n$$

where the polynomials q_n have the obvious interpretation. However, this is not a simple Taylor series, in general. Thus, if we have $p_n \to f$ and write $p_n(x) = \sum_{i=0}^{n} c_{ni} x^i$, it is *not* legitimate to conclude that

$$f(x) = \lim_{n \to \infty} p_n(x) = \lim_{n \to \infty} \sum_{i=0}^{n} c_{ni} x^i = \sum_{i=0}^{\infty} c_i x^i$$

where $c_i = \lim_{n \to \infty} c_{ni}$. This last limit will not exist in most cases.

The expansion of a function in an orthogonal series has its own cautionary examples. Consider the orthonormal family of Legendre polynomials, $p_0, p_1, \ldots$ defined on the interval $[-1, 1]$. They have the property

$$\int_{-1}^{1} p_n(x) p_m(x) \, dx = \delta_{nm}$$

For any continuous function f defined on this same interval, we can construct its series in Legendre functions:

$$f \sim \sum_{k=0}^{\infty} a_k p_k \qquad a_k = \langle f, p_k \rangle = \int_{-1}^{1} f(x) p_k(x)$$

Here we write $\sim$ to remind us that equality may or may not hold. It is only asserted that each continuous function has a corresponding formal series in Legendre functions. Can we not appeal to the Weierstrass approximation theorem to conclude that the series does converge to f? By now, the reader must guess that the answer is "No". The reason is not at all obvious, but depends on a startling theorem in Analysis, quoted here. (A proof is to be found in [Ch].)

Theorem 4. The Kharshiladze–Lozinski Theorem.   *For each* $n = 0, 1, 2, \ldots$ *let* P_n *be a projection of the space* $C[-1, 1]$ *onto the subspace* Π_n *of polynomials of degree at most* n. *Then* $\|P_n\| \to \infty$.

It is readily seen that the equation

$$P_n(f) = \sum_{k=0}^{n} a_k(f) p_n$$

where the coefficients a_k are as above, defines a projection of the type appearing in Theorem 4. That is, P_n is a continuous linear idempotent map from $C[-1,1]$ onto Π_n. Hence, $\|P_n\| \to \infty$. By the Banach–Steinhaus Theorem (Chapter 1, Section 7, page 41) the set of f in $C[-1,1]$ for which the series above converges uniformly to f is of the first category (relatively small) in $C[-1,1]$.

One should think of this phenomenon in the following way. The space $C[-1,1]$ contains not only the nice familiar functions of elementary calculus, but also the bizarre unmanageable ones that we do not see unless we go searching for them. Most functions are of the latter type. See Example 1 on page 42 to be convinced of this. To guarantee convergence of the series under discussion, one must make further smoothness assumptions about f. For example, if f is an analytic function of a complex variable in an ellipse that contains the line segment $[-1,1]$, then the Legendre series for f will converge uniformly to f on that segment. For results about these series, consult [San].

Interchanging the order of two integrals in a double integral can also involve difficulties. The Fubini Theorem in Chapter 8 addresses this issue. Here we offer an example of a double integral in a discrete measure space, where the integrals become sums. This is adapted from [MT].

Example. Consider a function of two positive integers defined by the formula

$$f(n,m) = \begin{cases} 0 & \text{if } m > n \\ -1 & \text{if } m = n \\ 2^{m-n} & \text{if } m < n \end{cases}$$

The two possible sums can be calculated in a straightforward way, and they turn out to be different:

$$\sum_m \sum_n f(n,m) = \sum_{m=1}^{\infty} \sum_{n=m}^{\infty} f(n,m) = \sum_{m=1}^{\infty} \left[-1 + \frac{1}{2} + \frac{1}{4} + \cdots \right] = \sum_m 0 = 0$$

$$\sum_n \sum_m f(n,m) = \sum_{n=1}^{\infty} \sum_{m=1}^{n} f(n,m) = f(1,1) + \sum_{n=2}^{\infty} \sum_{m=1}^{n} f(n,m)$$

$$= -1 + \sum_{n=2}^{\infty} \left[2^{1-n} + 2^{2-n} + \cdots + 2^{-1} - 1 \right]$$

$$= -1 + \sum_{n=2}^{\infty} -2^{1-n} = -1 - 1 = -2$$

The difficulty here is not to be attributed to the fact that our domain $\mathbb{N} \times \mathbb{N}$ stretches infinitely far to the right and upwards in the 2-dimensional plane. One can make this example work on the unit square in the plane by the following construction. Define intervals $I_n = [1/(n+1), 1/n]$. On each rectangle $I_n \times I_m$ define a function F whose integral over that rectangle is $f(n,m)$, as defined previously. We then find, from the above calculations, that

$$\int_0^1 \int_0^1 F(x,y)\, dx\, dy = \sum_m \sum_n f(n,m) = 0$$

whereas

$$\int_0^1 \int_0^1 F(x,y)\, dy\, dx = \sum_n \sum_m f(n,m) = -2$$

By referring to the Fubini Theorem, page 426, we see that our functions f and F do not satisfy the essential hypothesis of that theorem: They are not integrable over the Cartesian domain. The function $|f|$, for example, has an infinite number of values $+1$, and so cannot yield a finite integral over $\mathbb{N} \times \mathbb{N}$. Had we wished to apply the Tonelli Theorem, the crucial missing hypothesis would have been that $f \geqslant 0$ or $F \geqslant 0$.

Let us return to the functions defined by infinite series, for such functions are truly ubiquitous in Mathematics. We can ask, "How does integration interact with the summation process? Can integration be interchanged with summation?" The answer is that the conditions for this to be valid are less stringent than those for differentiation. This is to be expected, for (in general) integration is a smoothing process, whereas differentiation is the opposite: It emphasizes or magnifies the variations in a function. The relevant theorem, again conveniently accessible in [Wi], is as follows.

Theorem 5.   *If the functions g_n are continuous on $[a,b]$, and if the series $\sum_n g_n$ converges uniformly, then*

$$\int_a^b \sum_{n=1}^\infty g_n(x)\, dx = \sum_{n=1}^\infty \int_a^b g_n(x)\, dx$$

This theorem is often used to obtain Taylor series for troublesome functions. For example, if a Taylor series is needed for the Arctan function, we can start with the relationship

$$\frac{d}{dt} \text{Arctan}(t) = \frac{1}{1+t^2} = \sum_{n=0}^\infty (-t^2)^n$$

This is valid for $t^2 < 1$. Then for $x^2 < 1$,

$$\text{Arctan}(x) = \int_0^x \sum_{n=0}^\infty (-t^2)^n = \sum_{n=0}^\infty (-1)^n x^{2n+1}/(2n+1)$$

The interchange of differentiation and integration is another common technique in analysis. Here there are various theorems that apply, for example, the following one, given in [Wi].

Theorem 6.   *Let $f(x) = \int_a^b g(x,t)\, dt$, where g and $\partial g/\partial x$ are continuous on the rectangle $[A, B] \times [a, b]$ in the xt–plane. Then $f'(x) = \int_a^b \partial g(x,t)/\partial x\, dt$*

(Amusing anecdotes about differentiating under the integral sign occur in Feynman's memoirs [Feyn], pages 86, 87, and 110.) For more general situations, involving an arbitrary measure μ, we can still raise the question of whether

(3) $$\frac{d}{dx} \int_T g(x,t)\, d\mu(t) = \int_T \frac{\partial g}{\partial x}(x,t)\, d\mu(t)$$

The setting is as follows. A measure space $(T, \mathcal{A}, \mu)$ is prescribed. Thus T is a set, $\mathcal{A}$ is a σ-algebra of subsets of T, and $\mu : \mathcal{A} \to [0, \infty]$ is a measure. An open interval (a, b) is also prescribed. The function g is defined on $(a, b) \times T$ and takes values in $\mathbb{R}$. Select a point x_0 in (a, b) where $(\partial g / \partial x)(x_0, t)$ exists for all t. What further assumptions are needed in order that Equation (3) shall be true at a prescribed point x_0? Let us assume that:

(A) For each x in (a, b), the function $t \mapsto g(x, t)$ belongs to $L^1(T, \mathcal{A}, \mu)$.
(B) There exists a function $G \in L^1(T, \mathcal{A}, \mu)$ such that

$$\left| \frac{g(x, t) - g(x_0, t)}{x - x_0} \right| \leqslant G(t) \qquad (t \in T, \ a < x < b, \ x \neq x_0)$$

Theorem 7.   *Under the hypotheses given above, Equation (3) is true for the point $x = x_0$.*

Proof.   By Hypothesis (A) we are allowed to define

$$f(x) = \int_T g(x, t) \, d\mu(t)$$

The derivative $f'(x_0)$ exists if and only if for each sequence $[x_n]$ converging to x_0 we have

$$f'(x_0) = \lim_{n \to \infty} \frac{f(x_n) - f(x_0)}{x_n - x_0} = \lim_{n \to \infty} \int_T \frac{g(x_n, t) - g(x_0, t)}{x_n - x_0} \, d\mu(t)$$

By Hypothesis (B), the integrands in the preceding equation are bounded in magnitude by the single L^1-function G. The Lebesgue Dominated Convergence Theorem (see Chapter 8, page 406) allows an interchange of limit and integral. Hence

$$f'(x_0) = \int_T \lim_{n \to \infty} \frac{g(x_n, t) - g(x_0, t)}{x_n - x_0} \, d\mu(t) = \int_T \frac{\partial g}{\partial x}(x_0, t) \, d\mu(t) \qquad \blacksquare$$

This proof is given by Bartle [Bart1]. A related theorem can be found in McShane's book [McS]. A useful corollary of Theorem 5 is as follows.

Theorem 8.   *Let $(T, \mathcal{A}, \mu)$ be a measure space such that $\mu(T) < \infty$. Let $g : (a, b) \times T \to \mathbb{R}$. Assume that for each n, $(\partial^n g / \partial x^n)(x, t)$ exists, is measurable, and is bounded on $(a, b) \times T$. Then*

(4) $$\frac{d^n}{dx^n} \int_T g(x, t) \, d\mu(t) = \int_T \frac{\partial^n g}{\partial x^n}(x, t) \, d\mu(t) \qquad (n = 1, 2, \ldots)$$

Proof.   Since $\mu(T) < \infty$, any bounded measurable function on T is integrable. To see that Hypothesis (B) of the preceding theorem is true, use the mean value theorem:

$$\left| \frac{g(x, t) - g(x_0, t)}{x - x_0} \right| = \left| \frac{\partial g}{\partial x}(\xi, t) \right| \leqslant M$$

where M is a bound for $|\partial g / \partial x|$ on $(a, b) \times T$. By the preceding theorem, Equation (4) is valid for $n = 1$. The same argument can be repeated to give an inductive proof for all n. $\blacksquare$

Chapter 8

Measure and Integration

8.1   Extended Reals, Outer Measures, Measurable Spaces   381

8.2   Measures and Measure Spaces   386

8.3   Lebesgue Measure   391

8.4   Measurable Functions   394

8.5   The Integral for Nonnegative Functions   399

8.6   The Integral, Continued   404

8.7   The L^p-Spaces   408

8.8   The Radon–Nikodym Theorem   413

8.9   Signed Measures   417

8.10   Product Measures and Fubini's Theorem   420

8.1 Extended Reals, Outer Measures, Measurable Spaces

This chapter gives, in as brief a form as possible, the main features of measure theory and integration. The presentation is sufficiently general to cover Lebesgue measures and measures that arise in studying the continuous linear functionals on Banach spaces.

Since measures are employed to assign a size to sets, they are often allowed to assume infinite values. The **extended real number system** is designed to assist in this situation. This is the set $\mathbb{R}^* = \mathbb{R} \cup \{\infty\} \cup \{-\infty\}$. The two new points, ∞ and $-\infty$, that have been adjoined to $\mathbb{R}$ are required to behave as follows:

(1) $(-\infty, \infty) = \mathbb{R}$

(2) $x + \infty = \infty$ for $x \in (-\infty, \infty]$

(3) $x\infty = \infty$ for $x \in (0, \infty]$

(4) $0\infty = 0$

From these rules various others follow, such as $x - \infty = -\infty$ when $x \in [-\infty, \infty)$. One advantage of $\mathbb{R}^*$ is that every subset of $\mathbb{R}^*$ has a supremum and an infimum in $\mathbb{R}^*$. For example, the equation $\sup A = \infty$ means that for each $x \in \mathbb{R}$, A contains an element a such that $a > x$. Note that certain expressions, such as $-\infty + \infty$, must remain undefined.

Definition.   Let X be an arbitrary set. An **outer measure** "on X" is a function $\mu : 2^X \to \mathbb{R}^*$ such that:

(1) $\mu(\varnothing) = 0$ ($\varnothing$ is the empty set).

(2) If $A \subset B$, then $\mu(A) \leqslant \mu(B)$.

(3) $\mu\left[\bigcup_{i=1}^{\infty} A_i\right] \leqslant \sum_{i=1}^{\infty} \mu(A_i)$.

Of course, in these postulates, $A, B, \ldots$ are arbitrary subsets of X. Notice that (1) and (2) together imply that $\mu(A) \geqslant 0$ for all A. Let us look at some examples.

Example 1.   Let X be any set. If $A \subset X$, define $\mu(A) = 0$.   ∎

Example 2.   Let X be any set. Define $\mu(\varnothing) = 0$, and for any nonempty set A, put $\mu(A) = +\infty$.   ∎

Example 3.   Let X be any set. For a finite subset A, let $\mu(A) = \#(A)$, the number of elements in A. For all other sets, put $\mu(A) = \infty$. This is called **counting measure**.   ∎

Example 4.   Let X be any set and let x_0 be any point in X. Define $\mu(A)$ to be 1 if $x_0 \in A$ and to be 0 if $x_0 \notin A$.   ∎

Example 5.   Let X be any infinite set. Let $\{x_1, x_2, \ldots\}$ be a countable set of (distinct) points in X. Let λ_n be positive numbers ($n = 1, 2, \ldots$). Define $\mu(A) = \sum\{\lambda_n : x_n \in A\}$, and $\mu(\varnothing) = 0$.   ∎

Example 6.   Lebesgue Outer Measure. Let X be the real line. Define

$$\mu(A) = \inf\left\{\sum_{i=1}^{\infty}(b_i - a_i) \; : \; A \subset \bigcup_{i=1}^{\infty}(a_i, b_i) \, , \; a_i < b_i\right\}$$ ∎

Example 7.   Lebesgue Outer Measure on $\mathbb{R}^n$. In $\mathbb{R}^n$ define the "unit cube" to be the set Q of all points $(\xi_1, \ldots, \xi_n)$ whose components lie in the interval $[0, 1]$. If $x \in \mathbb{R}^n$ and $\lambda \in \mathbb{R}$, define $x + \lambda Q = \{x + \lambda v : v \in Q\}$. For $A \subset \mathbb{R}^n$, define

$$\mu(A) = \inf\left\{\sum_{i=1}^{\infty}|\lambda_i|^n \; : \; A \subset \bigcup_{i=1}^{\infty}(x_i + \lambda_i Q)\right\}$$ ∎

Example 8.   Lebesgue–Stieltjes Outer Measure. Let X be the real line, and select a monotone nondecreasing function $\gamma : \mathbb{R} \to \mathbb{R}$. Define

$$\mu(A) = \inf\left\{\sum_{i=1}^{\infty}\left[\gamma(b_i) - \gamma(a_i)\right] \; : \; A \subset \bigcup_{i=1}^{\infty}(a_i, b_i) \, , \; a_i < b_i\right\}$$

Notice that Lebesgue outer measure (in Example 6) is a special case, obtained when $\gamma(x) = x$.   ∎

   In order to see that Examples 6, 7, and 8 are bona fide outer measures, one can appeal to the following theorem.

Theorem 1.   *Let X be an arbitrary set, and $\mathcal{C}$ a collection of subsets of X, countably many of which cover X. Let β be a function from $\mathcal{C}$ to $\mathbb{R}^*$ such that*

$$\inf\{\beta(C) : C \in \mathcal{C}\} = 0$$

Then the equation

$$\mu(A) = \inf\left\{\sum_{i=1}^{\infty} \beta(C_i) \ : \ A \subset \bigcup_{i=1}^{\infty} C_i, \ C_i \in \mathcal{C}\right\}$$

defines an outer measure on X.

Proof.   Assume all the hypotheses. There are now three postulates for an outer measure to be verified. Our assumption about β implies that $\beta(C) \geqslant 0$ for all $C \in \mathcal{C}$. Therefore, $\mu(A) \geqslant 0$ for all A. Since $\varnothing \subset C$ for all $C \in \mathcal{C}$, $\mu(\varnothing) \leqslant \beta(C)$ for all C. Taking an infimum yields $\mu(\varnothing) \leqslant 0$.

If $A \subset B$ and $B \subset \bigcup_{i=1}^{\infty} C_i$, then $A \subset \bigcup_{i=1}^{\infty} C_i$ and $\mu(A) \leqslant \sum_{i=1}^{\infty} \beta(C_i)$. Taking an infimum over all countable covers of B, we have $\mu(A) \leqslant \mu(B)$.

Let $A_i \subset X$ $(i \in \mathbb{N})$ and let $\varepsilon > 0$. By the definition of $\mu(A_i)$ there exist $C_{ij} \in \mathcal{C}$ such that $A_i \subset \bigcup_{j=1}^{\infty} C_{ij}$ and $\sum_{j=1}^{\infty} \beta(C_{ij}) \leqslant \mu(A_i) + \varepsilon/2^i$. Since $\bigcup_{i=1}^{\infty} A_i \subset \bigcup_{i,j=1}^{\infty} C_{ij}$, we obtain

$$\mu\left(\bigcup_{i=1}^{\infty} A_i\right) \leqslant \sum_{i,j} \beta(C_{ij}) \leqslant \sum_{i=1}^{\infty}\left[\mu(A_i) + \varepsilon/2^i\right] = \varepsilon + \sum_{i=1}^{\infty} \mu(A_i)$$

Since this is true for each positive ε, we obtain $\mu(\bigcup_{i=1}^{\infty} A_i) \leqslant \sum_{i=1}^{\infty} \mu(A_i)$.   ∎

The postulates for an outer measure do not include all the desirable attributes that are needed for integration. For example, an essential property is additivity:

$$A \cap B = \varnothing \quad \Longrightarrow \quad \mu(A \cup B) = \mu(A) + \mu(B)$$

This property cannot be deduced from the axioms for an outer measure. See Problem 6 for a simple example. Even Lebesgue outer measure is not additive, although it seems to be a natural or intrinsic definition for "the measure of a set" in $\mathbb{R}$. If we concentrate for the moment on this all-important example, we ask, How can additivity be obtained? We could change the definition. But if the measure of an interval is to be its length, changing the definition will not succeed. The only solution is to reduce the domain of μ from $2^{\mathbb{R}}$ (or 2^X in general) to a smaller class of sets. This is the brilliant idea of Lebesgue (1901) that leads to Lebesgue measure on $\mathbb{R}$. The procedure for accomplishing this domain reduction is dealt with in the theorem of Carathéodory, proved later. First, we describe in the abstract the sort of domain that will be used for measures.

Definition. A **measurable space** is a pair $(X, \mathcal{A})$ in which X is a set and $\mathcal{A}$ is a nonempty family of subsets of X such that:

(i) If $A \in \mathcal{A}$, then $X \smallsetminus A \in \mathcal{A}$.

(ii) If $A_1, A_2, \ldots \in \mathcal{A}$, then $\bigcup_{i=1}^{\infty} A_i \in \mathcal{A}$.

In brief, $\mathcal{A}$ is closed under complementation and forming countable unions. A family $\mathcal{A}$ having these two properties is said to be a σ-**algebra**. If a measurable space $(X, \mathcal{A})$ is prescribed, we call the sets in $\mathcal{A}$ the **measurable** subsets of X.

Example 9. Let X be any set, and let $\mathcal{A}$ consist solely of X and $\varnothing$. Then $(X, \mathcal{A})$ is a measurable space. In fact, this is the smallest nonempty σ-algebra of subsets of X. ∎

Example 10. Let X be an arbitrary set, and let $\mathcal{A} = 2^X$ (the family of all subsets of X). Then $(X, \mathcal{A})$ is a measurable space and $\mathcal{A}$ is the largest σ-algebra of subsets of X. ∎

Example 11. Let X be any set and let A be a particular subset of X. Define $\mathcal{A} = \{X, \varnothing, A, X \smallsetminus A\}$. This is the smallest σ-algebra containing A. We are observing that, as long as $\mathcal{A}$ is nonempty, the set X and the empty set $\varnothing$ must belong to $\mathcal{A}$. In other words, these two sets will always be measurable. ∎

Example 12. Let X be any set, and let $\mathcal{A}$ consist of all countable subsets of X and their complements. Then $\mathcal{A}$ is the smallest σ-algebra containing all finite subsets of X. ∎

Lemma 1. *If $(X, \mathcal{A})$ is a measurable space, then X and $\varnothing$ belong to $\mathcal{A}$. Furthermore, $\mathcal{A}$ is closed under countable intersections and set difference.*

Proof. This is left to the reader as a problem. ∎

Lemma 2. *For any subset $\mathcal{F}$ of 2^X there is a smallest σ-algebra containing $\mathcal{F}$.*

Proof. As Example 10 shows, there is certainly one σ-algebra containing $\mathcal{F}$. The smallest one will be the intersection of all the σ-algebras $\mathcal{A}_\nu$ containing $\mathcal{F}$. It is only necessary to verify that $\bigcap \mathcal{A}_\nu$ is a σ-algebra. If $A_i \in \bigcap \mathcal{A}_\nu$, then $A_i \in \mathcal{A}_\nu$ for all ν. Since $\mathcal{A}_\nu$ is a σ-algebra, $\bigcup_{i=1}^{\infty} A_i \in \mathcal{A}_\nu$. Since this is true for all ν, $\bigcup_{i=1}^{\infty} A_i \in \bigcap \mathcal{A}_\nu$. A similar proof is needed for the other axiom. ∎

In any topological space X, the smallest σ-algebra containing the topology (i.e., containing all the open sets) is called the σ-algebra of **Borel** sets, or the "Borel σ-algebra."

Suggested references for this chapter are [Bart2], [Berb3], [Berb4], [DS], [Frie2], [Hal4], [HS], [Jo], [KS], [Loo], [OD], [RN], [Roy], [Ru3], [Tay3], and [Ti2].

Problems 8.1

1. Does the extended real number system $\mathbb{R}^*$ become a topological space if a neighborhood of ∞ is defined to be any set that contains an interval of the form $(a, \infty]$, and similarly for $-\infty$?

2. Why, in defining Lebesgue outer measure, do we not "approximate from within" and define

$$\mu(A) = \sup \left\{ \sum_{i=1}^{\infty} (b_i - a_i) : (a_i, b_i) \subset A, \text{ intervals mutually disjoint} \right\} ?$$

3. Prove that if μ is an outer measure and if $\mu(B) = 0$, then $\mu(A \cup B) = \mu(A)$.

4. An outer measure on a group is said to be **invariant** if $\mu(x + A) = \mu(A)$ for all x and A. Prove that Lebesgue outer measure has this property.

5. Under what conditions is Lebesgue–Stieltjes outer measure invariant, as defined in Problem 4?

6. Let $X = \{1, 2\}$, and define $\mu(\varnothing) = 0$, $\mu(X) = 2$, $\mu(\{1\}) = 1$ and $\mu(\{2\}) = 2$. Show that μ is an outer measure but is not additive.

7. Prove that the Lebesgue outer measure of each countable subset of $\mathbb{R}$ is 0.

8. How many outer measures having range in $\{0, 1, \ldots, n\}$ are there on a set of n elements?

9. Prove that the Lebesgue outer measure of the interval $[a, b]$ is $b - a$.

10. Let μ be an outer measure on X, and let $Y \subset X$. Define $\nu(A) = \mu(A)$ when $A \subset Y$. Is ν an outer measure on Y?

11. Does an outer measure necessarily obey this equation?

$$\mu(A \cup B) + \mu(A \cap B) \leqslant \mu(A) + \mu(B)$$

12. Let μ be an outer measure on X, and let $Y \subset X$. Define $\nu(A) = \mu(Y \cap A)$. Is ν an outer measure on X?

13. Let μ and ν be outer measures on X. Define $\theta(A) = \max\{\mu(A), \nu(A)\}$ for all $A \subset X$. Is θ an outer measure on X?

14. Are the outer measures in Examples 3 and 5 additive?

15. What is the Lebesgue outer measure of the set of irrational numbers in $[0, 1]$?

16. Prove Lemma 1.

17. Prove that every countable set in $\mathbb{R}$ is a Borel set.

18. Let $(X, \mathcal{A})$ be a measurable space, and let A and B be two subsets of X. If A is measurable and $B \smallsetminus A$ is not, what conclusions can be drawn about $A \cup B$ and $A \smallsetminus B$?

19. Let $(X, \mathcal{A})$ be a measurable space, and let $Y \in \mathcal{A}$. Define $\mathcal{B}$ to be the family of all sets $Y \cap A$, where A ranges over $\mathcal{A}$. Prove that $(Y, \mathcal{B})$ is a measurable space.

20. Does there exist a countably infinite σ-algebra?

21. Prove that $\mathcal{A}$ in Example 12 is a σ-algebra.

22. Is there an example of a σ-algebra containing exactly 5 sets?

8.2 Measures and Measure Spaces

Let $(X, \mathcal{A})$ be a measurable space, as defined in the preceding section. A function $\mu : \mathcal{A} \to \mathbb{R}^*$ is called a **measure** if:

(a) $\mu(\varnothing) = 0$.

(b) $\mu(A) \geqslant 0$ for all A in $\mathcal{A}$.

(c) $\mu(\bigcup_{i=1}^{\infty} A_i) = \sum_{i=1}^{\infty} \mu(A_i)$ if $\{A_1, A_2, \ldots\}$ is a disjoint sequence in $\mathcal{A}$.

Notice that the additivity property discussed in the preceding section is now being assumed in a strong form. It is called **countable additivity**. On the other hand, the domain of μ is **not** assumed to be 2^X but is instead the σ-algebra $\mathcal{A}$.

Example 1. Let X be any set and let $\mathcal{A} = \{X, \varnothing\}$. Define $\mu(\varnothing) = 0$ and let $\mu(X)$ be any number in $[0, \infty]$. ∎

Example 2. Let X be any set and $\mathcal{A} = 2^X$. Define $\mu(A)$ to be the number of elements in A if A is finite, and define $\mu(A) = \infty$ otherwise. ∎

Example 3. If $\mathcal{A} = 2^X$, let $\mu(\varnothing) = 0$ and $\mu(A) = \infty$ if $A \neq \varnothing$. ∎

Example 4. Let A be a subset of X such that $A \neq \varnothing$ and $A \neq X$. Let $\mathcal{A} = \{\varnothing, A, X \smallsetminus A, X\}$ and define $\mu(\varnothing) = 0$, $\mu(A) = 1$, $\mu(X \smallsetminus A) = 1$, $\mu(X) = 2$. ∎

Example 5. Let $\mathcal{A} = 2^X$, and let $\{x_1, x_2, \ldots\}$ be a countable subset of X. Select numbers $\lambda_i \in [0, \infty]$, $i \in \mathbb{N}$, and define $\mu(A) = \sum\{\lambda_i : x_i \in A\}$, $\mu(\varnothing) = 0$. ∎

If $(X, \mathcal{A})$ is a measurable space and μ is a measure defined on $\mathcal{A}$, then $(X, \mathcal{A}, \mu)$ is called a **measure space**.

Lemma 1.   *If $(X, \mathcal{A}, \mu)$ is a measure space, then:*

(1) $\mu(A) \leqslant \mu(B)$ *if* $A \in \mathcal{A}$, $B \in \mathcal{A}$, *and* $A \subset B$.

(2) $\mu\left(\bigcup_{i=1}^{\infty} A_i\right) \leqslant \sum_{i=1}^{\infty} \mu(A_i)$ *if* $A_i \in \mathcal{A}$.

Proof. For (1), write

$$\mu(B) = \mu[A \cup (B \smallsetminus A)] = \mu(A) + \mu(B \smallsetminus A) \geqslant \mu(A)$$

For (2), we create a disjoint sequence of sets B_i by writing $B_1 = A_1$, $B_2 = A_2 \smallsetminus A_1$, $B_3 = A_3 \smallsetminus (A_1 \cup A_2)$, and so on. (Try to remember this little trick.) These sets are in $\mathcal{A}$ by Lemma 1 in the preceding section. Also, $B_i \subset A_i$ and

$$\mu\left(\bigcup_{i=1}^{\infty} A_i\right) = \mu\left(\bigcup_{i=1}^{\infty} B_i\right) = \sum_{i=1}^{\infty} \mu(B_i) \leqslant \sum_{i=1}^{\infty} \mu(A_i)$$

∎

Lemma 2.   Let $(X, \mathcal{A}, \mu)$ be a measure space. If $A_i \in \mathcal{A}$ for $i \in \mathbb{N}$ and $A_1 \subset A_2 \subset \cdots$, then $\mu(A_n) \uparrow \mu(\bigcup_{i=1}^{\infty} A_i)$.

Proof. Let $A_0 = \varnothing$ and observe that $A_n = \bigcup_{i=1}^{n}(A_i \smallsetminus A_{i-1})$. It follows from the disjoint nature of the sets $A_i \smallsetminus A_{i-1}$ that $\mu(A_n) = \sum_{i=1}^{n} \mu(A_i \smallsetminus A_{i-1})$. Hence

$$\mu(A_n) \uparrow \sum_{i=1}^{\infty} \mu(A_i \smallsetminus A_{i-1}) = \mu\left[\bigcup_{i=1}^{\infty}(A_i \smallsetminus A_{i-1})\right] = \mu\left(\bigcup_{i=1}^{\infty} A_i\right) \qquad \blacksquare$$

Definition.   A measure space $(X, \mathcal{A}, \mu)$ is said to be **complete** if the conditions $A \subset B$, $B \in \mathcal{A}$, $\mu(B) = 0$ imply that $A \in \mathcal{A}$.

We now arrive at the point where we want to create a measure from an outer measure μ. This is to be done by restricting the domain of μ from 2^X to a suitable σ-algebra. A remarkable theorem of Carathéodory accomplishes this in one stroke:

Theorem 1 (Carathéodory Theorem).   Let μ be an outer measure on a set X. Let $\mathcal{A}$ be the family of all subsets A of X having the property

(1) $$\mu(S) = \mu(S \cap A) + \mu(S \smallsetminus A) \qquad \text{(for all } S \in 2^X)$$

Then $(X, \mathcal{A}, \mu)$ is a complete measure space.

Proof. (In the conclusion of the theorem, it is understood that μ is restricted to $\mathcal{A}$, although we refrain from writing $\mu|\mathcal{A}$.) It is to be shown that $\mathcal{A}$ is a σ-algebra, that μ is a measure on $\mathcal{A}$, and that the newly created measure space is complete. There are six tasks to be undertaken.

I. If $A \in \mathcal{A}$, then $X \smallsetminus A \in \mathcal{A}$, because for an arbitrary set S,

$$\mu(S) = \mu(S \cap A) + \mu(S \smallsetminus A) = \mu[S \smallsetminus (X \smallsetminus A)] + \mu[(S \cap (X \smallsetminus A)]$$

II. We prove that $\mathcal{A}$ is closed under the formation of finite unions. It suffices to consider the union of two sets A and B in $\mathcal{A}$. We have, for any S,

$$\begin{aligned}
\mu(S) &= \mu(S \cap A) + \mu(S \smallsetminus A) \\
&= \mu(S \cap A) + \mu[(S \smallsetminus A) \cap B] + \mu[(S \smallsetminus A) \smallsetminus B] \\
&\geqslant \mu[(S \cap A) \cup ((S \smallsetminus A) \cap B)] + \mu[S \smallsetminus (A \cup B)] \\
&= \mu[S \cap (A \cup B)] + \mu[S \smallsetminus (A \cup B)] \\
&\geqslant \mu[(S \cap (A \cup B)) \cup (S \smallsetminus (A \cup B))] = \mu(S)
\end{aligned}$$

III. Let $A_i \in \mathcal{A}$. Here we prove that $\bigcup_{i=1}^{\infty} A_i \in \mathcal{A}$. Define $B_i = A_1 \cup \cdots \cup A_i$ and $C_i = B_i \smallsetminus B_{i-1}$, where $B_0 = \varnothing$. By Parts I and II, together with the equation $C_i = X \smallsetminus [B_{i-1} \cup (X \smallsetminus B_i)]$, we see that C_i and B_i are in $\mathcal{A}$. For any

S, $\mu(S \cap B_n) \geqslant \sum_{i=1}^{n} \mu(S \cap C_i)$. This is proved by induction. For n it is trivial, because $B_1 = C_1$. If it is true for $n - 1$, then

$$\mu(S \cap B_n) = \mu(S \cap B_n \cap C_n) + \mu[(S \cap B_n) \smallsetminus C_n]$$
$$= \mu(S \cap C_n) + \mu(S \cap B_{n-1})$$
$$\geqslant \mu(S \cap C_n) + \sum_{i=1}^{n-1} \mu(S \cap C_i) = \sum_{i=1}^{n} \mu(S \cap C_i)$$

With this inequality available, we have, with $A = \bigcup_{i=1}^{\infty} A_i$,

$$(2) \qquad \mu(S) = \mu(S \cap B_n) + \mu(S \smallsetminus B_n) \geqslant \sum_{i=1}^{n} \mu(S \cap C_i) + \mu(S \smallsetminus A)$$

Since this is true for each n, we can write an inequality to show that $A \in \mathcal{A}$:

$$\mu(S) \geqslant \sum_{i=1}^{\infty} \mu(S \cap C_i) + \mu(S \smallsetminus A) \geqslant \mu \left[\bigcup_{i=1}^{\infty} (S \cap C_i) \right] + \mu(S \smallsetminus A)$$
$$= \mu \left[S \cap \bigcup_{i=1}^{\infty} C_i \right] + \mu(S \smallsetminus A) = \mu(S \cap A) + \mu(S \smallsetminus A) \geqslant \mu(S)$$

IV. The postulate $\mu(\varnothing) = 0$ is true for μ on $\mathcal{A}$ because it is a postulate for outer measures. That $\mu(A) \geqslant 0$ follows from the postulates of an outer measure:

$$0 = \mu(\varnothing) \leqslant \mu(A)$$

because $\varnothing \subset A$.

V. For the countable additivity of μ, look at the proof in Part III. If the sequence $A_1, A_2, \ldots$ is disjoint, then $C_i = A_i$, and Equation (2) will read, with $A = \bigcup_{i=1}^{\infty} A_i$,

$$\mu(S) \geqslant \sum_{i=1}^{n} \mu(S \cap A_i) + \mu(S \smallsetminus A)$$

In this equation, let $S = A$ and let $n \to 0$ to conclude that

$$\mu(A) \geqslant \sum_{i=1}^{\infty} \mu(A_i) \geqslant \mu \left(\bigcup A_i \right) = \mu(A)$$

VI. That the measure space $(X, \mathcal{A}, \mu)$ is complete follows from the more general fact that $A \in \mathcal{A}$ if $\mu(A) = 0$. Indeed, if $\mu(A) = 0$, then for $S \in 2^X$,

$$\mu(S) = \mu(A) + \mu(S) \geqslant \mu(S \cap A) + \mu(S \smallsetminus A) \qquad \blacksquare$$

If μ is an outer measure on X, then the sets A that have the property in Equation (1) of Carathéodory's Theorem are said to be μ-**measurable**. Carathéodory's Theorem thus asserts that the family of μ-measurable sets (μ being an outer measure) is a σ-algebra. If this σ-algebra is denoted by $\mathcal{A}$, then the concepts of μ-measurable and $\mathcal{A}$-measurable are the same (by the definition of $\mathcal{A}$). However, there can be other σ-algebras present (for the same space X), and there can be different kinds of measurability.

One example of this situation occurs when μ is Lebesgue outer measure, as defined in Example 6 of the preceding section (page 382). The σ-algebra $\mathcal{A}$ that arises from Carathéodory's Theorem is called the σ-algebra of **Lebesgue measurable sets**. A smaller σ-algebra is the family $\mathcal{B}$ of all Borel sets. This is the smallest σ-algebra containing the open sets. It turns out that $\mathcal{B}$ is a proper subset of $\mathcal{A}$. In some situations one uses the measure space $(\mathbb{R}, \mathcal{B}, \mu)$ in preference to $(\mathbb{R}, \mathcal{A}, \mu)$. It is convenient to use μ without indicating notationally whether its domain is $2^{\mathbb{R}}$, or $\mathcal{A}$, or $\mathcal{B}$. Remember, however, that μ on $2^{\mathbb{R}}$ is only an outer measure, and countable additivity can fail for sets not in $\mathcal{A}$.

We have seen that every outer measure leads to a measure via Carathéodory's Theorem. There is a converse theorem asserting, roughly, that every measure can be obtained in this way.

Theorem 2.   Let $(X, \mathcal{A}, \mu)$ be a measure space. For $S \in 2^X$ define

$$\mu^*(S) = \inf\{\mu(A) : S \subset A \in \mathcal{A}\}$$

Then μ^* is an outer measure whose restriction to $\mathcal{A}$ is μ. Furthermore, each set in $\mathcal{A}$ is μ-measurable.

Proof.   I. Since μ is nonnegative, so is μ^*. Since $\varnothing \in \mathcal{A}$, we have $0 \leqslant \mu^*(\varnothing) \leqslant \mu(\varnothing) = 0$.

II. If $S \subset T$, then $\{A : S \subset A \in \mathcal{A}\}$ contains $\{A : T \subset A \in \mathcal{A}\}$. Hence

$$\mu^*(S) = \inf\{\mu(A) : S \subset A \in \mathcal{A}\} \leqslant \inf\{\mu(A) : T \subset A \in \mathcal{A}\} = \mu^*(T)$$

III. If $S_i \in 2^X$ and $\varepsilon > 0$, select $A_i \in \mathcal{A}$ so that $S_i \subset A_i$ and $\mu^*(S_i) \geqslant \mu(A_i) - \varepsilon/2^i$. Then $\bigcup_{i=1}^{\infty} S_i \subset \bigcup_{i=1}^{\infty} A_i \in \mathcal{A}$. Consequently,

$$\mu^*\left(\bigcup_{i=1}^{\infty} S_i\right) \leqslant \mu\left(\bigcup_{i=1}^{\infty} A_i\right) \leqslant \sum_{i=1}^{\infty} \mu(A_i) \leqslant \sum_{i=1}^{\infty}\left[\mu^*(S_i) + \frac{\varepsilon}{2^i}\right]$$

Since ε can be any positive number, $\mu^*(\bigcup_{i=1}^{\infty} S_i) \leqslant \sum_{i=1}^{\infty} \mu(S_i)$.

IV. If $S \in \mathcal{A}$ and $S \subset A \in \mathcal{A}$, then $\mu^*(S) \leqslant \mu(S) \leqslant \mu(A)$. Taking an infimum for all choices of A, we get $\mu^*(S) \leqslant \mu(S) \leqslant \mu^*(S)$. This proves that μ^* is an extension of μ.

V. To prove that each A in $\mathcal{A}$ is μ^*-measurable, let S be any subset of X. Given $\varepsilon > 0$, we find $B \in \mathcal{A}$ such that $\mu^*(S) \geqslant \mu(B) - \varepsilon$ and $S \subset B$. Then

$$\mu^*(S) + \varepsilon \geqslant \mu(B) = \mu(B \cap A) + \mu(B \smallsetminus A) = \mu^*(B \cap A) + \mu^*(B \smallsetminus A)$$
$$\geqslant \mu^*(S \cap A) + \mu^*(S \smallsetminus A) \geqslant \mu^*(S)$$

This calculation used Parts III and IV of the present proof. Since ε was arbitrary, $\mu^*(S) = \mu^*(S \cap A) + \mu^*(S \smallsetminus A)$ for all S. Hence A is μ^*-measurable. ∎

The construction of μ^* in the preceding theorem yields some additional information. First, we define the concept of regularity for an outer measure. An outer measure μ on 2^X is said to be **regular** if for each $S \in 2^X$ there corresponds a μ-measurable set A such that $S \subset A$ and $\mu(S) = \mu(A)$.

Theorem 3.   *Under the same hypotheses as in Theorem 2, the*
outer measure μ^* *is regular.*

Proof. Let S be any subset of X. For each $n \in \mathbb{N}$ select $A_n \in \mathcal{A}$ so that
$S \subset A_n$ and $\mu^*(S) \geqslant \mu(A_n) - 1/n$. Put $A = \bigcap_{n=1}^{\infty} A_n$. Since $\mathcal{A}$ is a σ-algebra,
$A \in \mathcal{A}$. (See Lemma 1 in the preceding section, page 384.) From the inclusion
$S \subset A \subset A_n$ we get

$$\mu^*(S) \leqslant \mu^*(A) = \mu(A) \leqslant \mu(A_n) < \mu^*(S) + 1/n$$

Since this is true for all n, $\mu^*(S) = \mu^*(A)$. By the preceding theorem, A is
μ^*-measurable.   ∎

Problems 8.2

1. Let μ be Lebesgue outer measure. Let $\mathbb{Q}$ be the set of all rational numbers. Prove that
   if $\mathbb{Q} \cap [0,1]$ is contained in $\bigcup_{i=1}^{n}(a_i, b_i)$, then $\sum_{i=1}^{n}(b_i - a_i) \geqslant 1$. Show that this is not
   true if we permit a countable number of intervals (a_i, b_i).

2. Is there an example of a set X and a measure μ on 2^X such that $\mu(X) = 1$ and $\mu(\{x\}) = 0$
   for all points x in X?

3. In $\mathbb{R}$, is the smallest σ-algebra containing all singletons $\{x\}$ the same as the σ-algebra of
   all Borel sets?

4. Let $(X, \mathcal{A}, \mu)$ be a measure space, and let μ^* be the outer measure defined in Theorem 2
   (page 389). Define the **inner measure** μ_* induced by μ via the equation

   $$\mu_*(S) = \sup\{\mu(A) : S \supset A \in \mathcal{A}\}$$

   Prove these properties of μ_*: (i) $\mu_*(S) \leqslant \mu^*(S)$; (ii) $\mu_*(S) \geqslant 0$; (iii) $\mu_*(S) \leqslant \mu_*(T)$
   when $S \subset T$; (iv) $\mu_*(\varnothing) = 0$; (v) $\mu_*(A) = \mu(A)$ if $A \in \mathcal{A}$.

5. Prove that an outer measure μ on 2^X is a measure on 2^X if and only if every set in 2^X
   is μ-measurable.

6. Let $(X, \mathcal{A}, \mu)$ be a measure space, and let $A_i \in \mathcal{A}$. Prove that $\mu(\bigcup_{i=1}^{\infty} A_i) = \lim_{n \to \infty} \mu(\bigcup_{i=1}^{n} A_i)$.

7. The symmetric difference of two sets A and B is $A \bigtriangleup B = (A \smallsetminus B) \cup (B \smallsetminus A)$. Prove
   that for measurable sets A and B in a measure space, the condition $\mu(A \bigtriangleup B) = 0$ implies
   that $\mu(A) = \mu(B)$.

8. Let $(X, \mathcal{A}, \mu)$ be an incomplete measure space. Show how to enlarge $\mathcal{A}$ and extend μ so
   that a complete measure space is obtained.

9. Let $(X, \mathcal{A}, \mu)$ be a measure space. Let $\mathcal{B}$ be the family of all sets $B \in 2^X$ such that
   $A \cap B \in \mathcal{A}$ whenever $A \in \mathcal{A}$ and $\mu(A) < \infty$. Show that $\mathcal{B}$ is a σ-algebra containing $\mathcal{A}$.

10. Prove that if $(X, \mathcal{A}, \mu)$ and $(X, \mathcal{A}, \nu)$ are measure spaces, then so is $(X, \mathcal{A}, \mu + \nu)$. Gen-
    eralize.

11. If $(X, \mathcal{A}, \mu)$ and $(X, \mathcal{A}, \nu)$ are measure spaces such that $\mu \geqslant \nu$, is there a measure θ (on
    $\mathcal{A}$) such that $\nu + \theta = \mu$? (Caution: $\infty - \infty$ is not defined in $\mathbb{R}^*$.)

12. Let $(X, \mathcal{A}, \mu)$ be a measure space. Suppose that $A_n \in \mathcal{A}$ and $A_{n+1} \subset A_n$ for all n. Does
    it follow that $\mu(\bigcap_{n=1}^{\infty} A_n) = \lim_{n \to \infty} \mu(A_n)$?

13. Let X be an uncountable set and $\mathcal{A} = 2^X$. Define $\mu(A) = 0$ if A is countable and
    $\mu(A) = \infty$ otherwise. Is $(X, \mathcal{A}, \mu)$ a measure space?

14. Prove that for any outer measure μ and any set A such that $\mu(A) = 0$, A is μ-measurable.

15. Let $(X, \mathcal{A}, \mu)$ be a measure space, and let $B \in \mathcal{A}$. Define ν on $\mathcal{A}$ by writing $\nu(A) = \mu(A \cap B)$. Prove that $(X, \mathcal{A}, \nu)$ is a measure space.

16. Let $(X, \mathcal{A}, \mu)$ be a measure space. Let $A_n \in \mathcal{A}$ and $\sum_{n=1}^{\infty} \mu(A_n) < \infty$. Prove that the set of x belonging to infinitely many A_n has measure 0.

17. Let $(X, \mathcal{A}, \mu)$ be a measure space for which $\mu(X) < \infty$. Let A_n be a sequence of measurable sets such that $A_1 \subset A_2 \subset \cdots$ and $X = \bigcup A_n$. Show that $\mu(X \setminus A_n) \downarrow 0$.

8.3 Lebesgue Measure

In this section μ will denote both Lebesgue outer measure and Lebesgue measure. Both are defined for subsets of $\mathbb{R}$ by the equation

$$(1) \qquad \mu(S) = \inf \left\{ \sum_{i=1}^{\infty} |b_i - a_i| \; : \; S \subset \bigcup_{i=1}^{\infty} (a_i, b_i) \right\}$$

The outer measure is defined for all subsets of $\mathbb{R}$, while the measure μ is the restriction of the outer measure to the σ-algebra described in Carathéodory's Theorem (page 387). The sets in this σ-algebra are called the Lebesgue measurable subsets of $\mathbb{R}$. It is a very large class of sets, bigger than the σ-algebra of Borel sets. The latter has cardinality c, while the former has cardinality 2^c.

Theorem 1.   *The Lebesgue outer measure of an interval is its length.*

Proof. Consider first a compact interval $[a, b]$. Since $[a, b] \subset (a - \varepsilon, b + \varepsilon)$, we conclude from the definition (1) that $\mu([a, b]) \leqslant b - a + 2\varepsilon$ for every positive ε. Hence $\mu([a, b]) \leqslant b - a$. Suppose now that $\mu([a, b]) < b - a$. Find intervals (a_i, b_i) such that $[a, b] \subset \bigcup_{i=1}^{\infty}(a_i, b_i)$ and $\sum_{i=1}^{\infty} |b_i - a_i| < b - a$. We can assume $a_i < b_i$ for all i. By compactness and renumbering we can get $[a, b] \subset \bigcup_{i=1}^{n}(a_i, b_i)$. It follows that $\sum_{i=1}^{n}(b_i - a_i) < b - a$. By renumbering again we can assume $a \in (a_1, b_1)$, $b_1 \in (a_2, b_2)$, $b_2 \in (a_3, b_3)$, and so on. There must exist an index $k \leqslant n$ such that $b < b_k$. Then we reach a contradiction:

$$b - a > \sum_{i=1}^{n}(b_i - a_i) \geqslant \sum_{i=1}^{k}(b_i - a_i) = b_k - a_1 + \sum_{i=1}^{k-1}(b_i - a_{i+1}) > b_k - a_1 > b - a$$

If J is a bounded interval of the type (a, b), $(a, b]$, or $[a, b)$, then from the inclusions

$$[a + \varepsilon, b - \varepsilon] \subset J \subset [a - \varepsilon, b + \varepsilon]$$

we obtain $b - a - 2\varepsilon \leqslant \mu(J) \leqslant b - a + 2\varepsilon$ and $\mu(J) = b - a$.

Finally, if J is an unbounded interval, then it contains intervals $[a, b]$ of arbitrarily great length. Hence $\mu(J) = \infty$. ∎

Theorem 2.   *Every Borel set in* $\mathbb{R}$ *is Lebesgue measurable.*

Proof.   (S.J. Bernau) The family of Borel sets is the smallest σ-algebra containing all the open sets. The Lebesgue measurable sets form a σ-algebra. Hence it suffices to prove that every open set is Lebesgue measurable.

Recall that every open set in $\mathbb{R}$ can be expressed as a countable union of open intervals (a, b). Thus it suffices to prove that each interval (a, b) is Lebesgue measurable. We begin with an interval of the form (a, ∞), where $a \in \mathbb{R}$.

To prove that the open interval (a, ∞) is measurable, we must prove, for any set S in $\mathbb{R}$, that

$$(2) \qquad \mu(S) \geqslant \mu\Big[S \cap (a, \infty)\Big] + \mu\Big[S \smallsetminus (a, \infty)\Big]$$

Let us use the notation $|I|$ for the length of an interval I. Given $\varepsilon > 0$, select open intervals I_n such that $S \subset \bigcup_{n=1}^{\infty} I_n$ and $\sum |I_n| < \mu(S) + \varepsilon$. Define $J_n = I_n \cap (a, \infty)$, $K_n = I_n \cap (-\infty, a)$, and $K_0 = (a - \varepsilon, a + \varepsilon)$. Then we have

$$S \cap (a, \infty) \subset \bigcup_{n=1}^{\infty} J_n$$

$$S \smallsetminus (a, \infty) = S \cap (-\infty, a] \subset \bigcup_{n=0}^{\infty} K_n$$

$$J_n \cup K_n \subset I_n \quad \text{and} \quad J_n \cap K_n = \varnothing$$

Consequently,

$$\mu\big[S \cap (a, \infty)\big] + \mu\big[S \smallsetminus (a, \infty)\big] \leqslant \sum_{n=1}^{\infty} \{|J_n| + |K_n|\} + |K_0|$$

$$\leqslant \sum_{n=1}^{\infty} |I_n| + 2\varepsilon < \mu(S) + 3\varepsilon$$

Because ε was arbitrary, this establishes Equation (2). Since the measurable sets make up a σ-algebra, each set of the form $(-\infty, b] = \mathbb{R} \smallsetminus (b, \infty)$ is measurable. Hence the set $(-\infty, b) = \bigcup_{n=1}^{\infty} (-\infty, b - \frac{1}{n}]$ is measurable and so is $(a, b) = (-\infty, b) \cap (a, \infty)$. ∎

Theorem 3.   *Lebesgue outer measure is invariant on the group* $(\mathbb{R}, +)$.

Proof.   The statement means that $\mu(S) = \mu(v + S)$ for all $S \in 2^{\mathbb{R}}$ and all $v \in \mathbb{R}$. The translate $v + S$ is defined to be $\{v + x : x \in S\}$. Notice that the condition $S \subset \bigcup_{i=1}^{\infty} (a_i, b_i)$ is equivalent to the condition $x + S \subset \bigcup_{i=1}^{\infty} (x + a_i, x + b_i)$. Since the length of $(x + a_i, x + b_i)$ is the same as the length of (a_i, b_i), the definition of μ gives equal values for $\mu(S)$ and $\mu(x + S)$. ∎

Lemma.   Let $\{r_1, r_2, \ldots\}$ be an enumeration of the rational numbers in $[-1, 1]$. There exists a set P contained in $[0, 1]$ such that the sequence of sets $r_i + P$ is disjoint and covers $[0, 1]$.

Proof.   Let $\mathbb{Q}$ be the set of all rational numbers. Consider the family $\mathcal{F}$ of all sets of the form $x + \mathbb{Q}$, where $0 \leqslant x \leqslant 1$. Although our description of $\mathcal{F}$ involves many sets being listed more than once, the family $\mathcal{F}$ is, in fact, disjoint. To verify this, suppose that $x + \mathbb{Q}$ and $y + \mathbb{Q}$ have a point t in common. Then $t = x + q_1 = y + q_2$, for appropriate $q_i \in \mathbb{Q}$. Consequently, $x - y = q_2 - q_1 \in \mathbb{Q}$, from which it follows easily that $x + \mathbb{Q} = y + \mathbb{Q}$.

The family of sets $(x + \mathbb{Q}) \cap [0, 1]$, where $0 \leqslant x \leqslant 1$, is also disjoint, and each of these sets is nonempty. By Zermelo's Postulate ([Kel], page 33), there exists a set $P \subset [0, 1]$ such that P contains one and only one point from each set in the family.

Now we want to prove that the family $\{r_i + P\}$ is disjoint. Suppose that $t \in (r_i + P) \cap (r_j + P)$. Then $r_i + p = r_j + p'$ for suitable p and p' in P. We have $p = p' + (r_j - r_i)$, whence $p \in P \cap (p' + \mathbb{Q})$. Since $0 \in \mathbb{Q}$, we have $p' \in P \cap (p' + \mathbb{Q})$. By the properties of P, $p = p'$. From an equation above, $r_i = r_j$ and then $i = j$.

Finally, we show that $[0, 1] \subset \bigcup_{i=1}^{\infty}(r_i + P)$. If $0 \leqslant x \leqslant 1$, then P contains an element p in $x + \mathbb{Q}$. By the definition of P, $0 \leqslant p \leqslant 1$. We write $p = x + r$ for a suitable r in $\mathbb{Q}$. Then $r = p - x \in [-1, 1]$ and $-r \in [-1, 1]$. Thus $-r = r_i$ for some i. It follows that $x = p - r = p + r_i \in r_i + P$. ∎

Theorem 4.   There exists no translation-invariant measure ν defined on $2^{\mathbb{R}}$ such that $0 < \nu([0, 1]) < \infty$. Consequently, there exist subsets of $\mathbb{R}$ that are not Lebesgue measurable.

Proof.   The second assertion follows from the first because if every set of reals were Lebesgue measurable, then Lebesgue measure would contradict the first assertion.

To prove the first assertion, suppose that a measure ν exists as described. By the preceding lemma, the set P given there has the property

$$[0, 1] \subset \bigcup_{i=1}^{\infty}(r_i + P) \subset [-1, 2]$$

Also by the lemma, the sequence of sets $r_i + P$ is disjoint. Consequently,

$$0 < \nu([0, 1]) \leqslant \nu\left[\bigcup_{i=1}^{\infty}(r_i + P)\right] = \sum_{i=1}^{\infty}\nu(r_i + P) = \sum_{i=1}^{\infty}\nu(P)$$

Therefore, $\nu(P) > 0$, $\sum \nu(P) = \infty$, and we have the contradiction

$$\infty = \nu\left[\bigcup_{i=1}^{\infty}(r_i + P)\right] \leqslant \nu([-1, 2]) \leqslant 3\nu([0, 1])$$ ∎

Problems 8.3

1. Zermelo's Postulate states that if $\mathcal{F}$ is a disjoint family of nonempty sets, then there is a set that contains exactly one element from each set in the family $\mathcal{F}$. Prove this, using the Axiom of Choice.

2. Prove that the set P in the lemma is not Lebesgue measurable.

3. Prove that Lebesgue measure restricted to the Borel sets is not complete.

4. An F_σ-set is any countable union of closed sets, and a G_δ-set is any countable intersection of open sets. Prove that both types of sets are Borel sets.

5. Prove that the Cantor "middle-third" set is an uncountable Borel set of Lebesgue measure 0. This set is defined in Problem 1.7 26, page 46.

6. Prove that the infimum in the definition of Lebesgue outer measure is attained if the set is bounded and open.

7. Prove that the set $\mathbb{Q}$ of all rational numbers is a Borel set of measure 0.

8. Prove that for any Lebesgue measurable set A of finite measure and for any $\varepsilon > 0$ there are an open set G and a closed set F such that $F \subset A \subset G$ and $\mu(G) \leqslant \mu(F) + \varepsilon$.

9. Let S be a subset of $\mathbb{R}$ such that for each $\varepsilon > 0$ there is a closed set F contained in S for which $\mu(S \smallsetminus F) < \varepsilon$. Prove that S is Lebesgue measurable.

10. Prove that a set of Lebesgue measure 0 cannot contain a nonmeasurable set, but every set of positive measure does contain a nonmeasurable set.

11. Under what set operations is $2^{\mathbb{R}} \smallsetminus \mathcal{B}$ closed? Here $\mathcal{B}$ is the σ-algebra of Borel sets.

12. Prove that if $S \subset \mathbb{R}$ and for every $\varepsilon > 0$ there is an open set G containing S and satisfying $\mu(G \smallsetminus S) < \varepsilon$, then S is Lebesgue measurable.

13. In Theorem 4, is the result valid when the domain of ν is a subset of $2^{\mathbb{R}}$?

8.4 Measurable Functions

In the study of topological spaces, continuous functions play an important rôle. Analogously, in the study of measurable spaces, measurable functions are important. In fact, they are perhaps the principal reason for creating measurable spaces. Our considerations here are *general*, i.e., not restricted to Lebesgue measure.

Consider an arbitrary measurable space $(X, \mathcal{A})$, as defined in Section 8.1 (page 384). A function $f : X \to \mathbb{R}^*$ is said to be **$\mathcal{A}$-measurable** (or simply **measurable**) if $f^{-1}(B) \in \mathcal{A}$ whenever B is a Borel subset of $\mathbb{R}^*$.

Theorem 1.   Let $(X, \mathcal{A})$ be a measurable space. A function f from X to the extended reals $\mathbb{R}^*$ is measurable if it has any one of the following properties:

         a. $f^{-1}\big((a, \infty]\big) \in \mathcal{A}$ for each $a \in \mathbb{R}^*$
         b. $f^{-1}\big([a, \infty]\big) \in \mathcal{A}$ for each $a \in \mathbb{R}^*$
         c. $f^{-1}\big([-\infty, a)\big) \in \mathcal{A}$ for each $a \in \mathbb{R}^*$
         d. $f^{-1}\big([-\infty, a]\big) \in \mathcal{A}$ for each $a \in \mathbb{R}^*$

e. $f^{-1}\big((a,b)\big) \in \mathcal{A}$ for all a and b in $\mathbb{R}^*$

f. $f^{-1}(O) \in \mathcal{A}$ for each open set O in $\mathbb{R}^*$

Proof. We shall prove that each condition implies the one following it, and that **f** implies that f is measurable. That **a** implies **b** follows from the equation $f^{-1}\big([a,\infty]\big) = \bigcap_{n=1}^{\infty} f^{-1}\big((a - \frac{1}{n}, \infty]\big)$ and from the properties of a σ-algebra. That **b** implies **c** follows in the same manner from the equation $f^{-1}\big([-\infty, a)\big) = X \smallsetminus f^{-1}\big([a, \infty]\big)$. That **c** implies **d** follows from the equation $f^{-1}\big([-\infty, a]\big) = \bigcap_{n=1}^{\infty} f^{-1}\big([-\infty, a + \frac{1}{n})\big)$. That **d** implies **e** follows from writing $f^{-1}\big((a,b)\big) = \bigcup_{n=1}^{\infty} f^{-1}\big([-\infty, b - \frac{1}{n})\big) \smallsetminus f^{-1}\big([-\infty, a]\big)$. That **e** implies **f** is a consequence of the theorem that each open set in $\mathbb{R}^*$ is a countable union of intervals of the form (a, b), where a and b are in $\mathbb{R}^*$. To complete the proof, assume condition **f**. Let $\mathcal{S}$ be the family of all sets S contained in $\mathbb{R}^*$ such that $f^{-1}(S) \in \mathcal{A}$. It is straightforward to verify that $\mathcal{S}$ is a σ-algebra. By hypothesis, each open set in $\mathbb{R}^*$ belongs to $\mathcal{S}$. Hence $\mathcal{S}$ contains the σ-algebra of Borel sets. Consequently, $f^{-1}(B) \in \mathcal{A}$ for each Borel set B, and f is measurable. ∎

Our next goal is to study, for a given measurable space $(X, \mathcal{A})$, the class of measurable functions. First, we define the **characteristic function** of a set A to be the function $\mathfrak{X}_A$ given by

$$\mathfrak{X}_A(x) = \begin{cases} 1 & \text{if } x \in A \\ 0 & \text{if } x \notin A \end{cases}$$

Theorem 2. Let $(X, \mathcal{A})$ be a measurable space. The family of all measurable functions contains the characteristic function of each measurable set and is closed under these operations:

a. $f + g$ (provided that there is no point x where $f(x)$ and $g(x)$ are infinite and of opposite signs).

b. λf $(\lambda \in \mathbb{R}^*)$

c. fg

d. $\sup f_i$ $(i \in \mathbb{N})$

e. $\inf f_i$ $(i \in \mathbb{N})$

f. $\liminf f_i$ $(i \in \mathbb{N})$

g. $\limsup f_i$ $(i \in \mathbb{N})$

Proof. If $A \in \mathcal{A}$, then the characteristic function $\mathfrak{X}_A$ is measurable because $\mathfrak{X}_A^{-1}\big((a, \infty]\big) = \{x : \mathfrak{X}_A(x) > a\}$, and this last set is either X, A, or $\varnothing$. These three sets are measurable, and Theorem 1 applies.

Now suppose that f and g are measurable functions. Let $r_1, r_2, \ldots$ be an enumeration of all the rational numbers. Then

$$(f+g)^{-1}\big((a,\infty]\big) = \{x : f(x) + g(x) > a\}$$

$$= \bigcup_{i=1}^{\infty} \{x : f(x) > r_i \text{ and } g(x) > a - r_i\}$$

$$= \bigcup_{i=1}^{\infty} \Big[f^{-1}\big((r_i,\infty]\big) \cap g^{-1}\big((a - r_i,\infty]\big)\Big]$$

To verify this, notice that $f(x) + g(x) > a$ if and only if $a - g(x) < f(x)$, and this last inequality is true if and only if $a - g(x) < r_i < f(x)$ for some i. The last term in the displayed equation is a countable union of measurable sets, because f and g are measurable.

If f is measurable, then so is λf, because $(\lambda f)^{-1}\big((a,\infty]\big)$ is either $\varnothing$ (when $\lambda = 0$ and $a \geqslant 0$), or X (when $\lambda = 0$ and $a < 0$), or $f^{-1}\big((a/\lambda,\infty]\big)$ (when $\lambda > 0$), or $f^{-1}\big([-\infty, a/\lambda)\big)$ (when $\lambda < 0$).

If f is measurable, then so is f^2, because $(f^2)^{-1}\big((a,\infty]\big)$ is X when $a < 0$, and it is $f^{-1}\big((\sqrt{a},\infty]\big) \cup f^{-1}\big([-\infty, -\sqrt{a})\big)$ when $a \geqslant 0$. From the identity $fg = \frac{1}{4}(f+g)^2 - \frac{1}{4}(f-g)^2$ it follows that fg is measurable if f and g are measurable.

If f_i are measurable and if $g(x) = \sup_i f_i(x)$, then g is measurable because $g^{-1}\big((a,\infty]\big) = \bigcup_{i=1}^{\infty} f_i^{-1}\big((a,\infty]\big)$. A similar argument applies to infima, if we use an interval $[-\infty, a)$.

If f_i are measurable and if $g(x) = \limsup f_i(x)$, then g is measurable, because $g(x) = \lim_{n\to\infty} \sup_{i>n} f_i(x) = \inf_n \sup_{i>n} f_i(x)$. A similar argument applies to the limit infimum. ∎

Consider now a measure space $(X, \mathcal{A}, \mu)$. Let f and g be functions on X taking values in $\mathbb{R}^*$. If the set $\{x : f(x) \neq g(x)\}$ belongs to $\mathcal{A}$ and has measure 0, then we say that $f(x) = g(x)$ **almost everywhere**. This is an equivalence relation if the measure space is complete (Problem 1). More generally, if $P(x)$ is a proposition, for each x in X, then we say that P is **true almost everywhere** if the set $\{x : P(x) \text{ is false}\}$ is a measurable set of measure 0. The abbreviation **a.e.** is used for "almost everywhere." The French use **p.p.** for "presque partout."

Theorem 3.   *Let $(X, \mathcal{A}, \mu)$ be a complete measure space, as defined in Section 8.2, page 387. If f is a measurable function and if $f(x) = g(x)$ almost everywhere, then g is measurable.*

Proof. Define $A = \{x : f(x) \neq g(x)\}$. Then A is measurable and $\mu(A) = 0$. Also, $X \smallsetminus A$ is measurable. For $a \in \mathbb{R}^*$ we write

$$g^{-1}\big((a,\infty]\big) = \Big\{g^{-1}\big((a,\infty]\big) \cap (X \smallsetminus A)\Big\} \cup \Big\{g^{-1}\big((a,\infty]\big) \cap A\Big\}$$

On the right side of this equation we see the union of two sets. The first of these is measurable because it is $f^{-1}\big((a,\infty)\big) \smallsetminus A$. The second set is measurable because it is a subset of a set of measure 0, and the measure space is complete. ∎

Let $(X, \mathcal{A}, \mu)$ be a measure space, and let $f, f_1, f_2, \ldots$ be measurable functions. We say that $f_n \to f$ **almost uniformly** if to each positive ε there corresponds a measurable set of measure at most ε on the complement of which $f_n \to f$ uniformly.

> **Theorem 4 (Egorov's Theorem).**   *Let $(X, \mathcal{A}, \mu)$ be a measure space such that $\mu(X) < \infty$. For a sequence of finite-valued measurable functions $f, f_1, f_2, \ldots$ these properties are equivalent:*
> **a.** $f_n \to f$ *almost everywhere*
> **b.** $f_n \to f$ *almost uniformly*

Proof.   Assume that **b** is true. For each m in $\mathbb{N}$ there is a measurable set A_m such that $\mu(A_m) < 1/m$, and on $X \smallsetminus A_m$, $f_n(x) \to f(x)$ uniformly. Define $A = \bigcap_{m=1}^{\infty} A_m$. Then $\mu(A) = 0$ because $A \subset A_m$ for all m. Also, $f_n(x) \to f(x)$ on $X \smallsetminus A$ because $f_n(x) \to f(x)$ for $x \in X \smallsetminus A_m$ and $X \smallsetminus A = X \smallsetminus \bigcap A_m = \bigcup (X \smallsetminus A_m)$. Thus **a** is true.

Now assume that **a** is true. Let $g_n = f - f_n$. By altering g_n on a set of measure 0, we can assume that $g_n(x) \to 0$ everywhere. Next, we define $A_n^m = \{x : |g_i(x)| \leqslant 1/m \text{ for } i \geqslant n\}$. Thus $A_1^m \subset A_2^m \subset \cdots$ For each x there is an index n such that $x \in A_n^m$; in other words, $x \in \bigcup_{n=1}^{\infty} A_n^m$ and $X \subset \bigcup_{n=1}^{\infty} A_n^m$. Since X has finite measure, $\mu(X \smallsetminus A_n^m) \to 0$ as $n \to \infty$. (See Lemma 2 in Section 8.2, page 387.) Let $\varepsilon > 0$. For each m, let n_m be an integer such that $\mu(X \smallsetminus A_{n_m}^m) < \varepsilon/2^m$. Define $A = \bigcap_{m=1}^{\infty} A_{n_m}^m$. Then

$$\mu(X \smallsetminus A) = \mu\left(X \smallsetminus \bigcap_{m=1}^{\infty} A_{n_m}^m \right) = \mu\left(\bigcup_{m=1}^{\infty} (X \smallsetminus A_{n_m}^m) \right) \leqslant \sum_{m=1}^{\infty} \mu(X \smallsetminus A_{mn}^m) < \varepsilon$$

On A, $g_i(x) \to 0$ uniformly. Indeed, for $x \in A$ we have (for all m)

$$i \geqslant n_m \implies |g_i(x)| \leqslant \tfrac{1}{m} \qquad\qquad ∎$$

Let $(X, \mathcal{A})$ be a measurable space. A **simple function** is a measurable function $f : X \to \mathbb{R}^*$ whose range is a finite subset of $\mathbb{R}^*$. Then f can be written in the form $f = \sum_{i=1}^{n} \lambda_i \mathcal{X}_{A_i}$, where the λ_i can be taken to be distinct elements of $\mathbb{R}^*$, and A_i can be the set $\{x : f(x) = \lambda_i\}$. It then turns out that each A_i is measurable, that these sets are mutually disjoint, and their union is X. (Problem 2)

> **Theorem 5.**   *Let $(X, \mathcal{A})$ be a measurable space, and f any nonnegative measurable function. Then there exists a sequence of nonnegative simple functions g_n such that $g_n(x) \uparrow f(x)$ for each x. If f is bounded, this sequence can be constructed so that $g_n \uparrow f$ uniformly.*

Proof. ([HewS], page 159.) Define

$$A_i^n = \left\{ x \in X \; : \; \frac{i}{2^n} \leqslant f(x) < \frac{i+1}{2^n} \right\} \qquad (0 \leqslant i < n2^n)$$

$$B^n = \{ x \in X \; : \; f(x) \geqslant n \}$$

$$g_n = \sum_i \frac{i}{2^n} \mathfrak{X}_{A_i^n} + n \mathfrak{X}_{B^n}$$

The sets A_i^n and B^n are measurable, by Theorem 1. Hence g_n is a simple function. The definition of g_n shows directly that $g_n \leqslant f$. In order to verify that $g_n(x)$ converges to $f(x)$ for each x, consider first the case when $f(x) \neq \infty$. For large n and a suitable i, $x \in A_i^n$. Then $f(x) - g_n(x) < \frac{i+1}{2^n} - \frac{i}{2^n} = \frac{1}{2^n}$. On the other hand, if $f(x) = +\infty$, then $g_n(x) = n \to f(x)$.

For the monotonicity of $g_n(x)$ as a function of n for x fixed, first verify (Problem 3) that (for $i < n2^n$) $A_i^n = A_{2i}^{n+1} \cup A_{2i+1}^{n+1}$. If $x \in A_{2i}^{n+1}$, then $g_{n+1}(x) = 2i/2^{n+1} = i/2^n = g_n(x)$. If $x \in A_{2i+1}^{n+1}$, then $g_{n+1}(x) = (2i+1)/2^{n+1} \geqslant 2i/2^{n+1} = g_n(x)$. If $x \in B^n$, then $f(x) \geqslant n$, and therefore $x \in \bigcup_{i \geqslant n2^{n+1}} A_i^{n+1} \cup B^{n+1}$. It follows that $g_{n+1}(x) \geqslant n = g_n(x)$.

Finally, if f is bounded by m, then for $n \geqslant m$ we have $0 \leqslant f(x) - g_n(x) \leqslant 2^{-n}$. In this case the convergence is uniform.   ∎

Problems 8.4

1. Prove that the relation of two functions being equal almost everywhere is an equivalence relation if the underlying measure space is complete.

2. Prove the assertions about the sets A_i that were mentioned in the definition of a simple function.

3. In the proof of Theorem 5, verify that $A_i^{(n)} = A_{2i}^{(n+1)} \cup A_{2i+1}^{(n+1)}$.

4. Let $(X, \mathcal{A})$ be a measurable space, and let $r_1, r_2, \ldots$ be an enumeration of the rational numbers. Prove that a function $f : X \to \mathbb{R}^*$ is measurable if and only if all the sets $f^{-1}((r_i, \infty])$ are measurable.

5. Prove that every Borel set in $\mathbb{R}^*$ is one of the four types B, $B \cup \{\infty\}$, $B \cup \{-\infty\}$, $B \cup \{+\infty, -\infty\}$, where B is a Borel set in $\mathbb{R}$.

6. Prove that in order for f to be measurable it is necessary and sufficient that $f^{-1}(O)$ be measurable for all open sets O in $\mathbb{R}$ and that $f^{-1}(\{\infty\})$ and $f^{-1}(\{-\infty\})$ be measurable.

7. Prove that if f and g are measurable functions, then the sets $\{x : f(x) = g(x)\}$, $\{x : f(x) \geqslant g(x)\}$, and $\{x : f(x) > g(x)\}$ are measurable.

8. Prove that if f and g are measurable functions and if $(f+g)(x)$ is assigned some constant value on the set where $f(x)$ and $g(x)$ are infinite and of opposite sign, then $f + g$ is measurable.

9. Let $(\mathbb{R}, \mathcal{A})$ be the measurable space in which $\mathcal{A}$ is the family of all Lebesgue measurable sets. Give an example of a nonmeasurable function in this setting.

10. Let P be the σ-algebra of Borel sets in $\mathbb{R}$. Let $\mathcal{A}$ be the σ-algebra of Lebesgue measurable sets. . ove that $\mathcal{B}$ is a proper subset of $\mathcal{A}$.

11. Let $(X, \mathcal{A}, \mu)$ be a measure space for which $\mu(X) < \infty$. Prove that if f is a measurable function that is finite-valued almost everywhere, then for each $\varepsilon > 0$ there is an M such that $\mu(\{x : |f(x)| > M\}) < \varepsilon$.

12. Let $(X, \mathcal{A})$ be a measurable space and f a measurable function. What can you say about the following set?
$$\{S : S \subset \mathbb{R} \ \text{and} \ f^{-1}(S) \in \mathcal{A}\}$$

13. Prove that the composition $f \circ g$ of two Borel measurable functions on $\mathbb{R}$ is Borel measurable.

14. Let $(X, \mathcal{A}, \mu)$ be a measure space and f a measurable function. For each Borel set B in $\mathbb{R}^*$ define $\nu(B) = \mu(f^{-1}(B))$. Show that ν is a Borel measure, i.e., a measure on $\mathcal{B}$, the σ-algebra of Borel sets.

15. If $|f|$ is measurable, does it follow that f is measurable?

16. Show that the composition of two Lebesgue measurable functions need not be Lebesgue measurable.

17. Prove that if f is a real-valued Lebesgue measurable function, then there is a Borel measurable function equal to f almost everywhere.

18. Let $X = \mathbb{N}$, $\mathcal{A} = 2^X$, and let μ be counting measure, as defined in Example 3 in Section 8.1, page 382. Let f_n be the characteristic function of the set $\{1, 2, \ldots, n\}$. Prove that the sequence $[f_n]$ has property (a) but not property (b) in Egorov's Theorem. Resolve the apparent contradiction.

19. Refer to Problem 7 in Section 8.2, page 390, for the definition of the symmetric difference of two sets. Prove that $|\mathcal{X}_A - \mathcal{X}_B| = \mathcal{X}_{A \triangle B}$.

20. Prove that a monotone function $f : \mathbb{R} \to \mathbb{R}$ is Borel measurable.

21. Prove that the set of points where a sequence of measurable functions converges is a measurable set.

8.5 The Integral for Nonnegative Functions

With any measure space $(X, \mathcal{A}, \mu)$ there is associated (in a certain standard way) an **integral**. It will be a linear functional on the space of all measurable functions from X into $\mathbb{R}^*$. The motivation for an appropriate definition arises from our wish that the integral of the characteristic function of a measurable set should be the measure of the set:

(1)
$$\int \mathcal{X}_A = \mu(A) \qquad (A \in \mathcal{A})$$

The requirement that the integral act linearly leads to the definition of the integral of a simple function f:

(2)
$$\int f = \int \sum_{i=1}^n \alpha_i \mathcal{X}_{A_i} = \sum_{i=1}^n \alpha_i \mu(A_i)$$

In Equation (2) we assume that the sets $A_i, \ldots, A_n$ are mutually disjoint and that the α_i are distinct. Such a representation $f = \sum \alpha_i \mathcal{X}_{A_i}$ is called **canonical**.

Lemma 1.   Let $f = \sum_{i=1}^n \alpha_i \mathcal{X}_{A_i}$, where we assume only that the sets A_i are mutually disjoint measurable sets. Then $\int f = \sum_{i=1}^n \alpha_i \mu(A_i)$.

Proof.   The function f is simple, and its range contains at most n elements. Let $\{\beta_1, \ldots, \beta_k\}$ be the range of f, and let $B_i = f^{-1}(\{\beta_i\})$. Then $f = \sum_{i=1}^k \beta_i \mathcal{X}_{B_i}$,

400 Chapter 8 Measure and Integration

and this representation is canonical; i.e., it conforms to the requirements of Equation (2). Putting $J_i = \{j : \alpha_j = \beta_i\}$, we have

$$\int f = \sum_{i=1}^{k} \beta_i \mu(B_i) = \sum_{i=1}^{k} \beta_i \mu\left(\bigcup_{j \in J_i} A_j\right) = \sum_{i=1}^{k} \sum_{j \in J_i} \beta_i \mu(A_j)$$

$$= \sum_{i=1}^{k} \sum_{j \in J_i} \alpha_j \mu(A_j) = \sum_{j=1}^{n} \alpha_j \mu(A_j) \qquad \blacksquare$$

Lemma 2. If g and f are simple functions such that $g \leqslant f$, then $\int g \leqslant \int f$.

Proof. Start with canonical representations, as described following Equation (2):

$$g = \sum_{i=1}^{n} \alpha_i \mathcal{X}_{A_i} \qquad f = \sum_{j=1}^{k} \beta_j \mathcal{X}_{B_j}$$

Then we have (non-canonical) representations conforming to Lemma 1:

$$g = \sum_{i=1}^{n} \sum_{j=1}^{k} \alpha_i \mathcal{X}_{A_i \cap B_j} \qquad f = \sum_{j=1}^{k} \sum_{i=1}^{n} \beta_j \mathcal{X}_{A_i \cap B_j}$$

Since $g \leqslant f$, we have $\alpha_i \leqslant \beta_j$ whenever $A_i \cap B_j \neq \emptyset$. By Lemma 1

$$\int g = \sum_{i=1}^{n} \sum_{j=1}^{k} \alpha_i \mu(A_i \cap B_j) \qquad \int f = \sum_{j=1}^{k} \sum_{i=1}^{n} \beta_j \mu(A_j \cap B_j)$$

Hence

$$\int f - \int g = \sum_{i=1}^{n} \sum_{j=1}^{k} (\beta_j - \alpha_i) \mu(A_i \cap B_j) \geqslant 0 \qquad \blacksquare$$

The next step in the process involves the approximation of nonnegative measurable functions by simple functions, as addressed in Theorem 5 of Section 8.4, page 397. Suppose, then, that $g_1, g_2, \ldots$ are nonnegative simple functions such that $g_n \uparrow f$. Then we want the integral of f to be the limit of $\int g_n$. For technical reasons this is best accomplished by defining

$$(3) \qquad \int f = \sup\left\{ \int g : g \text{ simple and } g \leqslant f \right\}$$

In this equation we continue to assume $f \geqslant 0$.

At this juncture we have two definitions for the integral of a nonnegative simple function, namely, Equations (2) and (3). Let us verify that these definitions are not in conflict with each other.

Lemma 3.   If f is a nonnegative simple function, then its integral as given in Equation (2) equals its integral as given in Equation (3).

Proof. Since f itself is simple, the expression on the right of Equation (3) is at least $\int f$. On the other hand, if g is simple and if $g \leqslant f$, then by Lemma 2, $\int g \leqslant \int f$. By taking a supremum, we see that the right side of Equation (3) is at most $\int f$.   ∎

Lemma 4.   If f and g are nonnegative simple functions, then $\int(f+g) = \int f + \int g$.

Proof. Proceed exactly as in the proof of Lemma 2. Then

$$g + f = \sum_{i=1}^{n} \sum_{j=1}^{k} (\alpha_i + \beta_j) \mathcal{X}_{A_i \cap B_j}$$

By the disjoint nature of the family $\{A_i \cap B_j\}$ we have

$$\int (g + f) = \sum_{i=1}^{n} \sum_{j=1}^{k} (\alpha_i + \beta_j) \mu(A_i \cap B_j)$$

This is the same as $\int g + \int f$, as we see from an equation in the proof of Lemma 2.   ∎

Lemma 5.   For two measurable functions f and g, the condition $0 \leqslant f \leqslant g$ implies $0 \leqslant \int f \leqslant \int g$.

Proof. Since 0 is a simple function, the definition of $\int f$ in Equation (3) gives $\int f \geqslant \int 0 = 0$. If h is a simple function such that $h \leqslant f$, then $h \leqslant g$ and $\int h \leqslant \int g$ by the definition of $\int g$. In this last inequality, take the supremum in h to get $\int f \leqslant \int g$.   ∎

We now arrive at the first of the celebrated convergence theorems for the integral. It is these theorems that distinguish the integral defined here from other integrals, such as the Riemann integral.

Theorem 1.   Monotone Convergence Theorem.   If $[f_n]$ is a sequence of measurable functions such that $0 \leqslant f_n \uparrow f$, then $0 \leqslant \int f_n \uparrow \int f$.

Proof. (Rudin) Since $0 \leqslant f_n \leqslant f_{n+1} \leqslant f$, we have $0 \leqslant \int f_n \leqslant \int f_{n+1} \leqslant \int f$ by Lemma 5. Hence $\lim \int f_n$ exists and is no greater than $\int f$. For the reverse inequality, let $0 < \theta < 1$ and let g be a simple function satisfying $0 \leqslant g \leqslant f$. Put $A_n = \{x : f_n(x) \geqslant \theta g(x)\}$. If $f(x) = 0$, then $g(x) = f_n(x) = 0$, and $x \in A_n$ for all n. If $f(x) > 0$, then eventually $f_n(x) \geqslant \theta g(x)$. Hence $x \in \bigcup_{n=1}^{\infty} A_n$ and $X = \bigcup_{n=1}^{\infty} A_n$. Also, we have $A_n \subset A_{n+1}$ for all n. By Lemma 2 in Section 8.2, page 387, we have, for any measurable set E,

(4) $\mu(A_n \cap E) \uparrow \mu(E)$

From this it is easy to prove that $\int g\mathcal{X}_{An} \uparrow \int g$. Indeed, we write $g = \sum_{i=1}^{m} \lambda_i \mathcal{X}_{E_i}$ (E_i being mutually disjoint) and observe that

$$\int g\mathcal{X}_{An} = \int \sum_{i=1}^{m} \lambda_i \mathcal{X}_{An}\mathcal{X}_{E_i} = \int \sum_{i=1}^{m} \lambda_i \mathcal{X}_{An \cap E_i} = \sum_{i=1}^{m} \lambda_i \mu(A_n \cap E_i)$$

As $n \uparrow \infty$, we have $\mu(A_n \cap E_i) \uparrow \mu(E_i)$ by Equation (4). Since the coefficients λ_i are nonnegative, $\int g\mathcal{X}_n \uparrow \sum_{i=1}^{m} \lambda_i \mu(E_i) = \int g$. We have proved that

$$\theta \int g = \lim_n \int \theta g\mathcal{X}_{An} \leqslant \lim_n \int f_n$$

Since this is true for any θ in $(0,1)$, one concludes that $\int g \leqslant \lim_n \int f_n$. In this inequality take a supremum over all simple g for which $0 \leqslant g \leqslant f$, arriving at $\int f \leqslant \lim_n \int f_n$. ∎

Theorem 2.   *For nonnegative measurable functions f and g we have $\int(f+g) = \int f + \int g$.*

Proof. By Theorem 5 in Section 8.4, page 397, there exist nonnegative simple functions $f_n \uparrow f$ and $g_n \uparrow g$. Then $f_n + g_n \uparrow f + g$. By Theorem 1 (the Monotone Convergence Theorem) and Lemma 4 above, we have

$$\int(f+g) = \lim_n \int (f_n + g_n) = \lim \left[\int f_n + \int g_n \right] = \int f + \int g$$ ∎

Theorem 3.   *Let f be nonnegative and measurable. The conditions $\int f = 0$ and $f(x) = 0$ almost everywhere are equivalent.*

Proof. Let $A = \{x : f(x) > 0\}$ and $B = X \smallsetminus A$. If $f(x) = 0$ almost everywhere, then $\mu(A) = 0$. Hence

$$\int f = \int (f\mathcal{X}_A + f\mathcal{X}_B) = \int f\mathcal{X}_A + \int f\mathcal{X}_B$$
$$\leqslant \int \infty \mathcal{X}_A + \int 0\mathcal{X}_B = \infty\mu(A) + 0\mu(B) = 0$$

For the other implication, assume $\int f = 0$. Define $A_n = \{x : f(x) > \frac{1}{n}\}$. Then $A = \bigcup_{n=1}^{\infty} A_n$. Since $\frac{1}{n}\mathcal{X}_{An}$ is a simple function bounded above by f we have

$$0 = \int f \geqslant \int \frac{1}{n}\mathcal{X}_{An} = \frac{1}{n}\mu(A_n)$$

Thus $\mu(A_n) = 0$ for all n and $\mu(A) = 0$ by Lemma 1 in Section 8.2, page 386. ∎

Theorem 4. Fatou's Lemma.   *For a sequence of nonnegative measurable functions,* $\int (\liminf f_n) \leqslant \liminf \int f_n$.

Proof.   Recall that the limit infimum of a sequence of real numbers $[x_n]$ is defined to be $\lim_{n\to\infty} \inf_{i \geqslant n} x_i$. The limit infimum of a sequence of real-valued functions is defined pointwise: $(\liminf f_n)(x) = \liminf f_n(x) = \lim g_n(x)$, where $g_n(x) = \inf_{i \geqslant n} f_i(x)$. Observe that $g_{n-1}(x) \leqslant g_n(x) \leqslant f_n(x)$ and that $g_n \uparrow \liminf f_n$. Hence by Theorem 1 (The Monotone Convergence Theorem)

$$\int (\liminf f_n) = \int \lim g_n = \lim \int g_n = \liminf \int g_n \leqslant \liminf \int f_n \qquad \blacksquare$$

Theorem 5.   *If f and g are nonnegative measurable functions that are equal almost everywhere, then $\int f = \int g$.*

Proof.   Let $A = \{x : f(x) = g(x)\}$ and $B = X \smallsetminus A$. Then

$$0 \leqslant \int f \mathcal{X}_B \leqslant \int \infty \mathcal{X}_B = \infty \mu(B) = \infty 0 = 0$$

Similarly, $\int g \mathcal{X}_B = 0$. Hence

$$\int f = \int (f \mathcal{X}_X + f \mathcal{X}_B) = \int f \mathcal{X}_A = \int g \mathcal{X}_A = \int (g \mathcal{X}_A + g \mathcal{X}_B) = \int g \qquad \blacksquare$$

Theorem 5 states that $\int f$ is not affected if f is altered on a set of measure 0, while retaining measurability.

Problems 8.5

1. Give an example in which strict inequality occurs in Fatou's Lemma (Theorem 4).

2. Show that a monotone convergence theorem for decreasing sequences is not true. For example, consider f_n as the characteristic function of the interval $[n, \infty)$.

3. Prove that if f is nonnegative and Lebesgue integrable on $\mathbb{R}$ and if $F(x) = \int_{-\infty}^{x} f$, then F is continuous.

4. Define $f_n(x)$ to be n if $|x| \leqslant 1/n$ and to be 0 otherwise. What are $\int \lim f_n$ and $\lim \int f_n$?

5. Prove or disprove: If $(X, \mathcal{A}, \mu)$ is a measure space and if A and B are measurable sets, then $\int |\mathcal{X}_A - \mathcal{X}_B| = \mu(A \triangle B)$. Recall that $A \triangle B = (A \smallsetminus B) \cup (B \smallsetminus A)$.

6. Let f be Lebesgue measurable on $[0, 1]$, and define $\varphi(t) = \mu(f^{-1}((-\infty, t)))$. Find the salient properties of φ. For example, is it continuous from the right or left? Is it monotone? Is it measurable? Is it invertible? What are $\lim_{t\to\infty} \varphi(t)$ and $\lim_{t\to-\infty} \varphi(t)$?

7. (Continuation). Define $f^*(x) = \sup\{t : \varphi(t) \leqslant x\}$. Prove that $\varphi(t) \leqslant x$ if and only if $t \leqslant f^*(x)$. Prove that the sets $\{x : f(x) < t\}$ and $\{x : f^*(x) < t\}$ have equal measure. Hence, f^* is called an **equimeasurable nondecreasing rearrangement** of f. Prove that the sets $\{x : f(x) \geqslant t\}$ and $\{x : f^*(x) \geqslant t\}$ have the same measure. Prove that the sets $\{x : f(x) > t\}$ and $\{x : f^*(x) > t\}$ have the same measure. Prove that $f^*(\varphi(x)) \geqslant x \geqslant \varphi(f^*(x))$.

8. Give an example to show that the nonnegativity hypothesis cannot be dropped from Fatou's Lemma (Theorem 4).

9. Let f_n be nonnegative measurable functions (on any measure space). Prove that if $f_n \to f$ and $f \geqslant f_n$ for all n, then $\int f_n \to \int f$.

10. Prove, for any sequence in $\mathbb{R}$, that $\liminf(-x_n) = -\limsup x_n$.

11. Let $X = [0, 1]$. Is there a Borel measure μ on X that assigns the same positive measure to each open interval $(0, 1/n)$, $n = 1, 2, 3, \ldots$?

12. If f is a bounded function, then there is a sequence of simple functions converging uniformly to f. (The domain of f can be any set, and no measurability assumptions are needed.)

13. Let f_n be measurable functions such that $f_n \geqslant 0$ a.e. ("almost everywhere") and $f_n \uparrow f$ a.e. Prove that $\int f_n \uparrow \int f$.

14. Let f_n be nonnegative and measurable. Prove that $\int \sum_{n=1}^{\infty} f_n = \sum_{n=1}^{\infty} \int f_n$.

15. Let f be nonnegative and measurable. Prove that $\int f = \int_A f$, where $A = \{x : f(x) > 0\}$.

16. Let f_n be measurable functions such that $f_n \geqslant f_{n+1} \geqslant 0$ for all n and $\int f_n \downarrow 0$. Prove that $f_n \downarrow 0$ a.e.

17. Let f be the characteristic function of the set of irrational points in $[0, 1]$. Is f measurable? What are the Riemann and the Lebesgue integrals of f?

18. Prove that if $\{A_n\}$ is a disjoint sequence of measurable sets and if $X = \bigcup_{n=1}^{\infty} A_n$, then $\int f = \sum_{n=1}^{\infty} \int_{A_n} f$.

19. Prove that if f_n are nonnegative measurable functions for which $\sum_{n=1}^{\infty} \int f_n \leqslant 0$, then $f_n \to 0$ a.e.

20. Give an example of a sequence of Riemann integrable functions such that the inequalities $0 \leqslant f_n \leqslant f_{n+1} \leqslant 1$ hold, yet $\lim f_n$ is not Riemann integrable.

21. Give an example of a sequence of simple functions f_n converging pointwise to a simple function f, and yet $\int |f_n - f| \nrightarrow 0$.

22. Find a sequence of simple functions f_n converging uniformly to 0, yet $\int |f_n| \nrightarrow 0$.

8.6 The Integral, Continued

In the preceding section, the integral for nonnegative functions was developed in the general setting of an arbitrary measure space $(X, \mathcal{A}, \mu)$. Next on the agenda is the extension of the integral to "arbitrary" functions.

Definition.   Let $(X, \mathcal{A}, \mu)$ be a measure space, and let $f : X \to \mathbb{R}^*$. We define

$$\int f = \int f^+ - \int f^-$$

where $f^+ = \max(f, 0)$ and $f^- = \max(-f, 0)$. Note that $\int f$ remains undefined if $\int f^+ = \int f^- = \infty$.

The lattice operators max and min are defined for functions in a pointwise manner. Thus, $f^+(x) = \max(f(x), 0)$. Notice that $f = f^+ - f^-$, that $f^+ \geqslant 0$, that $f^- \geqslant 0$, and that $|f| = f^+ + f^-$.

The general definition just given for the integral is in harmony with the previous definition, Equation (3), Section 8.5, page 400, in the cases where both definitions are applicable. Indeed, if $f \geqslant 0$, then $f^+ = f$ and $f^- = 0$.

Definition.   A function $f : X \to \mathbb{R}^+$ is said to be **integrable** if it is measurable and if $\int |f| < \infty$. The set of all integrable functions on the given measure space is denoted by $L^1(X, \mathcal{A}, \mu)$, or simply by L^1 if there can be no ambiguity about the underlying measure space.

Lemma 1.   *A function f is integrable if and only if its positive and negative parts, f^+ and f^-, are integrable.*

Proof.   Assume that f is integrable. Then it is measurable, and the measurability of f^+ follows from the fact that $\{x : f^+(x) \geqslant a\}$ is X when $a \leqslant 0$ and is $\{x : f(x) \geqslant a\}$ when $a > 0$. The finiteness of the integral of $|f^+|$ is immediate from the inequality $|f^+(x)| \leqslant |f(x)|$. The remainder of the proof involves similar elementary ideas.   ∎

Theorem 1.   *The set $L^1(X, \mathcal{A}, \mu)$ is a linear space, and the integral is a linear functional on it.*

Proof.   Let f and g be members of L^1. To show that $f + g \in L^1$, write $h = f + g$, and

$$h^+ - h^- = h = f^+ - f^- + g^+ - g^-$$

From this it follows that

$$h^+ + f^- + g^- = h^- + f^+ + g^+$$

Since these are all nonnegative functions, Theorem 2 of Section 8.5, page 402, is applicable, and

$$\int h^+ + \int f^- + \int g^- = \int h^- + \int f^+ + \int g^+$$

Therefore, by Lemma 1,

$$\int (f + g) = \int h = \int h^+ - \int h^- = \int f^+ - \int f^- + \int g^+ - \int g^- = \int f + \int g$$

With this equation now established, we use Lemma 5 in Section 8.5 (page 401) to write

$$\int |f + g| \leqslant \int (|f| + |g|) = \int |f| + \int |g| < \infty$$

For scalar multiplication, observe first that if $\lambda \geqslant 0$ and $f \geqslant 0$, then the definition of the integral in Equation (3) of Section 8.5 (page 400) gives $\int \lambda f = \lambda \int f$. If $f \geqslant 0$ and $\lambda < 0$, then

$$\int \lambda f = \int (\lambda f)^+ - \int (\lambda f)^- = -\int (\lambda f)^- = -\int (-\lambda f^+) = \lambda \int f^+ = \lambda \int f$$

In the general case, we use what has already been proved:

$$\int \lambda f = \int \left[\lambda f^+ + (-\lambda) f^-\right] = \int \lambda f^+ + \int -\lambda f^- = \lambda \int f^+ - \lambda \int f^-$$
$$= \lambda \left[\int f^+ - \int f^-\right] = \lambda \int f$$

The finiteness of the integral is now trivial:

$$\int |\lambda f| = \int |\lambda|\,|f| = |\lambda| \int |f| < \infty \qquad \blacksquare$$

The second of the celebrated convergence theorems in the theory can now be given.

Theorem 2.   Dominated Convergence Theorem.   *Let $g, f_1, f_2, \ldots$ be functions in $L^1(X, \mathcal{A}, \mu)$ such that $|f_n| \leqslant g$. If the sequence $[f_n]$ converges pointwise to a function f, then $f \in L^1$ and $\int f_n \to \int f$.*

Proof. The functions $f_n + g$ are nonnegative. By Fatou's Lemma (Theorem 4 in Section 8.5, page 403) and by the preceding theorem,

$$\int g + \int f = \int (g + f) = \int \liminf (g + f_n) \leqslant \liminf \int (g + f_n)$$
$$= \liminf \left[\int g + \int f_n\right] = \int g + \liminf \int f_n$$

Since $\int g < \infty$, we conclude that $\int f \leqslant \liminf \int f_n$. Since $-f$ and $-f_n$ satisfy the hypotheses of our theorem, the same conclusion can be drawn for them: $\int -f \leqslant \liminf \int -f_n$. This is equivalent to $-\int f \leqslant -\limsup \int f_n$ and to $\int f \geqslant \limsup \int f_n$. Putting this all together produces

$$\liminf \int f_n \leqslant \limsup \int f_n \leqslant \int f \leqslant \liminf \int f_n \qquad \blacksquare$$

A **step function** is a function on $\mathbb{R}$ that is a simple function $\sum_{i=1}^n c_i \mathcal{X}_{A_i}$ in which the sets A_i are intervals, mutually disjoint.

Theorem 3.   *Let f be Lebesgue integrable on the real line. For any positive ε there exist a simple function g, a step function h and a continuous function k having compact support such that*

$$\int |f - g| < \varepsilon \qquad \int |f - h| < \varepsilon \qquad \int |f - k| < \varepsilon$$

Proof. By Lemma 1, f^+ and f^- are integrable. By the definition of the integral, Equation (3) in Section 8.5, page 400, there exist simple functions g_1

and g_2 such that $g_1 \leqslant f^+$, $g_2 \leqslant f^-$, $\int f^+ < \int g_1 + \varepsilon$, and $\int f^- < \int g_2 + \varepsilon$. Then $g_1 - g_2$ is a simple function such that

$$\int |f - g_1 + g_2| \leqslant \int |f^+ - g_1| + \int |f^- - g_2| = \int (f^+ - g_1) + \int (f^- - g_2) < 2\varepsilon$$

In order to establish the second part of the theorem, it now suffices to prove it in the special case that f is an integrable simple function. It is therefore a linear combination of characteristic functions of measurable sets of finite measure. It then suffices to prove this part of the theorem when $f = \mathfrak{X}_A$ for some measurable set A having finite measure. By the definition of Lebesgue measure, there is a countable family of open intervals $\{I_n\}$ that cover A and satisfy $\mu(A) \leqslant \sum_{n=1}^{\infty} \mu(I_n) < \mu(A) + \varepsilon$. There is no loss of generality in assuming that the family $\{I_n\}$ is disjoint, because if two of these intervals have a point in common, their union is a single open interval. Since the series $\sum \mu(I_n)$ converges, there is an index m such that $\sum_{n=m+1}^{\infty} \mu(I_n) < \varepsilon$. Put $B = \bigcup_{n=1}^{m} I_n$, $E = \bigcup_{n=m+1}^{\infty} I_n$, $h = \mathfrak{X}_B$, and $\varphi = \mathfrak{X}_E$. Then h is a step function. Since $A \subset B \cup E$, we have $f \leqslant h + \varphi$. Then

$$|h - f| \leqslant |h + \varphi - f| + |\varphi| = (h + \varphi - f) - \varphi$$

Consequently,

$$\int |h - f| \leqslant \int (h + \varphi - f) + \int \varphi$$
$$= \mu(B) + \mu(E) - \mu(A) + \mu(E) \leqslant 2\varepsilon$$

For the third part of the proof it suffices to consider an f that is an integrable step function. For this, in turn, it is enough to prove that the characteristic function of a single compact interval can be approximated in L^1 by a continuous function that vanishes outside that interval. This can certainly be done with a piecewise linear function.   ∎

The linear space $L^1(X, \mathcal{A}, \mu)$ becomes a pseudo-normed space upon introducing the definition $\|f\| = \int |f|$. Since a function that is equal to 0 almost everywhere will satisfy $\|f\| = 0$, we will not have a true norm unless we interpret each f in L^1 as an equivalence class consisting of all functions equal to f almost everywhere. This manner of proceeding is eventually the same as introducing the null space of the norm, $N = \{g \in L^1 : \|g\| = 0\}$, and considering the quotient space L^1/N. The elements of this space are cosets $f + N$, and the norm of a coset is defined to be $\|f + N\| = \inf\{\|f + g\| : g \in N\}$. This is the same as $\|f\|$.

A consequence of these considerations is that for f in L^1, the expression $f(x)$ is meaningless. After all, f stands for a class of functions that can differ from each other on sets of measure 0. The single point x is a set of measure zero, and we can change the value of f at x without changing f *as a member of* L^1. The conventional notation $\int f(x)\, dx$ should always be interpreted as $\int f$. Remember that the integral of f is not affected by changing the values of f on any set of measure 0, such as the set of all rational points on the line!

Problems 8.6

1. Define the lattice operations $\vee$ and $\wedge$ by writing

$$(f \vee g)(x) = \max\left(f(x), g(x)\right) \qquad (f \wedge g)(x) = \min\left(f(x), g(x)\right)$$

Prove that the set of measurable functions on a given measurable space is closed under these lattice operations. Prove the same assertion for L^1.

2. Complete the proof of Lemma 1.

3. Let $X = (0, 1)$, let $\mathcal{A}$ be the σ-algebra of Lebesgue measurable subsets of $(0, 1)$, and let μ be Lebesgue measure on $\mathcal{A}$. Which of these functions are in $L^1(X, \mathcal{A}, \mu)$: (a) $f(x) = x^{-1}$, (b) $g(x) = x^{-1/2}$, (c) $h(x) = \exp(-x^{-1})$, (d) $k(x) = \log x$?

4. If $f = g$ almost everywhere, does it follow that $f^+ = g^+$ and $f^- = g^-$ almost everywhere? What can be said of the converse?

5. Prove or disprove: If $\{f_n\}$ is a sequence of measurable functions such that $f_n \uparrow f$, then $\int f_n \uparrow \int f$.

6. Prove that if $f \in L^1(X, \mathcal{A}, \mu)$, then $|f| \in L^1$ and $\left|\int f\right| \leqslant \int |f|$. Verify that $|f| = f^+ + f^-$, that $f = f^+ - f^-$, that $0 \leqslant f^+ \leqslant |f|$, and $0 \leqslant f^- \leqslant |f|$.

7. Show that from the five hypotheses f_n integrable, h integrable, g measurable, $f_n \to f$, $|f_n| \leqslant h$ one cannot draw the conclusion $\int f_n g \to \int fg$. Find an appropriately weak additional hypothesis that makes the inference valid.

8. This problem and the next four involve **convergence in measure**. If $f, f_1, f_2, \ldots$ are measurable functions on a measure space $(X, \mathcal{A}, \mu)$ and if $\lim_n \mu\{x : |f_n(x) - f(x)| > \varepsilon\}$ is 0 for each $\varepsilon > 0$, then we say that $f_n \to f$ **in measure**. Prove that if $f_n \to f$ almost uniformly, then $f_n \to f$ in measure. (Almost uniform convergence is defined in Section 8.4, page 397.)

9. Consider the following sequence of intervals: $A_1 = [0, 1]$, $A_2 = [0, 1/2]$, $A_3 = [1/2, 1]$, $A_4 = [0, 1/4]$, $A_5 = [1/4, 1/2]$, $A_6 = [1/2, 3/4]$, $A_7 = [3/4, 1]$, $A_8 = [0, 1/8], \ldots$ Let f_n denote the characteristic function of A_n. Prove that $f_n \to 0$ in measure but f_n does not converge almost everywhere.

10. Using Lebesgue measure, test the sequence $f_n = \mathcal{X}_{[n-1,n]}$ for pointwise convergence, convergence almost everywhere, convergence almost uniformly, and convergence in measure.

11. Let $(X, \mathcal{A}, \mu)$ be a measure space such that $\mu(X) < \infty$. Let $f, f_1, f_2, \ldots$ be real-valued measurable functions such that $f_n \to f$ almost everywhere. Prove that $f_n \to f$ in measure.

12. Prove that the Monotone Convergence Theorem (page 401), Fatou's Lemma (page 403), and the Dominated Convergence Theorem (page 406) are valid for sequences of functions converging in measure.

13. Prove that if A is a Lebesgue measurable set of finite measure, then for each $\varepsilon > 0$ there is a finite union B of open intervals such that $\mu(A \triangle B) < \varepsilon$.

14. Prove that if f is Lebesgue measurable and finite-valued on a compact interval, then there is a sequence of continuous functions g_n defined on the same interval such that $g_n \to f$ almost uniformly.

15. Prove **Lusin's Theorem**: If f is Lebesgue measurable and finite-valued on $[a, b]$ and if $\varepsilon > 0$, then there is a continuous function g defined on $[a, b]$ that has the property $\mu\{x : f(x) \neq g(x)\} < \varepsilon$.

8.7 The L^p-Spaces

Throughout this section, a fixed measure space $(X, \mathcal{A}, \mu)$ is the setting. For each $p > 0$, the notation $L^p(X, \mathcal{A}, \mu)$, or just L^p, will denote the space of all measurable functions f such that $\int |f|^p < \infty$. The case when $p = 1$ has been considered in the preceding section. We write

$$(1) \qquad \|f\|_p = \left(\int |f|^p \right)^{1/p}$$

although this equation generally does not define a norm (nor even a seminorm if $p < 1$). The case $p = \infty$ will be included in our discussion by making two special definitions. First, $f \in L^\infty$ shall mean that for some M, $|f(x)| \leq M$ almost everywhere. Second, we define

$$(2) \qquad \|f\|_\infty = \inf\{M : |f(x)| \leq M \text{ a.e.}\}$$

The functions in L^∞ are said to be **essentially bounded**, and $\|f\|_\infty$ is called the **essential supremum** of $|f|$, written as $\|f\|_\infty = \text{ess sup} |f(x)|$.

When the equation $\dfrac{1}{p} + \dfrac{1}{q} = 1$ appears, it is understood that q will be ∞ when $p = 1$, and vice versa.

Theorem 1. Hölder's Inequality.   *Let* $1 \leq p \leq \infty$, $\dfrac{1}{p} + \dfrac{1}{q} = 1$,
$f \in L^p$, *and* $g \in L^q$. *Then* $fg \in L^1$ *and*

$$(3) \qquad \int |fg| = \|fg\|_1 \leq \|f\|_p \|g\|_q$$

Proof.   The seminorms involved here are homogeneous: $\|\lambda f\| = |\lambda| \, \|f\|$. Consequently, it will suffice to establish Equation (3) in the special case when $\|f\|_p = \|g\|_q = 1$. At first, let $p = 1$ and $q = \infty$. Since $g \in L^\infty$, we have $|g(x)| \leq M$ a.e. for some M. From this it follows that $\int |fg| \leq M \int |f| = M\|f\|_1$. By taking the infimum for all M, we obtain $\|fg\|_1 \leq \|f\|_1 \|g\|_\infty$.

Suppose now that $p > 1$. We prove first that if $a > 0$, $b > 0$, and $0 \leq t \leq 1$, then $a^t b^{1-t} \leq ta + (1-t)b$. The accompanying Figure 8.1 shows the functions of t on the two sides of this inequality (when $a = 2$ and $b = 12$). It is clear that we should prove convexity of the function $\varphi(t) = a^t b^{1-t}$. This requires that we prove $\varphi''(t) \geq 0$. Since $\log \varphi(t) = t \log a + (1-t) \log b$, we have

$$\frac{\varphi'(t)}{\varphi(t)} = \log a - \log b = c$$

whence $\varphi''(t) = c\varphi'(t) = c^2 \varphi(t) \geq 0$.

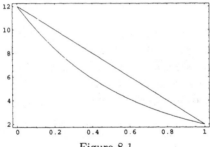

Figure 8.1

Now let $a = |f(x)|^p$, $b = |g(x)|^q$, $t = 1/p$, $1-t = 1/q$. Our inequality yields then $|f(x)g(x)| \leqslant \frac{1}{p}|f(x)|^p + \frac{1}{q}|g(x)|^q$. By hypothesis, the functions on the right in this inequality belong to L^1. Hence by integrating we obtain

$$\|fg\|_1 = \int |fg| \leqslant \frac{1}{p}\int |f|^p + \frac{1}{q}\int |g|^q = \frac{1}{p} + \frac{1}{q} = 1 = \|f\|_p \|g\|_q \qquad \blacksquare$$

Theorem 2. Minkowski's Inequality   Let $1 \leqslant p \leqslant \infty$. If f and g belong to L^p, then so does $f + g$, and

$$\|f + g\|_p \leqslant \|f\|_p + \|g\|_p$$

Proof.   The cases $p = 1$ and $p = \infty$ are special. For the first of these cases, just write

$$\int |f + g| \leqslant \int (|f| + |g|) = \int |f| + \int |g|$$

For $p = \infty$, select constants M and N for which $|f(x)| \leqslant M$ a.e. and $|g(x)| \leqslant N$ a.e. Then $|f(x) + g(x)| \leqslant M + N$ a.e. This proves that $f + g \in L^\infty$ and that $\|f + g\|_\infty \leqslant M + N$. By taking infima we get $\|f + g\|_\infty \leqslant \|f\|_\infty + \|g\|_\infty$.

Now let $1 < p < \infty$. From the observation that $|f + g| \leqslant 2\max\{|f|, |g|\}$, we have

$$|f + g|^p \leqslant 2^p \max\left\{|f|^p, |g|^p\right\} \leqslant 2^p\left(|f|^p + |g|^p\right)$$

This establishes that $f + g \in L^p$. Next, write

$$|f + g|^p = |f + g|\,|f + g|^{p-1} \leqslant |f|\,|f + g|^{p-1} + |g|\,|f + g|^{p-1}$$

Since $|f + g| \in L^p$, we can infer that $|f + g|^{p-1} \in L^q$ (where $\dfrac{1}{p} + \dfrac{1}{q} = 1$) because

$$\int |f + g|^{(p-1)q} = \int |f + g|^p < \infty$$

By the homogeneity of Minkowski's inequality, we may assume that $\|f+g\|_p = 1$. Observe now that Hölder's Inequality is applicable to the product $|f|\,|f + g|^{p-1}$ and to the product $|g|\,|f + g|^{p-1}$. Consequently,

$$1 = \int |f + g|^p \leqslant \int |f|\,|f + g|^{p-1} + \int |g|\,|f + g|^{p-1}$$
$$\leqslant \|f\|_p\|\,|f + g|^{p-1}\|_q + \|g\|_p\|\,|f + g|^{p-1}\|_q$$

This is equivalent to

$$1 \leqslant \left\{\|f\|_p + \|g\|_p\right\}\|f + g\|_p^{p/q} = \|f\|_p + \|g\|_p \qquad \blacksquare$$

Theorem 3.   The Riesz–Fischer Theorem.   *Each space*
$L^p(X, \mathcal{A}, \mu)$, *where* $1 \leqslant p \leqslant \infty$, *is complete.*

Proof.   The case $p = \infty$ is special and is addressed first. Let $[f_n]$ be a Cauchy sequence in L^∞. Define

$$E_{nm} = \left\{ x \; : \; |f_n(x) - f_m(x)| > \|f_n - f_m\|_\infty \right\}$$

By Problem 1, these sets all have measure 0. Hence the same is true of their union, E. If $x \in X \smallsetminus E$, then $|f_n(x) - f_m(x)| \leqslant \|f_n - f_m\|_\infty$, and thus $[f_n(x)]$ is a Cauchy sequence in $\mathbb{R}$ for each $x \in X \smallsetminus E$. This sequence converges to a number that we may denote by $f(x)$. Define $f(x) = 0$ for $x \in E$. On $X \smallsetminus E$, $|f(x)| = \lim |f_n(x)| \leqslant \lim \|f_n\|_\infty < \infty$. (Use the fact that a Cauchy sequence in a metric space is bounded.) Thus, $f \in L^\infty$. To prove that $\|f_n - f\|_\infty \to 0$, let $\varepsilon > 0$ and select N so that $\|f_n - f_m\|_\infty < \varepsilon$ when $n > m > N$. Then $|f_n(x) - f_m(x)| < \varepsilon$ on $X \smallsetminus E$, and $|f(x) - f_m(x)| \leqslant \varepsilon$ for $m > N$.

All the cases when $1 \leqslant p < \infty$ can be done together. Let $[f_n]$ be a Cauchy sequence in L^p. For each $k = 1, 2, 3, \ldots$ there exists a least index n_k such that the following implication is valid:

$$i, j \geqslant n_k \quad \Longrightarrow \quad \|f_i - f_j\|_p < 2^{-k}$$

It follows that $n_1 \leqslant n_2 \leqslant \cdots$ and that $\|f_{n_{k+1}} - f_{n_k}\|_p < 2^{-k}$. Let $g_0 = 0$ and $g_k = f_{n_{k+1}} - f_{n_k}$ for $k \geqslant 1$. Then

$$\sum_{k=0}^{\infty} \|g_k\|_p < \sum_{k=1}^{\infty} 2^{-k} = 1$$

Define $h_n = \sum_{k=0}^{n} |g_k|$ and $h = \lim h_n$. By Minkowski's Inequality (Theorem 2), $\|h_n\|_p \leqslant \sum_{k=0}^{n} \|g_k\|_p < 1$, and thus by Fatou's Lemma (Theorem 4 in Section 8.5, page 403)

$$\int h^p = \int \lim_n h_n^p \leqslant \lim_n \int h^p \leqslant 1$$

This proves that $h \in L^p$. Consequently, the set A on which $h(x) = \infty$ is of measure 0. On $X \smallsetminus A$ the two series $\sum_{k=0}^{\infty} |g_k(x)|$ and $\sum_{k=0}^{\infty} g_k(x)$ converge. Therefore, we can define $f(x) = \sum_{k=0}^{\infty} g_k(x)$ for $x \in X \smallsetminus A$ and let $f(x) = 0$ on A. Since

$$\sum_{k=0}^{i} g_k = f_{n_1} + (f_{n_2} - f_{n_1}) + (f_{n_3} - f_{n_2}) + \cdots + (f_{n_{i+1}} - f_{n_i}) = f_{n_{i+1}}$$

we have $f_{n_i}(x) \to f(x)$ a.e. Since $|f| \leqslant \sum_{k=0}^{\infty} |g_k| = h$, we conclude that $f \in L^p$. It remains to be shown that $\|f - f_n\|_p \to 0$. By the definition of n_k, if $j \geqslant n_k$, then

$$\|f - f_j\|_p = \|\lim_i f_{n_i} - f_j\|_p = \lim_i \|f_{n_i} - f_j\|_p \leqslant 2^{-k} \qquad \blacksquare$$

Problems 8.7

1. Prove that if $f \in L^\infty$, then the set $\{x : |f(x)| > \|f\|_\infty\}$ has measure 0.

2. Let X be any set, and take $\mathcal{A}$ to be 2^X and μ to be counting measure. In this setting, the space $L^p(X, \mathcal{A}, \mu)$ is often denoted by $\ell^p(X)$. Prove that for each $f \in \ell^p(X)$ the support of f is countable. Here, the support of f is defined to be $\{x \in X : f(x) \neq 0\}$.

3. (Continuation) Prove that if X is a set of n points, then $\dim \ell^p(X)$ is n.

4. (Continuation) For $n = 2$, draw the set $\{f \in \ell^p(X) : \|f\|_p = 1\}$ using $p = 1, 2, 10, \infty$.

5. Let $f_n \in L^\infty$ and $f_n \geqslant 0$. Prove that $\sup \|f_n\|_\infty = \|\sup f_n\|_\infty$.

6. In $L^p(X, \mathcal{A}, \mu)$, write $f \equiv g$ if $\|f - g\|_p = 0$. Prove that $f \equiv g$ if and only if $f = g$ a.e. (Thus the equivalence relation is independent of p.) Prove that the equivalence relation is "consistent" with the other structure in L^p by establishing that the conditions $f_1 \equiv f_2$ and $g_1 \equiv g_2$ imply that $f_1 + g_1 \equiv f_2 + g_2$, $\lambda f_1 \equiv \lambda f_2$, and $\|f_1\| = \|f_2\|$.

7. The space $\ell^p(\mathbb{N})$ of Problem 2 is usually written simply as ℓ^p, and if $f \in \ell^p$, we usually write f_n instead of $f(n)$. Show that if $f \in \ell^p$, $g \in \ell^q$, $1/p + 1/q = 1$, then $fg \in \ell^1$ and $\sum_{n=1}^\infty |f_n g_n| \leqslant (\sum_{n=1}^\infty |f_n|^p)^{1/p} (\sum_{n=1}^\infty |g_n|^q)^{1/q}$.

8. Let $(E, \| \ \|)$ be a pseudo-normed linear space. Let $M = \{f \in E : \|f\| = 0\}$. Prove that M is a linear subspace of E. In the quotient space E/M the elements are cosets $f + M$. Define $\|f + M\| = \inf\{\|f + g\| : g \in M\}$. Show that this defines a norm in E/M.

9. Let f and f_n belong to $L^\infty(X, \mathcal{A}, \mu)$. Show that $\|f - f_n\|_\infty \to 0$ if and only if $f_n \to f$ almost uniformly. (See the definition in Section 8.4, page 397.)

10. Let $(X, \mathcal{A}, \mu)$ be a measure space for which $\mu(X) < \infty$. Show that if $0 < \alpha < \beta \leqslant \infty$, then $L^\beta \subset L^\alpha$. Show that the hypothesis $\mu(X) < \infty$ cannot be omitted.

11. Prove that if $0 < \alpha < \beta \leqslant \infty$, then $\ell^\alpha \subset \ell^\beta$. (See the definition in Problem 7.)

12. Show that in the proof of the Riesz–Fischer Theorem (Theorem 3), the sequence $[f_n]$ need not converge to f almost everywhere. Consider, for example, the characteristic functions of the intervals $[0, 1]$, $[0, \frac{1}{2}]$, $[\frac{1}{2}, 1]$, $[0, \frac{1}{3}]$, $[\frac{1}{3}, \frac{2}{3}]$, $[\frac{2}{3}, 1]$, ... Show that $\|f_n\|_p \to 0$ but $f_n(x)$ is divergent for each x in $[0, 1]$.

13. Let $(X, \mathcal{A}, \mu)$ be a measure space for which $\mu(X) < \infty$. Prove that for each $f \in L^\infty$, $\lim_{p \to \infty} \|f\|_p = \|f\|_\infty$.

14. Prove that for any measure space, if $0 < \alpha < \beta \leqslant \infty$, then $L^\infty \cap L^\alpha \subset L^\infty \cap L^\beta$.

15. Prove for any measure space: If $0 < \alpha < \beta < \gamma < \infty$, then $L^\alpha \cap L^\gamma \subset L^\beta \cap L^\gamma$.

16. Let $f(x) = [x \log^2(1/x)]^{-1}$ and prove that f is in $L^1[0, \frac{1}{2}]$ but is not in $\bigcup_{p>1} L^p[0, \frac{1}{2}]$.

17. Prove that if $[f_n]$ is a Cauchy sequence in L^p, then it has a subsequence that converges almost everywhere.

18. Let f and f_n belong to L^p. If $\|f_n - f\|_p \to 0$ and $f_n \to g$ a.e., what relationship exists between f and g?

19. Let $(X, \mathcal{A}, \mu)$ be a measure space for which $\mu(X) = 1$. (Such a space is a **probability space**.) Prove that if f and g are positive, measurable, and satisfy $fg \geqslant 1$, then the inequality $\int f \cdot \int g \geqslant 1$ holds.

20. Prove that if $f_n \in L^1(X, \mathcal{A}, \mu)$ and $\sum_{n=1}^\infty \|f_n\|_1 < \infty$, then $f_n \to 0$ a.e.

21. If $0 < \int |f| < \infty$, then there is a continuous function g having compact support such that $\int fg \neq 0$.

22. Prove that if $f \in L^p(X, \mathcal{A}, \mu)$ for all sufficiently large values of p, and if the limit of $\|f\|_p$

exists when $p \to \infty$, then the value of the limit is $\|f\|_\infty$.

23. Show that in general, $L^\infty(X, \mathcal{A}, \mu) \neq \bigcap_{p>1} L^p(X, \mathcal{A}, \mu)$. Are there cases when equality occurs?

24. If $\mathcal{A} = 2^X$ and μ is counting measure on $\mathcal{A}$, what is $L^\infty(X, \mathcal{A}, \mu)$?

25. Prove that for $f \in L^1(X, \mathcal{A}, \mu)$ we have $|\int f| \leqslant \int |f|$. When does equality occur here?

26. Let $1 < p < \infty$, $1/p + 1/q = 1$, and $f \in L^p$. Prove that $|f|^p \in L^1$, that $|f|^{p-1} \in L^q$, and that for $r \neq 0$, $|f|^r \in L^{p/r}$.

8.8 The Radon–Nikodym Theorem

In elementary calculus, the expression $\int_a^x f(t)\, dt$ is called an "indefinite integral." It is a function of the two arguments f and x, or of f and the set $[a, x]$. Therefore, in general integration theory the analogous concept is an integral $\int_A f$ depending on the two arguments f and A. Recall that our notation is as follows:

$$\int_A f = \int f \mathcal{X}_A$$

where $\mathcal{X}_A$ is the characteristic function of the set A. The set A and the function f should be measurable with respect to the underlying measure space $(X, \mathcal{A}, \mu)$.

Now suppose that a second measure ν is defined on the σ-algebra $\mathcal{A}$. If $\nu(A) = 0$ whenever $\mu(A) = 0$, we say that ν is **absolutely continuous** with respect to μ, and we write $\nu \ll \mu$.

One easy way to produce such a measure ν is given in the next theorem.

Theorem 1.   If $(X, \mathcal{A}, \mu)$ is a measure space, and if f is a nonnegative measurable function, then the equation

$$(1) \qquad\qquad \nu(A) = \int_A f\, d\mu \qquad (A \in \mathcal{A})$$

defines a measure ν that is absolutely continuous with respect to μ.

Proof. The postulates for a measure are quickly verified.

(a) $\nu(\varnothing) = \int_\varnothing f = \int f \mathcal{X}_\varnothing = \int 0 = 0$

(b) $\nu(A) \geqslant 0$ because $f \geqslant 0$

(c) If $[A_i]$ is a disjoint sequence of measurable sets, then

$$\nu\left(\bigcup_{i=1}^\infty A_i\right) = \int_{\cup A_i} f = \int f \mathcal{X}_{\cup A_i} = \int f \sum \mathcal{X}_{A_i} = \int \sum f \mathcal{X}_{A_i}$$

$$= \int \lim_n \sum_{i=1}^n f \mathcal{X}_{A_i} = \lim_n \int \sum_{i=1}^n f \mathcal{X}_{A_i} = \lim_n \sum_{i=1}^n \int f \mathcal{X}_{A_i}$$

$$= \sum_{i=1}^\infty \nu(A_i)$$

This calculation used the Monotone Convergence Theorem (Section 8.5, page 401). The absolute continuity of ν is clear: if $\mu(A) = 0$, then $\nu(A) = \int_A f = 0$. ∎

It is natural to seek a converse for this theorem. Thus we ask whether each measure that is absolutely continuous with respect to μ must be of the form in Equation (1). The answer is a qualified "Yes." It is necessary to make a slight restriction. Consider a general measure space $(X, \mathcal{A}, \mu)$. We say that X (or μ) is σ-**finite** if X can be written as a countable union of measurable sets, each having finite measure. For example, the real line with Lebesgue measure is σ-finite, since we can write $\mathbb{R} = \bigcup_{n=1}^{\infty}[-n, n]$.

> **Theorem 2. Radon–Nikodym Theorem.**   *Let μ and ν be σ-finite measures on a measurable space $(X, \mathcal{A})$. If ν is absolutely continuous with respect to μ, then there exists a nonnegative measurable function h, determined uniquely up to a set of μ-measure 0, such that $\nu(A) = \int_A h\,d\mu$ for all $A \in \mathcal{A}$.*

Proof.   We prove the theorem first under the assumption that $\mu(X) < \infty$ and $\nu(X) < \infty$. Consider the Hilbert space $L^2 = L^2(X, \mathcal{A}, \mu + \nu)$. For any f in L^2, define $\Phi(f) = \int f\,d\mu$. It is easily verified that Φ is a linear functional on L^2. Furthermore it is bounded (continuous) because by the Hölder Inequality (Theorem 1 in Section 8.7, page 409)

$$|\Phi(f)| = \left| \int f \cdot 1\,d\mu \right| \leqslant \int |f| \cdot 1\,d(\mu + \nu) \leqslant \|f\|_2 \, \|1\|_2$$

By the Riesz Representation Theorem for Hilbert space (Section 2.3, page 81) there exists an element h_0 in L^2 such that

$$\Phi(f) = \int f h_0\,d(\mu + \nu) \qquad f \in L^2$$

This means that $\int f\,d\mu = \int f h_0\,d(\mu + \nu)$, whence

$$\int f(1 - h_0)\,d\mu = \int f h_0\,d\nu$$

Let $B = \{x : h_0(x) \leqslant 0\}$. Then $1 - h_0 \geqslant 1$ on B, and consequently

$$0 \leqslant \mu(B) \leqslant \int \mathcal{X}_B (1 - h_0)\,d\mu = \int \mathcal{X}_B h_0\,d\nu \leqslant 0$$

Thus $\mu(B) = 0$ and $h_0(x) > 0$ a.e. (with respect to μ). Since $\nu \ll \mu$, we have $\nu(B) = 0$ also. Hence for any $A \in \mathcal{A}$,

$$\nu(A) = \int \mathcal{X}_A\,d\nu = \int h_0^{-1} \mathcal{X}_A h_0\,d\nu = \int h_0^{-1} \mathcal{X}_A (1 - h_0)\,d\mu$$
$$= \int_A h_0^{-1}(1 - h_0)\,d\mu = \int_A h\,d\mu \qquad (h = h_0^{-1}(1 - h_0))$$

To see that $h \geqslant 0$ a.e., with respect to μ, write $A = \{x : h(x) < 0\}$, so that $0 \leqslant \nu(A) = \int_A h \, d\mu \leqslant 0$, whence $\mu(A) = 0$.

For the second half of the proof we assume only that μ and ν are σ-finite. Then $X = \bigcup_{n=1}^{\infty} A_n = \bigcup_{n=1}^{\infty} B_n$, where A_n and B_n are measurable sets such that $\mu(A_n) < \infty$ and $\nu(B_n) < \infty$ for each n. Write the doubly-indexed family $A_i \cap B_j$ as a sequence C_n. Then $X = \bigcup C_n$, $\mu(C_n) < \infty$, and $\nu(C_n) < \infty$. With no loss of generality we assume that the sequence $[C_n]$ is disjoint. Define measures ν_n and μ_n by putting $\nu_n(A) = \nu(A \cap C_n)$ and $\mu_n(A) = \mu(A \cap C_n)$. Since $\nu \ll \mu$, we have $\nu_n \ll \mu_n$ for all n. By the first half of the proof there exist functions h_n such that $\nu_n(A) = \int_A h_n \, d\mu_n$, for all $A \in \mathcal{A}$. Since the C_n-sequence is disjoint, we can define h on X by specifying that $h(x) = h_n(x)$ for $x \in C_n$. Then we have

$$\int_A h \, d\mu = \int_A \sum h \mathcal{X}_{C_n} \, d\mu = \sum \int_A h_n \, d\mu_n = \sum \nu_n(A) = \sum \nu(A \cap C_n) = \nu(A)$$

For the uniqueness of h, suppose that $\int_A h \, d\mu = \int_A h' \, d\mu$ for all $A \in \mathcal{A}$. Letting $A = \{x : h(x) > h'(x)\}$, we have

$$\int_A (h - h') \, d\mu = 0 \qquad h > h' \text{ on } A$$

It follows that $\mu(A) = 0$. By symmetry, the set where $h'(x) > h(x)$ is also of measure 0. Hence $h = h'$ a.e. (μ).   ∎

The preceding paragraphs have involved the concept of absolute continuity of one measure with respect to another. The antithesis of this is "mutual singularity." Two measures μ and ν on the same measure space are said to be **mutually singular** if there is a measurable set B such that $\mu(B) = \nu(X \smallsetminus B) = 0$. This relation is written symbolically as $\mu \perp \nu$. It is obviously a symmetric relation.

Theorem 3.   Lebesgue Decomposition Theorem.   If μ and ν are σ-finite measures on the measurable space $(X, \mathcal{A})$, then there exist unique measures ν_1 and ν_2 on $\mathcal{A}$ such that $\nu = \nu_1 + \nu_2$, $\nu_1 \ll \mu$, and $\nu_2 \perp \mu$.

Proof. By the Radon–Nikodym Theorem (Theorem 2, above)—or indeed by the first half of its proof—there exists a measurable function h such that

$$\mu(A) = \int_A h \, d(\mu + \nu) \qquad (A \in \mathcal{A})$$

Define the set $B = \{x : h(x) = 0\}$. Next, define

$$\nu_1(A) = \nu(A \smallsetminus B) \qquad \nu_2(A) = \nu(A \cap B) \qquad (A \in \mathcal{A})$$

Obviously, $\nu_1 + \nu_2 = \nu$. By Problem 8.2.15, page 391, ν_1 and ν_2 are measures. Let us prove that $\nu_2 \perp \mu$. Since $h = 0$ on B, we have $\mu(B) = \int_B h \, d(\mu + \nu) = 0$. On the other hand, $\nu_2(X \smallsetminus B) = \nu((X \smallsetminus B) \cap B) = \nu(\varnothing) = 0$. Next, we prove that $\nu_1 \ll \mu$. Suppose that $\mu(A) = 0$. Then $\int_A h\, d(\mu + \nu) = 0$, $\int_A h\, d\nu = 0$, and $\int_{A \smallsetminus B} h\, d\nu = 0$. But $h > 0$ on $A \smallsetminus B$, and hence $\nu(A \smallsetminus B) = 0$, $\nu_1(A) = 0$. Finally, we prove the uniqueness of our decomposition. Suppose that another decomposition is given: $\nu = \nu_3 + \nu_4$, where $\nu_3 \ll \mu$ and $\nu_4 \perp \mu$. Then there exists a set C such that $\mu(C) = \nu_4(X \smallsetminus C) = 0$. (This set C is akin to B in the first part of our proof.) If $D = B \cup C$, then $0 = \mu(B) + \mu(C) \geqslant \mu(D)$ and $\mu(D) = 0$. It follows, for any measurable set A, that $\nu_1(A \cap D) \leqslant \nu_1(D) = 0$. Hence

$$\nu(A \cap D) = (\nu_1 + \nu_2)(A \cap D) = \nu_2(A \cap D)$$
$$= \nu_2(A \cap D) + \nu_2(A \smallsetminus D) = \nu_2(A)$$

The same argument will prove that $\nu(A \cap D) = \nu_4(A)$. Hence $\nu_2 = \nu_4$. Since $\nu = \nu_1 + \nu_2 = \nu_3 + \nu_4$, one is tempted to conclude outright that $\nu_1 = \nu_3$. However, if A is a set for which $\nu_2(A) = \nu_4(A) = \infty$, we cannot perform the necessary subtraction. Using the σ-finite property of the space, we find a disjoint sequence of measurable sets X_n such that $X = \bigcup X_n$ and $\nu(X_n) < \infty$. Then $\nu_1(X_n \cap A) = \nu_3(X_n \cap A)$ for all n and for all A. It follows that $\nu_1(A) = \nu_3(A)$ and that $\nu_1 = \nu_3$. ∎

Problems 8.8

1. Is the relation of absolute continuity (for measures) reflexive? What about symmetry and transitivity? Is it a partial order? A linear order? A well-ordering? Give examples to support each conclusion.

2. Solve Problem 1 for the relation of mutual singularity.

3. The function h in the Radon–Nikodym Theorem is often denoted by $\dfrac{d\nu}{d\mu}$. Prove that
$$\frac{d\nu}{d\theta} = \frac{d\nu}{d\mu}\frac{d\mu}{d\theta} \text{ if } \nu \ll \mu \ll \theta.$$

4. Refer to Problem 3 and prove that $\dfrac{d(\nu + \theta)}{d\mu} = \dfrac{d\nu}{d\mu} + \dfrac{d\theta}{d\mu}$ if $\nu \ll \mu$ and $\theta \ll \mu$.

5. Refer to Problem 3 and prove that $\dfrac{d\mu}{d\nu}\dfrac{d\nu}{d\mu} = 1$ if $\nu \ll \mu \ll \nu$.

6. Refer to Problem 3 and prove that $\int f\, d\nu = \int f\dfrac{d\nu}{d\mu}\, d\mu$ if $\nu \ll \mu$.

7. Let $X = [0,1]$, let $\mathcal{A}$ be the family of all Lebesgue measurable subsets of X, let ν be Lebesgue measure, and let μ be counting measure. Show that $\nu \ll \mu$. Show that there exists no function h for which $\nu(A) = \int_A h\, d\mu$. Explain the apparent conflict with the Radon–Nikodym Theorem.

8. Prove that in the Radon–Nikodym Theorem, $h(x) < \infty$ for all x. Show also that if $\nu(X) < \infty$, then $h \in L^1(X, \mathcal{A}, \mu)$.

9. *Extension of Radon–Nikodym Theorem.* Let μ and ν be measures on a measurable space $(X, \mathcal{A})$. Suppose that there exists a disjoint family $\{B_\alpha\}$ of measurable sets having these properties:

    (i) $\mu(B_\alpha) < \infty$ for all α.

    (ii) $\mu(A) = 0$ if $A \in \mathcal{A}$ and $\mu(A \cap B_\alpha) = 0$ for all α.

    (iii) $\nu(A) = 0$ if $A \in \mathcal{A}$ and $\nu(A \cap B_\alpha) = 0$ for all α.

If $\nu \ll \mu$, then there is an h as in the Radon–Nikodym Theorem, but it may be measurable only with respect to the σ-algebra

$$\mathcal{B} = \{B : B \subset X, B \cap A \in \mathcal{A} \text{ when } A \in \mathcal{A} \text{ and } \mu(A) < \infty\}$$

10. If there corresponds to each positive ε a positive δ such that

$$[A \in \mathcal{A} \text{ and } \mu(A) < \delta] \implies \nu(A) < \varepsilon$$

then $\nu \ll \mu$, and conversely.

11. Let $(X, \mathcal{A}, \mu)$ be a measure space. Fix $B \in \mathcal{A}$ and define $\nu(A) = \mu(B \cap A)$ for all $A \in \mathcal{A}$. Is ν absolutely continuous with respect to μ? Is $\mu - \nu$ a measure? Is the equation $\int_A f \, d\nu = \int_{A \cap B} f \, d\mu$ true for measurable f and A? Is $\mu - \nu$ singular with respect to μ? Give examples.

12. If $\nu \ll \mu$ and $\lambda \perp \mu$ then $\nu \perp \lambda$.

13. If $\mu \perp \nu \ll \mu$, then $\nu = 0$.

14. Give an example of the Radon–Nikodym Theorem in which the function h fails to be bounded a.e.

15. Let $(X, \mathcal{A}, \mu)$ be a σ-finite measure space. Let ν be a measure on $\mathcal{A}$. The existence of a constant c for which $\nu \leqslant c\mu$ is equivalent to the existence of a bounded measurable nonnegative function h such that $\nu(A) = \int_A f \, d\mu$.

8.9 Signed Measures

In this section we examine the consequences of relaxing the nonnegativity requirement on a measure. Let $(X, \mathcal{A})$ be a measurable space. A function μ from $\mathcal{A}$ to $\mathbb{R}^*$ is called a **signed measure** if

   (i) The range of μ does not include both $+\infty$ and $-\infty$.

   (ii) $\mu(\varnothing) = 0$

   (iii) $\mu(\bigcup_{i=1}^{\infty} A_i) = \sum_{i=1}^{\infty} \mu(A_i)$ when $\{A_i\}$ is a disjoint sequence in $\mathcal{A}$.

The reason for the first requirement is that we want to avoid the meaningless expression $\infty - \infty$. Thus, if $\mu(A) = \infty$ and $\mu(B) = -\infty$, then from the equation $A = (A \cap B) \cup (A \smallsetminus B)$ we see that one of the terms $\mu(A \cap B)$ and $\mu(A \smallsetminus B)$ must be $+\infty$. Likewise, one of $\mu(A \cap B)$ and $\mu(B \smallsetminus A)$ must be $-\infty$. Hence the right side of the equation

$$\mu(A \cup B) = \mu(A \cap B) + \mu(A \smallsetminus B) + \mu(B \smallsetminus A)$$

is meaningless.

Theorem 1.   Jordan Decomposition.   *The difference of two measures (defined on the same σ-algebra), one of which is finite, is a signed measure. Conversely, every signed measure μ is the difference of two measures μ^+ and μ^-, one of which is finite. Furthermore, we* *may require these two measures to be mutually singular, and in that case they are uniquely determined by μ.*

Proof.   For the first assertion, let μ_1 and μ_2 be measures, and suppose that μ_1 is finite. Put $\mu = \mu_1 - \mu_2$. To see that μ is a signed measure, note first that μ does not assume the value $+\infty$. Next, we have $\mu(\varnothing) = 0$ since μ_1 and μ_2 have this property. Finally, let $\{A_i\}$ be a disjoint sequence of measurable sets. Then

$$\mu\left(\bigcup_{i=1}^{\infty} A_i\right) = \mu_1\left(\bigcup_{i=1}^{\infty} A_i\right) - \mu_2\left(\bigcup_{i=1}^{\infty} A_i\right)$$

$$= \sum_{i=1}^{\infty} \mu_1(A_i) - \sum_{i=1}^{\infty} \mu_2(A_i)$$

$$= \lim_n \sum_{i=1}^{n} \mu_1(A_i) - \lim_n \sum_{i=1}^{n} \mu_2(A_i)$$

$$= \lim_n \left[\sum_{i=1}^{n} \mu_1(A_i) - \sum_{i=1}^{n} \mu_2(A_i)\right]$$

$$= \lim_n \sum_{i=1}^{n} \left[\mu_1(A_i) - \mu_2(A_i)\right] = \sum_{i=1}^{\infty} \mu(A_i)$$

Notice that on the second line of this calculation the first sum is finite, although the second may be infinite.

For the other half of the proof, let μ be a signed measure that does not assume the value $+\infty$. In an abuse of language, we say that a (measurable) set S is *positive* if $\mu(A) \geqslant 0$ for all measurable subsets A in S. Define

$$\theta = \sup\{\mu(S) : S \text{ is a positive set}\}$$

Let S_n be a sequence of positive sets such that $\mu(S_n) \uparrow \theta$, and define $P = \bigcup_{n=1}^{\infty} S_n$. Let us prove that P is a positive set. If $A \subset P$, we write

$$A_n = (A \cap S_n) \smallsetminus (S_1 \cup \cdots \cup S_{n-1})$$

Since $A_n \subset S_n$, we have $\mu(A_n) \geqslant 0$. Since A is the union of the disjoint family $\{A_n\}$, it follows that

$$\mu(A) = \sum_{n=1}^{\infty} \mu(A_n) \geqslant 0$$

Since P is a positive set,

$$\mu(P) = \mu(P \smallsetminus S_n) + \mu(S_n) \geqslant \mu(S_n)$$

whence $\mu(P) \geqslant \theta$ and $\theta < \infty$.

Now we wish to prove that $\mu(A) \leqslant 0$ whenever $A \subset X \smallsetminus P$. Suppose, on the contrary, that $A \subset X \smallsetminus P$ and $\mu(A) > 0$. If A contains a positive set B of positive measure, then $P \cup B$ is a positive set for which $\mu(P \cup B) = \mu(P) + \mu(B) > \theta$, in contradiction to the definition of θ. Thus, A contains no positive set of positive measure. Define sets $A_1, A_2, \ldots$ as follows. Let n_1 be the first positive integer such that there exists a set A_1 satisfying

$$A_1 \subset A \qquad \mu(A_1) < -\frac{1}{n_1}$$

Since $0 < \mu(A) = \mu(A \smallsetminus A_1) + \mu(A_1)$, we see that $A \smallsetminus A_1$ is a subset of A having positive measure. It is therefore not a positive set. Hence there is a first positive integer n_2 and a set A_2 such that

$$A_2 \subset A \smallsetminus A_1 \qquad \mu(A_2) < -\frac{1}{n_2}$$

Continue in this manner, finding at the kth step a set A_k and an integer n_k such that

$$A_k \subset A \smallsetminus (A_1 \cup \cdots \cup A_{k-1}) \qquad \mu(A_k) < -\frac{1}{n_k}$$

Define $B = A \smallsetminus \bigcup_{k=1}^{\infty} A_k$. By the same argument used earlier, B has positive measure. It is actually a positive set. To verify this, suppose on the contrary that there exists a set $C \subset B$ such that $\mu(C) < 0$. Let m be the first positive integer such that $\mu(C) < -1/m$. Since $C \subset A \smallsetminus (A_1 \cup \cdots \cup A_{k-1})$ for every k, we have $n_k \leqslant m$ for all k. Hence $\mu(\bigcup_{k=1}^{\infty} A_k) = -\infty$ and $\mu(B) = +\infty$, a contradiction.

Now define μ^+ and μ^- by writing, for $A \in \mathcal{A}$,

$$\mu^+(A) = \mu(A \cap P) \qquad \mu^-(A) = -\mu(A \smallsetminus P)$$

We see that $\mu^+ \perp \mu^-$ because $\mu^+(X \smallsetminus P) = 0 = \mu^-(P)$.

Our last task is to prove the uniqueness of this decomposition. Suppose that $\mu = \mu_1 - \mu_2 = \nu_1 - \nu_2$, where these are measures such that $\nu_1 \perp \nu_2$ and $\mu_1 \perp \mu_2$. Then there exists a set Q such that $\nu_1(X \smallsetminus Q) = 0 = \nu_2(Q)$. We can prove that $\nu_1 \leqslant \mu_1$ by writing

$$\nu_1(A) = \nu_1(A \cap Q) + \nu_1(A \smallsetminus Q) = \nu_1(A \cap Q) = (\mu + \nu_2)(A \cap Q)$$
$$= \mu(A \cap Q) = (\mu_1 - \mu_2)(A \cap Q) \leqslant \mu_1(A \cap Q) \leqslant \mu_1(A)$$

By the symmetry in this situation, we can prove $\mu_1 \leqslant \nu_1$. Hence $\mu_1 = \nu_1$ and $\mu_2 = \nu_2$. ∎

Theorem 2. Radon–Nikodym Theorem for Signed Measures.
Let $(X, \mathcal{A}, \mu)$ be a σ-finite measure space. If ν is a finite-valued signed measure that is absolutely continuous with respect to μ, then there is a measurable function h such that for all $A \in \mathcal{A}$, $\nu(A) = \int_A h \, d\mu$.

Proof. By the preceding theorem, there exist measures ν^+ and ν^- such that $\nu = \nu^+ - \nu^-$ and $\nu^+ \perp \nu^-$. Consequently, there exists a measurable set P for which $\nu^+(X \smallsetminus P) = 0 = \nu^-(P)$. If A is a measurable set satisfying $\mu(A) = 0$, then $\mu(A \cap P) = 0$ and $\nu(A \cap P) = 0$, by the absolute continuity. Hence

$$\nu^+(A) = \nu^+(A \cap P) + \nu^+(A \smallsetminus P) = \nu^+(A \cap P)$$
$$= (\nu + \nu^-)(A \cap P) = \nu(A \cap P) = 0$$

This establishes that ν^+ is absolutely continuous with respect to μ. It follows that ν^- is also absolutely continuous with respect to μ. By the earlier Radon–Nikodym Theorem (Theorem 2 in Section 8.8, page 414), there exist nonnegative measurable functions h_1 and h_2 such that for A in $\mathcal{A}$,

$$\nu^+(A) = \int_A h_1 \, d\mu \qquad \nu^-(A) = \int_A h_2 \, d\mu$$

It follows that h_1 and h_2 are finite almost everywhere. Thus, there is nothing suspicious in the equation

$$\nu(A) = \nu^+(A) - \nu^-(A) = \int_A h_1 \, d\mu - \int_A h_2 \, d\mu = \int_A (h_1 - h_2) \, d\mu \qquad ∎$$

Theorem 3. The Hahn Decomposition.   *If μ is a signed measure on the measurable space $(X, \mathcal{A})$, then there is a decomposition of X into a disjoint pair of measurable sets N and P such that $\mu(A) \geqslant 0$ when $A \subset P$ and $\mu(A) \leqslant 0$ when $A \subset N$.*

Proof.   Left as a problem.   ∎

Problems 8.9

1. Use the Jordan decomposition theorem to prove the Hahn decomposition theorem.

2. Prove that μ^+ in Theorem 1 has the property

$$\mu^+(A) = \sup\{\mu(S) : S \in \mathcal{A} \text{ and } S \subset A\}$$

3. Prove that a signed measure μ is monotone on a positive set. Thus, if $A \subset B \subset S$, where S is a positive set, then $\mu(A) \leqslant \mu(B)$.

4. If μ is a signed measure, does it follow that $-\mu$ is also a signed measure? Are sums and differences of signed measures signed measures?

5. Let $(X, \mathcal{A}, \mu)$ be a measure space, and let h be a nonnegative measurable function. Define $\nu(A) = \int_A h \, d\mu$. Prove that $\int f \, d\nu = \int f h \, d\mu$ for all measurable f.

6. Let μ and ν be measures on the measurable space $(X, \mathcal{A})$. Suppose that 0 is the only measurable function such that $\nu(A) \geqslant \int_A f \, d\mu$, for all $A \in \mathcal{A}$. Prove that $\nu \perp \mu$.

7. Let $(X, \mathcal{A}, \mu)$ be a measure space such that each singleton $\{x\}$ is measurable. Define $\nu(A)$ to be the sum of all $\mu(\{x\})$ as x ranges over A. Does this define a measure on $\mathcal{A}$?

8. Is the function h in Theorem 2 unique?

8.10 Product Measures and Fubini's Theorem

Suppose that two measure spaces are given: $(X, \mathcal{A}, \mu)$ and $(Y, \mathcal{B}, \nu)$. Is there a suitable way of making the Cartesian product $X \times Y$ into a measure space? In particular, can this be done in such a way that

$$\int_{X \times Y} f(x, y) = \int_X \int_Y f(x, y) \, d\nu(y) \, d\mu(x) \ ?$$

We begin by forming the class of all sets of the form $A \times B$, where $A \in \mathcal{A}$ and $B \in \mathcal{B}$. Such sets are called **measurable rectangles** or simply **rectangles**. The family of all rectangles is not a σ-algebra. For example, $(A_1 \times B_1) \cup (A_2 \times B_2)$ will not be a rectangle, in general. To understand this, observe that if points (x_1, y_1) and (x_2, y_2) belong to a rectangle, then (x_1, y_2) and (x_2, y_1) also belong to that rectangle.

The next step, therefore, is to construct the σ-algebra $\mathcal{A} \otimes \mathcal{B}$ generated by the rectangles. (Refer to Lemma 2 in Section 8.1, page 384). For any subset E of the Cartesian product $X \times Y$, we define cross-sections

$$E_x = \{y \in Y : (x, y) \in E\}$$
$$E^y = \{x \in X : (x, y) \in E\}$$

Lemma 1.   Let $(X, \mathcal{A})$ and $(Y, \mathcal{B})$ be two measurable spaces. If $E \in \mathcal{A} \otimes \mathcal{B}$, then $E_x \in \mathcal{B}$ for all $x \in X$ and $E^y \in \mathcal{A}$ for all $y \in Y$.

Proof.   Define

$$\mathcal{M} = \{E : E \subset X \times Y \text{ and } E^y \in \mathcal{A} \text{ for all } y \in Y\}$$

We shall prove that $\mathcal{M}$ is a σ-algebra containing all rectangles. From this it will follow that $\mathcal{M} \supset \mathcal{A} \otimes \mathcal{B}$, since the latter is the smallest σ-algebra containing all rectangles. Then, if $E \in \mathcal{A} \otimes \mathcal{B}$, we can conclude that $E \in \mathcal{M}$ and that $E^y \in \mathcal{A}$ for each y. Now consider any rectangle $E = A \times B$. If $y \in B$, then $E^y = A \in \mathcal{A}$. If $y \notin B$, then $E^y = \varnothing \in \mathcal{A}$. Thus in all cases $E^y \in \mathcal{A}$ and $E \in \mathcal{M}$. Next, let E be any member of $\mathcal{M}$. The equation

$$(1) \qquad [(X \times Y) \smallsetminus E]^y = X \smallsetminus E^y$$

shows that $(X \times Y) \smallsetminus E$ belongs to $\mathcal{M}$. If $E_i \in \mathcal{M}$, then by the equation

$$(2) \qquad \left[\bigcup_{i=1}^{\infty} E_i \right]^y = \bigcup_{i=1}^{\infty} E_i^y$$

we see that $\bigcup E_i \in \mathcal{M}$. ∎

An **algebra** of subsets of a set X is a collection $\mathcal{C}$ such that if A and B belong to $\mathcal{C}$ then $X \smallsetminus A$ and $A \cup B$ belong also to $\mathcal{C}$.

Lemma 2.   The collection of all unions of finite disjoint families of rectangles constructed from a pair of σ-algebras is an algebra.

Proof.   Let $\mathcal{C}$ be the collection referred to, and let E and F be members of $\mathcal{C}$. Then E and F have expressions $E = \bigcup_{i=1}^{n}(A_i \times B_i)$ and $F = \bigcup_{j=1}^{m}(C_j \times D_j)$, both being unions of disjoint families. Since

$$(3) \qquad E \cap F = \bigcup_{i=1}^{n} \bigcup_{j=1}^{m} \left[(A_i \cap C_j) \times (B_i \cap D_j) \right]$$

we see that $E \cap F \in \mathcal{C}$, and that $\mathcal{C}$ is closed under the taking of intersections. From the equation

(4) $$(X \times Y) \smallsetminus (A \times B) = \left[(X \smallsetminus A) \times B\right] \cup \left[X \times (Y \smallsetminus B)\right]$$

we get

$$(X \times Y) \smallsetminus E = (X \times Y) \smallsetminus \bigcup_{i=1}^{n}(A_i \times B_i) = \bigcap_{i=1}^{n}\left[(X \times Y) \smallsetminus (A_i \times B_i)\right]$$

$$= \bigcap_{i=1}^{n}\left\{\left[(X \smallsetminus A_i) \times B_i\right] \cup \left[X \times (Y \smallsetminus B_i)\right]\right\}$$

This shows that the complement of E belongs to $\mathcal{C}$, because $\mathcal{C}$ is closed under finite intersections. By the de Morgan identities, $\mathcal{C}$ is closed under unions.   ∎

Lemma 3.   *In any measure space* $(X, \mathcal{A}, \mu)$ *the following are true for measurable sets* A_i :

(1) *If* $A_1 \subset A_2 \subset \cdots$, *then* $\mu\left(\bigcup_{i=1}^{\infty} A_i\right) = \lim_n \mu(A_n)$

(2) *If* $A_1 \supset A_2 \supset \cdots$ *and* $\mu(A_1) < \infty$, *then* $\mu\left(\bigcap_{i=1}^{\infty} A_i\right) = \lim_n \mu(A_n)$

Proof.   Assume the hypothesis in (1), and define $B_n = A_n \smallsetminus A_{n-1}$. The sequence $\{B_n\}$ is disjoint, and consequently,

$$\mu\left(\bigcup_{i=1}^{\infty} A_i\right) = \mu\left(\bigcup_{i=1}^{\infty} B_i\right) = \sum_{i=1}^{\infty} \mu(B_i) = \lim_{n \to \infty} \sum_{i=1}^{n} \mu(B_i)$$

$$= \lim_{n \to \infty} \mu\left(\bigcup_{i=1}^{n} B_i\right) = \lim_{n \to \infty} \mu(A_n)$$

To establish (2), assume its hypothesis. Then $\{A_1 \smallsetminus A_n\}$ is an increasing sequence, and by part (1) we have

$$\mu(A_1) - \mu\left(\bigcap_{i=1}^{\infty} A_i\right) = \mu\left(A_1 \smallsetminus \bigcap_{i=1}^{\infty} A_i\right) = \mu\left(\bigcup_{i=1}^{\infty}(A_1 \smallsetminus A_i)\right)$$

$$= \lim_{n \to \infty} \mu(A_1 \smallsetminus A_n) = \lim_{n \to \infty} \left(\mu(A_1) - \mu(A_n)\right)$$

$$= \mu(A_1) - \lim_{n \to \infty} \mu(A_n) \qquad \blacksquare$$

A **monotone** class of sets is a family $\mathcal{M}$ having these two properties:

(1) If $A_i \in \mathcal{M}$ and $A_1 \subset A_2 \subset \cdots$ then $\bigcup_{i=1}^{\infty} A_i \in \mathcal{M}$

(2) If $A_i \in \mathcal{M}$ and $A_1 \supset A_2 \supset \cdots$ then $\bigcap_{i=1}^{\infty} A_i \in \mathcal{M}$

If X is a set, then 2^X is a monotone class. Also, every σ-algebra of sets is a monotone class (easily verified). If $\mathcal{A}$ is any family of sets, then there exists a smallest monotone class containing $\mathcal{A}$. This assertion depends on the easy fact that the intersection of a collection of monotone classes is also a monotone class.

Lemma 4.   *Let C be an algebra of sets, as defined above. Then the monotone class generated by C is identical with the σ-algebra generated by C.*

Proof. Let $\mathcal{M}$ and S be respectively the monotone class and the σ-algebra generated by C. Since every σ-algebra is a monotone class, we have $\mathcal{M} \subset S$. The rest of the proof is devoted to showing that $\mathcal{M}$ is a σ-algebra (so that $S \subset \mathcal{M}$).

For any set F in the monotone class $\mathcal{M}$ we define

$$\mathcal{K}_F = \{A : \text{ the sets } A \smallsetminus F, \ F \smallsetminus A, \text{ and } A \cup F \text{ belong to } \mathcal{M}\}$$

Assertion 1 $\mathcal{K}_F$ is a monotone class.

There are two properties to verify, one of which we leave to the reader. Suppose that $A_i \in \mathcal{K}_F$ and $A_1 \subset A_2 \subset \cdots$ Let $A = \bigcup_{i=1}^{\infty} A_i$. Then $A_i \smallsetminus F$, $F \smallsetminus A_i$, and $A_i \cup F$ all belong to $\mathcal{M}$ and form monotone sequences. Since $\mathcal{M}$ is a monotone class, we have

$$A \smallsetminus F = \bigcup_{i=1}^{\infty} (A_i \smallsetminus F) \in \mathcal{M}$$

$$F \smallsetminus A = \bigcap_{i=1}^{\infty} (F \smallsetminus A_i) \in \mathcal{M}$$

$$F \cup A = \bigcup_{i=1}^{\infty} (F \cup A_i) \in \mathcal{M}$$

These calculations establish that $A \in \mathcal{K}_F$.

Assertion 2   If $F \in C$, then $C \subset \mathcal{K}_F$.

To prove this let E be any element of C. Since C is an algebra, we have $E \smallsetminus F$, $F \smallsetminus E$, and $E \cup F$ all belonging to C and to $\mathcal{M}$. By the definition of $\mathcal{K}_F$, $E \in \mathcal{K}_F$.

Assertion 3   If $F \in C$, then $\mathcal{M} \subset \mathcal{K}_F$.

To prove this, note that $\mathcal{K}_F$ is a monotone class containing C, by Assertions 1 and 2. Hence $\mathcal{K}_F \supset \mathcal{M}$, since $\mathcal{M}$ is the smallest monotone class containing C.

Assertion 4   If $F \in C$ and $E \in \mathcal{M}$, then $E \in \mathcal{K}_F$.

This is simply another way of expressing Assertion 3.

Assertion 5   If $F \in C$ and $E \in \mathcal{M}$, then $F \in \mathcal{K}_E$.

This is true because the statement $E \in \mathcal{K}_F$ is logically equivalent to $F \in \mathcal{K}_E$.

Assertion 6   If $E \in \mathcal{M}$, then $C \subset \mathcal{K}_E$.

This is a restatement of Assertion 5.

Assertion 7   If $E \in \mathcal{M}$, then $\mathcal{M} \subset \mathcal{K}_E$.

This follows from Assertions 6 and 1 because $\mathcal{K}_E$ is a monotone class containing C, while $\mathcal{M}$ is the smallest such monotone class.

Assertion 8 $\mathcal{M}$ is an algebra.

To prove this, let E and F be members of $\mathcal{M}$. Then $F \in \mathcal{K}_E$ by Assertion 7. Hence $E \smallsetminus F$, $F \smallsetminus E$, and $E \cup F$ all belong to $\mathcal{M}$.

Assertion 9 $\mathcal{M}$ is a σ-algebra.

To prove this, let $A_i \in \mathcal{M}$ and define $B_n = A_1 \cup \cdots \cup A_n$. By Assertion 8, $\mathcal{M}$ is an algebra. Hence $B_n \in \mathcal{M}$ and $B_1 \subset B_2 \subset \cdots$ Since $\mathcal{M}$ is a monotone class, $\bigcup_{n=1}^{\infty} B_n \in \mathcal{M}$. It follows that $\bigcup_{n=1}^{\infty} A_n \in \mathcal{M}$. ∎

Theorem 1. First Fubini Theorem.   If $(X, \mathcal{A}, \mu)$ and $(Y, \mathcal{B}, \nu)$
are σ-finite measure spaces, and if $E \in \mathcal{A} \otimes \mathcal{B}$, then

(1) The function $y \mapsto \mu(E^y)$ is measurable.
(2) The function $x \mapsto \nu(E_x)$ is measurable.
(3) $\int_X \nu(E_x)\, d\mu(x) = \int_Y \mu(E^y)\, d\nu(y)$

Proof.   Let $\mathcal{M}$ be the family of all sets E in $\mathcal{A} \otimes \mathcal{B}$ for which the assertion in
the theorem is true. Our task is to show that $\mathcal{M} = \mathcal{A} \otimes \mathcal{B}$.

We begin by showing that every measurable rectangle belongs to $\mathcal{M}$. Let
$E = A \times B$, where $A \in \mathcal{A}$ and $B \in \mathcal{B}$. Since $E^y = A$ or $E^y = \varnothing$, depending on
whether $y \in B$ or $y \in Y \smallsetminus B$, we have $\mu(E^y) = \mathcal{X}_B(y)\mu(A)$. This is a measurable
function of y. Furthermore,

$$\int_Y \mu(E^y)\, d\nu(y) = \int_Y \mathcal{X}_B(y)\mu(A)\, d\nu(y) = \mu(A)\nu(B)$$

We can carry out the same argument for $\nu(E_x)$ to see that $E \in \mathcal{M}$.

In the second part of the proof, let $\mathcal{C}$ denote the class of all sets in $\mathcal{A} \otimes \mathcal{B}$
that are unions of finite disjoint families of rectangles. By Lemma 2, $\mathcal{C}$ is an
algebra. We shall prove that $\mathcal{C} \subset \mathcal{M}$. Let $E \in \mathcal{C}$. Then $E = \bigcup_{i=1}^{n} E_i$ where
$E_1, \ldots, E_n$ is a disjoint set of rectangles. Hence

$$\mu(E^y) = \mu\left(\left(\bigcup_{i=1}^{n} E_i\right)^y\right) = \mu\left(\bigcup_{i=1}^{n} E_i^y\right) = \sum_{i=1}^{n} \mu(E_i^y)$$

This shows that $y \mapsto \mu(E^y)$ is a measurable function. By the symmetry in this
situation, $x \mapsto \nu(E_x)$ is measurable and $\nu(E_x) = \sum_{i=1}^{n} \nu((E_i)_x)$. Since $E_i \in \mathcal{M}$
by the first part of our proof, we have

$$\int_X \nu(E_x)\, d\mu(x) = \sum_{i=1}^{n} \int_X \nu((E_i)_x)\, d\mu(x) = \sum_{i=1}^{n} \int_Y \mu(E_i^y)\, d\nu(y)$$

$$= \int_Y \sum_{i=1}^{n} \mu(E_i^y)\, d\nu(y) = \int_Y \mu(E^y)\, d\nu(y)$$

This establishes that $E \in \mathcal{M}$ and that $\mathcal{C} \subset \mathcal{M}$.

In the third segment of the proof, we show that $\mathcal{M}$ is closed under the taking
of unions of increasing sequences of sets. Let $E_i \in \mathcal{M}$ and $E_1 \subset E_2 \subset \cdots$ Define
$E = \bigcup_{i=1}^{\infty} E_i$. Then by Lemmas 1 and 3, $\mu(E^y) = \mu(\bigcup_{i=1}^{\infty} E_i^y) = \lim_n \mu(E_n^y)$.
Hence, $y \mapsto \mu(E^y)$ is a measurable function. Also, since $E_n \in \mathcal{M}$,

$$\int_Y \mu(E^y)\, d\nu(y) = \lim_n \int_Y \mu(E_n^y)\, d\nu(y) = \lim_n \int_X \nu((E_n)_x)\, d\mu(x) = \int_X \nu(E_x)\, d\mu(x)$$

by the Monotone Convergence Theorem (Theorem 1 in Section 8.5, page 401).
This shows that $E \in \mathcal{M}$.

In the fourth part of the proof we establish that $\mathcal{M}$ is closed under taking
intersections of decreasing sequences of sets. Since X and Y are σ-finite, there

exist $A_n \in \mathcal{A}$ and $B_n \in \mathcal{B}$ such that $X = \bigcup_{n=1}^{\infty} A_n$, $Y = \bigcup_{n=1}^{\infty} B_n$, $\mu(A_n) < \infty$, and $\nu(B_n) < \infty$. We may suppose further that $A_1 \subset A_2 \subset \cdots$ and that $B_1 \subset B_2 \subset \cdots$ Let $\{E_i\}$ be a decreasing sequence of sets in $\mathcal{M}$, and set $E = \bigcap_{n=1}^{\infty} E_n$. We want to prove that $E \in \mathcal{M}$. Since $E = \bigcup_{n=1}^{\infty}[E \cap (A_n \times B_n)]$ and since $\mathcal{M}$ is closed under "increasing unions," it suffices to prove that $E \cap (A_n \times B_n) \in \mathcal{M}$ for each n. We therefore define

$$\mathcal{F} = \{F : F \cap (A_n \times B_n) \in \mathcal{M} \quad \text{for} \quad n = 1, 2, \ldots\}$$

Now it is to be proved that $E \in \mathcal{F}$. Since $E = \bigcap_{i=1}^{\infty} E_i$, it will be sufficient to prove that $\mathcal{F}$ is a monotone class and that $E_i \in \mathcal{F}$ for each i. Since $E_i \in \mathcal{M} \subset \mathcal{A} \otimes \mathcal{B}$, we have only to prove that $\mathcal{F}$ is a monotone class containing $\mathcal{A} \otimes \mathcal{B}$. By Lemma 4, this will follow if we can show that $\mathcal{F}$ is a monotone class containing $\mathcal{C}$. That $\mathcal{F} \supset \mathcal{C}$ can be verified as follows. Since $A_n \times B_n \in \mathcal{C}$, and $\mathcal{C}$ is an algebra, we have the implications

$$F \in \mathcal{C} \Longrightarrow F \cap (A_n \times B_n) \in \mathcal{C} \subset \mathcal{M} \Longrightarrow F \in \mathcal{F}$$

To prove that $\mathcal{F}$ is a monotone class, let $\{F_i\}$ be an increasing sequence in $\mathcal{F}$, and set $F = \bigcup_{i=1}^{\infty} F_i$. The equation

$$F \cap (A_n \times B_n) = \bigcup_{i=1}^{\infty} \left[F_i \cap (A_n \times B_n) \right]$$

shows that $F \cap (A_n \times B_n) \in \mathcal{M}$, since $\mathcal{M}$ is closed under "increasing unions." Hence $F \in \mathcal{F}$. Next, take a decreasing sequence $\{F_i\}$ in $\mathcal{F}$, and let $F = \bigcap_{i=1}^{\infty} F_i$. Let n be fixed. For each i, the set $G_i = F_i \cap (A_n \times B_n)$ belongs to $\mathcal{M}$. Let $G = \bigcap_{i=1}^{\infty} G_i$. For each $y \in Y$, $G_1^y \subset A_n$, whence $\mu(G_1^y) \leqslant \mu(A_n) < \infty$. It follows from Lemma 3 that $\mu(G^y) = \lim_i \mu(G_i^y)$. This proves that $\mu(G^y)$ is a measurable function of y. Since

$$\int_Y \mu(G_1^y)\, d\nu(y) \leqslant \int_Y \mu(A_n)\mathcal{X}_{B_n}(y)\, d\nu(y) = \mu(A_n)\nu(B_n) < \infty$$

the Dominated Convergence Theorem (Theorem 2 in Section 8.6, page 406) implies that $\int_Y \mu(G^y)\, d\nu(y) = \lim_i \int_Y \mu(G_i^y)\, d\nu$. Similarly, $\int_X \nu(G_x)\, d\mu(x) = \lim_i \int_X \nu((G_i)_x)\, d\mu(x)$. But for each i, $\int_X \nu((G_i)_x)\, d\mu(x) = \int_Y \mu(G_i^y)\, d\nu(y)$ because $G_i \in \mathcal{M}$. Hence $\int_Y \mu(G^y)\, d\nu(y) = \int_X \nu(G_x)\, d\mu(x)$. Thus $G \in \mathcal{M}$. Since $G = F \cap (A_n \times B_n)$, $F \in \mathcal{F}$.

We are now at the point where $\mathcal{M}$ is a monotone class containing $\mathcal{C}$. By Lemma 4, $\mathcal{M}$ contains the σ-algebra generated by $\mathcal{C}$. Thus $\mathcal{M} \supset \mathcal{A} \otimes \mathcal{B}$. ∎

The preceding theorem enables us to define a measure ϕ on $\mathcal{A} \otimes \mathcal{B}$ by the equation

$$\phi(E) = \int_Y \mu(E^y)\, d\nu(y) = \int_X \nu(E_x)\, d\mu(x)$$

This measure ϕ is called the **product measure** of μ and ν. It is often denoted by $\mu \otimes \nu$.

Lemma 5.   If $(X, \mathcal{A}, \mu)$ and $(Y, \mathcal{B}, \nu)$ are σ-finite measure spaces, then so is $(X \times Y, \mathcal{A} \otimes \mathcal{B}, \mu \otimes \nu)$.

Proof.   It is clear that the set function ϕ has the property $\phi(\varnothing) = 0$ and the property $\phi(E) \geqslant 0$. If $\{E_i\}$ is a disjoint sequence of sets in $\mathcal{A} \otimes \mathcal{B}$, then $\{E_i^y\}$ is a disjoint sequence in $\mathcal{A}$. Hence, by the Dominated Convergence Theorem,

$$\phi\left(\bigcup_{i=1}^{\infty} E_i\right) = \int_Y \mu\left(\left(\bigcup_{i=1}^{\infty} E_i\right)^y\right) d\nu(y) = \int_Y \mu\left(\bigcup_{i=1}^{\infty} E_i^y\right) d\nu(y)$$

$$= \int_Y \sum_{i=1}^{\infty} \mu(E_i^y) \, d\nu(y) = \sum_{i=1}^{\infty} \int_Y \mu(E_i^y) \, d\nu(y) = \sum_{i=1}^{\infty} \phi(E_i)$$

Thus ϕ is a measure. For the σ-finiteness, observe that if $X = \bigcup_{n=1}^{\infty} A_n$ and $Y = \bigcup_{n=1}^{\infty} B_n$, where $A_1 \subset A_2 \subset \cdots$ and $B_1 \subset B_2 \subset \cdots$, then $X \times Y = \bigcup_{n=1}^{\infty}(A_n \times B_n)$. If, further, $\mu(A_n) < \infty$ and $\nu(B_n) < \infty$ for all n, then we have $\phi(A_n \times B_n) = \mu(A_n)\nu(B_n) < \infty$. ∎

Theorem 2.   Second Fubini Theorem.   Let $(X, \mathcal{A}, \mu)$ and $(Y, \mathcal{B}, \nu)$ be two σ-finite measure spaces. Let f be a nonnegative function on $X \times Y$ that is measurable with respect to $(X \times Y, \mathcal{A} \otimes \mathcal{B})$. Then

(1) For each x, $y \mapsto f(x, y)$ is measurable

(2) For each y, $x \mapsto f(x, y)$ is measurable

(3) $y \mapsto \int_X f(x, y) \, d\mu(x)$ is measurable

(4) $x \mapsto \int_Y f(x, y) \, d\nu(y)$ is measurable

(5) $\int_{X \times Y} f(x, y) \, d\phi = \int_X \int_Y f(x, y) \, d\nu \, d\mu = \int_Y \int_X f(x, y) \, d\mu \, d\nu$

Proof.   If f is the characteristic function of a measurable set E, then (1) is true because $f(x, y) = \mathcal{X}_{E_x}(y)$. Part (2) is true by the symmetry in the situation. Since

$$\int_X f(x, y) \, d\mu(x) = \int_X \mathcal{X}_E(x, y) \, d\mu(x) = \int_X \mathcal{X}_{E^y}(x) \, d\mu(x) = \mu(E^y)$$

the preceding lemma asserts that (3) is also true in this case. Part (4) is true by symmetry. For part (5), write

$$\int_{X \times Y} f(x, y) \, d\phi = \phi(E) = \int_Y \mu(E^y) \, d\nu(y)$$

$$= \int_Y \int_X f(x, y) \, d\mu(x) \, d\nu(y)$$

The other equality is similar. Thus, Theorem 2 is true when f is the characteristic function of a measurable set.

If f is a simple function, then f has properties (1) to (5) by the linearity of the integrals.

If f is an arbitrary nonnegative measurable function, then there exist simple functions f_n such that $f_n \uparrow f$. Since the limit of a sequence of measurable functions is measurable, f has properties (1) to (4). By the Monotone Convergence Theorem, property (5) follows for f. ∎

An extension of the Fubini Theorem exists for a more general class of measures, namely signed measures and even complex-valued measures. Since the complex-valued measures include the real-valued measures, we describe them first. Let $(X, \mathcal{A})$ be a measurable space. A **complex measure** "on X" is a function $\mu : \mathcal{A} \to \mathbb{C}$ such that:

(I) $\sup_{A \in \mathcal{A}} |\mu(A)| < \infty$

(II) $\mu(\bigcup A_i) = \sum_{i=1}^{\infty} \mu(A_i)$ for any disjoint sequence of measurable sets A_i.

In an abuse of notation, we define $|\mu|$ by the equation

$$|\mu|(A) = \sup \sum_{i=1}^{\infty} |\mu(A_i)|$$

where the supremum is over all partitions of A into a disjoint sequence of measurable sets. It is clear that $|\mu(A)| \leqslant |\mu|(A)$, because $\{A\}$ is a competing partition of A. The theory goes on to establish that $|\mu|$ is an ordinary (i.e., nonnegative) measure and $|\mu|(X) < \infty$. This feature distinguishes the theory of complex or signed measures from the traditional nonnegative measures. References: [DS], [Roy], [Ru3], [HS], [Berb3], [Berb4].

The Fubini Theorem in this new setting is as follows:

Theorem 3. Fubini's Theorem for Complex Measures.   *Let $(X, \mathcal{A})$ and $(Y, \mathcal{B})$ be two measurable spaces, and let μ and ν be complex measures on X and Y, respectively. Let f be a complex-valued measurable function on $X \times Y$. If $\int_Y \int_X |f(x, y)| \, d|\mu| \, d|\nu| < \infty$, then*

$$\int_Y \int_X f(x, y) \, d\mu(x) \, d\nu(y) = \int_X \int_Y f(x, y) \, d\nu(y) \, d\mu(x)$$

This theorem is to be found in [DS].

Problems 8.10

1. Verify Equations (1) and (2).

2. Verify Equations (3) and (4). Show that the two sets on the right side of Equation (4) are mutually disjoint.

3. Prove that every σ-algebra is a monotone class. Prove that the intersection of a family of monotone classes is also a monotone class.

4. Prove that if a monotone class is an algebra, then it is a σ-algebra. Can you get this conclusion from weaker hypotheses?

References

[AS] Abramowitz, M. and I. Stegun, *Handbook of Mathematical Functions with Formulas, Graphs, and Mathematical Tables*, U.S. Department of Commerce, National Bureau of Standards, 1964. Reprint, Dover Publications, New York.

[Ad] Adams, R. A., *Sobolev Spaces*, Academic Press, New York, 1975. (Vol. 65 in the series Pure and Applied Mathematics.)

[Agm] Agmon, S., *Lectures on Elliptic Boundary Value Problems*, Van Nostrand, New York, 1965.

[AG] Akhiezer, N. I. and I. M. Glazman, *Theory of Linear Operators in Hilbert Space*, Ungar, New York, 1963.

[ATS] Alekseev, V. M., V.M. Tikhomirov, and S. V Fomin, *Optimal Control*, Consultants Bureau, New York, c1987.

[AY] Alexander, J. C. and J. A. Yorke, "The homotopy continuation method: Numerically implemented topological procedures," *Trans. Amer. Math. Soc.* **242** (1978), 271–284.

[AlG] Allgower, E. and K. Georg, "Simplicial and continuation methods for approximating fixed points and solutions to systems of equations," *SIAM Review* **22** (1980), 28–85.

[AGP] Allgower, E. L., K. Glasshoff, and H.-O. Peitgen, eds., *Numerical Solution of Nonlinear Equations*, Lecture Notes in Math., vol. 878, Springer-Verlag, New York, 1981.

[Ar] Aronszajn, N., *Introduction to the Theory of Hilbert Spaces*, Research Foundation, Oklahoma State University, Stillwater, Oklahoma, 1950.

[At] Atkinson, K. E., *A Survey of Numerical Methods for the Solution of Fredholm Integral Equations of the Second Kind*, SIAM Publications, Philadelphia, 1976.

[Au] Aubin, J. P., *Applied Functional Analysis*, 2nd ed., Wiley, New York, 1999.

[Av1] Avez, A., *Introduction to Functional Analysis, Banach Spaces, and Differential Calculus*, Wiley, New York, 1986.

[Av2] Avez, A., *Differential Calculus*, Wiley, New York, 1986.

[Ax] Axelsson, O., "On Newton Type Continuation Methods," *Comm. Applied Analysis* 4 (2000), 575–595.

[BN] Bachman, G. and L. Narici, *Functional Analysis*, Academic Press, New York, 1966.

[Bac] Bachman, G., *Elements of Abstract Harmonic Analysis*, Academic Press, New York, 1964.

[Bak] Baker, C. T. H., *The Numerical Treatment of Integral Equations*, Oxford University Press, 1977.

[Ban] Banach, S., *Théorie des Opérations Linéaires*, Hafner, New York, 1932.

[Barb] Barbeau, E. J., *Mathematical Fallacies, Flaws, and Flimflam*, Mathematical Association of America, Washington, 2000.

[Bar] Barnes, E. R., "A variation on Karmarkar's algorithm for solving linear programming problems," *Mathematical Programming* **36** (1986), 174–182.

[Bart] Bartle, R. G., *The Elements of Real Analysis*, 2nd ed., Wiley, New York, 1976.

[Bart2] Bartle, R. G., *Elements of Integration Theory*, Wiley, New York, 1966. Revised edition, 1995.

[Bea] Beauzamy, B., *Introduction to Banach Spaces and their Geometry*, North Holland, Amsterdam, 1985.

[Bec] Beckner, W., "Inequalities in Fourier Analysis," *Ann. Math.* **102** (1975), 159–182.

[Berb] Berberian, S. K., *Introduction to Hilbert Space*, Chelsea Publishing Co., New York, 1976. Reprint by American Mathematical Society, Providence, RI.

[Berb2] Berberian, S. K., *Notes on Spectral Theory*, Van Nostrand, New York, 1966.

[Berb3] Berberian, S. K., *Fundamentals of Real Analysis*, Springer-Verlag, New York, 1999.

[Berb4] Berberian, S. K., *Measure and Integration*, Macmillan, New York, 1965.

[Berez] Berezanski, Ju. M., *Expansions in Eigenfunctions of Self-Adjoint Operators*, Amer. Math. Soc., Providence, RI, 1968.

[Bes] Besicovitch, A. S., *Almost Periodic Functions*, Dover, New York, 1954.

[Bl] Bliss, G. A., *Calculus of Variations*, Mathematical Association of America, 1925.

[Bo] Bolza, O., *Lectures on the Calculus of Variations*, 1904. Chelsea Publishing Co., Reprint, 1973.

[Br] Bracewell, R. N., *The Fourier Transform and Its Applications*, McGraw-Hill, New York, 1986.

[Brae] Braess, D., *Finite Elements*, Cambridge University Press, Cambridge, 1997.

[Brez] Brezinski, C., *Biorthogonality and its Application to Numerical Analysis*, Dekker, Basel, 1991.

[BrS] Brophy, J. F. and P. W. Smith, "Prototyping Karmarkar's algorithm using MATH-PROTAN," *Directions* **5** (1988), 2–3. IMSL Corp. Houston.

[Cann1] Cannell, D. M., "George Green: An enigmatic mathematician," *Amer. Math. Monthly* **106** (1999), 137–151.

[Cann2] Cannell, D. M., *George Green, Mathematician and Physicist, 1793–1841*, SIAM, Philadelphia, 2000.

[Car] Carathéodory, C., *Calculus of Variations and Partial Differential Equations of the First Order*, Holden-Day Publishers, 1965. Second edition, Chelsea, New York, 1982.

[Cart] Cartan, H., *Cours de Calcul Différentiel*, Hermann, Paris 1977.

[CCC] Chen, M. J., Z. Y. Chen, and G. R. Chen, *Approximate Solutions of Operator Equations*, World Scientific, Singapore, 1997.

[Ch] Cheney, E.W., *Introduction to Approximation Theory*, McGraw-Hill, New York, 1966. 2nd ed., Chelsea Publ. Co., New York, 1985. American Mathematical Society, 1998.

[CL] Cheney, E. W. and W. A. Light, *A Course in Approximation Theory*, Brooks/Cole, Pacific Grove, CA, 1999.

[CMY] Chow, S. N., J. Mallet-Paret, J. A. Yorke, "Finding zeroes of maps: Homotopy methods that are constructive with probability one," *Math. Comp.* **32** (1978), 887–899.

[Cia] Ciarlet, P., *The Finite Element Method for Elliptic Problems*, North Holland, New York, 1978.

[Coc] Cochran, J. E., *Applied Mathematics*, Wadsworth, Belmont, CA, 1982.

[Coh] Cohen, P. J., "The independence of the continuum hypothesis," *Proc. Nat. Acad. Sci., U.S.A.* **50** (1963), 1143–1148 and **51** (1964), 105–110.

[Col] Collatz, L., *Functional Analysis and Numerical Mathematics*, Springer-Verlag, New York, 1966.

[Con] Constantinescu, F., *Distributions and Their Applications in Physics*, Pergamon Press, Oxford, Eng., 1980.

[Cor] Corduneanu, C., *Integral Equations and Applications*, Cambridge University Press, 1991.

[CS] Corwin, L., and R. Szczarla, *Calculus in Banach Spaces*, Dekker, New York, 1979.

[Cou] Courant, R., *Calculus of Variations*, Lecture Notes from the Courant Institute, 1946.

[CH] Courant, R. and D. Hilbert, *Methods of Mathematical Physics, Vol. I, II*, Interscience, New York, 1953, 1962.

[CP] Curtain, R.F. and A.J. Pritchard, *Functional Analysis in Modern Applied Mathematics*, Academic Press, New York, 1977.

[Dav] Davis, H. T., *Introduction to Nonlinear Differential and Integral Equations*, Dover Publications, New York, 1962.

[Davp] Davis, P. J., *Interpolation and Approximation*, Blaisdell, New York, 1963. Reprint, Dover Publications, New York.

[Day] Day, M. M., *Normed Linear Spaces*, Academic Press, New York, 1962. Reprint, Springer-Verlag, Berlin.

[DM] Debnath, L. and P. Mikusinski, *Introduction to Hilbert Spaces with Applications*, Academic Press, New York, 1990.

[Deb] Debnath, L., *Integral Transforms and Their Applications*, CRC Press, Boca Raton, FL., 1995.

[DMo] Delves, L. M. and J. L. Mohamed, *Computational Methods for Integral Equations*, Cambridge University Press, 1985.

[Det] Dettman, J. W., *Mathematical Methods in Physics and Engineering*, McGraw-Hill, New York, 1962. Reprint, Dover Publications, New York, 1988.

[DiS] Diaconis, P. and M. Shahshahani, "On nonlinear functions of linear combinations," *SIAM J. Sci. Statis. Comput.* **5** (1984), 175–191.

[Dies] Diestel, J., *Sequences and Series in Banach Spaces*, Springer-Verlag, New York, 1984.

[Dieu] Dieudonné, J., *Foundations of Modern Analysis*, Academic Press, New York, 1960.

[Dono] Donoghue, W. F., *Distributions and Fourier Transforms*, Academic Press, New York, 1969. Vol. 32 in the series Pure and Applied Mathematics.

[Dug] Dugundji, J., *Topology*, Allyn and Bacon, Boston, 1965.

[DS] Dunford, N. and J. T. Schwartz, *Linear Operators, Part I, General Theory*, Interscience, New York, 1958.

[DM] Dym, H. and H. P. McKean, *Fourier Series and Integrals*, Academic Press, New York, 1972.

[Dzy] Dzyadyk, V. K., *Approximation Methods for Solutions of Differential and Integral Equations*, VSP Publishers, Zeist, The Netherlands, 1995.

[Eav] Eaves, B. C., "A short course in solving equations with PL homotopies," *SIAM-AMS Proceedings* **9** (1976), 73–144.

[Edw] Edwards, R. E., *Functional Analysis*, Holt Rinehart and Winston, New York, 1965.

[Egg] Eggleston, H. G., *Convexity*, Cambridge University Press, 1958.

[Els] Elsgolc, L. E., *Calculus of Variations*, Pergamon, London, 1961.

[Ev] Evans, L. C., *Partial Differential Equations*, Amer. Math. Soc., Providence, RI, 1998.

[Ewi] Ewing, G. W., *Calculus of Variations with Applications*, W.W. Norton, 1969. Reprint, Dover Publications, 1985.

[Fef] Feferman, S., "Does mathematics need new axioms?," *Amer. Math. Monthly* **106** (1999), 99-111.

[Fern] Fernandez, L. A., "On the limits of the Lagrange multiplier rule," *SIAM Rev.* **39** (1997), 292-297.

[Feyn] Feynman, R. P., *Surely You're Joking, Mr. Feynman*, W. W. Norton, New York, 1985.

[Fic] Ficken, F. A., "The continuation method for functional equations," *Comm. Pure Appl. Math.* **4** (1951), 435-456.

[Fla] Flanders, H., *Differential Forms*, Academic Press, New York, 1963.

[Fol] Folland, G. B., *Fourier Analysis and Its Applications*, Wadsworth-Brooks-Cole, Pacific Grove, Calif., 1992.

[Fox] Fox, C., *An Introduction to the Calculus of Variations*, Oxford University Press, 1963. Reprint, Dover Publications, 1987.

[Fri] Friedlander, F. G., *Introduction to the Theory of Distributions*, Cambridge University Press, 1982.

[Frie1] Friedman, A., *Generalized Functions and Partial Differential Equations*, Prentice-Hall, Englewood Cliffs, N.J., 1963.

[Frie2] Friedman, A., *Foundations of Modern Analysis*, Holt Rinehart and Winston, New York 1970. Reprint, Dover Publications, New York, 1982.

[Fried] Friedman, B., *Principles and Techniques of Applied Mathematics*, Wiley, New York, 1956. Reprint, Dover Publications.

[FM] Furi, M. and M. Martelli, "On the mean-value theorem, inequality, and inclusion," *Amer. Math. Monthly* **98** (1991), 840-846.

[Gar] Garabedian, P. R., *Partial Differential Equations*, Wiley, New York, 1964. Reprint, Chelsea Publications, New York.

[GG] Garcia, C. B. and F. J. Gould, "Relations between several path-following algorithms and local and global Newton methods," *SIAM Rev.* **22** (1980), 263-274.

[GZ1] Garcia, C. B. and W. I. Zangwill, "An approach to homotopy and degree theory," *Math. Oper. Res.* **4** (1979), 390-405.

[GZ2] Garcia, C. B. and W. I. Zangwill, "Finding all solutions to polynomial systems and other systems of equations," *Math. Programming* **16** (1979), 159–176.

[GZ3] Garcia, C. B. and W. I. Zangwill, *Pathways to Solutions, Fixed Points and Equilibria*, Prentice Hall, Englewood Cliffs, N.J., 1981.

[GF] Gelfand, I. M. and S. V. Fomin, *Calculus of Variations*, Prentice-Hall, Englewood Cliffs, N.J., 1963.

[GV] Gelfand, I. M. and N. Ya. Vilenkin, *Generalized Functions*, 4 volumes, Academic Press, 1964. (Vol. 1 is by Gelfand and G.E. Shilov.)

[Go] Gödel, K., *The Consistency of t*ɪe *Axiom of Choice and of the Generalized Continuum Hypothesis with the Axioms of Set Theory*, Princeton University Press, 1940.

[GP] Goffman, C. and G. Pedrick, *First Course in Functional Analysis*, Chelsea Publishing Co., New York. Reprint, American Mathematical Society.

[Gol] Goldberg, R. R., *Fourier Transforms*, Cambridge University Press, 1970.

[Gold] Goldstein, A. A., *Constructive Real Analysis*, Harper and Row, New York, 1967.

[Gr] Graves, L. M., *The Theory of Functions of Real Variables*, McGraw-Hill, New York, 1946.

[Green] Green, G., *Mathematical Papers of George Green*, edited by N.M. Ferrers, Amer. Math. Soc., Providence, RI, 1970.

[Gre] Greenberg, M. D., *Foundations of Applied Mathematics*, Prentice Hall, Englewood Cliffs, NJ, 1978.

[Gri] Griffel, D. H., *Applied Functional Analysis*, John Wiley, New York, 1981.

[Gro] Groetsch, C. W., *Elements of Applicable Functional Analysis*, Marcel Dekker, New York, 1980.

[Hal1] Halmos, P. R., "What does the spectral theorem say?," *Amer. Math. Monthly* **70** (1963), 241–247.

[Hal2] Halmos, P. R., *A Hilbert Space Problem Book*, van Nostrand, Princeton, 1967.

[Hal3] Halmos, P. R., *Introduction to Hilbert Space*, Chelsea Publishing Co., New York, 1951.

[Hal4] Halmos, P. R., *Measure Theory*, Van Nostrand, New York, 1950. Reprint, Springer-Verlag, New York.

[Hel] Helson, H., *Harmonic Analysis*, Addison-Wesley, London, 1983.

[Hen] Henrici, P. *Discrete Variable Methods in Ordinary Differential Equations*, Wiley, New York, 1962,

[Hes1] Hestenes, M. R., *Calculus of Variations and Optimal Control Theory*, Wiley, New York, 1965.

[Hes2] Hestenes, M. R., "Elements of the Calculus of Variations" pp. 59-91 in *Modern Mathematics for the Engineer*, E. F. Beckenback, ed., McGraw-Hill, New York, 1956.

[HS] Hewitt, E. and K. Stromberg, *Real and Abstract Analysis*, Springer-Verlag, New York, 1965.

[HP] Hille, E. and R. S. Phillips, *Functional Analysis and Semigroups*, Amer. Math. Soc., Providence, RI 1957.

[HS] Hirsch, M. W. and S. Smale, "On algorithms for solving $f(x) = 0$," *Comm. Pure Appl. Math.* **32** (1979), 281–312.

[Hol] Holmes, R. B., *Geometric Functional Analysis and its Applications*, Springer-Verlag, New York, 1975.

[Ho] Hörmander, L., *The Analysis of Linear Partial Differential Operators I*, Springer-Verlag, Berlin, 1983.

[Horv] Horváth, J., *Topological Vector Spaces and Distributions*, Addison-Wesley, London, 1966.

[Hu] Huet, D., *Distributions and Sobolev Spaces*, Lecture Note #6, Department of Mathematics, University of Maryland, 1970.

[Hur] Hurley, J. F., *Multivariate Calculus*, Saunders, Philadelphia, 1981.

[In] Ince, E. L., *Ordinary Differential Equations*, Longmans Green, London, 1926. Reprint, Dover Publications, New York, 1948.

[IK] Isaacson, E. and H. B. Keller, *Analysis of Numerical Methods*, Wiley, New York, 1966.

[Ja1] James, R., "Weak compactness and reflexivity," *Israel J. Math.* **2** (1964), 101–119.

[Ja2] James, R., "A non-reflexive Banach space isometric with its second conjugate space," *Proc. Nat. Acad. Sci. U.S.A.* **37** (1951), 174–177.

[Jam] Jameson, G. J. O., *Topology and Normed Spaces*, Chapman and Hall, London, 1974.

[JKP] Jaworowski, J., W. A. Kirk, and S. Park, *Antipodal Points and Fixed Points*, Lecture Note Series, Number 28, Seoul National University, Seoul 1995.

[Jon] Jones, D. S., *The Theory of Generalised Functions*, McGraw-Hill, 1966. 2nd. Edition, Cambridge University Press, 1982.

[Jo] Jones, F., *Lebesgue Integration on Euclidean Space*, Jones and Bartlet, Boston, 1993.

[JLJ] Jost, J., and X. Li-Jost, *Calculus of Variations*, Cambridge University Press, 1999.

[KA] Kantorovich, L. V. and G. P. Akilov, *Functional Analysis in Normed Spaces*, Pergamon Press, London, 1964.

[KK] Kantorovich, L.V. and V.I Krylov, *Approximate Methods of Higher Mathematics*, Interscience, New York, 1964.

[Kar] Karmarkar, N., "A new polynomial-time algorithm for linear programming," *Combinatorica* **4** (1984), 373–395.

[Kat] Katznelson, Y., *An Introduction to Harmonic Analysis*, Wiley, New York, 1968. Reprint, Dover Publications, New York.

[Kee] Keener, J. P., *Principles of Applied Mathematics*, Addison-Wesley, New York, 1988.

[Kel] Kelley, J. L., *General Topology*, D. Van Nostrand, New York, 1955. Reprint, Springer-Verlag, New York.

[KN] Kelley, J. L., I. Namioka, et al., *Linear Topological Spaces*, D. Van Nostrand, New York, 1963.

[KS] Kelley, J. L. and T. P. Srinivasen, *Measure and Integral*, Springer-Verlag, New York, 1988.

[Kello] Kellogg, O. D., *Foundations of Potential Theory*, Dover, New York.

[Ken] Keener, J. P., *Principles of Applied Mathematics*, Perseus Books Group, Boulder, CO, 1999.

[KC] Kincaid, D. and Cheney, W., *Numerical Analysis*, 3nd ed., Brooks/Cole, Pacific Grove, CA., 2001.

[KF] Kolmogorov, A. N. and S. V. Fomin, *Introductory Real Analysis*, Dover Publications, New York, 1975.

[Ko] Körner, T. W., *Fourier Analysis*, Cambridge University Press, 1988.

[Kras] Krasnoselski, M.A., *Topological Methods in the Theory of Nonlinear Integral Equations*, Pergamon, New York, 1964.

[Kr] Kress, R. *Linear Integral Equations*, Springer-Verlag, Berlin, 1989. 2nd edition, 1999.

[Kre] Kreysig, E., *Introductory Functional Analysis with Applications*, Wiley, New York, 1978.

[KRN] Kuratowski, K. and C. Ryll-Nardzewski, "A general theorem on selectors", *Bull. Acad. Polonaise Sciences, Serie des Sciences Math. Astr. Phys.* **13** (1965), 397-403.

[Lanc] Lanczos, C., *Applied Mathematics*, Dover Publications, New York, 1988.

[Lan1] Lang, S., *Analysis II*, Addison-Wesley, London, 1969.

[Lan2] Lang, S., *Introduction to Differentiable Manifolds*, Interscience, New York, 1962.

[Las] Lass, H., *Vector and Tensor Analysis*, McGraw Hill, New York, 1950.

[Lax] Lax, P. D., "Change of variables in multiple integrals," *Amer. Math. Monthly* **106** (1999), 497–501.

[LSU] Lebedev, N. N., I. P. Skalskaya, and Y. S. Uflyand, *Worked Problems in Applied Mathematics*, Reprint, Dover Publications, New York, 1979.

[Leis] Leis, R., *Initial Boundary Value Problems in Mathematical Physics*, Wiley, New York, 1986.

[Li] Li, T. Y. "Solving polynomial systems," *Math. Intelligencer* **9** (1987), 33–39.

[LL] Lieb, E. H. and M. Loss, *Analysis*, Amer. Math. Soc., Providence, 1997.

[LT] Lindenstrauss, J. and L. Tzafriri, *Classical Banach Spaces I*, Springer-Verlag, Berlin.

[LM] Lions, J.L. and E. Magenes, *Nonhomogeneous Boundary Value Problems and Applications*, Springer-Verlag, New York, 1972.

[Lo] Logan, J. D., *Applied Mathematics: A Contemporary Approach*, Wiley, New York, 1987.

[Lov] Lovett, W. V., *Linear Integral Equations*, McGraw-Hill, New York, 1924. Reprint, Dover Publications, New York, 1950.

[Loo] Loomis, L. H., *An Introduction to Abstract Harmonic Analysis*, Van Nostrand, New York, 1953.

[Lue1] Luenberger, D. G., *Introduction to Linear and Nonlinear Programming*, Addison-Wesley, London, 1965.

[Lue2] Luenberger, D. G., *Optimization by Vector Space Methods*, Wiley, New York, 1969.

[MT] Marsden, J. E. and A. J. Tromba, *Vector Calculus* (2nd ed.), W.H. Freeman, San Francisco, 1981.

[Mar] Martin, J. B., *Plasticity: Fundamentals and General Results*, MIT Press, Cambridge, MA, 1975.

[Mas] Mason, J., *Methods of Functional Analysis for Applications in Solid Mechanics*, Elsevier, Amsterdam, 1985.

[Maz] Mazja, V. G., *Sobolev Spaces*, Springer-Verlag, Berlin, 1985.

[McK] McKinsey, J. C. C., *Introduction to the Theory of Games*, McGraw-Hill, New York, 1952.

[Mey] Meyer, G. H., "On solving nonlinear equations with a one-parameter operator embedding," *SIAM J. Numer. Analysis* **5** (1968), 739–752.

[Mich1] Michael, E., "Continuous Selections," *Ann. Math.* **63** (1956), 361–382.

[Mich2] Michael, E., "Selected Selection Theorems," *Amer. Math. Monthly* **63** (1956), 233–238.

[Mil] Milne, W. E., *Numerical Solution of Differential Equations*, Dover, New York.

[Moo] Moore, R. E., *Computational Functional Analysis*, Wiley, New York, 1985.

[Mor1] Morgan, A. "A homotopy for solving polynomial systems," *Applied Math. and Comp.* **18** (1986), 87–92.

[Mor2] Morgan, A. *Solving Polynomial Systems Using Continuation for Engineering and Scientific Problems*, Prentice Hall, Englewood Cliffs, N.J., 1987.

[Morr] Morris, P., *Introduction to Game Theory*, Springer-Verlag, New York, 1994.

[NaSn] Naylor, A.W. and G.R. Snell, *Linear Operator Theory in Engineering and Science*, Springer-Verlag, New York, 1982.

[Naz1] Nazareth, J. L., "Homotopy techniques in linear programming," *Algorithmica* **1** (1986), 529–535.

[Naz2] Nazareth, J. L., "The implementation of linear programming algorithms based on homotopies," *Algorithmica* **15** (1996), 332–350.

[Nel] Nelson, E., *Topics in Dynamics, Vol 1: Flows*, Princeton University Press (1969).

[NSS] Nickerson, H. K., D. C. Spencer, and N. E. Steenrod, *Advanced Calculus*, van Nostrand, New York, 1959.

[OD] Oden, J. T. and L. F. Demkowicz, *Applied Functional Analysis*, CRC Press, New York, 1996.

[OdR] Oden, J. T. and J. N. Reddy, *An Introduction to the Mathematical Theory of Finite Elements*, Wiley, New York, 1976.

[Ol] Olver, F. W. J., *Asymptotics and Special Functions*, Academic Press, New York, 1974.

[OR] Ortega, J. M. and W. C. Rheinboldt, *Iterative Solution of Nonlinear Equations in Several Variables*, Academic Press, New York, 1970.

[Par] Park, Sehie, "Eighty years of the Brouwer fixed point theorem," in *Antipodal Points and Fixed Points* by J. Jaworowski, W. A. Kirk, and S. Park. Lecture Notes Series, No. 28, Seoul National University, 1995. 55–97.

[Part] Parthasarathy, T., *Selection Theorems and Their Applications*, Lecture Notes in Math. No. 263, Springer-Verlag, New York, 1972.

[Ped] Pedersen, M., *Functional Analysis in Applied Mathematics and Engineering*, CRC, Boca Raton, FL, 1999.

[Pet] Petrovskii, I. G., *Lectures on the Theory of Integral Equations*, Graylock Press, Rochester, NY, 1957.

[PM] Polyanin, A. and A. V. Manzhirov, *Handbook of Integral Equations*, CRC Press, Boca Raton, FL, 1998.

[PBGM] Pontryagin, L. S., V. G. Boltyanskii, R. V. Gamkrelidze, E. F. Mishchenko, *The Mathematical Theory of Optimal Processes*, Interscience Publ., New York, 1962.

[Pry] Pryce, J. D., *Numerical Solution of Sturm–Liouville Problems*, Oxford University Press, 1993.

[Red] Reddy, J.N., *Applied Functional Analysis and Variational Methods in Engineering*, McGraw-Hill, New York, 1986.

[RS] Reed, M. and B. Simon, *Methods of Modern Mathematical Physics*, Vol. I, Academic Press, New York, 1980.

[Rh1] Rheinboldt, W. C., *Numerical Analysis of Parameterized Nonlinear Equations*, Wiley, New York, 1986.

[Rh2] Rheinboldt, W. C., "Solution fields of nonlinear equations and continuation methods," *SIAM J. Numer. Analysis* **17** (1980), 221–237.

[Ri] Richtmyer, R. D., *Principles of Advanced Mathematical Physics*, 2 volumes, Springer-Verlag, New York, 1978.

[RN] Riesz, F. and B. Sz.-Nagy, *Functional Analysis*, Frederick Ungar, 1955. Reprint, Dover Publications, New York, 1991.

[Rie] Riesz, T., *Perturbation Theory for Linear Operators*, Springer-Verlag, New York, 1966.

[Ro] Roach, G. F., *Green's Functions*, 2nd ed., Cambridge University Press, 1982.

[Ros] Rosenbloom, P. C., "The method of steepest descent" in *Numerical Analysis*, J. H. Curtiss, ed., Symposia in Applied Math., vol.VI, 1956, 127–176.

[Roy] Royden, H. L., *Real Analysis*, Macmillan, New York, 1968.

[Rub] Rubin, H. and J. E. Rubin, *Equivalents of the Axiom of Choice*, North Holland Publ. Co., Amsterdam, 1985.

[Ru1] Rudin, W. *Functional Analysis*, McGraw-Hill, New York, 1973.

[Ru2] Rudin, W., *Fourier Analysis on Groups*, Interscience, New York, 1963.

[Ru3] Rudin, W., *Real and Complex Analysis*, 2nd ed., McGraw-Hill, New York, 1974.

[Sa] Saaty, T. L., *Modern Nonlinear Equations*, McGraw-Hill, New York, 1967. Reprint, Dover Publications, New York, 1981.

[Sag] Sagan, H., *Introduction to the Calculus of Variations*, McGraw-Hill Book Co., 1969. Reprint, Dover Publications, 1992.

[San] Sansone, G. *Orthogonal Functions*, Interscience, New York, 1959.

[Schu] Schur, I., "Über lineare Transformationen in der Theorie der unendlichen Reihen," *J. Reine Angew. Math.* **151** (1920), 79–111.

[Schj] Schwartz, J. T., *Non-Linear Functional Analysis*, Gordon and Breach, New York, 1969.

[Schl] Schwartz, L., *Mathematics for the Physical Sciences*, Addison-Wesley, London, 1966.

[Schl2] Schwartz, L., *Théorie des Distributions, I, II*, Hermann et Cie, Paris, 1951.

[Sem] Semadeni, Z., *Schauder Bases in Banach Spaces of Continuous Functions*, Lecture Notes in Mathematics, vol. 918, Springer-Verlag, New York, 1982.

[Sho] Showalter, R. E., *Hilbert Space Methods for Partial Differential Equations*, Pitman, London, a977. (Available on-line from http://ejde.math.swt.edu//mono-toc.html.)

[Sim] Simmons, G. F., *Introduction to Topology and Modern Analysis*, McGraw-Hill, 1963.

[Sing] Singer, I., *Bases in Banach Spaces* (2 volumes), Springer-Verlag, Berlin. 1970, 1981.

[Sm] Smale, S., "Algorithms for solving equations," *Proceedings of the International Congress of Mathematicians*, 1986.

[Sma] Smart, D. R., *Fixed Point Theorems*, Cambridge University Press, 1974.

[Smi] Smith, K. T., *Primer of Modern Analysis*, Bogden and Quigley, Belmont, CA, 1971. Springer-Verlag, Berlin, 1983.

[So] Sobolev, S. L., *Applications of Functional Analysis in Mathematical Physics*, Amer. Math. Soc. Translations Series, 1963.

[Sta] Stakgold, I., *Green's Functions and Boundary Value Problems*, Wiley, New York, 1979.

..

[SW] Stein, E. M. and G. Weiss, *Introduction to Fourier Analysis on Euclidean Spaces*, Princeton University Press, 1971.

[StWi] Stoer, J. and C. Witzgall, *Convexity and Optimization in Finite Dimensions*, Springer-Verlag, New York, 1970.

[Str1] Strang, G., *Linear Algebra and Its Applications*, 3rd ed., Harcourt Brace Jovanovich, San Diego, 1988.

[Str2] Strang, G., *Introduction to Applied Mathematics*, Wellesley-Cambridge, Wellesley, MA, 1986.

[Sz] Szegő, G., *Orthogonal Polynomials*, American Mathematical Society Colloquium Publications, vol. 23, 1959.

[Tay1] Taylor, A. E., *Advanced Calculus*, Ginn, New York, 1955.

[Tay2] Taylor, A. E., *Introduction to Functional Analysis*, Wiley, New York, 1958. Reprint, Dover Publications.

[Tay3] Taylor, A. E. *General Theory of Functions and Integration*, Blaisdell, New York, 1965. Reprint, Dover Publications, New York.

[Ti1] Titchmarsh, E. C., *Introduction to the Theory of Fourier Integrals*, Oxford University Press, 1937. Reprinted by Chelsea Publ. Co., New York, 1986.

[Ti2] Titchmarsh, E. C., *The Theory of Functions*, Oxford University Press, 1939.

[Tod] Todd, M. J., "An introduction to piecewise linear homotopy algorithms for solving systems of equations" in *Topics in Numerical Analysis*, P. R. Turner, ed., Lecture Notes in Mathematics, vol. 965, Springer-Verlag, New York, 1982, 147–202.

[Tri] Tricomi, F. G., *Integral Equations*, Interscience, New York, 1957. Reprint, Dover Publications, New York, 1985.

[Vic] Vick, J. W., *Homology Theory*, Academic Press, New York, 1973.

[Wac] Wacker, H. G., ed., *Continuation Methods*, Academic Press, New York, 1978.

[Wag] Wagner, D. H., "Survey of measurable selection theorems: an update," in *Measure Theory Oberwolfach 1979*, D. Kölzow, ed., Lecture Notes in Mathematics, vol. 794, Springer-Verlag, Berlin, 1980,

[Wal] Walter, G. G., *Wavelets and Other Orthogonal Systems with Applications*, CRC Press, Boca Raton, FL, 1994.

[Was] Wasserstrom, E., "Numerical solutions by the continuation method," *SIAM Review* **15** (1973), 89–119.

[Wat] Watson, L. T., "A globally convergent algorithm for computing fixed points of C^2 maps," *Appl. Math. Comput.* **5** (1979), 297–311.

[Wein] Weinstock, R., *Calculus of Variations, with Applications to Physics and Engineering*, McGraw-Hill, New York, 1952. Reprint, Dover Publications 1974.

[West] Westfall, R. S., *Never at Rest: A Biography of Isaac Newton*, Cambridge University Press, 1980.

[Whi] Whitehead, G. W., *Homotopy Theory*, MIT Press, Cambridge, Massachusetts, 1966.

[Wid1] Widder, D. V., *Advanced Calculus*, 2nd ed., Prentice-Hall, Englewood Cliffs, NJ, 1961. Reprint, Dover Publications, New York.

[Wie] Wiener, N., *The Fourier Integral and Certain of Its Applications*, Cambridge University Press, Cambridge, 1933. Reprint, Dover Publications, New York, 1958.

[Wilf] Wilf, H. S., *Mathematics for the Physical Sciences*, Dover Publications, New York, 1978.

[Will] Williamson, J. H., *Lebesgue Integration*, Holt, Rinehart and Winston, New York, 1962.

[Yo] Yosida, K., *Functional Analysis*, 4th ed., Springer-Verlag, Berlin, 1974.

[Youl] Young, L. C., *Lectures on the Calculus of Variations and Optimal Control Theory*, Chelsea Publishing Co., 1980.

[Youn] Young, N., *An Introduction to Hilbert Space*, Oxford University Press, 1988.

[Ze] Zeidler, E., *Applied Functional Analysis*, Springer-Verlag, New York, 1995.

[Zem] Zemanian, A. H., *Distribution Theory and Transform Analysis*, Dover Publications, New York, 1987.

[Zie] Ziemer, W. P., *Weakly Differentiable Functions*, Springer, New York, 1989.

[Zien] Zienkiewicz, O. C. and K. Morgan, *Finite Elements and Approximation*, Wiley, New York, 1983.

[Zy] Zygmund, A., *Trigonometric Series*, 2nd ed., Cambridge University Press, 1959.

Index

A-orthogonal, 233
A-orthonormal, 233
Absolute continuity, 413
Absolutely convergent, 14, 17
Accumulation point, 12
Adjoint of an operator, 50, 82–83
Adjoint space, 34
Affine map, 120
Alaoglu Theorem, 370
Alexander's Theorem, 365
Algebra of sets, 421
Almost everywhere, 396
Almost periodic functions, 76, 77
Almost uniformly, 397
Angle between vectors, 67
Annihilator, 36
Approximate inverse, 188
Arzelà–Ascoli Theorems, 347ff
Autocorrelation, 293
Axiom of Choice, 31
Babuška–Lax–Milgram Theorem,
    201
Baire Theorem, 40
Banach limits, 37
Banach space, 10
Banach–Alaoglu Theorem, 370
Banach–Steinhaus Theorem, 41
Bartle–Graves Theorem, 342
Base for a topology, 362
Base, 5
Basin of attraction, 135
Bernoulli, J., 153
Bessel functions, 179
Bessel's Inequality, 72
Best approximation, 192
Bilinear functional, 201
Binomial Theorem, 262
Binomial coefficients, 261
Biorthogonal System, 82, 192
Bohl's Theorem, 339
Borel Sigma-algebra, 384
Borel sets, 392
Bounded above, 6

Bounded functional, 81
Bounded map, 25
Bounded set, 20, 368
Brachistochrone Problem, 153,
    157ff
Brouwer's Theorem, 333
Calculus of Variations, 152
Canonical embedding, 58
Cantor set, 46
Carathéodory's Theorem, 387
Category argument, 45, 46, 47, 48
Category, 41
Catenary, 153, 156, 169
Cauchy sequence, 10
Cauchy–Riemann equations, 199
Cauchy–Schwarz Inequality, 62
Césarò means, 13
Chain Rule, 121
Chain, 31
Characteristic function of a set,
    395
Characters, 288
Chebyshev polynomials, 214
Closed Graph Theorem, 49
Closed Range Theorem, 50
Closed graph, 47
Closed mapping, 47
Closed set, 16
Closure of a set, 16, 363
Cluster point, 12
Collocation methods, 213ff
Compact operator, 85, 351
Compact set, 8
Compactness in the weak
    topologies, 369
Compactness, 19, 20, 364
Complete measure space, 387
Completeness, 9, 10, 15, 21
Completion of a space, 15, 60
Composition operator, 252
Condensation of singularities, 46
Conjugate direction methods, 232
Conjugate gradient method, 235

Conjugate space, 34
Conjugate-linear map, 90
Connectedness, 124
Continuation methods, 238
Continuity, 15
Contraction Mapping Theorem,
    177, 333
Contraction, 132, 176
Convergence in measure, 408
Convergence of distributions, 257
Convergence of test functions, 249
Convergence, 8, 11, 17
Convex functional, 231
Convex hull, 12
Convex set, 6
Convolution of distributions, 285
Convolution, 269ff, 290ff
Coset, 29
Cosine transform, 300, 321
Countable additivity, 386
Countably compact, 227
Counting measure, 382
Cycloid, 153, 158
Degenerate kernel, 176, 357
Dense set, 14, 28, 36
Derivative of a distribution, 253
Descent methods, 225
Diaconis–Shahshahani Theorem,
    361
Diagonal dominance, 172ff
Diameter of a set, 185
Differentiable, 115
Differential operator, 24, 273
Dini's Theorem, 350
Dirac distribution, 250, 256, 260,
    268, 283
Direct sum, 80, 143
Directed set, 363
Directional derivative, 227
Dirichlet Problem, 167, 198
Discrete space, 46
Discrete topology, 362
Discretization, 170
Distance function, 9, 19, 23, 34,
    64
Distributions, 246, 249
Dominate, 32

Dominated convergence theorem,
    406
Dual space, 34
Eberlein–Smulyan Theorem, 59
Egorov's Theorem, 397
Eigenvalue, 91
Eigenvector, 92
Elliptic, 211
Embedding theorems, 330ff
Equimeasurable rearrangement,
    403
Equivalence, 4
Equivalent norms, 23, 27,39
Essential supremum, 409
Euclidean norm, 4
Euler Equation, 155ff, 164
Euler–Lagrange Equation, 155
Extended real number system, 381
Extension of a function, 31
Extremum problems, 145
Fatou's Lemma, 403
Feasible set, 243
Fermat's Principle, 162, 164
Finite dimensional, 5
Fixed point of Fourier transform,
    301
Fixed-Point Theorems, 140, 333
Formal adjoint, 279, 280
Fourier coefficients, 72
Fourier projections, 42
Fourier series, 42, 167
Fourier transform table, 292
Fourier transform, 24, 287ff
Fréchet derivative, 115
Fréchet–Kolmogorov Theorem,
    350
Fredholm Alternative, 351ff
Fredholm integral equation, 175,
    178, 190
Fredholm theory, 356
Fubini Theorems, 424, 426, 427
Fundamental solution of an
    operator, 273
Fundamental set, 36
G_δ set, 46
Gödel's Theorem, 30
Gâteaux derivative, 120, 228
Galerkin method, 198

Game theory, 345
Gamma function, 293
Gaussian elimination, 172
Gaussian function, 318
Gaussian quadrature, 223
Generalized Cauchy–Schwarz
    Inequality, 84
Generalized function, 246
Generalized sequence, 364
Geodesic, 13, 164ff
Geometrical optics, 162
Goldschmidt solution, 157
Gradient, 117
Gram matrix, 197
Gram–Schmidt process, 75
Greatest lower bound, 6
Green's Identity, 277
Green's Theorem, 161, 200, 203,
    205, 210
Green's functions, 107ff, 215
Hölder Inequality, 55, 409
Hahn decomposition, 420
Hahn–Banach Theorem, 32
Half-space, 38
Hamel base, 32
Hammerstein Equation, 225
Harmonic function, 199
Harmonic series, 18
Hausdorff space, 362
Hausdorff–Young Theorem, 309
Heat equation, 318ff
Heaviside distribution, 250, 254,
    256, 257, 283
Heine–Borel Theorem, 19
Helmholtz equation, 320
Hermite functions, 309
Hermitian matrices, 104
Hermitian operator, 83
Hilbert cube, 351
Hilbert space, 61, 63
Hilbert–Schmidt operator, 83, 96,
    98
Homotopy, 237ff
Hyperplane, 38
Idempotent operator, 189, 191
Implicit Function Theorems, 135ff
Infimum, 6
Initial-value problem, 179ff

Inner measure, 390
Inner product, 61
Integrable function, 405
Integral equations, 131, 141, 357
Integral operator, 24
Integration, 399ff
Interior Mapping Theorem, 48
Interior of a set, 363
Invariant measure, 385, 392
Inverse Fourier transform, 301ff
Inverse Function Theorems, 139,
    140
Invertible, 28
Isolated point, 47
Isometric, 35
Isoperimetric Problem, 159, 161
Iteration, 176
Iterative refinement, 187, 188
Jacobian, 118
James' Theorem, 60
Jordan decomposition, 417
Kantorovich Theorem, 127, 130
Kernel, 26
Kharshiladze–Lozinski Theorem,
    377
Kuratowski–Ryll–Nardzewski
    Theorem, 342
Lagrange interpolation, 193
Lagrange multipliers, 145, 148,
    152, 159
Laplace transform, 24, 287
Laplacian, 198, 275, 297
Laurent's Theorem, 315
Least upper bound, 6
Lebesgue Decomposition
    Theorem, 415
Lebesgue measurable set, 389, 391
Lebesgue measure, 391
Lebesgue outer measure, 382
Lebesgue space, 4
Lebesgue–Stieltjes outer measure,
    382
Legendre polynomials, 76, 77, 377
Leibniz formula, 265
Limit in the mean, 308
Linear functional, 24
Linear independence, 4
Linear inequalities, 344

Linear mapping, 24
Linear operator, 24
Linear programming, 243
Linear space, 2
Linear topological spaces, 367ff
Linear transformation, 24
Lion, 154
Lipschitz condition, 120, 178, 180
Local integrability, 251
Locally-convex space, 370
Locally-finite covering, 345
Lower semicontinuity, 22, 226, 340
Lusin's Theorem, 408
Malgrange–Ehrenpreis Theorem, 273
Mathematica, 126, 205, 230
Maximal element, 32
Mazur's Theorem, 372
Mean-Value Theorem, 122, 123
Measurable functions, 394ff
Measurable rectangle, 421
Measurable sets, 384
Measurable space, 384
Measure space, 386
Measure, 386
Metric space, 8, 13
Meyers–Serrin Theorem, 330
Michael Selection Theorem, 341
Min-Max Theorem, 346
Minimizing sequence, 226
Minimum deviation, 192
Minkowski Inequality, 55
Minkowski functional, 334, 343
Minkowski's Inequality, 410
Mollifier, 249
Monomial, 6, 261
Monotone class, 422
Monotone convergence theorem, 401
Monotone norm, 14
Moore's Theorem, 373, 375
Multi-index, 246
Multinomial Theorem, 263
Multiplication operator, 252, 268
Multivariate interpolation, 313
Mutually singular, 415
Natural embedding, 58
Neighborhood base, 367

Neighborhood, 17
Net, 364
Neumann Theorem, 28, 133, 186
Neural networks, 315
Newton's Method, 125
Newton, I., 154
Non-differentiable function, 13
Non-expansive, 19, 185
Norm, 3
Normal equations, 200
Normal operator, 100
Nowhere dense, 41
Null space, 26
o–notation, 119
Objective function, 243
Open set, 17
Order of a distribution, 253
Order of a multi-index, 247
Ordered vector space, 150, 152
Orthogonal complement, 65
Orthogonal projection, 72, 74, 193
Orthogonal set, 64, 70
Orthonormal base, 73
Orthonormal set, 71
Outer measure, 382
Paracompactness, 345
Parallelogram law, 61, 62
Partial derivative, 117, 118, 144
Partially ordered set, 31, 363
Partition of unity, 282
Pascal's triangle, 268
Picard iteration, 181
Plancherel Theorem, 305ff
Point-evaluation functional, 29, 193, 214
Pointwise convergence, 11
Poisson summation formula, 298
Poisson's Equation, 203, 210
Polar set, 370
Polygonal path, 13
Polynomial, 261
Positive cone, 150, 152
Positive sets, 418
Pre-Hilbert space, 61
Product measures, 420ff, 425
Product spaces, 365
Projection methods, 79, 191, 194
Pseudo-norm, 370

Pythagorean Law, 62, 70
Quadrature, 175, 219, 222
Radial projection, 19
Radiative transfer, 186
Radon–Nikodym Theorem, 413ff
Rank of an operator, 197
Rapidly decreasing function, 294
Rayleigh quotient, 149
Rayleigh–Ritz Method, 166ff, 205ff
Reflexive spaces, 58
Regular distribution, 252
Regular outer measure, 389
Relative topology, 363
Rellich–Kondrachov Theorem, 331
Residual set, 47
Residual vector, 188, 229
Residue calculus, 315ff
Riemann Sum, 218
Riemann integral, 43
Riemann's Theorem, 18
Riesz Representation Theorem, 81
Riesz's Lemma, 22
Riesz–Fischer Theorem, 63, 411
Rothe's Theorem, 338
Saddle point, 347
Schauder base, 38, 204
Schauder–Tychonoff Theorem, 334
Schur's Lemma, 56
Schwartz space, 294
Selection theorems, 339ff
Self-adjoint operator, 83
Seminorm, 370
Separable kernel, 176, 357
Separable space, 75
Separation theorem, 151, 342, 343
Sigma-Algebra, 384
Sigma-finite, 414
Signed measures, 417ff
Similarity, 103
Simple function, 397
Simplex, 345
Simpson's Rule, 223
Sinc-function, 289
Sine transform, 321
Singular-Value decomposition, 98
Skew-Hermitian operator, 101
Snell's Law, 163

Sobolev spaces, 325
Sobolev–Hilbert spaces, 332
Span, 5
Spectral Theorem, 93
Stable sequence, 80
Steepest Descent, 124, 228
Step function, 406
Stone–Weierstrass Theorem, 359
Strictly positive definite functions, 315
Sturm–Liouville problems, 105ff, 203
Subbase for a topology, 363
Subsequence, 8
Sup norm, 3
Support of a distribution, 282
Support of a function, 247
Supremum, 6
Surjective Mapping Theorem, 139, 142
Szegő's Theorem, 44
Tangent, 119
Tauber Theorem, 38
Tempered distributions, 321ff
Test function, 247
Topological spaces, 17, 361
Totally ordered set, 31
Translation of a distribution, 270
Translation operator, 38, 252, 328
Tridiagonal, 172
Two-point boundary value problem, 171, 208ff
Tychonoff Theorem, 366
Uncertainty Principle, 310
Uniform Boundedness Theorem, 42
Uniform continuity, 16
Uniform convergence, 11
Unit ball, 7
Unit cell, 7
Unitary matrices, 104
Unitary operator, 101
Upper bound, 6, 31
Upper semicontinuity, 208
Variance of a function, 310
Vector space, 2
Volterra integral equation, 141, 182, 183, 185, 189

Weak Cauchy property, 88
Weak convergence in Hilbert
　　space, 87
Weak convergence, 53
Weak topology, 368
Weak* topology, 368
Weakly complete, 57
Weierstrass M-Test, 373

Weierstrass nondifferentiable
　　function, 259ff, 374
Weierstrass, 11
Wrońskian, 106
Young's Theorem, 332
Zarantonello, 183
Zermelo–Fraenkel Axioms, 31
Zorn's Lemma, 32

Symbols

t^*	Point-evaluation functional, 29, 43
L^*	The adjoint of a mapping L, 50
dim	Dimension, 5
$\mathbb{R}$	The real number field
$\mathbb{R}^*$	The extended real number system, 381
$\mathbb{C}$	The complex number field, 3
$\Pi_k(\mathbb{R}^n)$	Space of all polynomials of degree at most k in n variables, 263
Π	The space of all polynomials in one variable
$\|x\|_\infty$	Sup-norm on $\mathbb{R}^n$, 3
$\|x\|_1$	ℓ_1-norm on $\mathbb{R}^n$, 3
$\mathbb{R}^n$	n-Dimensional Euclidean space, 3–4
δ_{ij}	Kronecker delta (1 if $i = j$ and 0 otherwise), 71
$C(S)$	Space of continuous functions on a domain S, 3, 14, 348ff
$C^\infty(\mathbb{R})$	Space of all infinitely differentiable functions on $\mathbb{R}$, 247
$L|U$	Restriction of a map L to a set U, 59
$f \circ g$	The composition of functions, f with g, 27
ℓ^∞ or ℓ_∞	Space of bounded functions on $\mathbb{N}$ with sup-norm, 4, 12
ℓ_1	Space of summable functions on $\mathbb{N}$, 14, 34
ℓ_p	Space of p-th power summable sequences, 54
c_0	Space of sequences converging to zero, with sup-norm, 12
L^p	Space of p-th power integrable functions, 409
$\mathcal{R}(L)$	Range of operator L, 51, 191
$\mathcal{L}(X, Y)$	Space of bounded linear maps from X to Y, 25, 27
xy	Inner product in $\mathbb{R}^n$, 263, 288
$\langle x, y \rangle$	Inner product, 61
$\#$	Number of elements in a set, 39
$\mathbb{Z}$	Set of all integers
$\mathbb{Z}_+$	Set of all nonnegative integers
$\mathbb{Z}_+^n$	Set of n-tuples of nonnegative integers.
$\mathbb{N}$	The set of natural numbers $\{1, 2, \ldots\}$
dist	Distance from a point to a set, 9, 19, 23
Π_k	Space of polynomials of degree at most k in one variable.
$\twoheadrightarrow$	Surjective mapping, 49, 193
$\rightarrow$	Special convergence for test functions, 249
$\Longrightarrow$	Implication symbol
$*$	Convolution, 269

$\mathcal{X}$	Characteristic function of a set, 395
X^*	Conjugate Banach space, 34
$\perp$	Orthogonality symbol, 64
$\perp$	Annihilator symbol, 52, 65
$f^\wedge,\ \hat{f}$	Fourier transform of f, 288
$\mapsto$	Mapping symbol
$\rightharpoonup$	Symbol for weak convergence, 53
$\|A\|$	Norm of a quadratic form, 84
$\binom{n}{m}$	Binomial coefficient, 261
$\mathcal{D}$	Space of test functions, 247
$\mathcal{D}'$	Space of distributions, 249
$\mathcal{S}$	the Schwartz space, 294
$\varnothing$	Empty set, 17
∇^2	Laplacian, 198
$\exists$	"There exists"
$\forall$	"For all"
$\bigcap$	Intersection of a family of sets
$\bigcup$	Union of a family of sets
$\setminus$	Set difference, 22
$\&$	Logical AND
$W^{k,p}(\Omega)$	Sobolev space, 326
$V^{k,p}(\Omega)$	Sobolev space, 329
$C_b^m(\Omega)$	331
$H^k(\Omega)$	332

Graduate Texts in Mathematics

(continued from page ii)

66 WATERHOUSE. Introduction to Affine Group Schemes.
67 SERRE. Local Fields.
68 WEIDMANN. Linear Operators in Hilbert Spaces.
69 LANG. Cyclotomic Fields II.
70 MASSEY. Singular Homology Theory.
71 FARKAS/KRA. Riemann Surfaces. 2nd ed.
72 STILLWELL. Classical Topology and Combinatorial Group Theory. 2nd ed.
73 HUNGERFORD. Algebra.
74 DAVENPORT. Multiplicative Number Theory. 3rd ed.
75 HOCHSCHILD. Basic Theory of Algebraic Groups and Lie Algebras.
76 IITAKA. Algebraic Geometry.
77 HECKE. Lectures on the Theory of Algebraic Numbers.
78 BURRIS/SANKAPPANAVAR. A Course in Universal Algebra.
79 WALTERS. An Introduction to Ergodic Theory.
80 ROBINSON. A Course in the Theory of Groups. 2nd ed.
81 FORSTER. Lectures on Riemann Surfaces.
82 BOTT/TU. Differential Forms in Algebraic Topology.
83 WASHINGTON. Introduction to Cyclotomic Fields. 2nd ed.
84 IRELAND/ROSEN. A Classical Introduction to Modern Number Theory. 2nd ed.
85 EDWARDS. Fourier Series. Vol. II. 2nd ed.
86 VAN LINT. Introduction to Coding Theory. 2nd ed.
87 BROWN. Cohomology of Groups.
88 PIERCE. Associative Algebras.
89 LANG. Introduction to Algebraic and Abelian Functions. 2nd ed.
90 BRØNDSTED. An Introduction to Convex Polytopes.
91 BEARDON. On the Geometry of Discrete Groups.
92 DIESTEL. Sequences and Series in Banach Spaces.
93 DUBROVIN/FOMENKO/NOVIKOV. Modern Geometry—Methods and Applications. Part I. 2nd ed.
94 WARNER. Foundations of Differentiable Manifolds and Lie Groups.
95 SHIRYAEV. Probability. 2nd ed.
96 CONWAY. A Course in Functional Analysis. 2nd ed.
97 KOBLITZ. Introduction to Elliptic Curves and Modular Forms. 2nd ed.
98 BRÖCKER/TOM DIECK. Representations of Compact Lie Groups.
99 GROVE/BENSON. Finite Reflection Groups. 2nd ed.
100 BERG/CHRISTENSEN/RESSEL. Harmonic Analysis on Semigroups: Theory of Positive Definite and Related Functions.
101 EDWARDS. Galois Theory.
102 VARADARAJAN. Lie Groups, Lie Algebras and Their Representations.
103 LANG. Complex Analysis. 3rd ed.
104 DUBROVIN/FOMENKO/NOVIKOV. Modern Geometry—Methods and Applications. Part II.
105 LANG. $SL_2(\mathbf{R})$.
106 SILVERMAN. The Arithmetic of Elliptic Curves.
107 OLVER. Applications of Lie Groups to Differential Equations. 2nd ed.
108 RANGE. Holomorphic Functions and Integral Representations in Several Complex Variables.
109 LEHTO. Univalent Functions and Teichmüller Spaces.
110 LANG. Algebraic Number Theory.
111 HUSEMÖLLER. Elliptic Curves.
112 LANG. Elliptic Functions.
113 KARATZAS/SHREVE. Brownian Motion and Stochastic Calculus. 2nd ed.
114 KOBLITZ. A Course in Number Theory and Cryptography. 2nd ed.
115 BERGER/GOSTIAUX. Differential Geometry: Manifolds, Curves, and Surfaces.
116 KELLEY/SRINIVASAN. Measure and Integral. Vol. I.
117 SERRE. Algebraic Groups and Class Fields.
118 PEDERSEN. Analysis Now.
119 ROTMAN. An Introduction to Algebraic Topology.
120 ZIEMER. Weakly Differentiable Functions: Sobolev Spaces and Functions of Bounded Variation.
121 LANG. Cyclotomic Fields I and II. Combined 2nd ed.
122 REMMERT. Theory of Complex Functions. *Readings in Mathematics*
123 EBBINGHAUS/HERMES et al. Numbers. *Readings in Mathematics*
124 DUBROVIN/FOMENKO/NOVIKOV. Modern Geometry—Methods and Applications. Part III.
125 BERENSTEIN/GAY. Complex Variables: An Introduction.
126 BOREL. Linear Algebraic Groups. 2nd ed.
127 MASSEY. A Basic Course in Algebraic Topology.
128 RAUCH. Partial Differential Equations.
129 FULTON/HARRIS. Representation Theory: A First Course. *Readings in Mathematics*
130 DODSON/POSTON. Tensor Geometry.

131 LAM. A First Course in Noncommutative Rings. 2nd ed.
132 BEARDON. Iteration of Rational Functions.
133 HARRIS. Algebraic Geometry: A First Course.
134 ROMAN. Coding and Information Theory.
135 ROMAN. Advanced Linear Algebra.
136 ADKINS/WEINTRAUB. Algebra: An Approach via Module Theory.
137 AXLER/BOURDON/RAMEY. Harmonic Function Theory. 2nd ed.
138 COHEN. A Course in Computational Algebraic Number Theory.
139 BREDON. Topology and Geometry.
140 AUBIN. Optima and Equilibria. An Introduction to Nonlinear Analysis.
141 BECKER/WEISPFENNING/KREDEL. Gröbner Bases. A Computational Approach to Commutative Algebra.
142 LANG. Real and Functional Analysis. 3rd ed.
143 DOOB. Measure Theory.
144 DENNIS/FARB. Noncommutative Algebra.
145 VICK. Homology Theory. An Introduction to Algebraic Topology. 2nd ed.
146 BRIDGES. Computability: A Mathematical Sketchbook.
147 ROSENBERG. Algebraic K-Theory and Its Applications.
148 ROTMAN. An Introduction to the Theory of Groups. 4th ed.
149 RATCLIFFE. Foundations of Hyperbolic Manifolds.
150 EISENBUD. Commutative Algebra with a View Toward Algebraic Geometry.
151 SILVERMAN. Advanced Topics in the Arithmetic of Elliptic Curves.
152 ZIEGLER. Lectures on Polytopes.
153 FULTON. Algebraic Topology: A First Course.
154 BROWN/PEARCY. An Introduction to Analysis.
155 KASSEL. Quantum Groups.
156 KECHRIS. Classical Descriptive Set Theory.
157 MALLIAVIN. Integration and Probability.
158 ROMAN. Field Theory.
159 CONWAY. Functions of One Complex Variable II.
160 LANG. Differential and Riemannian Manifolds.
161 BORWEIN/ERDÉLYI. Polynomials and Polynomial Inequalities.
162 ALPERIN/BELL. Groups and Representations.
163 DIXON/MORTIMER. Permutation Groups.
164 NATHANSON. Additive Number Theory: The Classical Bases.
165 NATHANSON. Additive Number Theory: Inverse Problems and the Geometry of Sumsets.
166 SHARPE. Differential Geometry: Cartan's Generalization of Klein's Erlangen Program.
167 MORANDI. Field and Galois Theory.
168 EWALD. Combinatorial Convexity and Algebraic Geometry.
169 BHATIA. Matrix Analysis.
170 BREDON. Sheaf Theory. 2nd ed.
171 PETERSEN. Riemannian Geometry.
172 REMMERT. Classical Topics in Complex Function Theory.
173 DIESTEL. Graph Theory. 2nd ed.
174 BRIDGES. Foundations of Real and Abstract Analysis.
175 LICKORISH. An Introduction to Knot Theory.
176 LEE. Riemannian Manifolds.
177 NEWMAN. Analytic Number Theory.
178 CLARKE/LEDYAEV/STERN/WOLENSKI. Nonsmooth Analysis and Control Theory.
179 DOUGLAS. Banach Algebra Techniques in Operator Theory. 2nd ed.
180 SRIVASTAVA. A Course on Borel Sets.
181 KRESS. Numerical Analysis.
182 WALTER. Ordinary Differential Equations.
183 MEGGINSON. An Introduction to Banach Space Theory.
184 BOLLOBAS. Modern Graph Theory.
185 COX/LITTLE/O'SHEA. Using Algebraic Geometry.
186 RAMAKRISHNAN/VALENZA. Fourier Analysis on Number Fields.
187 HARRIS/MORRISON. Moduli of Curves.
188 GOLDBLATT. Lectures on the Hyperreals: An Introduction to Nonstandard Analysis.
189 LAM. Lectures on Modules and Rings.
190 ESMONDE/MURTY. Problems in Algebraic Number Theory.
191 LANG. Fundamentals of Differential Geometry.
192 HIRSCH/LACOMBE. Elements of Functional Analysis.
193 COHEN. Advanced Topics in Computational Number Theory.
194 ENGEL/NAGEL. One-Parameter Semigroups for Linear Evolution Equations.
195 NATHANSON. Elementary Methods in Number Theory.
196 OSBORNE. Basic Homological Algebra.
197 EISENBUD/HARRIS. The Geometry of Schemes.
198 ROBERT. A Course in p-adic Analysis.
199 HEDENMALM/KORENBLUM/ZHU. Theory of Bergman Spaces.
200 BAO/CHERN/SHEN. An Introduction to Riemann–Finsler Geometry.

201 HINDRY/SILVERMAN. Diophantine
     Geometry: An Introduction.
202 LEE. Introduction to Topological
     Manifolds.
203 SAGAN. The Symmetric Group:
     Representations, Combinatorial
     Algorithms, and Symmetric Functions.
     2nd ed.
204 ESCOFIER. Galois Theory.

205 FÉLIX/HALPERIN/THOMAS. Rational
     Homotopy Theory.
206 MURTY. Problems in Analytic Number
     Theory.
     Readings in Mathematics
207 GODSIL/ROYLE. Algebraic Graph Theory.
208 CHENEY. Analysis for Applied
     Mathematics.